Astrophysics Update 2

John W. Mason (Editor)

Astrophysics Update 2

Springer

Published in association with
Praxis Publishing
Chichester, UK

Editor
Dr John W. Mason
Olympus Mons
51 Orchard Way
Barnham
West Sussex PO22 0HX
UK

SPRINGER–PRAXIS BOOKS IN ASTRONOMY AND PLANETARY SCIENCES
SUBJECT *ADVISORY EDITOR*: Dr. Philippe Blondel, C.Geol., F.G.S., Ph.D., M.Sc., Senior Scientist, Department of Physics, University of Bath, Bath, UK; John Mason B.Sc., M.Sc., Ph.D.

ISBN-13 978-3-642-06778-5 e-ISBN-13 978-3-540-30313-8

Springer is part of Springer-Science + Business Media (springeronline.com)
Bibliographic information published by Die Deutsche Bibliothek
Die Deutsche Bibliothek lists this publication in the Deutsche Nationalbibliografie; detailed bibliographic data are available from the Internet at http://dnb.ddb.de

Softcover reprint of the hardcover 1st edition 2006

Cover design: Jim Wilkie
Completed in LaTex: EDV-Beratung, Germany

Editor's Preface

Important developments in astrophysics and cosmology continue at a frenetic pace, and it is increasingly difficult to keep up with progress in these fields. This second volume of *Astrophysics Update* comprises a collection of eleven topical reviews, each presented as a separate chapter, covering important areas of astrophysics and cosmology. The contributions have been written by scientists at the forefront of research in the selected areas, in a style which, we hope, will be accessible not only to advanced undergraduate students and beginning graduate students, but also to professional astronomers working in the field.

In this volume, we have attempted, once again, to highlight a diverse range of topics, from studies of the early universe to work in stellar astrophysics. It is clear that much of the new research is being driven by a combination of multi-wavelength observations with powerful new ground- and space-based telescopes together with new and improved facilities for modelling and simulation.

Ten years ago, in 1995, the discovery of Gliese 229B, the first bona fide 'brown dwarf', was announced. Since that time an increasing number of objects less massive and cooler than red M-dwarf stars, but more massive than giant planets such as Jupiter, have been discovered. It is, therefore, appropriate that in Chapter 1 of this volume Katharina Lodders and Bruce Fegley look at recent progress in the classification and chemistry of low mass dwarfs. The authors discuss the spectral classification of L and T dwarfs, objects which include the coolest known stars and substellar objects, and they examine the elements whose chemistry plays a key role in shaping the optical and infrared spectra of these so-called "failed" stars.

Neutron stars are the superdense remnants of Type II supernova explosions, produced in the collapse of massive stars after they have exhausted their available nuclear fuel. In Chapter 2, Tod Strohmayer examines neutron stars from the viewpoint that they provide a natural laboratory for the study of matter under the most extreme conditions of density and pressure, with the strongest magnetic fields known. The latest attempts to measure neutron star masses and radii are discussed as they provide a means to constrain important aspects of neutron star structure and fundamental physics. Recent observational efforts have concentrated on measurements of X-ray pulsation signals and observations of the thermal emission from neutron star surfaces.

The observations of the Chandra X-ray Observatory, launched into Earth orbit in 1999, have made outstanding contributions to many areas of astrophysics. In Chapter 3, Martin Weisskopf and John Hughes discuss how Chandra observations of supernova remnants (SNRs) have provided new insights into the supernova phenomenon. Here, Chandra observations have clarified the existence of a class of compact central objects associated with SNRs that are radio quiet and appear unpulsed in X-rays. Chandra has also isolated and resolved the structure of pulsar wind nebulae. In some cases pulsars have been clearly identified within SNRs, while in others Chandra has been unable to identify an associated compact central object.

The study of colliding galaxies has seen enormous progress in recent years, and in Chapter 4, Curtis Struck discusses a number of key issues in this field. These include the role of major and minor collisions and mergers in building galaxies, whether collisions differ in cluster, group or nearly isolated environments due to their differing relative velocities, and how the large-scale dynamics of collisions and interactions induce star formation and nuclear activity, both before and during mergers. The author also looks at the association of ultraluminous infrared galaxies (ULIRGs) with colliding galaxies, and examines to what degree long-term secular effects are triggered by mergers and feedback.

For some years it has been known that there is a close correlation between the mass of supermassive black holes (SMBHs) in galaxy nuclei and the velocity dispersion inside the effective radius of the bulge. This suggests a common formation scenario for galaxies and their central black holes. In Chapter 5, Françoise Combes discusses the various processes involved in the formation and evolution of SMBHs, particularly self-regulation and feedback which are key to this observed correlation. The author also describes a wide range of phenomena including black hole formation in the very early Universe, mini-quasars and reionisation, intermediate mass black holes, downsizing and lifetime of activity, and how mergers of binary SMBHs help to heat and destroy central stellar cusps. Bright quasars appear to grow rapidly at high redshift and their number density then decreases sharply, while the density of low-luminosity objects peaks more recently and then decreases smoothly.

The correlation between central black hole mass and velocity dispersion is picked up in Chapter 6, where Mark Lacy discusses quasars and their host galaxies. Two major new quasar surveys with tens of thousands of objects have recently appeared and have enabled the quasar luminosity function to be refined. Estimates of black hole masses have been used to try and improve our physical understanding of active galactic nuclei, such as in the case of quasar radio emission. One consequence of the large-scale quasar surveys has been an estimate of quasar clustering and its evolution with redshift. The number density of very high redshift quasars has also been used to estimate their contribution to reionisation. The author also describes significant developments in the study of broad absorption line quasars, advances in finding the obscured quasar population and quasar host galaxy studies.

In Chapter 7, Matthias Bartelmann looks at applications of gravitational lensing in cosmology. Since it appears that the Universe is dominated by dark matter, gravitational lensing has evolved into one of the most important tools for learning about the amount, composition and distribution of matter in the Universe. The author first summarises the theory and basic properties of gravitational lensing, and then applies this to isolated lenses such as compact objects, galaxies and galaxy clusters. Cosmologically important applications of gravitational lensing are described, including searches for MACHOs, time-delay measurements in multiple image systems which provide estimates of the Hubble Constant, and mapping the dark matter distribution in galaxy clusters. The concept of lensing by large-scale structures and its measurement are also introduced.

The techniques used to identify the oldest "living stars in the Universe are reviewed by Timothy Beers in Chapter 8. Such stars are inferred to be low mass objects that inherited their observed low metallicities from the earliest generations of massive (Population III) stars, but with lifetimes so long that they are still shining today. The author concentrates on the results of two modern surveys: the HK survey and the Hamburg/ESO survey. The distribution of measured stellar metallicities derived from these samples and some results of recent high-resolution spectroscopy of very metal-poor stars are described. Special attention is given to the nature of r-process enhanced metal-poor stars.

Ultraluminous infrared galaxies (ULIRGs) are mergers between gas-rich galaxies where the interaction triggers a dust-enshrouded starburst and AGN activity. These fascinating, immensely luminous objects are the focus of the review by Carol Lonsdale, Duncan Farrah and Harding Smith in Chapter 9. In the past decade, comprehensive multi-wavelength observations of ULIRGs have shown that instead of being merely local oddities, these objects are vastly more numerous at high redshift. The authors describe recent progress in understanding the physics and evolution of local ULIRGs, and the contribution of high redshift ULIRGs to the cosmic infrared background and history of star formation. They also show how ULIRGs may be used as diagnostics of the formation of massive galaxies and large-scale structures.

The role of galactic winds and outflows in galaxy formation are considered by Crystal Martin in Chapter 10. The author begins by describing the astrophysics of galaxy formation and the role of processes such as gas cooling and feedback. The amount of feedback from massive stars is discussed and the hydrodynamics of supersonic outflows from stellar winds and ensembles of supernovae are explained. The author demonstrates how our knowledge of galactic winds has been refined by observations, which have also constrained dynamical models of such outflows, and the measured properties of galactic outflows are described. The cosmological impact of galactic winds is also assessed with regard to shaping the galaxy luminosity function and dispersing heavy elements into the intergalactic medium.

Finally, in Chapter 11, Fred Taylor, Isobel Hook and Matthew Colless look at the exciting prospects for astronomical research which will result from the construction of extremely large telescopes (ELTs), instruments having apertures between 20m and 100m. The authors describe the mirror technology and optical designs required for such large telescopes, their mountings and enclosures, the selection of suitable observing sites for them, their instrumentation and the likely cost of their construction. They then review the main science drivers for such telescopes, which include the discovery and characterisation of Earth-like planets around other stars, star formation throughout the Universe, the growth of galaxies in the early Universe, and identification of the first objects and reionisation of the Universe.

This book has benefited from the support and assistance of a large number of people. I would like to offer my sincere thanks to all of the contributing authors for their considerable efforts, perseverance and enthusiasm for this project. I am also most grateful to Frank Herweg of Springer, Heidelberg for his invaluable assistance and advice in the preparation of the LaTeX files for this book, including his work on many of the illustrations prior to publication. Finally, I am indebted to Sue Peterkin of Praxis Publishing for her very considerable assistance at all stages in the organization and coordination of this project, and to Clive Horwood, Publisher, for his encouragement, advice and patience throughout.

Barnham, November 2005 John W. Mason

List of Contributors

Matthias Bartelmann
Zentrum für Astronomie der Universität Heidelberg
Institut für Theoretische Astrophysik
Albert Überle-Str. 2
69120 Heidelberg
Germany
mbartelmann@ita.uni-heidelberg.de

Timothy C Beers
Department of Physics and Astronomy and Joint Institute for Nuclear Astrophysics
Michigan State University
East Lansing
MI 48824
USA
beers@pa.msu.edu

Matthew Colless
Director
Anglo-Australian Observatory
P O Box 296
Epping
NSW 1710
Australia
colless@aao.gov.au

Françoise Combes
Observatoire de Paris
LERMA
61 Avenue de l'Observatoire
F- 75014 Paris
France
francoise.combes@obspm.fr

Duncan Farrah
Department of Astronomy
Cornell University
Space Sciences Building
Ithaca
NY 14853
USA
duncan@isc.astro.cornell.edu

Bruce Fegley Jr
Planetary Chemistry Laboratory
Department of Earth and Planetary Sciences
and McDonnell Center for the Space Sciences
Washington University
Campus Box 1169
St. Louis
MO 63130
USA
bfegley@wustl.edu

Isobel Hook
Department of Astrophysics
University of Oxford
Keble Road
Oxford
OX1 3RH
UK
imh@astro.ox.ac.uk

John P Hughes
Department of Physics and Astronomy
Rutgers, The State University of New Jersey
136 Frelinghuysen Road
Piscataway
NJ 08854-8019
USA
jph@physics.rutgers.edu

Mark Lacy
Spitzer Science Center
California Institute of Technology
MS 220-6
Pasadena
CA 91125
USA
mlacy@ipac.caltech.edu

Katharina Lodders
Planetary Chemistry Laboratory
Department of Earth and Planetary Sciences
and McDonnell Center for the Space Sciences
Washington University
Campus Box 1169
St. Louis
MO 63130
USA
lodders@wustl.edu

Carol J Lonsdale
Infrared Processing and Analysis Center
California Institute of Technology
MS 100-22
Pasadena
CA 91125
USA
cjl@ipac.caltech.edu

Crystal L Martin
Packard Fellow
Department of Physics
University of California, Santa Barbara
Santa Barbara
CA 93106
USA
cmartin@physics.ucab.edu

Harding E Smith
Center for Astrophysics and Space Science
University of California San Diego
9500 Gilman Drive
La Jolla
CA 92093
USA
hsmith@ucsd.edu

Tod E Strohmayer
Exploration of the Universe Division
NASA Goddard Space Flight Center
Greenbelt
MD 20771
USA
tod.e.strohmayer@nasa.gov

Curtis Struck
Department of Physics and Astronomy
Iowa State University
Ames
IA 50011
USA
curt@iastate.edu

Fred Watson
Anglo-Australian Observatory
Coonabarabran
NSW 2357
Australia
fgw@aao.gov.au

Martin C Weisskopf
Space Sciences Directorate
NASA Marshall Space Flight Center
XD01
Huntsville
AL 35812
USA
martin.weisskopf@msfc.nasa.gov

Contents

1 Chemistry of Low Mass Substellar Objects

K. Lodders and B. Fegley, Jr.

Abstract: "Brown dwarfs" is the collective name for objects more massive than giant planets such as Jupiter but less massive than M dwarf stars. This review gives a brief description of the classification and chemistry of low mass dwarfs. The current spectral classification of stars includes L and T dwarfs that encompass the coolest known stars and substellar objects. The relatively low atmospheric temperatures and high total pressures in substellar dwarfs lead to molecular gas and condensate chemistry. The chemistry of elements such as C, N, O, Ti, V, Fe, Cr, and the alkali elements play a dominant role in shaping the optical and infrared spectra of the "failed" stars. Chemical diagnostics for the sub-classifications are described.

1.1 Introduction

2005 marked the 10th anniversary of the first discovery of Gl 229B, a bona-fide brown dwarf – a substellar object too low in mass to sustain hydrogen fusion to shine as a star (Nakajima et al. 1995, Oppenheimer et al. 1995). In the 1960s, it was already recognized that a lower mass limit of >0.08 solar masses is required for dwarf stars to enter the main sequence and that such objects only release energy from gravitational contraction and evolve into "black dwarfs" (Kumar 1963). These low-mass objects became more colorful in 1975, when Tarter called them "brown dwarfs", a name still commonly used. However, since these low-mass and faint objects appear reddish purple to magenta instead of brown (Burrows et al. 2001), we use "low mass objects", "substellar objects" or "substellar dwarfs" in this review.

Since 1995, a plethora of objects less massive and cooler than true M dwarf stars, but more massive than giant planets such as Jupiter, has become known through the Two Micron All Sky Survey (2MASS), Deep Survey of the Southern Sky (DENIS), and Sloan Digital Sky Survey (SDSS) infrared sky searches. These objects extend the spectral sequence of stars to fainter and lower mass objects (e.g., reviews by Chabrier & Baraffe 2000, Basri 2000, Burrows et al. 1999, 2001). By 1999 two new classes, L and T, were added to the stellar spectral classification

(Kirkpatrick et al. 1999, Martin et al. 1999. Current counts are ~300 L dwarfs, and ~60 T dwarfs[1].

The L dwarfs more closely resemble M dwarfs in their spectra and include the lightest real stars and the heaviest substellar objects. The T dwarfs, characterized by methane and water absorption bands, have closer spectral resemblance to giant planets but are much more massive than ~13 Jupiter masses, which is the mass limit for deuterium burning (see below). The latest suggestion for addition to the spectral sequence is the letter "Y" for objects that lack water absorption bands because the outer atmospheres of Y dwarfs should be cool enough ($< \sim 500$ K) to sequester water into clouds (Ackerman & Marley 2001, Burrows et al. 2003). There are no direct observations yet of objects that are cooler than the coolest known T dwarfs ($T_{eff} \sim 700$ K) to complete the bridge to Jupiter (Teff ~125 K).

The detection of the dim substellar objects is challenging because they are small (about Jupiter's radius) and low in mass. The upper mass limit for substellar dwarfs is about 7–8% that of the Sun (for comparison Jupiter' mass is 0.095%), insufficient to sustain H-burning like in real stars. Substellar objects may burn deuterium if they exceed 13 Jupiter masses and burn ^{6}Li and ^{7}Li if they exceed 65 Jupiter masses (the mass limits for H and D burning are for objects with solar metallicity, see Burrows et al. 2001). In the most massive substellar objects, D-burning lasts up to $\sim 3 \times 10^7$ years and Li-burning up to $\sim 3 \times 10^8$ years, whereas H-burning operates several 10^9 years in dwarf stars. Most of the energy released by a substellar object over its lifetime is from gravitational energy gained during its formation and contraction as it sits in space and cools.

Substellar dwarfs never exceed ~3000 K near their surfaces. As blackbody curves show, the more an object cools, the less it is visible at optical wavelengths. M dwarf stars emit most strongly at red optical wavelengths (~0.75 μm) but maximum emissions of the cooler L ($\sim 1200 < T_{eff} < 2200$ K) and T dwarfs ($300\,K < T_{eff} < \sim 1200$ K) are shifted to longer wavelengths in the near infrared (1–2 μm). Substellar objects also appear intrinsically fainter (i.e., compared to an M dwarf star at the same distance) because of the significant decrease in the energy density at maximum in cooler blackbodies. Furthermore increased opacity from metal oxides, metal hydrides, water, carbon monoxide, methane, and ammonia at different near-infrared wavelengths may reduce the overall energy flux by so much that essentially nothing of an ideal blackbody intensity curve remains.

The molecular chemistry in the relatively dense and cool atmospheres of substellar objects leads to significant changes in the observed far-red and near-infrared spectra when compared to hotter dwarf stars. Chemical thermodynamics and kinetics determine the gases present at a given depth in the atmosphere as functions of temperature, pressure, and overall elemental composition. The

[1]see http://spider.ipac.caltech.edu/staff/davy/ARCHIVE;
http://www.astro.ucla.edu/~adam/homepage/research/tdwarf/ and
http://www.astro.ucla.edu/~mclean/BDSSarchive/

chemical speciation changes when a solar composition gas is cooled from high to low temperatures and is largely responsible for the observed differences in stellar and sub-stellar spectra. At high photospheric temperatures such as in K and early M dwarf stars, monatomic ions and neutral atoms dominate, but at the lower temperatures near the M/L transition and in T dwarfs, neutral atoms, particularly of the alkali elements, and molecules such as TiO, VO, FeH, CrH, CO, H_2O, and CH_4 become more and more important. Condensation of some elements from the gas into clouds also occurs in L and T dwarfs.

1.2 Classification Schemes

The spectral M, L, and T dwarf classifications make use of the temperature dependent appearance and disappearance of ions, atoms, and molecules in the optical and near IR spectra of the these objects e.g., Kirkpatrick et al. 1999, 2000 (=K99, K00), Martin et al. 1999 (=M99), Burgasser et al. 2002a (=B02), 2003, Geballe et al. 2002 (=G02), Leggett et al. 2002 (=L02), McLean et al. 2003 (=Mc03), Nakajima et al. 2004). Some of the major characteristics also used for classification are summarized in Table 1.1.

Spectral subtype numbers from zero for the hottest (e.g., L0, T0) to (currently) nine for the coolest (e.g., L9, T9) are set when certain spectral features appear or disappear (Table 1.1). Note that currently there are 2 scales for subtyping the L dwarfs that differ in the assignments of the higher L subtype numbers (M99, K99, K00). Here we follow the majority of the literature and use the Kirkpatrick et al. scale; however, the conversion of this subtype scale to that by Martin et al. is also given in Table 1.1.

Late M dwarfs ($>$ M6) are characterized by increasingly stronger TiO, VO H_2O, FeH, CrH, CaH, and MgH bands. Water bands grow stronger from mid-M dwarfs through the L and T dwarf sequence in all spectral ranges. The L dwarf sequence starts when TiO and VO bands begin to weaken while CrH, FeH, and water bands strengthen in the optical. In the near infrared, CO, CrH and FeH absorptions strengthen towards the mid-L subclasses, and then decline.

However, the FeH absorption at 0.9896 µm reappears and strengthens again in early T dwarfs and reaches a maximum in strength around T5 before completely disappearing at T8 (Burgasser et al. 2002a,b). The band of CrH at 0.9979 µm behaves similarly and could be a misidentified FeH band according to Cushing et al. (2003) who investigated and identified many FeH transitions. However, the assignment of the band at 0.9979 µm to CrH is confirmed by opacity computations for CrH by Burrows et al. (2002a) and FeH by Dulick et al. (2003). These studies also show that the FeH/CrH abundance ratio plays a critical role for the observability of CrH.

The L dwarfs show CO overtone bands in the 1–2.5 µm range and the band at 2.3 µm is traceable into the T dwarf sequence up to T3 (B02, Mc03). The strongest band of CH_4 at 3.3 µm already appears in L5 dwarfs (Noll et al. 2000), and in some L6.5 and L8 dwarfs very weak bands of CH_4 are seen at 2.2 and 1.6 µm (see Nakajima et al. 2004, McLean et al. 2003). The T type is defined by

Table 1.1. Spectral Classification of L and T Dwarfs

Spectral type M99	K99	Spectral characteristics	Chemistry
M8	**M8**	TiO (705.3, 843.2 nm) at max. (M99, (K99))	TiO at max.
M9	**M9**	VO (740, 790 nm) at max (M99, (K99))	VO at max
L0	**L0**	blueward TiO gone, redder TiO remains (K99,M99); Al I, Ca I disappear in J band (Mc03); Rb I (780.0, 794.8 nm); Cs I (852.1, 894.3 nm) weak (K99, M99)	Ca, Al, Ti cond.
L1	**L1**	VO bands (740, 790 nm) weaken; Na I (589.0 and 589.6 nm) doublet weakens; Rb I, Cs I increasing in strength (K99,M99)	V cond. $Cs > Cs^+$
L2	**L2**	all TiO bands gone except at 843.2 nm (K99); Ca I triplet still in K band (Mc03)	
L2.5	**L3**	Fe I at 1.189 μm still present (Mc03)	
	L3.5	Fe I at 1.189 μm gone (Mc03); K I (1.169, 1.197, 1.25 μm) peak in strength (R01)	Fe cond.
L3	**L4**	all VO gone (K99); CrH (861.1 nm) = FeH (869.2 nm); FeH at 869.2 nm and J band at max. (R01, B02, K99, M99, Mc03); K I doublets in J-band at max (B02, Mc03)	FeH ≈ CrH K at max.
L4	**L5**	CH_4 appears at 3.3 μm (Noll et al. 2000); CaH at 685 nm still seen (M99, Mc03); FeH (869.2 nm and J band) weakens (B02,K99,M99); CrH 861.1 nm at max. (B02, K99, M99); optical Rb I, Cs I strengthening, but weaker in J band (K99, Mc03)	
L5	**L6**	CrH weakens in optical (K99, M99) Li I (670.8 nm) still observed in some objects (K99, M99)	Cr cond. Li = LiCl
L5.5	**L7**	TiO 843.2 nm (K99, M99) essentially absent, ditto TiO 549.7, 559.7, 615.9, 615.9, 638.4 nm (M99); Li I most likely absent (K99,M99)	
L6	**L8**	subtle CH_4 band in K band (G02, N04); no or subtle CH_4 in H band (B02,G02,Mc03); Na I (589.0, 589.6 nm) very broad, Na I barely seen in optical and J band (K99,M99,Mc03); K I doublets at min. in J band (N04)	$CO = CH_4$
	T0	CH_4 appears in K and H bands (G02) FeH (989.6 nm) weaker (N04) or as strong as in L8 (B02b)	$CO = CH_4$
	T1	weak CH_4 in H, J, and K bands (B02,G02); strengths of $CO = CH_4$ in K band; K I (1.2432, 1.2522 μm) present in J band (B02)	$CO = CH_4$
	T2	strengths of CH_4 > CO in K band (B02); CrH (861.1 nm) and FeH (869.2 nm) disappear (B03); FeH (989.6 nm) weaker (N04) or as strong as in L8 (B02b); Cs I (852.1, 894.3 nm) peak in strength (B03)	
	T3	CO (2.295 μm) very weak or absent in K band (B02,Mc03,N04)	
	T4	FeH (989.6 nm), CrH (996.9 nm) increase again in strength (B02,B02b,B03,N04)	
	T5	No CO in K band (B02); FeH (989.6 nm) and CrH (996.9 nm) peak in strength (B02b,B03,N04); K I at maximum in J band (B02,Mc03,N04)	Na_2S cond.
	T6	K I at in J band begins to weaken (B02)	
	T7	K I in J band very weak (B02, Mc03)	
	T8	no FeH (989.6 nm) (B02, Mc03); no K I in J band (B02, Mc03); Cs I marginally detectable in optical (B03)	$(NH_3 = N_2)$?

B02: Burgasser et al. 2002a,b; G02: Geballe et al. 2002; K99: Kirkpatrick et al. 1999, 2000; M99: Martin et al. 1999; Mc03: McLean et al. 2003; N04: Nakajima et al. 2004, R01: Reid et al. 2001
H band centered at ~1.6 μm; J band centered at ~1.25 μm; K band centered at ~2.2 μm

the onset of methane absorption in the 1.6 and 2.2 μm photometric bandpasses (H and K bands). This causes the shift towards blue in the infrared J-H and H-K colors in the spectral sequence near the L/T transition (L02, Golimowski et al. 2004). Within the T sequence increasingly stronger H_2O and CH_4 absorptions appear between 1.0–2.2 μm (hence also the name "methane" dwarfs), and in addition, optical spectra of T dwarfs are shaped from collisionally induced absorption (CIA) by H2 (K99, Tokunaga & Kobayashi 1999, Liebert et al. 2000, B02, G02).

Throughout the L-sequence, the absorption resonance doublets of monatomic Na I (0.5890, 0.5896 μm) and K I (0.7665, 0.7699 μm) weaken and become strongly pressure broadened (see, K99, Tsuji et al. 1999, Burrows et al. 2000, 2002b, Liebert et al. 2000, Burrows & Volobuyev 2003). This continues within the T dwarf sequence, where the extremely broadened wings of the Na I and K I doublet absorptions dominate the slope of large portions in the red and near IR spectra so that most of the sub-typing of T dwarfs is done better with near infrared spectra (e.g., B02, G02). The near infrared K I doublets in the J band (Table 1.1) peak in strength around L3.5-L4 (Reid et al. 2001, B02, Mc03) and then decline but reach a second increase in strength around T5/T6 (e.g., B02), a behavior that is mimicked by FeH noted above.

The lines of Rb I (0.7800, 7948 μm) and Cs I (0.8521, 8943 μm) appear near the M/L transition, grow in strength throughout the L sequence, and reach a peak in strength at T2, after which their strength decreases (K99, B02). The presence of the resonance doublet of monatomic Li I (0.6708 μm) – a useful but limited test for the substellar nature of an object (see below) – appears to be restricted to earlier L subtypes of up to L6.

The introduction of the L and T classes is relatively new, and the definitions of the L and T subtypes are still work in progress. Characterization of the subtypes depends on the number of known and observed substellar dwarfs, the spectral wavelength range (optical vs. infrared) covered and resolution of the spectra, and the choice of spectral features used to parameterize the subclasses. Kirkpatrick et al. (1999, 2000) and Martin et al. (1999) utilized red optical spectra (0.63–1.01 μm) for M and L dwarfs, and near-infrared spectral features were used to classify T dwarfs (BO2, G02). Work on linking the optical and infrared observations into one scheme is ongoing and unified classification schemes for L and T dwarfs using spectral indices from the same wavelength ranges are under development (see, e.g., M99, K99, K00, Reid et al 2001, Tokunaga & Kobayashi 1999, Testi et al. 2001, B02, G02, Mc03, Nakajima et al. 2004). Geballe et al. (2002) combined optical (0.6–1.0 μm) and infrared (1.0–2.5 μm) spectra for M, L, and T dwarfs to find suitable infrared flux indices that also can be linked to the optical classification scheme of L dwarfs from K99 and M99. Apparently only the water band at 1.5 μm is useful for sub-typing the entire L and T sequence and for relating the infrared and optical classification schemes (but see also Tokunaga & Kobayashi 1999, Reid et al. 2001, Testi et al. 2001, Knapp et al. 2004). For objects of higher spectral type than L3 in the schemes of M99 and K99, the strength of the methane band at 2.2 μm provides an additional common index for the L and T classes (G02). McLean et al. (2003) suggested a classification scheme for L and T dwarfs based on several indicators such as the relative strengths of the atomic lines of Na, K, Fe, Ca, Al, and Mg and bands of water, carbon monoxide, methane, and FeH in near infrared spectra.

1.3 Effective Temperatures along the L and T Dwarf Sequence

The effective temperatures (T_{eff}) of substellar objects depend on their total mass, radius, and their age, and are generally below 3000 K (see Burrows et al. 2001). The effective temperatures are related to luminosity (L) by $L = 4\pi R^2 \sigma T_{eff}^4$, where R is the radius of a substellar dwarf and σ the Stefan–Boltzmann constant. The luminosities can be obtained from the bolometric magnitudes if bolometric (filter) corrections as well as parallaxes of the objects are known. Bolometric corrections were recently computed for a large set of L and T dwarfs (Golimowski et al. 2004) which combined with parallax measurements (Dahn et al. 2002, Vrba et al. 2004) and estimates of substellar object radii give T_{eff} estimates summarized in Table 1.2.

Other determinations of the effective temperature require detailed modeling of the chemistry, the atmospheric structure and synthetic spectra (e.g., Burrows et al. 2001). Ideally, the spectral subtypes should correspond to a regular temperature scale, but this goal has not yet been reached for L and T dwarfs. The spectral classification schemes utilize the temperature dependent chemistry of the different elements but the appearance or disappearance of different atoms and molecules does not happen in evenly spaced temperature intervals (see chemistry below). So using the presence/absence of chemistry features alone for subtyping cannot lead to a temperature scale that linearly correlates with spectral subtype. Defining the subtypes so that they correspond to evenly spaced differences in effective temperatures requires that one knows the absolute temperature at the beginning and the end of the L dwarf sequence, and similarly at the start and end of the T dwarf sequence. This requires synthetic spectra to calibrate effective temperatures to the strength of certain absorption features so that practical flux ratios for subtyping can be defined.

Table 1.2. Estimates of Effective Temperatures[a]

M9/L0[b]	L8-9/T0	T8	Source
2500	1390	760 (T8/9)	Vrba et al. 2004
2400	1360	–	Dahn et al. 2002
2300	1450	600–750 (T9)	Golimowski et al. 2004
2250	1230		Burgasser et al. 2002a
2200	1400	–	Stephens et al. 2001, Marley et al 2002
2200	1200	–	M99, Basri et al. 2000
2100	1300	800 (T8)	Geballe et al. 2001, 2002; Leggett et al. 2001, 2002
2030	1300	920 (T7.5)	Nakajima et al. 2004; Tsuji 2002
2000	1400	–	Schweitzer et al. 2001, 2002
2000	1300	–	K99, K00

[a] Typical uncertainties are ±100 K.
[b] Spectral type L according to K99

The two classification schemes for L dwarfs by K99, K00 and M99 agree from late M to L3 but the subtype classification and the assigned T_{eff} diverge significantly for later subclasses. K99 applied observed molecular and atomic features and chemical equilibrium calculations of Burrows & Sharp (1999) to estimate T_{eff} ranges of 2000 to 1500 K between L0 to L8. M99 defined each L spectral subtype to be exactly 100 K wide and they used a T_{eff} scale for a set of L dwarfs determined from optical lines of Cs I and Rb I by Basri et al. (2000).

These initially derived T_{eff} probably need some revision because at the time, the applied chemistry of Burrows & Sharp (1999) used by K99 & K00, and the atmosphere models employed by Basri et al. (2000) used by M99 assumed chemical equilibrium without sedimentation of condensates and also excluded condensate opacities. However, since condensates are unavoidably present in L and T dwarfs they affect the emerging spectral flux, either by blocking and scattering radiation or by having removed gas opacity (see e.g., Tsuji et al. 1996a,b, 1999, Jones & Tsuji 1997, Ackerman & Marley 2001, Marley et al. 2002, Tsuji 2002).

Estimates for the effective temperatures in the spectral class L are about 2500–2000 K at M9/L0 for the coolest "true" stars, to about 1500–1200 K for the substellar objects near the L/T transition (Table 1.2). Within the T dwarf sequence a few objects have well constrained effective temperatures. For Gl 229B (T6) and Gl 570D (T8) these are 940 ± 80 K and 805 ± 20 K, respectively (Saumon et al. 2000; Geballe et al. 2001). The coolest T dwarf currently known is 2MASS J04151954-0935066 and estimates for its T_{eff} are 600–750 K (Golimowski et al. 2004), and 760 K (Vrba et al. 2004). Overall, the uncertainties in the T_{eff} scales of L and T dwarfs remain large (e.g., references in Table 1.2).

1.4 The Case for Condensate Cloud Layers

Condensate formation is important for understanding the atmospheric chemistry and spectra of low mass stars and substellar objects. Condensate cloud formation in an aging, cooling atmosphere seems unavoidable, hence condensate cloud layers are to be expected in L and T dwarfs. Cloud layer formation leads to drastic compositional and morphological atmospheric changes by removal of potential absorbers from the observable atmosphere and by blocking and scattering emerging spectral flux depending on how close they are to the photosphere. So in addition to chemical effects (see below), the physical effects of clouds (number and sizes of cloud layers, size of cloud particles) must be included in the atmospheric models that are needed to compute spectra. Physical cloud models have been studied by Lunine et al. 1989, Tsuji et al. 1996a,b, 1999, 2004 Jones & Tsuji 1997, Chabrier et al. 2000, Ackerman & Marley 2001, Allard et al. 2001, Marley et al. 2002, Tsuji 2002, Cooper et al. 2003, Tsuji & Nakajima 2003, Woitke & Helling 2003).

Condensation in a planetary, substellar or stellar atmosphere proceeds differently than condensate formation in a low gravity environment such as the solar nebula (or other protoplanetary disks) and stellar outflows (such as from giant stars). In a bound atmosphere, condensates precipitating from the gas at high temperatures ('primary condensates') settle due to the influence of gravity and form relatively thin discrete cloud layers. Consequently, the primary condensates cannot react with gas at higher altitudes above the condensate clouds, and are out of equilibrium with the overlaying atmosphere. Thus there are no secondary condensates from gas-solid or gas-liquid reactions at lower temperatures.

The cloud-layer condensation approach (occasionally called rainout) works well for the Jovian planets where refractory elements (e.g., Ca, Al, Mg, Si, Fe, Ti, V) must be sequestered into high-temperature clouds deep in the atmospheres (e.g., Lewis 1969, Barshay & Lewis 1978, Fegley & Prinn 1985a,b, 1986, Fegley & Lodders 1994). Hence it is plausible to apply this approach to the atmospheres of more massive, substellar objects (Fegley & Lodders 1996). Giant planets, T and L dwarfs show a wide range in effective temperatures and the number of cloud layers depends on the volatility of the different elements, so giant planets, T, and L dwarfs will have different numbers of cloud layers (Fig. 1.1). Refractory oxide clouds such as corundum clouds already appear in the photospheres of late M dwarfs (Jones & Tsuji 1997). Silicates and liquid iron metal can form cloud layers in the hottest L dwarfs.

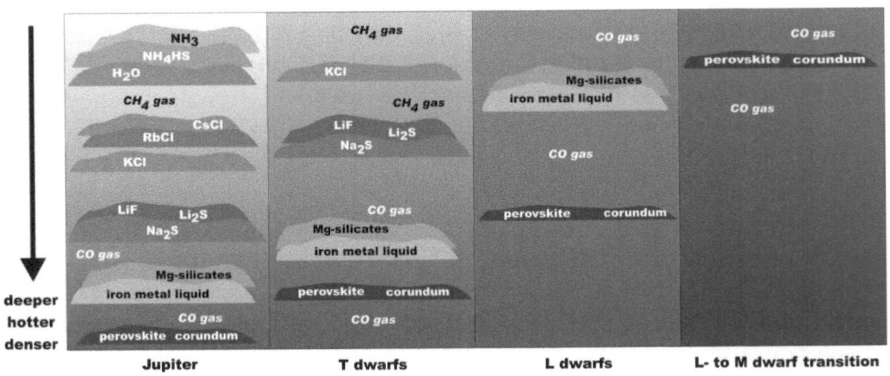

Fig. 1.1. The change in cloud layer structure from cool to hot objects

On the other hand, Jupiter has water, ammonium hydrosulfide (NH_4SH), and solid ammonia as top-level clouds (Fig. 1.1). Deeper inside, alkali halide and sulfide clouds follow, then silicate and iron cloud layers. The deepest Jovian cloud layer consists of refractory Ca-Al- and Ca-Ti oxides. In the hotter T and L dwarfs, the cloud layer structure shown for Jupiter is successively stripped off at the top, and only the cloud layer with the most refractory condensates may be left in the hottest L dwarfs. Thus, looking at hotter substellar objects is like looking at deeper and deeper regions of Jupiter's atmosphere after removing the cooler atmospheric regions. The net effect of condensate cloud formation is that the atmosphere above a cloud is depleted in gases that contained the elements condensed into the clouds.

The following examples show how observable chemical tracers can be used to constrain the cloud layer model. The absence of silane (SiH_4) and the presence of germane (GeH_4) in the atmospheres of Jupiter and Saturn is due to depletion of refractory Si, but not of volatile Ge, by magnesium silicate condensate cloud formation deep in their atmospheres (Fegley & Lodders 1994). Silicon is much more abundant than Ge in a solar composition gas (atomic Si/Ge ~8300) but SiH_4 is not observed on either Jupiter or Saturn (observational upper limits are $SiH_4/H_2 \sim 1 \times 10^{-9}$ by volume (1 ppbv)). For comparison, the solar Si/H_2 molar ratio is 7.09×10^{-5}, which is about 71,000 times larger than the observational upper limit on the silane abundance. In contrast, GeH_4 is observed with a GeH_4/H2 ratio ~0.7 ppbv on Jupiter and ~0.4 ppbv on Saturn. These values are less than the solar Ge/H_2 molar ratio of 8.5 ppbv because not all Ge in the atmospheres of Jupiter and Saturn is present as GeH_4 (Fegley & Lodders 1994). The presence of silicate clouds in T dwarfs is testable with searches for SiH_4 in T dwarfs and Si, SiO, SiS or SiH in L dwarfs. In the atmosphere above the Mg_2SiO_4 and $MgSiO_3$ clouds, Si-bearing gases should be highly depleted and be absent at the highest altitudes.

The *Galileo* entry probe mass spectrometer (GPMS) detected H_2S at about 3 times the solar S/H_2 ratio in Jupiter's atmosphere (Niemann et al. 1998). This is consistent with depletion of Fe-metal by Fe-cloud condensation (e.g., Lewis 1969, Barshay & Lewis 1978, Fegley & Lodders 1994). If Fe cloud formation did not occur, H_2S would be completely absent from the atmosphere of Jupiter because formation of FeS (troilite) from Fe metal grains with H_2S gas at ~700 K consumes all H_2S gas (solar Fe/S ~ 2). Hence the Fe-metal condensation cloud layer model is in accord with the GPMS observations of H_2S on Jupiter.

Monatomic K gas is present in the atmospheres of the T dwarfs Gl 229B and Gl 570D (e.g., Burrows et al. 2000, Geballe et al. 2001). The observed K abundances require that refractory rock-forming elements such as Al, Ca, and Si be removed by condensate cloud formation deep in the atmosphere of Gl 570D and Gl 229B (e.g., Geballe et al. 2001). Otherwise, potassium would condense into silicate minerals such as $KAlSi_3O_8$ (orthoclase) at high temperatures, and K gas would be depleted from the observable atmosphere (Lodders 1999, Burrows et al. 2000).

1.5 Chemistry of Selected Elements

This section summarizes the chemistry of C, N, O, the major rock-forming elements Ca, Al, Mg, Si, Fe, Cr, and of Li and the other alkali elements (Na, K, Rb, Cs) that significantly influence the spectra of low mass objects. The chemistry is derived from thermochemical equilibrium computations, and where necessary kinetic considerations (see Fegley & Lodders 1994, Lodders 1999, 2002, Lodders & Fegley 2002). The atmospheric chemistry for Jupiter and Saturn of essentially all naturally occurring elements is discussed by Fegley & Lodders (1994) and is a useful guide for the chemistry of other elements in the coolest low mass objects.

However, some differences may arise because Jupiter and Saturn are enriched in elements heavier than He relative to solar abundances.

Relatively recently the solar abundances of C and O were significantly revised downward and abundances of many other elements were updated (see Lodders 2003). The solar abundances are routinely used to model objects of solar metallicity hence the revisions in solar abundances change previous results for the temperature and pressure dependent distribution of the elements between gases and condensates. However, compared to previous results the new solar abundances do not change the types of gases and condensates that are present in substellar objects and only cause relative shifts of species abundances as functions of temperature and pressure. Here all results from thermochemistry are shown for the solar system abundances in Lodders (2003).

The major chemistry features as a function of total pressure (Ptot) and temperature are summarized in Figs. 1.2a,b and are described in the following subsections. The Figures also show model atmospheres for an M dwarf (T_{eff} = 2200 K, dust-free; Tsuji et al. 1996a), Gl 229B (T_{eff} = 960 K, Marley et al. 1996), and Jupiter to give some orientation which chemical features are relevant for objects of different mass.

1.5.1 Carbon, nitrogen, and oxygen chemistry

After hydrogen, C, N, and O are the next most abundant chemically reactive elements. The review of the C, N, and O chemistry is kept brief because a detailed description is in Lodders & Fegley (2002). Much of the chemistry and therefore the spectral appearance of substellar objects is governed by the distribution of C and O between water, methane, and carbon monoxide according to the net thermochemical reaction

$$CO + 3\,H_2 = CH_4 + H_2O$$

The distribution of C and O between these gases also determines the stability of many other gases and condensates containing other elements, which makes this equilibrium of fundamental importance for understanding the chemistry of substellar objects.

Figures 1.2a and b illustrate carbon equilibrium chemistry as a function of P and T. The dotted curve labeled $CO = CH_4$ is the equal abundance curve for these two gases. Carbon monoxide is more abundant than CH_4 in the lower pressure and higher temperature region to the left of the curve (the CO field). Methane is more abundant than CO in the higher pressure and lower temperature region to the right of the curve (the CH_4 field). The shift from CO as the major C-bearing gas to CH_4 as the major C-bearing gas is gradual and both gases are always present, even if one is overwhelmingly dominant. This point is illustrated by Figure 2 in Fegley and Prinn (1989) which shows logarithmic CO/CH_4 contours of ... 0.01, 0.1, 1, 10, 100 ... as a function of temperature and P_{tot} for solar elemental abundances of H:C:O = 1446:1:0.6.

Late M dwarfs with relatively high temperatures and low total pressures in their outer atmospheres plot in the CO field (Figs. 1.2a,b). In their observable

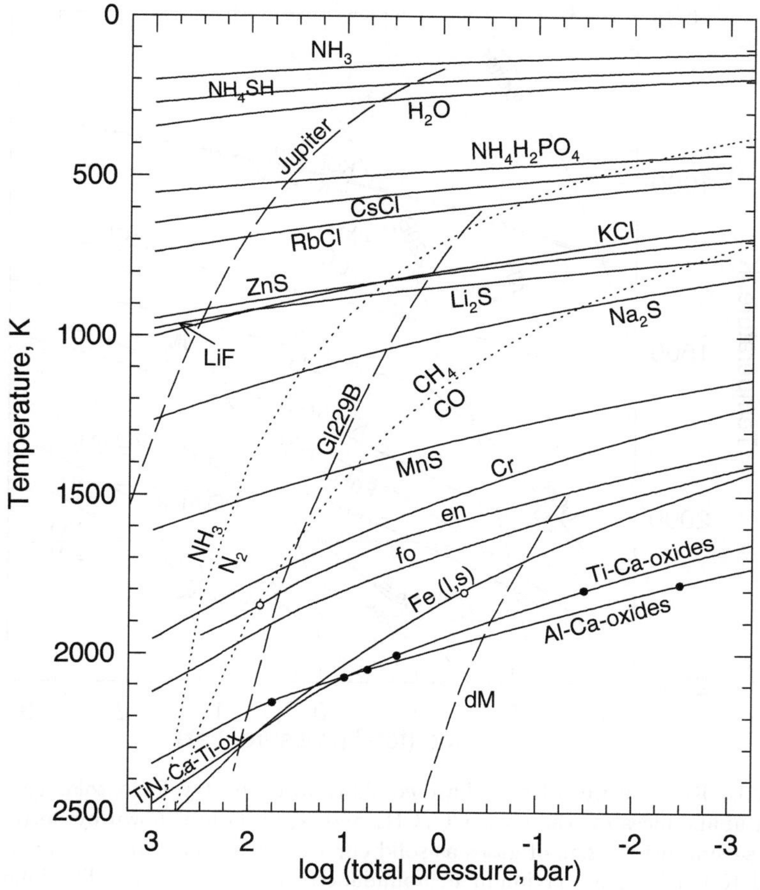

Fig. 1.2a. Major features of the equilibrium chemistry for a solar gas as a function of temperature and total pressure. The condensation temperatures of major element and alkali element condensates are the solid lines. The constituent condensate elements are removed from the gas above these curves. The curve labeled "Ca-Al-oxide" shows which Al-bearing condensate forms first as a function of total pressure. At low pressures, corundum forms. With increasing pressure hibonite, grossite, and gehlenite are the first Al-bearing condensates and the transitions are indicated by the black dots on the Ca-Al-oxide curve. Similarly, the curve "Ca-Ti-oxide" shows that at low total pressure, perovskite is the first Ti bearing condensate, then $Ca_4Ti_3O_{10}$, then $Ca_3Ti_2O_7$, and at the highest P_{tot}, TiN. The white dots on the enstatite and iron condensation curves indicate melting points. The long-dashed curves indicate atmospheric P-T conditions for Jupiter, Gl 229B ($T_{eff} = 960\,K$, Marley et al. 1996), and a M-dwarf ($T_{eff} = 2200\,K$, dust-free; Tsuji et al. 1996a). The dotted curves show equal gas abundances for the pairs CO/CH_4, and N_2/NH_3

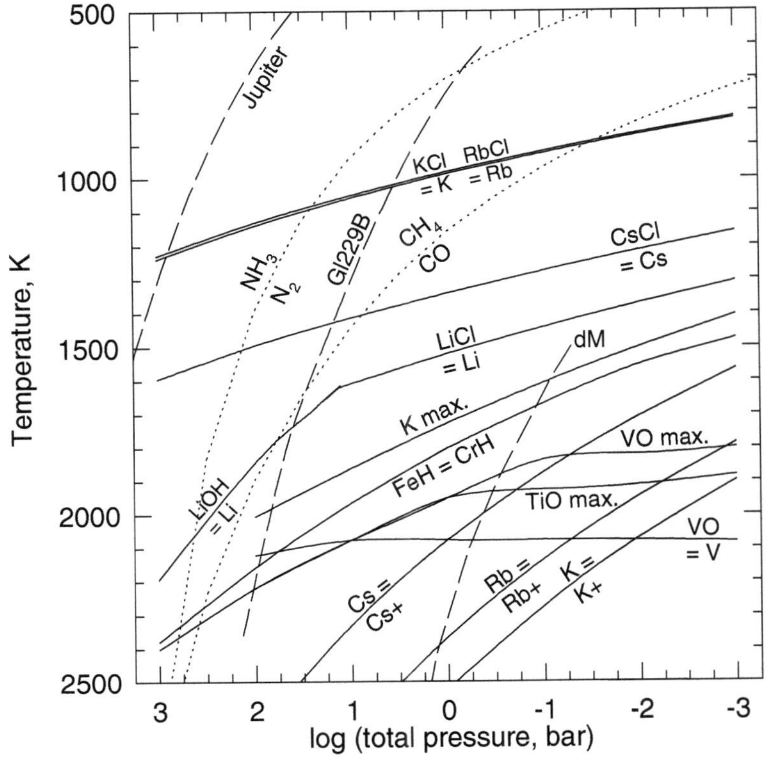

Fig. 1.2b. Key features of gas phase equilibrium chemistry for a solar gas. Curves of equal abundances for the pairs CO/CH_4 and N_2/NH_3 are shown by dotted curves and for several other pairs of gases as solid curves. Also shown are curves where TiO, VO and K reach their maximum in abundances (e.g., TiO max.) The long-dashed curves indicate atmospheric P-T conditions for Jupiter, Gl 229B, and an M-dwarf (see Fig. 1.2a).

atmospheres essentially all C is bound to CO and most of the rest of oxygen not in CO is bound to water. On the other hand, Jupiter's atmospheric P-T profile falls into the methane field and only a very small amount of CO is present (but also see kinetics below) and essentially all C is found as methane and O as water.

The P-T profile of the relatively late T dwarf Gl 229B (T6) with an effective temperature of 970 K falls into the methane field but crosses into the CO field at about 1600 K. Thus, CO is an abundant gas in the deeper atmosphere of Gl 229B and convective mixing of kinetically stable CO from the deeper interior into the observable atmosphere can occur. For example Noll et al. (1997) detected the fundamental band of CO (4.7 μm) in Gl 229B, revealing a CO/CH_4 ratio $\sim 10^3$ times larger than expected from chemical equilibrium (see Fegley & Lodders 1996, Griffith & Yelle 1999, Saumon et al. 2000, 2003, and kinetics below).

The L/T dwarfs transition is determined by the appearance of abundant methane and the $CO = CH_4$ boundary can provide an estimate for the effective temperature at the L/T transition. At the total pressures characteristic of substellar dwarf atmospheres methane is only more abundant than CO at temperatures below 1500–1200 K (Fig. 1.2b). This temperature range is in accord with temperature estimates at the L/T transitions listed in Table 1.2. However, the kinetics of CO destruction (see below) is also important in late L and early T dwarfs because it affects methane abundances which in turn affect the onset of methane bands. It also affects the atmospheric structure because methane is the second strongest absorber after H_2O in T dwarfs (e.g., Saumon et al. 2003).

Figure 1.2 also shows the curve of equal gas abundances for molecular nitrogen and ammonia ($N_2 = NH_3$, dotted curve). This is shifted to higher P_{tot} and lower temperatures than the $CO = CH_4$ boundary. This makes the N_2/NH_3 a promising temperature probe for the coolest T dwarfs.

Ammonia was first detected in the H and K bands of the near-IR spectrum of the T6 dwarf Gl 229B (Saumon et al. 2000). The P-T profile of Gl 229B stills falls into the field where N_2 is more abundant than NH_3 but approaches the $NH_3 = N_2$ boundary at lower T where NH_3 abundances may become large enough for detection. The models by Burrows et al. (2003) predict observable amounts of NH_3 at 1.5, 1.95 μm (H band) and 2.95 μm (K-band) for substellar dwarfs with effective temperatures below 600 K. These expectations are consistent with the observations by Knapp et al. (2004) and Golimowski et al. (2004) that NH3 absorptions are absent in the H and K bands for the coolest known T dwarf 2MASS J04151954-0935066 with $T_{eff} \sim 700$ K.

Ammonia absorption in T dwarfs is best detectable at mid-infrared wavelengths, and NH_3 absorptions at 7.8 and 10 μm were recently reported by Roellig et al. (2004) in spectra for the composite T1/T6 binary system ε Ind (Ba/Bb) with the Spitzer Space Telescope. Note that one of the binary components has essentially the same spectral type as Gl 229B so that the presence of NH_3 in T6 dwarfs seems certain. The mid-IR absorption features of NH_3 seem to be absent in the spectrum of a L8 dwarf (Roellig et al. 2004). Given that the L/T transition is marked by the onset of methane bands in the near IR and that there are considerable temperature differences between the $CO = CH_4$ and $N_2 = NH_3$ boundaries, large detectable amounts of ammonia are unlikely near the L/T transition. Instead, the appearance of ammonia in the near- and mid infrared bands will probably become a marker for the T/Y dwarf transition.

However, kinetics (see below) are also important for the N_2 to NH_3 conversion. For example, ammonia was found to be 50% depleted in the K band spectrum of Gl 229B which suggests that N_2-rich gas is transported upward in the atmosphere (Saumon et al. 2000). Observations of NH3 in mid-infrared spectra (such as reported by Roellig et al. 2004) probe further levels of the atmospheres. Quantitative analyses of NH3 bands at different atmospheric levels then should provide useful information on the mixing processes in substellar dwarf over several pressure scale heights (Saumon et al. 2000, 2003).

The larger abundances of O and N lead to massive cloud layers at the lowest temperatures when water, NH_4SH, and solid NH_3 condense (Fig. 1.2a) which effectively remove all N and O from the atmosphere above the clouds. However, these clouds are not expected within the atmosphere of the latest T dwarfs. As mentioned earlier, objects with water clouds will be members of a new spectral class (Y dwarfs) or larger planetary objects.

1.5.2 Refractory rock-forming elements: Al, Ca, Ti and V

The major rock forming elements (Al, Ca, Ti) can form high temperature condensate cloud layers. Although not a major element, V is on the list of elements here because it is observed in dwarfs and its chemistry is similar and related to that of Ti. The condensation temperatures of Ca-Al-oxides and Ca-Ti-oxides are shown in Fig. 1.2a by the solid lines. Curves for equal abundances of certain pairs of gases and curves for maximum abundances of a given gas are shown in Fig. 1.2b.

At temperatures below 2500 K but above temperatures where the first condensates appear (Figs. 1.2a,b), the major gases are monatomic Ca; Al, AlH, Al_2O, AlF, and AlCl; TiO; V and VO (at $P_{tot} = 1$ bar). The high temperature distribution of vanadium is of interest because the V/VO abundance ratio could serve as a good temperature indicator in M-dwarfs. High temperatures favor V as the dominant gas and low temperatures VO, and the boundary where V and VO reach equal abundances (at $T \sim 2080$) is relatively insensitive to P_{tot} (Fig. 1.2b).

The more pressure sensitive transition from Ti to TiO occurs at temperatures above 2500 K for the total pressures shown here. Once VO is the major V-bearing gas, VO_2 gas increases in abundance with decreasing temperature, and similarly, TiO_2 gas gains at the expense of TiO gas. The curves where TiO and VO gas abundances reach their maximum are shown in Fig. 1.2b. At $P_{tot} > 10$ bar, VO and TiO remain the dominant gases until all V and Ti gases are removed into condensates. This is the reason why the maximum TiO and VO abundances coincide with the condensation temperatures of the Ca-Ti-oxides.

When a condensate appears, its constituent elements are removed from the gas of the overlaying, cooler atmosphere above the condensate cloud. The abundances of *all* gases that contain the constituent element(s) of the condensate drop, e.g., once Ti condenses, the abundances of gaseous TiO, Ti, TiO_2, etc. are all reduced.

The most refractory major condensates contain Al-, Ca-, and Ti (Fig. 1.2a) and the specific mineralogy of the initial condensate depends on total pressure (Lodders 2002). The first Ca-Ti-bearing condensate as a function of P_{tot} are perovskite ($CaTiO_3$) below 0.03 bar, $Ca_4Ti_3O_{10}$ from 0.03 to 3.2 bar, $Ca_3Ti_2O_7$ from 3.2 to ∼10 bars, and above ∼10 bars, TiN (osbornite) replaces the Ca-Ti-oxides as the first Ti-bearing condensate. The P-T range at which a certain Ca-Ti-oxide is stable is indicated by black dots on the "Ca-Ti-oxide" condensation curve in Fig. 1.2a. Vanadium condenses into solid solution with Ca-Ti-oxides and can also form a solid solution with TiN at the highest total pressures.

The gradual disappearance of TiO and VO bands from the spectra of early L dwarfs is most plausibly explained by condensate formation. Once the refractory elements Ti and V are removed into cloud layers, they are no longer available to form gaseous TiO and VO (Fegley & Lodders 1994, 1996, Lodders 2002).

The mineralogy of the first Ca-Al-bearing condensates is also a function of total pressure (Lodders 2002) At the lowest Ptot ($< 3 \times 10^{-3}$ bar) corundum (Al_2O_3) is the initial condensate. With increasing P_{tot}, the initial condensates are hibonite ($CaAl_{12}O_{19}$) from 3×10^{-3} to 5.6 bars, grossite ($CaAl_4O_7$) from 5.6 to ~63 bar, and melilite (a solid solution of gehlenite and akermanite) above 63 bar. The condensation temperatures of the Ca-Al-oxides are similar to those of the Ca-titanates and formation of calcium aluminate clouds leads to the disappearance of the Ca I and Al I lines and CaH bands near the M to L dwarf transition (Table 1.1).

1.5.3 Abundant condensates from rock-forming elements: Fe, Cr, Mg, Si

The solar abundances of Fe, Mg and Si are much larger than those of the refractory rock-forming elements (Ca, Al, Ti) and condensates of Mg, Si and Fe form the major mass of high temperature condensates. Chromium is included in the list here because its chemistry is comparable to that of Fe so they are described together.

After condensation of the refractory elements Ca, Al, and Ti, iron metal forms the next cloud layer (at 1843 K at 1 bar). Before condensation occurs, monatomic Fe is the major Fe-bearing gas in dwarf atmospheres. Chromium behaves similarly – until condensation of Cr-metal (1518 K at 1 bar) the major Cr-bearing gas is monatomic Cr (note that condensation of Cr_2O_3 instead of Cr metal only occurs at lower Ptot $\sim 10^{-3}$ bar).

The hydrides FeH and CrH are always less abundant than the respective monatomic gases and the hydride gas abundances actually decrease with decreasing temperature (at constant Ptot) according to the equilibrium $M(g) + 0.5\, H_2 = MH(g)$ where M = Fe or Cr. We note that other metal hydrides such as NiH and CoH also appear in a similar manner.

Although less abundant than the monatomic gases, the hydrides of Fe and Cr strongly shape the spectra of L and T dwarfs. The major reason for increasingly stronger FeH and CrH bands within the L sequence up to spectral types L4-L6 is that other absorbers such as TiO and VO gradually disappear into condensates (see previous section) so that absorption features from other compounds become more apparent. The decline in FeH band strength starting with spectral type L5 must be related to the condensation of Fe metal which reduces the abundances of all Fe-bearing gases (i.e., monatomic Fe and FeH). Condensation of Cr metal reduces the abundances of monatomic Cr and CrH and occurs at lower temperatures than for Fe (Fig. 1.2a) which explains the decline of CrH band strength at a later spectral type (L6). After Fe metal condensation, abundances of FeH drop and at some point CrH and FeH abundances are equal. This is shown by the curve CrH = FeH in Fig. 1.2b. If spectroscopic properties of CrH and FeH are sufficiently similar, this curve approximates these equal abundances to equal band strengths, which is observed around spectral type L4 (Table 1.1).

At temperatures between Fe and Cr metal cloud condensation, a substantial cloud layer mass develops when Mg and Si condense as forsterite (Mg_2SiO_4) and enstatite ($MgSiO_3$) within a relatively narrow temperature interval (1688 and 1597 K, resp. at 1 bar). This removes opacity sources associated with the major Mg and Si-bearing gases (Mg, Si, SiO, SiS, SiH) from the overlaying gas.

A comparison of the condensation temperatures of the Mg-silicates and those of Fe and Cr and the behavior of the FeH and CrH bands with spectral type suggests that condensation of the Mg-silicate cloud layer should leave its spectral signature in the mid-L sequence (~L4-L6). So far, no direct chemical tracers have been observed that can confirm the presence of Mg-silicate clouds. These clouds may be traceable by the strength of neutral monatomic Mg and Si lines, or in the case of Si, observations of band strengths of SiO, SiS and various Si hydrides (depending on P_{tot}). However, the latter require near to mid IR observations and bands of the Si-bearing molecules could be swamped by those other strong absorbers (e.g., FeH, CrH, CO, H_2O) still present when the silicates form. The Mg I lines may be more prominent that Si I lines because Mg I is the dominant Mg gas whereas Si I gas is a minor fraction of all Si-bearing gases. Very weak Mg I lines can be seen in the J-band spectra of late M dwarfs and are probably present in L2 dwarfs (Mc03) but these weak lines are not useful for monitoring any systematic decrease of Mg through the L-sequence.

Any clouds close to or above the photosphere will leave their mark on infrared colors. The effect of clouds is also important for understanding the weakening of some spectral features in the mid-L sequences and the strengthening again of the same features (FeH, K I, see below) in the early T sequence. Detailed models suggest that the clouds are expected to be most optically thick around L4-L6 (see, Ackerman & Marley, 2001, Burrows et al. 2001, Tsuji 2002, Knapp et al. 2004). Only in cooler objects where the silicate clouds sit deeper below the photosphere does the observable part of the atmosphere become more transparent again.

Figure 1.2a shows that Fe, forsterite, and enstatite may condense as liquids at sufficiently high pressures. For example, Fe liquid forms at $T \geq 1809$ K, forsterite liquid at $T \geq 2163$ K, and enstatite liquid at $T \geq 1851$ K. Many models of dwarf atmospheres now include dust opacities but mainly of solid dust. However, for several objects the derived P-T conditions may fall into the stability fields of liquid condensates. The formation of dust particles or suspended liquid droplets (hazes or aerosols) can have different effects on atmospheric opacities and needs to be considered in atmospheric modeling. Liquid condensates may also dissolve other elements, have complex compositions, and form at lower temperatures.

1.5.4 The alkali elements Na, K, Rb, and Cs

Atomic alkali element lines are prominent in all L dwarfs and remain detectable in the hottest T dwarfs. With decreasing temperature, more of the alkali atoms convert into halide gases and also condense into sulfide and/or halide clouds. As before, we follow the chemistry from high to low temperatures in Figs. 1.2a,b and use 1 bar as reference total pressure for the discussion.

Monatomic Na and K are the major Na and K gases at the highest temperatures relevant for L and T dwarfs; in M dwarfs K^+ is important in addition to neutral K (Fig. 1.2b). Monatomic Na remains the dominant Na gas until Na_2S condensation at ~1000 K but the abundances of other Na-bearing gases such as NaCl steadily increase with decreasing temperature. In contrast, KCl becomes the dominant K-bearing gas at temperatures below the boundary at which monatomic K and KCl abundances are equal (K = KCl; ~980 K at 1 bar, see Fig. 1.2b). Formation of a KCl condensate cloud layer at ~800 K leads to removal of all K-bearing gases from the overlaying atmosphere. Thus, monatomic Na is mainly depleted through condensation of Na2S and monatomic K is depleted by conversion into KCl gas. These chemical changes are expected for temperatures below 1000 K (at 1 bar).

The strength of the Na I doublet (0.5890 & 0.5896 μm) begins to decline and to broaden at L1 and is barely perceptible at L8. The spectral region of the K I resonant doublet lines (0.7665 & 0.7699 μm) develops into a broad depression throughout the L sequence (K99). This behavior is due less to the chemical changes because above 1200 K, the lowest estimate for T_{eff} at the L/T transition (Table 1.2), monatomic Na and K are still the most abundant Na and K-bearing gases, although their chlorides become more abundant as temperature decreases. Instead, pressure-broadening of the Na I and K I resonance doublets diminishes the Na and K lines strengths.

Rubidium and Cs have the lowest ionization potentials of naturally occurring elements so Rb^+ and Cs^+ ions are the dominant gases at the highest temperatures shown in Fig. 1.2b. The abundances of Rb^+ and Cs^+ decrease with decreasing temperature and the curves labeled Rb^+ = Rb and Cs^+ = Cs indicate where the ion and neutral atom abundances are equal. At intermediate temperatures monatomic Rb and Cs are the most abundant Rb and Cs gases. RbCl and CsCl gas become more abundant at lower temperatures. The equimolar curves (Rb = RbCl and Cs = CsCl) are reached at 987 and 1342 K, respectively (at 1 bar; Fig. 1.2b). Further drops in temperature then allow condensation of solid RbCl (607 K at 1 bar) and CsCl (542 K at 1 bar) which depletes the atmosphere in gaseous RbCl and CsCl.

1.5.5 Lithium chemistry

Lithium is of particular interest because low mass dwarfs (<0.065 M_{sun}) are not expected to burn Li and their protosolar abundances of Li should be preserved. In that case, the substellar nature of a low-mass object could be confirmed by the observation of monatomic Li (Rebolo et al. 1992). However, this Li-test is of limited use in cooler dwarfs because the abundance of atomic Li (gas) is reduced below the bulk Li abundance when other Li-bearing gases such as LiOH and LiCl form and consume monatomic Li.

Lithium chemistry is more sensitive to P_{tot} than the chemistry of the other alkali elements, and several other gases (LiOH, LiH, LiF) gases are more important in addition to LiCl (Lodders 1999). Monatomic Li is the dominant Li-bearing gas down to 1520 K (at 1 bar) where LiCl becomes the major Li-bearing gas (Fig. 1.2b). At P_{tot} above ~30 bar, monatomic Li converts to LiOH which is then the most abundant Li gas. The Li = LiOH boundary (restricted to high

P_{tot}) is much more sensitive to total pressure than the Li = LiCl boundary at lower P_{tot}. Additional Li-bearing gases (not shown) become abundant at lower temperatures until condensation of Li_2S (at low P_{tot}) and LiF (at high P_{tot}) occurs (Fig. 1.2a.)

Lithium chemistry is influenced by the position of the $CO = CH_4$ boundary and LiOH is generally more important when CH_4 is more abundant than CO, and LiCl is generally more important when CO is more abundant than methane. The reason for this is that more H_2O is available when CH_4 forms via: $CO + 3H_2 = CH_4 + H_2O$ so that the reaction $LiCl + H_2O = LiOH + HCl$ also proceeds.

The increasing formation of Li-bearing molecules at lower temperatures places limits on the use of monatomic Li as the sole measurement of the Li content in a cool atmosphere. In order to obtain a measurement of the bulk Li abundance in L dwarfs the abundance of LiCl needs to be measured in addition to monatomic Li. Recently Weck et al. (2004) determined the spectroscopic properties of LiCl gas, which, in principle, would allow such determinations. However, as noted by Weck et al. (2004), suitable bands of LiCl are only in the mid infrared (15.8 μm) and these are most likely also blended by water absorptions.

1.5.6 Notes on a few other element condensates

For reference, we plotted some condensates of other, more abundant elements (Mn, Zn, P, S) expected in substellar dwarf atmospheres. Manganese sulfide (MnS) condenses between Cr-metal and Na_2S, and is expected to form a small cloud layer in late L-dwarfs (at 1340 K at 1 bar). Another sulfide, ZnS condenses at 800 K (at 1 bar), at similar temperatures as KCl in dwarfs around the L/T transition. A low temperature condensate that may appear already in very late T dwarfs is $NH_4H_2PO_4$ and is certainly expected for Y dwarfs. Another very low temperature condensate, NH_4SH, should also appear as a larger cloud layer in Y dwarfs since this condensate removes most of the abundant H_2S from the atmosphere. Details for the chemistry of these elements are not yet described for substellar dwarf atmospheres, but their chemistry in Jupiter and Saturn is described in Fegley & Lodders (1994).

1.6 Kinetics

Thermochemical reactions in the atmospheres of planets and low mass substellar objects take place in dynamic environments with only limited time available for reactions to reach chemical equilibrium. Furthermore, different chemical reactions take place at different rates, which generally decrease exponentially with decreasing temperature.

Since the mid 1970s, telescopic observations showed CO, PH_3, GeH_4, and AsH3 on Jupiter and Saturn at abundances much greater than expected from chemical equilibrium considerations. Vertical convective mixing brings these species to observable regions of Jupiter's and Saturn's atmospheres from deeper levels where their chemical equilibrium abundances are orders of magnitude larger (Fegley & Lodders 1994). Similar effects may be important for CO, N_2, PH_3, GeH_4, AsH_3, and HCN in the atmospheres of sub-stellar objects (Fegley & Lodders 1996). We briefly review the theory underlying thermochemical disequilibrium chemistry in planetary and sub-stellar atmospheres and then describe possible consequences for chemistry of C, N, Cs, and Fe.

Carbon monoxide is a good example to explain the basic theory. As mentioned earlier, CO is the dominant carbon gas at high temperatures and low pressures, while CH_4 is the dominant carbon gas at low temperatures and high pressures. The CO equilibrium abundance decreases strongly with decreasing temperature. The altitude range over which [CO], the CO number density, decreases by a factor of e is the CO chemical scale height $h_{\rm chem}$

$$h_{\rm chem} = -\frac{[{\rm CO}]/t_{\rm chem}}{\frac{\rm d}{{\rm d}z}([{\rm CO}]/t_{\rm chem})}$$

The $t_{\rm chem}$ in this equation is the chemical lifetime for CO destruction. It is defined by

$$t_{\rm chem} = -\frac{[{\rm CO}]}{{\rm d}[{\rm CO}]/{\rm d}t}$$

The conversion between CO and CH_4 occurs via the net thermochemical reaction

$$\rm CO + 3\,H_2 = CH_4 + H_2O$$

However, three H_2 molecules do not simultaneously collide with a CO molecule to yield one CH_4 and one H_2O molecule. Instead, the net reaction plausibly proceeds via a series of elementary reactions (i.e., the actual chemical steps that take place) such as that proposed by Prinn & Barshay (1977):

$$\begin{array}{ll}
\rm H_2 = H + H & \\
\rm CO + H_2 = H_2CO & \text{(formaldehyde)} \\
\rm H_2CO + H_2 \rightarrow CH_3 + OH & \\
\rm CH_3 + H + M = CH_4 + M & \\
\rm H + OH + M = H_2O + M & \\
\hline
\rm CO + 3\,H_2 = CH_4 + H_2O & \text{Net Reaction}
\end{array}$$

Here the M that appears in some of the equations is any third body, statistically H_2 or He in the atmosphere of a sub-stellar mass object, and it is necessary to absorb the energy released by forming the H_3C-H bond in methane and the H-OH bond in water. The = sign denotes reactions that are in equilibrium in the hot, deep atmospheres of gas giant planets and sub-stellar mass objects. The direction of a reaction is indicated by the arrow. This series of elementary reactions sums to $\rm CO + 3\,H_2 = CH_4 + H_2O$, but the actual reaction sequence is more complicated than suggested by that one reaction.

The chemical lifetime $t_{\rm chem}$ for CO destruction is computed from the rate of the slowest elementary reaction (the rate determining step), which is the reaction

$$\rm H_2CO + H_2 \rightarrow CH_3 + OH$$

The t_{chem} for CO destruction is

$$t_{chem} = \frac{[CO]}{[H_2CO][H_2]k}$$

The rate constant k for the rate-determining step is

$$k = 2.3 \times 10^{-10} \exp(-36{,}200/T)\,cm^3\,s^{-1}$$

The t_{chem} and h_{chem} for CO decrease dramatically with decreasing temperature in atmospheric regions where the activation energy factor ($E_a/R = 36{,}200$) is much larger than temperature T. Carbon monoxide is destroyed by conversion to $CH_4 + H_2O$ in gas parcels transported upward by convective mixing. The convective mixing time t_{mix} is

$$t_{mix} \sim H^2/K_{eddy}$$

The H and K_{eddy} in this equation are the pressure scale height and the vertical eddy diffusion coefficient. The pressure scale height is a function of atmospheric temperature, mean molecular weight (μ), and gravity

$$H = \frac{RT}{\mu g}$$

The vertical eddy diffusion coefficient is estimated from the convective heat flux and is typically 10^7–$10^9\,cm^2\,s^{-1}$ for the tropospheres of Jupiter and Saturn. Destruction of CO stops at the level where

$$t_{chem}(CO) = t_{mix} \sim H^2/K_{eddy}$$
$$h_{chem}^2 \ll H^2$$

This level is the quench level. For example, Prinn & Barshay (1977) calculated a quench level of 1100 K for CO on Jupiter where h_{chem} (CO) ~25 km versus $H \sim 220$ km.

Several other molecules such as HCN and N_2 are analogous to CO because they also have bonds with large dissociation energies (CO 11.09 eV, N_2 9.76 eV, CN 7.75 eV). The activation energy factors for destruction of N_2 ($E_a/R = 81{,}515$) and HCN ($E_a/R = 70{,}456$) are also much larger than T in the atmospheres of Jovian planets and sub-stellar mass objects. Thus, HCN and N_2 are also expected to be quenched at high temperature with the exact temperature depending upon the considerations above.

In contrast, several other species such as monatomic Cs and FeH are unquenchable in the atmospheres of gas giant planets and sub-stellar mass objects because the activation energy factors are effectively zero. For example, with decreasing temperature Cs is probably converted into CsCl by the elementary reaction

$$Cs + HCl \rightarrow CsCl + H \qquad \Delta H = -1.6\,kJ\,mol^{-1}$$

Clay & Hussain (1990) measured a rate constant $k = 5.1 \times 10^{-12}\,cm^3\,s^{-1}$ at 828 K. The estimated temperature dependence of the rate constant is

$$k = 5 \times 10^{-10} \exp(-3800/T)\,\mathrm{cm^3\,s^{-1}}$$

This estimate is based on the measured value of Clay and Hussain (1990) and the temperature dependences for the rate constants for the analogous conversions of Li to LiCl, Na to NaCl, and K to KCl. The chemical lifetime of Cs is

$$t_{\mathrm{chem}}(\mathrm{Cs}) = (k[\mathrm{HCl}])^{-1}$$

For example, calculations for an otherwise solar composition gas with a metallicity of -0.5 dex give t_{chem} (Cs) values ≤ 1 second at 1 bar for $T \geq 1100$ K. This shows that monatomic alkali to alkali halide conversions occur on very fast time scales that are much faster than plausible mixing timescales (such as on Jupiter). Griffith & Yelle (2000) also concluded that the observation of Cs in Gl 229B, where CsCl instead of Cs is expected at the low temperatures, cannot be caused by the presence of Cs from convection as initially proposed by Burrows & Sharp (1999). Instead it is more likely that the observations of Cs probe a deeper, hotter atmospheric level where Cs is the dominant Cs gas (see Fig. 1.2b).

No kinetic data are available for destruction of FeH gas. However, several plausible destruction reactions are highly exothermic at 1000 K and should have activation energy factors close to zero

$$\begin{array}{ll}
\mathrm{FeH + OH \rightarrow H_2O + Fe} & \Delta\mathrm{H} = -344\,\mathrm{kJ\,mol^{-1}} \\
\mathrm{FeH + CH_3 \rightarrow CH_4 + Fe} & \Delta\mathrm{H} = -288\,\mathrm{kJ\,mol^{-1}} \\
\mathrm{FeH + H \rightarrow H_2 + Fe} & \Delta\mathrm{H} = -282\,\mathrm{kJ\,mol^{-1}} \\
\mathrm{FeH + HS \rightarrow H_2S + Fe} & \Delta\mathrm{H} = -228\,\mathrm{kJ\,mol^{-1}} \\
\mathrm{FeH + FeH \rightarrow H_2 + 2Fe} & \Delta\mathrm{H} = -120\,\mathrm{kJ\,mol^{-1}}
\end{array}$$

These reactions may proceed at rates close to the gas kinetic rate (10^{-10}–$10^{-9}\,\mathrm{cm^3\,s^{-1}}$) and have chemical lifetimes much less than the vertical mixing times in the atmospheres of sub-stellar mass objects.

Thus, the observations of strengthening FeH bands within the T dwarf sequence cannot be related to convective upward mixing of FeH gas as favored by Nakajima et al. (2004). This possibility was already considered by Burgasser et al. (2002b) who concluded that the fragile molecule FeH (and, similarly CrH with low dissociation energies 1.63 eV and 1.93 eV, respectively) precludes quenching of the destruction reactions of FeH and CrH from high temperatures.

1.7 Summary: A Chemical Temperature Scale

Chemistry can be used to develop a chemical temperature scale', which should help to constrain effective temperatures within the L and T dwarf sequences. A chemical temperature scale is model-atmosphere independent and only uses the appearance (e.g., molecular gas formation) and disappearance (e.g., removal of gases by condensation) of certain compounds. However, the chemical temperature scale cannot easily be related to an effective temperature scale because observations at different spectral wavelength track chemical changes that occur

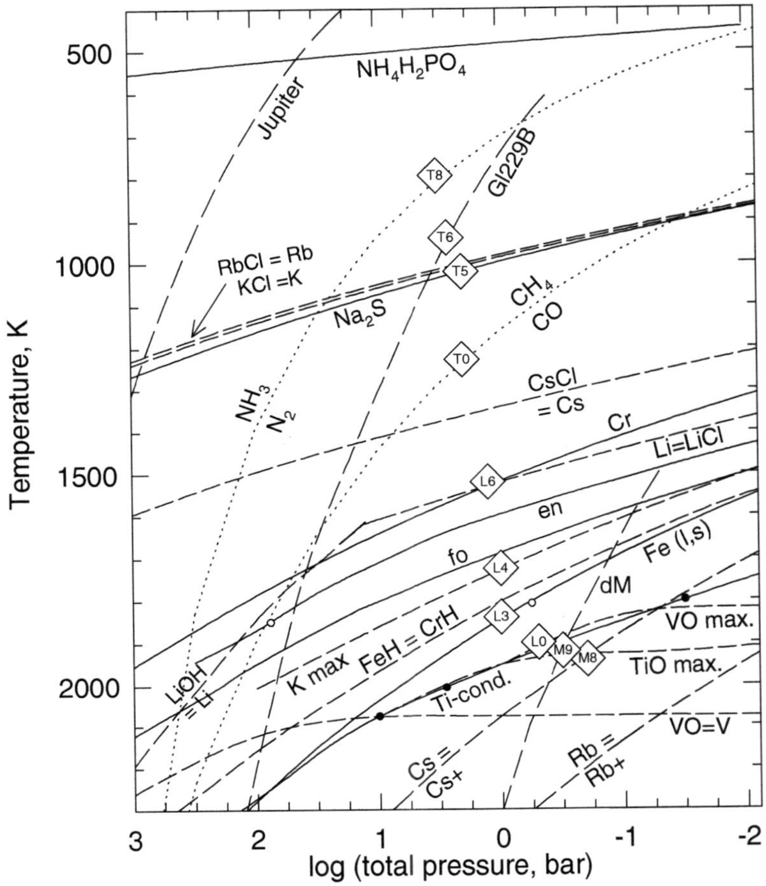

Fig. 1.3. A summary of the major chemistry features from Figs. 1.2a,b. The large diamonds indicate the temperature dependent chemical characteristics for a given spectral type. These "chemical" temperatures are only lower limits to the true effective temperatures. See the text for more explanations.

at different depth (and thus temperatures) in the atmospheres. Furthermore, the temperatures derived from chemistry must be lower limits and are probably 100–200 K less than the effective temperatures because the chemistry observed through the absorption features occurs within or above the photosphere. Keeping these limits in mind, we use the condensation temperatures and curves for gas abundance from Figs. 1.2a,b to relate the chemistry to spectral types. The major chemical features from Figs. 1.2a,b, relevant to the observations (Table 1.1) are shown in Fig. 1.3.

At the high temperature end, the V = VO boundary is located at ∼2080 K over a wide range of total pressures. Temperatures between the V = VO boundary and Ca-Ti-oxides and Ca-Al-oxide condensation curves should be characteristic for mid- to late M dwarfs because TiO and VO are major gases, and all neutral alkali atoms become more abundant. For these objects, the Li abundance derived from the Li I line is representative of the bulk Li abundance and can be used to decide an object's substellar nature.

Chemical temperatures at the L/M dwarf transition using the TiO (g) and VO (g) abundances from thermochemical computations can be derived as follows (see Lodders 2002). In M8 dwarfs TiO bands are at maximum strength (K99, M99), hence the chemical temperature for M8 dwarfs are placed on the maximum TiO gas abundance curve in Fig. 1.3. Similarly, VO bands peak in strength at M9, which places M9 onto the maximum VO gas curve. Total pressures should be lower than ~1 bar because otherwise the maxima of TiO and VO both coincide with the Ca-Ti condensation curve. The position on the total pressure axis is further constrained by the positions of the $Cs = Cs^+$ and $Rb = Rb^+$ boundaries. The lines of monatomic Cs and Rb strengthen beyond M9 in part because the monatomic gases become the most abundant Cs and Rb bearing gases (Figs. 1.2b, 1.3). The chemical temperature marker for M8 is placed in the vicinity where the TiO max curve is crossed by the $Cs = Cs^+$ boundary (Fig. 1.3). This leads to temperatures of 1940 K for M8 and 1920 K for M9.

In L0, TiO and VO bands start to decline and temperatures must be below the Ti- and V-bearing condensate curve (labeled Ca-Ti cond.). Temperatures at the beginning of the L sequence are <1920 K and are tentatively placed at 1900 K for L0. In addition to temperature, the TiO and VO abundances can also uniquely constrain the total atmospheric pressures at the M/L dwarf transition (see $KL0_2$).

A comparison with the relatively large range of estimates for the effective temperatures of 2000–2500 K at the M/L transition (Table 1.2) shows that the chemical temperature of ~1900 K is lower by 100–600 K.

The next major chemical change in the early L sequence is the condensation of metal, forsterite, and enstatite between 1840 and 1600 K (at 1 bar). Useful markers for the chemical temperatures are the curves for FeH = CrH and maximum K I abundances shown in Fig. 1.3. The K I lines peak in strengths and equal band strengths of FeH and CrH are seen in L4 dwarfs. At 1 bar, FeH = CrH occurs at 1800 K and the maximum in K abundance at 1725 K. For illustrative purposes, we place spectral type L3 onto the Fe condensation line since the Fe I lines disappear around L3 (Table 1.1) and type L4 onto the K maximum curve (Fig. 1.3). These temperatures are then also consistent with the position of the FeH = CrH abundance curve. As to be expected the chemical temperature range of 1725–1840 K for L3 and L4 is about 100 K lower than the effective temperatures of 1800–1950 K for L3 and L4 found by Vrba et al. (2004).

The next chemistry markers are at spectral type L6 where CrH bands start to weaken and which seems to be the lowest spectral type at which monatomic Li is still seen (Table 1.1). Interestingly, the condensation curve of Cr metal and the equal abundance curve for Li = LiCl are positioned very close together (Figs. 1.2, 1.3) and intersect at 1520 K (at ~1.3 bar), which is used as chemical temperature for L6. This temperature is only 50 K less than the effective temperature of 1570 K for L6 estimated by Vrba et al. (2004).

The chemical temperature at the L-T transition is related to the $CH_4 = CO$ boundary (Figs. 1.2, 1.3). In the total pressure range of 1.3–3 bar, the temperatures from the $CH_4 = CO$ boundary give 1180–1280 K. The lower bound in total pressure is that obtained for spectral type L6 above, and the upper bound comes from the positioning of the effective temperature of the T dwarf Gl 229B in Fig. 1.3. We plot the chemical temperature for T0 at 1230 K (at 2 bar). This chemical temperature coincides with the effective temperature at the L-T transition given by Burgasser et al. (2002a), but is generally 100–200 K lower than other estimates of the effective temperatures around the L-T transition (Table 1.2). However, only knowledge of methane and CO *abundance ratios* in late L and T dwarfs are good temperature diagnostics (Lodders & Fegley 2002) but this requires quantitative CO and methane abundance determinations which in turn require high-resolution model spectra for analyses.

There are no other direct chemical markers for the higher spectral types within the T sequence. The re-occurrence and peak in strengths of the K I lines and FeH bands around T5 (see Table 1.1) is most plausibly explained by effects from cloud physics and not by chemistry. Burgasser et al. (2002b) suggested that the cloud deck becomes more patchy within the T sequence, which allows to see deeper into the atmosphere so that K I and FeH bands are detectable again. These spectral features then begin to fade at T6 (Table 1.1) and we note that the effective temperature of T6 is lower than the Na_2S condensation temperature (see Fig. 1.3). Thus, the appearance of a new, small cloud deck from Na_2S around spectral types T5/T6 can block again the view into deeper atmospheric layers and is probably partly responsible that FeH and K are no longer seen in the spectra of the late T dwarfs. If so, the Na_2S condensation temperature of 1005–1035 (for P_{tot} of 1.3–3 bar) may reflect the chemical temperature for T5. Finally, effective temperatures of 700–800 K at the end of the T sequence (Table 1.2) are close to the $NH_3 = N_2$ boundary at the 3 bar level (Fig. 1.3).

Acknowledgements

Work supported in part by NSF grant AST-0406963 and NASA Planetary Atmospheres Program grant NAG5-11958.

References

Ackerman, A.S., & Marley, M.S. 2001, Precipitating condensation clouds in substellar atmospheres, Astrophys. J., **556**, 872-884

Allard, F., Hauschildt, P.H., Alexander, D.R., Tamanai, A., & Schweitzer, A. 2001, The limiting effects of dust in brown dwarf model atmospheres, Astrophys. J., **556**, 357-372

Barshay, S.S., & Lewis, J.S. 1978, Chemical structure of the deep atmosphere of Jupiter, Icarus, **33**, 593-611

Basri, G. 2000, Observations of brown dwarfs, ARAA, **38**, 485-519

Basri, G., Mohanty, S., Allard, F., Hauschildt, P.H., Delfosse, X., Martin, E.L., Forveille, T., & Goldman, B. 2000, An effective temperature scale for late-M and L dwarfs, from resonance absorption lines of Cs I and Rb I, Astrophys. J., **538**, 363-385

Burgasser, A.J., Kirkpatrick, J.D., Brown, M.E., Reid, I.N., Burrows, A., Liebert, J., Matthews, K., Gizis, J.E., Dahn, C.C., Monet, D.G., Cutri, R.M., & Skrutskie, M.F. 2002a, The spectra of T dwarfs. I. Near-infrared data and spectral classification, Astrophys. J., **564**, 421-451 (B02)

Burgasser, A.J., Marley, M.S., Ackerman, A.S., Saumon, D., Lodders, K., Dahn, C.C., Harris, H.C., & Kirkpatrick, J.D. 2002b, Evidence for cloud disruption in the L/T dwarf transition, Astrophys. J., **571**, L151-L154

Burgasser, A.J., Kirkpatrick, J.D., Liebert, J., & Burrows, A. 2003, The spectra of T dwarfs. II Red optical spectra, Astrophys. J., **594**, 510-524

Burrows, A., & Sharp, C.M. 1999, Chemical equilibrium abundances in brown dwarf and extrasolar giant planet atmospheres, Astrophys. J., **512**, 843-863

Burrows, A., Hubbard, W.B., Lunine, J.I., Marley, M.S., & Saumon, D. 1999, New ideas in the theory of extrasolar giant planets and brown dwarfs, in Protostars and Planets IV, V., Mannings, A. Boss, S. Russell, eds., (Univ. Arizona Press: Tucson), p. 1339

Burrows, A., Marley, M.S., & Sharp, C. 2000, The near-infrared and optical spectra of methane dwarfs and brown dwarfs, Astrophys. J., **531**, 438-446

Burrows, A., Hubbard, W.B., Lunine, J.I., Liebert, J. 2001, The theory of brown dwarfs and extrasolar giant planets, Rev. Mod. Phys., **73**, 719-765

Burrows, A., Ram, R.S., Bernath, P., Sharp, C.M., & Wilson, J.A. 2002a, New CrH opacities for the study of L and brown dwarf atmospheres, Astrophys. J., **577**, 986-992

Burrows, A., Burgasser, A.J., Kirkpatrick, J.D., Liebert, J., Milsom, J.A., Sudarsky, D., Hubeny, I. 2002b, Theoretical spectral models of T dwarfs at short wavelengths and their comparison with data, Astrophys. J., **573**, 394-417

Burrows, A., & Volobuyev, M. 2003, Calculations of the far-wing line profiles of sodium and potassium in the atmospheres of substellar-mass objects, Astrophys. J., **583**, 985-995

Burrows, A., Sudarsky, D., & Lunine, J.I. 2003, Beyond the T dwarfs: Theoretical spectra, colors, and detectability of the coolest brown dwarfs, Astrophys. J., **596**, 587-596

Chabrier, G., & Baraffe, I. 2000, Theory of low-mass stars and substellar objects, ARAA, **38**, 337-377

Chabrier, G., Baraffe, I., Allard, F., & Hauschildt, P.H. 2000, Evolutionary models for very low-mass stars and brown dwarfs with dusty atmospheres, Astrophys. J., **542**, 464-472

Clay, R.S., & Hussain, D. 1990, Determination of absolute rate data for reactions of ground state atomic caesium, Cs (6 $^2S_{1/2}$), by time-resolved atomic resonance absorption spectroscopy, J. Chem. Res. (S), **1990**, 384-385

Cooper, C.S., Sudarsky, D.M., John A., Lunine, J.I., & Burrows, A. 2003, Modeling the formation of clouds in brown dwarf atmospheres, Astrophys. J., **586**, 1320-1337

Cushing, M.C., Rayner, J.T., Davis, S.P., & Vacca, W.D. 2003, FeH absorption in the near-infrared spectra of late M and L dwarfs, Astrophys. J., **582**, 1066-1072

Dahn, C.C., Harris, H.C., Vrba, F.J., Guetter, H.H., Canzian, B., Henden, A.A., Levine, S.E., Luginbuhl, C.B., Monet, A.K.B., Monet, D.G., Pier, J.R., Stone, R.C., Walker, R.L., Burgasser, A.J., Gizis, J.E., Kirkpatrick, J.D., Liebert, J., &

Reid, I.N. 2002, Astrometry and photometry for cool dwarfs and brown dwarfs, Astron. J., **124**, 1170-1189
Dulick, M., Bauschlicher, C.W., Burrows, A., Sharp, C.M., Ram, R.S., & Bernath, P. 2003, Line intensities and molecular opacities of the FeH F 4Deltai-X 4Deltai Transition, Astrophys. J., **594**, 651-663
Fegley, B., & Lodders, K. 1994, Chemical models of the deep atmospheres of Jupiter and Saturn, Icarus, **110**, 117-154
Fegley, B., & Lodders, K. 1996, Atmospheric chemistry of the brown dwarf Gliese 229B: Thermochemical equilibrium predictions, Astrophys. J., **472**, L37-L39
Fegley, B., & Prinn, R.G. 1985a,Equilibrium and non-equilibrium chemistry of Saturn's atmosphere: Implications for the observability of PH3, N2, CO, and GeH_4, Astrophys. J., **299**, 1067-1078
Fegley, B., & Prinn, R.G. 1985b, Predicted chemistry of the deep atmosphere of Uranus prior to the Voyager 2 encounter, Nature, **318**, 48-50
Fegley, B., & Prinn, R.G. 1986, Chemical models of the deep atmosphere of Uranus, Astrophys. J., **307**, 852-865
Fegley, B., & Prinn, R.G. 1989, Solar nebula chemistry: Implications for volatiles in the solar system, in *The Formation and Evolution of Planetary Systems*, H. Weaver & L. Danly, eds. (Cambridge Univ. Press, Cambridge, UK), 171-211
Geballe, T.R., Saumon, D., Leggett, S.K., Knapp, G.R., Marley, M.S., & Lodders, K. 2001, Infrared observations and modeling of one of the coolest T dwarfs, Gl 570D, Astrophys. J. **556**, 373-379
Geballe, T.R., Knapp, G.R., Leggett, S.K., Fan, X., Golimowski, D.A., Anderson, S., Brinkmann, J., Csabai, I., Gunn, J.E., Hawley, S.L., & 19 coauthors 2002, Toward spectral classification of L and T dwarfs: Infrared and optical spectroscopy and analysis, Astrophys. J., **564**, 466-481 (G02)
Golimowski, D.A., Leggett, S.K., Marley, M.S., & many other co-authors 2004, L' and M' photometry of ultracool dwarfs, Astron. J, **127**, 5316-3536
Griffith, C.A. & Yelle, R.V. 1999, Disequilibrium chemistry in a brown dwarf's atmosphere: Carbon monoxide in Gliese 229B, Astrophys. J., **519**, L85-L88
Griffith, C.A. & Yelle, R.V. 2000, Equilibrium chemistry in a brown dwarf's atmosphere: Cesium in Gliese 229B, Astrophys. J., **532**, L59-L62
Jones, H.R.A., & Tsuji, T. 1997, Spectral evidence for dust in late-type M dwarfs, Astrophys. J., **480**, L39-L41
Kirkpatrick, J.D., Reid, I.N., Liebert, J., Cutri, R.M., Nelson, B., Beichman, C.A., Dahn, C.C., Monet, D.G., Gizis, J.E., Skrutskie, M.F. 1999, Dwarfs cooler than "M": The definition of spectral type "L" using discoveries from the 2 Micron All-Sky Survey (2MASS), Astrophys. J., **519**, 802-833 (K99)
Kirkpatrick, J.D., Reid, I.N., Liebert, J., Gizis, J.E., Burgasser, A.J., Monet, D.G., Dahn, C.C., Nelson B., & Williams, R.J. 2000, 67 Additional L dwarfs discovered by the Two Micron All Sky Survey, Astron. J., **120**, 447-472 (K00)
Knapp, G.R., Leggett, S.K., Fan, X., Marley, M.S., Geballe T.R., Golimowski, D.A., Finkbeiner, D., Gunn, J.E., & many other co-authors 2004, Near-infrared photometry and spectroscopy of L and T dwarfs: The effects of temperature, clouds, and gravity, Astron. J., **127**, 3553-3578
Kumar, S.S., 1963, The structure of stars of very low mass, Astrophys. J., **137**, 1121-1125
Leggett, S.K., Allard, F, Geballe, T, Hauschildt, P.H., & Schweitzer, A., 2001, Infrared spectral energy distributions of late M and L dwarfs, Astrophys. J., **548**, 908-918

Leggett, S.K., Golimowski, D.A., Fan, X., Geballe, T.R., Knapp, G.R., Brinkmann, J., Csabai, I., Gunn, J.E., Hawley, S.L., Henry, T.J., & 9 coauthors, 2002, Infrared photometry of late-M, L, and T dwarfs, Astrophys. J., **564**, 452-465 (L02)

Lewis, J.S. 1969, Observability of spectroscopically active compounds in the atmosphere of Jupiter, Icarus, **10**, 393-409

Liebert, J., Reid, I.N., Burrows, A., Burgasser, A.J., Kirkpatrick, J.D. & Gizis, J.E. 2000, An improved red spectrum of the methane or T dwarf SDSS 1624+0029: The role of the alkali metals, Astrophys. J., **533**, L155-L158

Lodders, K. 1999, Alkali element chemistry in cool stellar dwarf atmospheres, Astrophys. J., **519**, 793-801

Lodders, K. 2002, Titanium and vanadium chemistry in low-mass dwarf stars, Astrophys. J., **577**, 974-985

Lodders, K. 2003, Solar system abundances and condensation temperatures of the elements, Astrophys. J., **591**, 1220-1247

Lodders, K., & Fegley, B. 2002, Atmospheric chemistry in giant planets, brown dwarfs, and low-mass dwarf stars I. Carbon, nitrogen and oxygen, Icarus, **155**, 393-424

Lunine, J.I., Hubbard, W.B., Burrows, A., Wang, Y.P., & Garlow, K. 1989, The effect of gas and grain opacity on the cooling of brown dwarfs, Astrophys. J., **338**, 314-337

Marley, M.S., Saumon, D., Guillot, T., Freedman, R.S., Hubbard, W.B., Burrows, A., & Lunine, J.I. 1996, Atmospheric, evolutionary, and spectral models of the brown dwarf Gliese 229B, Science, **272**, 1919-1921

Marley, M.S., Seager, S., Saumon, D., Lodders, K., Ackerman, A.S., Freedman, R., & Fan, X. 2002, Clouds and chemistry: Brown dwarf atmospheric properties from optical and infrared colors, Astrophys. J., **568**, 335-342

Martin, E.L., Delfosse, X., Basri, G., Goldman, B., Forveille, T., & Zapatero Osorio, M.R. 1999, Spectroscopic classification of late-M and L field dwarfs, Astron. J., **118**, 2466-2482 (M99)

McLean, I.S., McGovern, M.R., Burgasser, A.J., Kirkpatrick, J.D., Prato, L, & Kim, S.S. 2003, The NIRSPEC brown dwarf spectroscopic survey. I. Low-resolution near-infrared spectra, Astrophys. J., **596**, 561-586

Nakajima, T., Oppenheimer, B.R., Kulkarni, S.R., Golimowski, D.A., Matthews, K., & Durrance, S.T. 1995, Discovery of a cool brown dwarf, Nature, **378**, 463-465

Nakajima, T., Tsuji, T., & Yanagisawa, K. 2004, Spectral classification and effective temperatures of L and T dwarfs based on near-infrared spectra, Astrophys. J., **607**, 499-510 (N04)

Niemann, H.B., Atreya, S.K., Carignan, G.R., Donahue, T.M., Haberman, J.A., Harpold, D.N., Hartle, R.E., Hunten, D.M., Kasprzak, W.T., Mahaffy, P.R., Owen, T.C., Way, S.H. 1998. The composition of the Jovian atmosphere as determined by the Galileo probe mass spectrometer, J. Geophys. Res., **103**, 22831-22845

Noll, K.S., Geballe, T.R., & Marley, M.S. 1997, Detection of abundant carbon monoxide in the brown dwarf Gliese 229B, Astrophys. J., **489**, L87-L91

Noll, K.S., Geballe, T.R., Leggett, S.K., & Marley, M.S. 2000, The onset of methane in L dwarfs, Astrophys. J., **541**, L75-L78

Oppenheimer, B.R., Kulkarni, S.R., Matthews, K., & Nakajima, T. 1995, Infrared spectrum of the cool brown dwarf Gl 229B, Science, **270**, 1478-1479

Prinn, R.G., & Barshay, S.S. 1977, Carbon monoxide on Jupiter and implications for atmospheric convection, Science, **198**, 1031-1034

Rebolo, R., Martin, E.L., & Magazzu, A. 1992, Spectroscopy of a brown dwarf candidate in the Alpha Persei open cluster, Astrophys. J., **389**, L83-L86

Reid, I.N, Burgasser, A.J., Cruz, K.L., Kirkpatrick, J.D., & Gizis, J.E.,, 2001, Near-infrared spectral classification of late M and L dwarfs, Astron. J., **121**, 1710-1721

Roellig, T.L., Van Cleve, J.E., Sloan, G.C., Wilson, J.C., Saumon, D., Leggett, S.K., Marley, M.S., Cushing, M.C., Kirkpatrick, J.D., Mainzer, A.K., & Houck, J.R. 2004, Spitzer Infrared Spectrograph (IRS) Observations of M, L, and T dwarfs, Astrophys. J. Suppl. Ser., **154**, 418-421

Saumon, D., Geballe, T.R., Leggett, S.K., Marley, M.S., Freedman, R.S., Lodders, K., Fegley, B., & Sengupta, S.K. 2000, Molecular abundances in the atmosphere of the T dwarf Gl 229B, Astrophys. J., **541**, 374-389

Saumon, D., Marley, M.S., Lodders, K., & Freedman, R.S. 2003, Non-equilibrium chemistry in the atmospheres of brown dwarfs. ASP Conf. Series IAU Symp. 211 on Brown Dwarfs (ed. Martin, E.L.) 345-353

Schweitzer, A., Gizis, J.E., Hauschildt, P.H., Allard, F., & Reid, I.N. 2001, Analysis of Keck HIRES spectra of early L-type dwarfs, Astrophys. J., **555**, 368-379

Schweitzer, A., Gizis, J.E., Hauschildt, P.H., Allard, F., Howard, E.M., & Kirkpatrick, J.D. 2002, Effective temperatures of late L dwarfs and the onset of methane signatures, Astrophys. J., **566**, 435-379

Stephens, D.C., Marley, M.S., Noll, K.S., & Chanover, N. 2001, L-band photometry of L and T dwarfs, Astrophys. J., **556**, L97-L101

Testi,L. D'Antona, F., Ghinassi, F., Licandro, J., Magazz, A., Maiolino, R., Mannucci, F., Marconi, A., Nagar, N., Natta, A., & Oliva, E. 2001, NICS-TNG low-resolution 0.85-2.45 micron spectra of L dwarfs: A near-infrared spectral classification scheme for faint dwarfs, Astrophys. J., **552**, L147-L150

Tokunaga, A.T., & Kobyashi, N. 1999, K-band spectra and narrowband photometry of DENIS field brown dwarfs, Astron. J., 117, 1010

Tsuji, T. 2002, Dust in the photospheric environment: Unified cloudy models of M, L, and T dwarfs, Astrophys. J., **575**, 264-290

Tsuji, T., & Nakajima, T. 2003, Transition from L to T dwarfs on the color-magnitude diagram, Astrophys. J., **585**, L151-L154

Tsuji, T., Ohnaka, K., & Aoki, W. 1996a, Dust formation in stellar photospheres: A case of very low mass stars and a possible resolution on the effective temperature scale of M dwarfs, Astron. Astrophys., **305**, L1-L4

Tsuji, T., Ohnaka, K., Aoki, W., & Nakajima, T. 1996b, Evolution of dusty photospheres through red to brown dwarfs: how dust forms in very low mass objects, Astron. Astrophys., **308**, L29-L32

Tsuji, T., Ohnaka, K., & Aoki, W. 1999, Warm dust in the cool brown dwarf Gliese 229B and spectroscopic diagnosis of dusty photospheres, Astrophys. J., **520**, L119-L122

Tsuji, T., Nakajima, T., & Yanagisawa, K. 2004, Dust in the photospheric environment. II. Effect on the near-infrared spectra of L and T dwarfs, Astrophys. J., **607**, 511-529

Vrba, F.J., Henden, A.A., Luginbuhl, C.B., Guetter, H.H., Munn, J.A., Canzian, B., Burgasser, A.J., Kirkpatrick, J.D., Fan, X., Geballe, T.R., & 5 coauthors, 2004, Preliminary parallaxes of 40 L and T dwarfs from the US Naval Observatory Infrared Astrometry Program, Astron. J., **127**, 2948-2968

Woitke, P., & Helling, C. 2003, Dust in brown dwarfs. Part 2. The coupled problem of dust formation and sedimentation, Astron. Astrophys., **399**, 297-313

2 Neutron Star Structure and Fundamental Physics

T.E. Strohmayer

Abstract: Neutron stars provide a natural laboratory for the study of matter under the most extreme conditions of density, pressure and magnetic fields known. Precise measurements of neutron star masses and radii may provide one of the only means to constrain important aspects of dense matter physics as well as fundamental particle physics. Moreover, neutron stars may contain states of matter not present anywhere else in the observable universe, for example, an exotic color-flavor-locked (CFL), superfluid quark matter phase. Here, I discuss and review recent efforts to measure neutron star masses and radii in order to use them as probes of fundamental physics. Such efforts have focused on measurements of X-ray pulsation signals from both isolated and accreting neutron stars, as well as observations of the thermal emission from neutron star surfaces. Indeed, recent X-ray observations have provided the first evidence of discrete, redshifted spectral features from neutron star surfaces, which may eventually provide precise mass and radii estimates, however, uncertainties remain in their interpretation. Fast X-ray timing observations of X-ray bursting neutron stars have revealed spin modulation pulsations, and recent studies of "magnetar" giant flares have found fast oscillations which may be signatures of global vibrations of neutron star crusts. I will describe how these observations are yielding new insights on neutron star structure.

2.1 Introduction

In some sense neutron stars don't get enough respect. In the public eye they often get second billing to their more compact cousins, the black holes, which have deservedly captured many headlines over the past decade. While black holes are famously said to have no "hair," it is the very fact that neutron stars have lots of "hair", that is, an actual physical surface and interior, rather than an abstract singularity, that makes them such important objects for both astronomers and physicists.

Neutron stars are formed in a Type II supernova explosion, the final collapse of a massive ($\approx 10M_{\odot}$) star after it has exhausted all of its nuclear fuel. The physics of neutron star formation is extremely complex, and it would be premature to claim that we understand it completely, perhaps even crudely. A number of important pieces of fundamental physics; such as the equation of state (EOS) of dense matter, extensive libraries of nuclear and neutrino reaction cross sections, etc. are essential to the story, not to mention the influence of macroscopic processess such as convection, rotation and magnetic fields (see, for example, Fryer & Warren 2004; Prakash et al. 2001). Although the details of exactly how neutron stars form are still uncertain, we are certain that they do form. The theoretical existence of collapsed neutron stars, and their possible association with supernovae, was first discussed in the late '30s by Zwicky and Baade (Baade & Zwicky 1934; Zwicky 1938). The existence of "massive neutron cores" was explored theoretically by Oppenheimer & Volkoff (1939), who showed that an object of $\sim 1M_\odot$ with a diameter of about 20 km could be supported by the Fermi degeneracy pressure of neutrons, but it was not until the discovery of radio pulsars in the late '60s that these objects actually came under observational scrutiny (Hewish et al. 1968).

Today, neutron stars are detected over an impressive range of the electromagnetic spectrum; from radio to gamma-ray energies. More than a thousand of the estimated 10^8 neutron stars thought to inhabit the Galaxy have been observed. Most of these are "ordinary" radio pulsars, but a substantial fraction are primarily observed at X-ray wavelengths; the isolated and accreting neutron stars. Why are neutron stars of such interest to astronomers and physicists? You could call it the "X-factor," meaning that many of the physical parameters which describe matter are at their most extreme in neutron stars. For example, given a mass $M = 1.4M_\odot$, and radius $R = 10$ km, we find a mean density, $\bar{\rho} = 3M/4\pi R^3 = 6.7 \times 10^{14}$ g cm^{-3}, that is comparable to nuclear densities. Neutron stars are easily the most dense objects known. Indeed, at such high density the fundamental theory of particle and nuclear interactions, Quantum Chromodynamics (QCD), is not sufficiently constrained to allow us to calculate the exact state of matter. Perhaps the only way to constrain the theory is to make precise measurements of both the masses and radii of neutron stars.

In addition to extreme densities, neutron stars are also the sites of the strongest magnetic fields accessible to observation. Their dipole field strengths span an impressive range; from less than 10^8 G, in the old, accreting, fast spinning neutron stars (see, for example, Psaltis & Chakrabarty 1999), to more than 10^{15} G in the young, isolated and slowly spinning "magnetars," the strongest magnets known. The term "magnetars" was coined by Duncan & Thompson (1992; Thompson & Duncan 1995), who suggested that strongly magnetized neutron stars might be the sources of the "soft gamma-ray repeaters (SGRs), the prototype of which is the source of the famous March 5, 1979 gamma-ray flare (Mazets et al. 1979; Cline et al. 1980). Currently, four SGR sources are known, three of which have produced similar giant flares, and a second class of objects, the Anomalous X-ray Pulsars (AXPs), are also thought to be a manifestation of extreme neutron star magnetism (see Woods & Thompson 2004 for a recent review).

The global structure of a neutron star is described by the general relativistic equations of hydrostatic equilibrium, also know as the Tolman-Oppenheimer-Volkoff (TOV) equations;

$$\frac{dP}{dr} = -\frac{G\left(m(r) + 4\pi r^3 P/c^2\right)\left(\rho + P/c^2\right)}{r\left(r - 2Gm(r)/c^2\right)}, \tag{2.1}$$

$$\frac{dm(r)}{dr} = 4\pi\rho r^2 . \tag{2.2}$$

Here, P, ρ, and $m(r)$ are the pressure, mass-energy density, and gravitational mass inside radius r. These equations show that the overall structure of a neutron star, it's mass and radius, depends directly on the run of pressure and density in the star, that is, the EOS of matter in the star's interior, which in turn, depends on the fundamental particle interactions of matter. Thus, by accurately measuring neutron star masses and radii we obtain fundamental information about the structure of matter. In practice one usually computes a mass–radius relation, $M(R)$, for a particular theoretical EOS. Figure 2.1 shows a number of theoretical mass–radius relations from different EOS from the recent study by Lattimer & Prakash (2001). The left panel shows several modern EOSs with "normal" (ie. nucleonic constituents), whereas the right panel displays several relations incorporating more exotic components, such as kaon condensates and quark matter. Quark matter configurations differ from standard stars in that they are "self-bound," that is, supported by the strong interactions themselves, therefore the density does not drop to zero at their surfaces. In principle, the mass–radius relation for these stars has a sufficiently different functional form to distinguish them from standard configurations, if the mass–radius relation can be measured with sufficient precision. Indeed, one of the fundamental observational goals of neutron star astrophysics – some might say the "holy grail" – is to accurately place real neutron stars on such a plot of mass versus radius. Although the masses of some neutron stars, particularly pulsars in binary systems, have been very accurately determined, there is no accurate radius information for this sample (Thorsett & Chakrabarty 1999; Nice et al. 2004).

The process of gravitational collapse which forms a neutron star leaves it incredibly hot. Peak temperatures in the core at formation can reach 5×10^{11} K. Most of this heat energy is lost quickly – in a matter of minutes – through emission of energetic neutrinos from the star's core, however, a small fraction is retained and is only more slowly radiated from the surface over time. The loss of heat from the star, in the form of neutrinos and photons, depends on the detailed properties of matter, including the EOS as well as interesting superfluid properties. Thus, measuring the surface temperatures and ages of neutron stars provides another avenue for probing dense matter physics (see, for example Page et al. 2004).

In addition to their dependence on the properties of ultra-dense matter, neutron stars also require extreme gravitational physics to be described accurately. This can be illustrated in a number of ways, for example, the escape velocity from a neutron star surface is $v_{esc} \sim (GM/R)^{1/2} = 1.4 \times 10^5$ km s^{-1}, about 1/2 the speed of light! Moreover, an object of mass m falling onto a neutron star from infinity will attain a kinetic energy $E \approx GMm/R = mc^2(GM/c^2R) \approx 0.42mc^2$, which represents almost half of the rest-mass energy of the object. This explains why accreting neutron stars (and black holes), are such powerful energy sources.

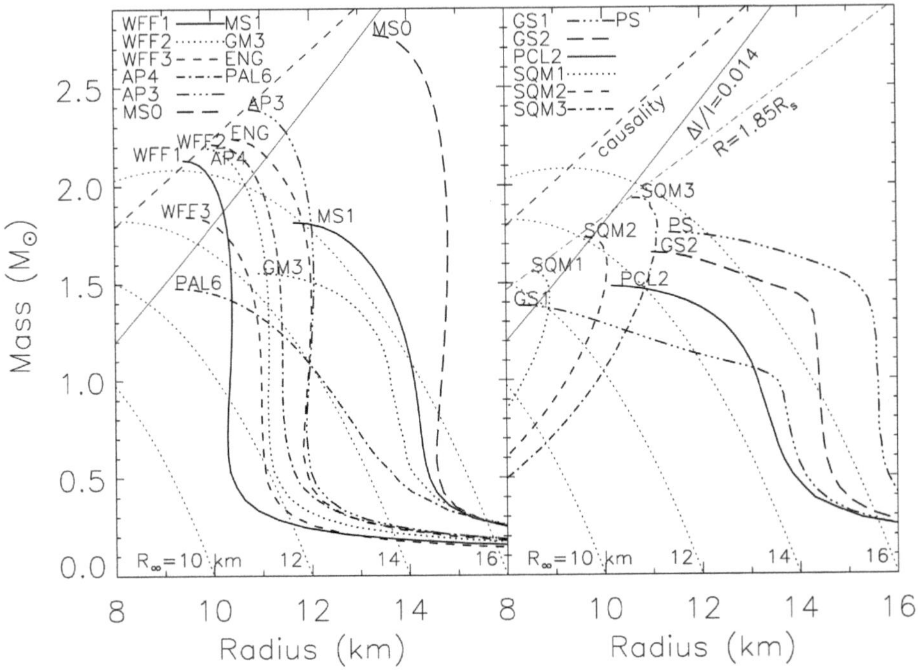

Fig. 2.1. Mass–radius relations for a sample of theoretical EOS from the study of Lattimer & Prakash (2001). The left panel summarizes stars which have nucleonic and/or hyperonic constituents, while the right panel shows relations for stars with more exotic components, such as pion and kaon condensates, or quark matter. See Lattimer & Prakash (2001) for additional details.

Although the theory of general relativity has passed all tests to date, there are predicted effects which can only be tested when the gravitational fields are very strong, such as in the vicinity of a neutron star. Among these effects are the existence of an inner-most stable circular orbit (ISCO), and the dragging of inertial reference frames close to the compact object. By studying emission from neutron star surfaces it may be possible to make inferences about these effects.

One of the most important gravitational effects in the context of studying neutron star structure is the redshift. In order to escape a neutron star's powerful gravitational field photons must give up some of their energy. The amount of energy lost, $\Delta E = E_0[(1 - 2GM/c^2R)^{-1/2} - 1]$ called the gravitational redshift, provides a direct measure of the mass to radius ratio, $\beta = GM/c^2R$, also called the compactness, of the neutron star. So, an observation of, for example, redshifted spectral lines from the surface of a neutron star would provide important information on its structure. However, a redshift measurement alone does not by itself allow a unique determination of both the mass and radius separately. To do that requires additional information. For example, rotation of a neutron star produces a Doppler shift which broadens spectral lines emitted at the surface. If the spin rate is known, then the width of the lines contains information about the radius of the star. Moreover, effects such as Stark broadening, which depends on the surface density (essentially M/R^2) will also broaden spectral lines (Paerels 1997). By accurately measuring both the gravitational redshift and the line profile (how broad the lines are), it should be possible to determine both the mass and radius of the star.

In Sect. 2.2 we will summarize some of the specific fundamental particle physics questions that we hope to address with accurate measurements of neutron star properties. In Sect. 2.3 we move to a brief summary of the recent observational advances which are providing new insights into neutron star structure. This includes recent advances in fast X-ray timing and high resolution X-ray spectroscopy. In Sect. 2.4 we discuss in more detail particular studies using the new methods based on fast X-ray timing, and the implications they have for neutron star structure. In Sect. 2.5 we describe the recent results from and prospects for high resolution X-ray spectroscopy. We conclude in Sect. 2.6 with a discussion of what future observational capabilities might offer in regard to cracking the remaining mysteries of neutron star structure.

2.2 Neutron Star Interiors: Fundamental Physics

In traversing the 10 km or so from the surface to its center the state of matter within a neutron star ranges from what we would call normal (near the surface) to exotic, or just downright bizarre, at the center. If the star has been accreting matter in a binary system, then it will have a surface "ocean" of light elements at the top and progressively heavier nuclei which are the products of complex nuclear burning processes (see Schatz et al. 2001). At higher densities a crystalline crust forms, and nuclei become more neutron-rich. Above a density of $\approx 4 \times 10^{11}$ gm cm^{-3}, known as the neutron drip density, neutrons begin to "leak" out of the nuclei, and the nucleonic lattice coexists with a, possibly superfluid, neutron fluid (Baym & Pethick 1975). The existence of such a superfluid is inferred from observations of sudden jumps (glitches) in the spin rates of pulsars (see Anderson & Itoh 1975; Link, Epstein & Lattimer 1999). As we proceed deeper, towards the core, the crustal lattice eventually dissolves, forming a stew of nucleons, electrons and muons. It is the core of the star, where most of the mass resides and where the densities are highest, that we would most like to probe, for a wide range of theoretical possibilities exist, and a narrowing down of these would tell us much about fundamental physics. For example, exotic species such as pions and kaons may condense out of the mix, and it is possible that a combined phase of hadronic and quark matter forms (Glendenning 1992). The existence of previously unknown states of quark matter is a particularly exciting possibility.

In terrestrial nuclei the number of neutrons is never very much larger than the number of protons, that is, the matter is nearly symmetric. In neutron stars, however, the ratio of neutrons to protons can be much larger. Thus, terrestrial experiments have difficulty probing the full range of the nuclear symmetry energy (the isospin dependence of nuclear forces). A number of studies have recently explored the relationship between the nuclear symmetry energy, the EOS and bulk properties of neutron stars (see Steiner et al. 2005; Lattimer & Prakash 2001). For example, Lattimer & Prakash (2001) demonstrated that the radius of a neutron star is largely dependent on the pressure of matter in the vicinity of nuclear saturation density, about 0.16 baryons fm^{-3}. Since the pressure at this density is largely determined by the density dependence of the nuclear symmetry energy, accurate measurements of neutron star radii can place important constraints on the isospin dependence of nuclear forces. Steiner et al. (2005) have also explored the dependence of the symmetry energy on a number of neutron star and nuclear observables; including neutron star radii and neutron skin thicknesses of heavy nuclei. They find that a number of these observables could also be used to constrain the symmetry energy.

In the past decade there has been a great deal of theoretical work done toward elucidating the phase diagram of matter above supra-nuclear densities. In particular, there is now substantial evidence to support the idea that at high enough densities the ground states of QCD form a family of so-called color-superconducting phases (see Alford et al. 2001; Rajagopal & Wilczek 2001). In these states the quarks form Cooper pairs in a manner analogous to the pairing of electrons in a BCS superconductor. In such theories an important parameter (or parameters) are the gap energies, Δ, which specify the energy of formation of a pair. The QCD gap energies are not precisely known, but are probably in the range from 10 - 100 MeV. At the highest densities, where differences between the quark masses become negligible, the pairing is completely antisymmetric in quark color and flavor, resulting in the so-called color-flavor locked (CFL) phase (Alford, Rajagopal & Wilczek 1999). The CFL state has some remarkable properties, it is a color superconductor, and is electrically neutral, having no electrons in the bulk phase. At somewhat lower densities quark masses become important, in particular the strange quark mass, and one finds additional color-superconducting phases. One of these, the gapless CFL phase (gCFL) has recently been studied by Alford et al. (2005), and has been found to have an unusual heat capacity and neutrino emissivity which modifies the cooling of the star at late times. This would have the effect of keeping stars with this quark matter phase in their cores, and older than about 10 million years, substantially warmer than "standard" cooling calculations would suggest. Thus, a detection of an old, warm neutron star could be an indication of the presence of quark matter in its interior. At present it is not known with any certainty at what densities these phases actually appear, and thus whether they actually occur in real neutron stars, however, further theoretical studies are warranted and might eventually yield a robust observational signature of quark matter.

2.3 An Overview of Recent Observational Advances

There has been enormous growth in both our observational and theoretical understanding of neutron stars and their astrophysical settings in the past decade.

The launch of NASA's Rossi X-ray Timing Explorer (RXTE) about a decade ago opened up a new field of millisecond X-ray timing studies of neutron stars. In addition, the launch of NASA's *Chandra* observatory and ESA's XMM-Newton satellite have provided new, high resolution X-ray spectroscopic capabilities. In this section I will provide a brief, selective overview of some of the recent observational results which have direct implications for our understanding of neutron star structure. In subsequent sections I will discuss in more detail the specific constraints on neutron star properties made possible by these new phenomena.

2.3.1 Fast X-ray timing observations

Observations of accreting neutron star binary systems with RXTE have found a number of millisecond timing signatures which have important implications for neutron star structure. These include the discovery of kilohertz (kHz) quasiperiodic oscillations (QPO) in the X-ray flux from accreting neutron stars (see van der Klis 2005 for a recent review), and the discovery of coherent oscillations during thermonuclear bursts (so called "burst oscillations") from accreting neutron stars (see Strohmayer & Bildsten 2004 for a recent review). These systems, generically known as low mass X-ray binaries (LMXB), comprise a neutron star accreting matter via roche lobe overflow from a low mass ($< 1M_{\odot}$) companion.

The kHz QPOs were first discovered in observations of the accreting neutron star systems 4U 1728-34 and Sco X-1 (Strohmayer et al. 1996; van der Klis 1996). Since then the phenomenon has been identified in an additional 19 objects. Figure 2.2 (left) shows an example of kHz QPOs observed from 4U 1728-34. These oscillations have a rich phenomenology, and have been measured at frequencies above 1300 Hz (van Straaten et al. 2000), close to the orbital frequency in the vicinity of the neutron star surface, so there is little doubt that they are generated close to the star. A comprehensive model of these oscillations is not yet agreed upon, however, it appears likely that the highest observed frequency in a given source is related to the orbital frequency at a location in the inner accretion disk, perhaps the ISCO predicted by general relativity (Miller, Lamb & Psaltis 1998; Stella & Vietri 1998; Zhang, Strohmayer & Swank 1997). Since the orbital frequency at a particular radius depends on the stellar mass, and a particle cannot orbit any closer than at the stellar radius, measurements of the orbital frequencies near the star can, in principle, be used to place constraints on the mass and radius.

Many accreting neutron stars exhibit X-ray bursts; flashes of quasi-thermal X-ray emission produced by unstable thermonuclear burning of matter accumulated on their surfaces (see Strohmayer & Bildsten 2004 for a recent review). These X-ray bursts are powerful probes of neutron stars because they provide a source of X-rays directly from the surface. High frequency oscillations during X-ray bursts were discovered shortly after RXTE observations began (Strohmayer et al. 1996). These "burst oscillations" were first discovered as discrete peaks in Fourier power spectra of time series accumulated during X-ray bursts from some neutron star binaries. For details of their observational characteristics see Strohmayer & Bildsten (2004). An example of a burst oscillation at 363 Hz

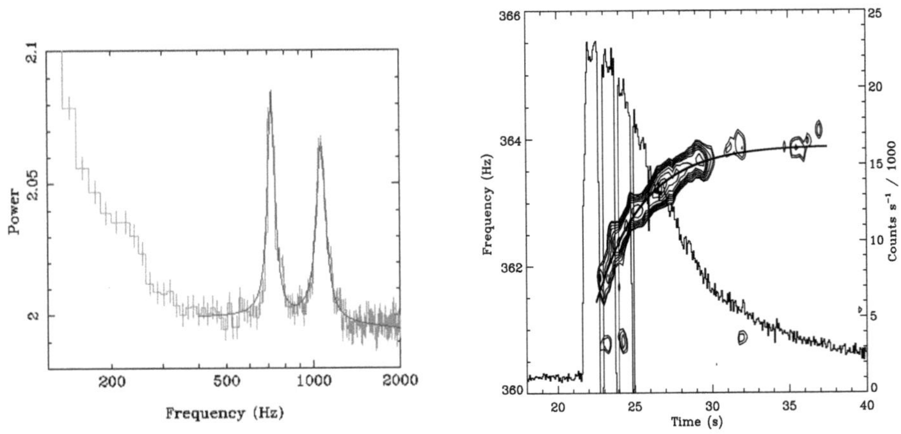

Fig. 2.2. Power spectrum of X-ray flux variations showing high frequency (kHz) quasiperiodic oscillations (QPO) in the low mass X-ray binary 4U 1728-34 (left), and burst oscillations at $\approx$ 363 Hz (ie. the spin frequency) in a thermonuclear burst from the same source (right). The right panel display contours of constant Fourier power versus time and frequency. The separation in frequency between the kHz peaks is approximately the spin frequency. The burst intensity profile is also shown.

from 4U 1728-34 is shown in Figure 2.2 (right). In their discovery paper Strohmayer et al. (1996) suggested that the burst oscillations result from spin modulation of the thermal burst flux, and their is now extensive evidence to support this conclusion. Some of the first pieces of evidence included; the large modulation amplitudes at the onset of bursts, the time evolution of the pulsed amplitude during the rise of bursts (Strohmayer, Zhang & Swank 1997), the coherence of the oscillations (Smith, Morgan & Bradt 1997; Strohmayer & Markwardt 1999; Muno et al. 2000), and the long term stability of the oscillation frequencies (Strohmayer et al. 1998). In the last few years the observations of highly coherent, orbitally modulated pulsations in a "superburst" from 4U 1636-53 (Strohmayer & Markwardt 2002), and burst oscillations at the known spin frequencies of two accreting millisecond pulsars; SAX J1808.4-3658 (Chakrabarty et al. 2003), and XTE J1814-338 (Strohmayer et al. 2003; Watts, Strohmayer & Markwardt 2005) have solidified the spin modulation scenario.

So, we are confident that burst oscillations result from a non-uniform brightness pattern on the neutron star surface. Now, the emission and propagation of photons from the surfaces of rapidly rotating neutron stars are strongly dependent on relativistic effects. For example, the amplitude of pulsations is affected by gravitational light deflection which depends on the compactness, $\beta = GM/c^2R$ (Pechenick, Ftaclas & Cohen 1983; Miller & Lamb 1998; Strohmayer et al. 1998), and the shape (harmonic content) of the pulses is influenced by the rotational velocity, $v_{\rm rot} = 2\pi R \nu_{\rm spin} \sin i$, where $\nu_{\rm spin}$ is the neutron star spin frequency. If the spin frequency is known (for example, from burst oscillations or persistent pulsations), then the velocity depends directly on the stellar radius, R, and the binary system inclination, i. Thus, the properties of observed burst oscillation pulses encode information about the structure of the neutron stars and, for example, detailed theoretical modelling of the pulse profiles can be used to constrain neutron star masses and radii. Moreover, similar arguments hold for the pulses observed in the persistent X-ray flux from accreting millsecond pulsars.

The giant, hard X-ray flares produced by highly magnetized neutron stars ("magnetars") are some of the most energetic stellar events known. The recent December 27, 2004 flare from SGR 1806-20 liberated more than 10^{46} erg in a few mintues (Palmer et al. 2005; Terasawa et al. 2005), and was the brightest cosmic transient yet observed, in terms of energy flux received at Earth. It is thought that these flares result from a catastrophic rearrangement of the global magnetic field structure of the star, fracturing the crust of the star in the process (Duncan 1998; Thompson & Duncan 2001). Recently, Israel et al. (2005) discovered with RXTE that the X-ray flux during a portion of the flare was oscillating at about 92 Hz. They also found evidence for lower frequency QPOs at $\approx$ 18 and 30 Hz. A neutron star is thought to have a solid crust (somewhat analogous to Earth's crust), which can support twisting, or torsional, motions (McDermott, van Horn & Hansen 1988). Israel et al. (2005) suggested that these oscillations might be caused by just such motions. Subsequently, Strohmayer & Watts (2005) examined RXTE data from the August 27, 1998 giant flare from SGR 1900+14, and found that it also contained high frequency QPOs. They found strong oscillation signals at 53.5, 84, and 155.1 Hz, with perhaps a fourth feature at 28 Hz. In both flares these fast oscillations are only detected over a unique range of spin phase of each neutron star. This provides strong evidence that they are produced close to or, indeed, at the stellar surface, and are likely associated with a particular site on the star, perhaps the location of a crust fracture. If the observed frequencies are indeed those of global torsional oscillation modes, whose frequencies depend on the interior structure, then they could become a new, unique probe of neutron star structure.

2.3.2 High resolution X-ray spectroscopy

It has been known for some time that continuum spectral analysis of Eddington limited X-ray bursts can, in principle, provide information on neutron star masses and radii (for a discussion see the reviews by Lewin, van Paradijs & Taam 1993; and Strohmayer & Bildsten 2004). The basic idea is that the maximum luminosity, the Eddington limit, from a neutron star surface is a function of the stellar mass, radius and composition of the atmosphere. This means that the observed effective temperature of the atmosphere at the Eddington limit is a unique function of mass and radius. However, the method has historically suffered from several systematic uncertainties; the unknown atmospheric composition, uncertainties in the intrinsic spectrum (leading to errors in deriving the effective temperature from the observed color temperature), as well as uncertainty in what fraction of the neutron star surface is emitting. With the discovery of burst oscillations, which provide a direct indication for brightness asymmetries, this concern has taken on added importance. Although some of these problems remain, work in this area with the higher signal to noise RXTE data has continued (see, for example Shaposhnikov, Titarchuk & Haberl 2003; Shaposhnikov & Titarchuk 2004). A related method of deriving radius information from continuum spectra is to fit accurate neutron star atmosphere models to the thermal flux from quiescent neutron stars. This essentially capitalizes on the definition of the flux, $F = R^2\sigma T_{\rm eff}^4/d^2$, where $T_{\rm eff}$, and d are the effective temperature of the neutron star atmosphere, and the source distance, respectively. One must also take account of general relativistic corrections (see Rutledge et al. 2001; Walter et al. 2004). Difficulties with carrying out this program include continuing uncertainties in the atmosphere models (Zavlin & Pavlov 2002), as well as distance uncertainties, although these can be mitigated by finding quiescent neutron stars in globular clusters, for example (Rutledge et al. 2002).

Perhaps a more straightforward method of measuring neutron star parameters is via the detection of spectral features (lines and edges) from their surface atmospheres. An observation of an identified spectral line gives the gravitational redshift, $1 + z = (1 - 2\beta)^{-1/2}$, at the neutron star surface, which provides a direct measurement of the compactness, β. Unfortunately, it has been difficult to detect any atmospheric lines from neutron stars. Recent observations of isolated or quiescent neutron stars with the high resolution capabilities of *Chandra* and XMM/Newton have been frustratingly devoid of spectral features (see Walter & Lattimer 2002; Burwitz et al. 2001; Drake et al. 2002; Pavlov et al. 2001). A potential problem for non-accreting neutron stars is the high surface gravity which can remove via sedimentation the heavy, line forming metals on a rather short timescale (see Bildsten, Chang & Paerels 2003). Because of this, accreting neutron stars are arguably more promising sources for line searches. An encouraging recent result is the detection of absorption features in bursts from the LMXB 0748-676 with XMM-Newton (Cottam, Paerels & Mendez 2002). They found ≈ 10 eV equivalent width absorption features by co-adding 28 bursts seen with the Reflection Grating Spectrometers (RGS) spectrometers. Figure 2.3 shows their RGS spectra and best model. Their proposed identification with Fe XXVI $n = 2 - 3$ transitions gives a redshift of $z = 0.35$ from the neutron star surface. Although extremely exciting, this remains so far a single detection, and the ubiquity of such features in X-ray bursts in general remains to be established with future observations.

In addition to providing a direct measure of β, additional mass–radius information is encoded in the line profile. If the line width is dominated by rotation of the neutron star, then a measurement of it constrains the stellar radius through the surface velocity $v_{\rm rot} = 2\pi\nu_{\rm spin}R\sin i$. For the burst oscillation sources, with known spin frequencies in the 45–620 Hz range, rotation should be the dominant broadening mechanism as long as the system inclination is not too small (see for example, Ozel & Psaltis 2003; Villareal & Strohmayer 2004). Rotationally dominated line profiles also contain information on the fraction of the neutron star surface that is involved in the line formation. For example, emission from a fraction of the neutron star surface produces a characteristic "double-horned" line profile. Indeed, the relative strengths of the red and blue wings is sensitive to

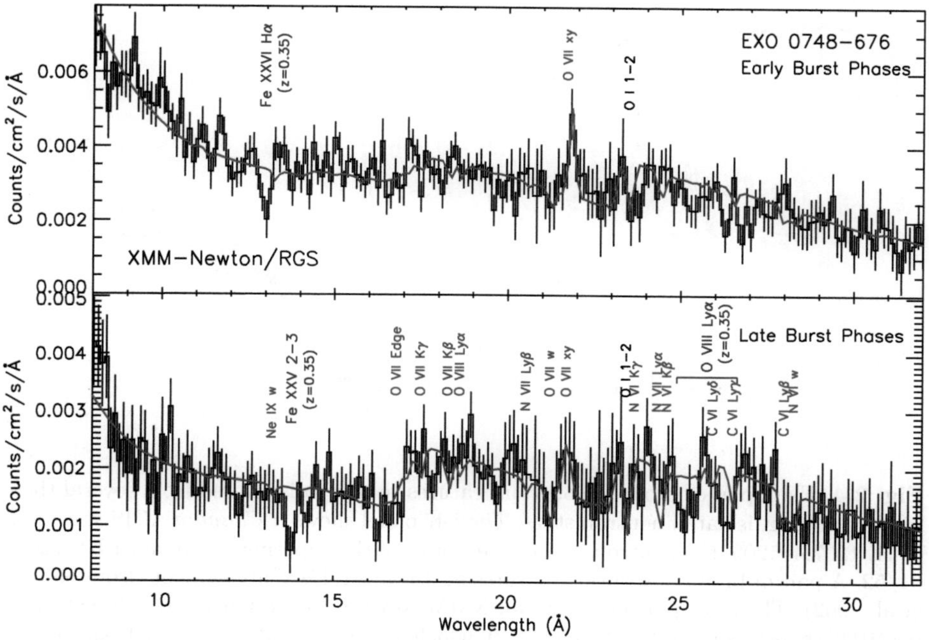

Fig. 2.3. High resolution X-ray spectra of bursts from the LMXB EXO 0748–676. XMM-Newton Reflection Grating Spectrometer (RGS) data from the early (top) and late (bottom) phases of the sum of 28 X-ray bursts are shown. The solid curve shows the best model of continuum and line features due to ionized gas in the immediate environs of the neutron star. The dips near 13 (top) and 14 (bottom) Angstroms are consistent with redshifted lines of Fe XXVI from the neutron star surface. Reprinted by permission from Nature (Cottam, Paerels & Mendez 2002) copyright (2002) Macmillan Publishers Ltd.

relativistic gravitational effects, such as frame dragging (Bhattacharyya, Miller & Lamb 2004).

In addition to the accreting neutron stars a smaller number of thermally radiating, isolated neutron stars have been observed over the past decade (see Haberl 2004 for a recent review). These objects have quasi-thermal spectra characterized by black body temperatures in the 30–70 eV range. It was hoped that deep, high resolution spectroscopic observations of these objects with *Chandra* and XMM-Newton would reveal atmospheric absorption lines with which to probe both the atmospheric composition as well as the neutron star mass and radius. Although ubiquitous, narrow spectral lines have not been found, in several sources broad absorption features have been discovered in the 0.2 to 2 keV energy band (Sanwal et al. 2002; van Kerkwijk et al. 2004; Haberl et al. 2004). Figure 2.4 shows two examples of these broad absorption lines. A full theoretical understanding of these spectral features is not yet in hand, however, the current thinking is that they may result from atomic transitions of atoms in very strong magnetic fields ($> 10^{12}$ G), or perhaps are formed by proton or ionic cyclotron lines in

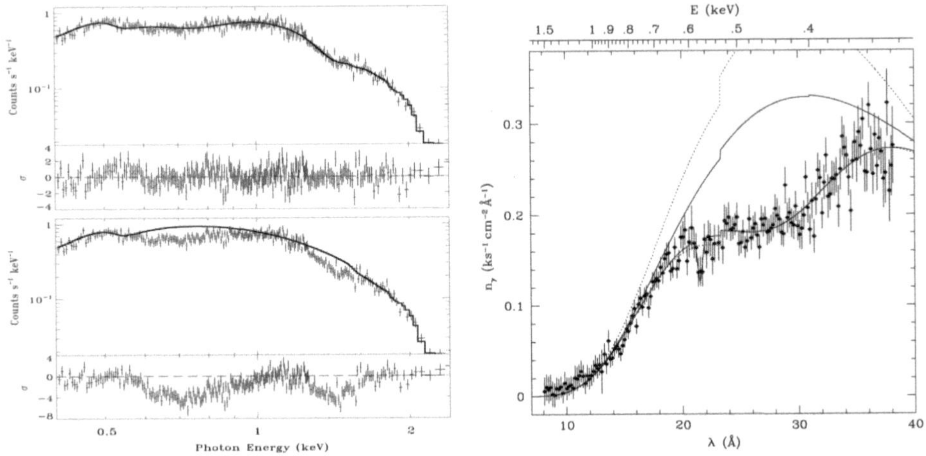

Fig. 2.4. Examples of broad absorption features recently discovered from several thermally emitting, isolated neutron stars. The left panel shows a *Chandra* ACIS spectrum of 1E 1207.4–5209, the neutron star in the center of the supernova remnant PKS 1209–51/52. A pair of broad absorption features at 0.7 and 1.4 keV are evident (from Sanwal et al. 2002). The right panel displays an XMM-Newton Reflection Grating Spectrometer (RGS) spectrum of RX J1605.3+3249, which shows a broad dip near 0.45 keV (from van Kerkwijk et al. 2004).

a similarly strong field (see for example, Hailey & Mori 2002; Ho & Lai 2004; van Kerkwijk et al. 2004). In either case, more theoretical work is needed to understand the effects of strong magnetic fields on the formation of atmospheric spectra. If such work leads to a clearer understanding of these features, then it should become possible to use them as detailed probes of neutron stars

2.4 Constraints from Fast X-ray Timing

2.4.1 Burst oscillations

As alluded to above, the shape of rotational modulation pulses are a unique function of a number of neutron star and binary orbital parameters. The pulsation amplitude depends most importantly on the size and shape of the surface area emitting X-rays, the viewing geometry, and the stellar compactness, β. The sharpness of the pulse profile (sometimes expressed in terms of the strength of harmonics) depends on relativistic beaming induced by the rotational motion. The maximum surface velocity, $v_{\rm rot}$, is given by $2\pi\nu_{\rm spin}R\sin i$. Since the spin frequencies of burst oscillation sources are known, the velocity is directly related to the stellar radius and the inclination angle, i, which is often not strongly constrained. Indeed, pulse profiles should provide a unique signature of these different quantities if they can be measured with sufficient precision.

Since the discovery of burst oscillations a number of studies have been undertaken to attempt to constrain neutron star properties using these ideas. Miller & Lamb (1998) explored theoretically the strength and shape of pulses from point-like hot spots on neutron stars. They demonstrated that the angular and spectral dependence of the surface emissivity can be crucial in obtaining accurate constraints. Nath, Strohmayer & Swank (2002) attempted to model bolometric pulse profiles seen during the rising portion of X-ray bursts from 4U 1636-53. They concluded that models with a single hot spot did not yet strongly constrain the mass and radius. They modelled the emission from a circular, linearly growing hotspot, and included light deflection in a Schwarzschild spacetime. Weinberg, Miller & Lamb (2001) computed theoretical pulse profiles produced by rotating neutron stars, including the rotational Doppler effects and aberration of the surface emissivity. They concluded that pulse profile fitting is superior to other indirect measures of the harmonic content, such as Fourier amplitudes. Muno, Ozel, & Charkrabarty (2002) have explored the amplitude evolution and harmonic content of cooling phase burst oscillations from a number of different sources. They used the observed limits on harmonic content to constrain the location and size of the hot spot or spots responsible for the observed modulations.

An observational limitation of these earliest studies was the inability to detect any harmonic signals in burst oscillations. However, studies of bursts from the accreting millisecond pulsar XTE J1814-338 have recently led to the first detection of significant harmonic structure in such oscillations (Strohmayer et al. 2003). In addition, the large number of bursts from this source and the good signal to noise achievable by co-adding bursts makes this object a particulary important source in the context of constraining neutron star structure using burst oscillations. Bhattacharyya et al. (2005) have fit the burst oscillation light curves from XTE J1814-338 in several energy bands using fully general relativistic models that include light deflection and frame dragging in a Kerr metric. They also allow for beaming of the emission from the surface and have also numerically computed the neutron star structure using realistic EOSs. They use a simple model of a circular hot spot for the emission region on the surface. They find that this model can adequately describe the pulse profiles seen from XTE J1814-338. For their two representative EOS they find from pulse profile fitting that $\beta < 0.24$, and they also find a beaming solution which is similar to that expected for electron scattering in a plane parallel atmosphere, consistent with simple expectations for neutron star atmospheres during bursts (see, for example London, Howard & Taam 1986). Figure 2.5 shows the pulse profile observed during bursts from XTE J1814–338 (left), as well as the constraints in the mass–radius plane for two different EOS from pulse profile fitting.

Although these studies are very promising they have not yet reached the level of sophistication required to obtain precise mass–radius measurements. A number of present uncertainties will have to be resolved before this can happen. For example, the form of the asymmetry on the stellar surface (put in a simple way, the shape of the hot spot) is very important in the modelling. Incorrect assumptions about the form of the asymmetry will lead to systematic errors

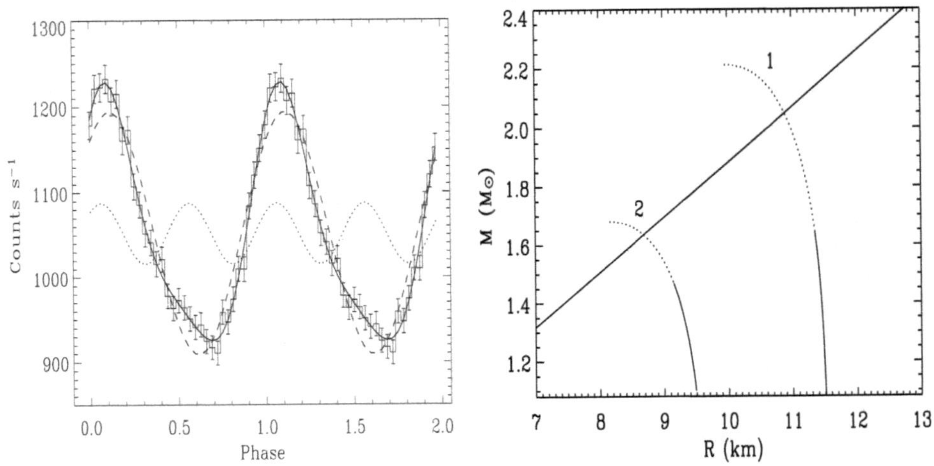

Fig. 2.5. Observed pulse profile showing significant harmonic structure during X-ray bursts from the accreting millisecond pulsar XTE J1814-338 (left; from Strohmayer et al. 2003), as well as constraints on the neutron star mass and radius for two representative EOS from pulse profile fitting (right; from Bhattacharyya et al. 2005). For both EOSs, the solid portion of each curve is the allowed region. See the text for further discussion of the compactness constraints.

in the derived mass and radius. Although this limitation can in principle be overcome by looking at burst rises – where the asymmetry is thought to take the form of a relatively small hot spot – this comes with the introduction of another limitation, inadequate signal. Fortunately, there are theoretical efforts underway to understand the nature of the asymmetries which produce burst oscillations. Some initial work was focused on the possibility of oscillation modes being responsible, particularly r-modes in the surface layers (Heyl 2004; Lee 2004; Lee & Strohmayer 2005; Heyl 2005; Piro & Bildsten 2005). Spitkovsky, Levin & Ushomirsky (2002) investigated the importance of the Coriolis force in understanding how thermonuclear burning fronts ignite and spread on a neutron star. They find that pressure gradients can drive zonal flows, some of which may be unstable to the formation of vortices, which could provide an asymmetry. An exciting, recent idea is that unstable modes associated with dynamical shear instabilities may be induced during X-ray bursts (Cumming 2005).

2.4.2 Accreting millisecond pulsars

The pulsations in the accretion-driven X-ray flux from millisecond pulsars can also be used to constrain neutron star structure. An advantage of such studies over burst oscillation pulses is that the pulse profiles can generally be measured with much greater precision. This is simply because the pulses are detected all the time (when the sources are active) rather than just during X-ray bursts. A drawback is that the emission mechanism is likely more complicated than the relatively simple thermal emission thought to dominate during bursts. This

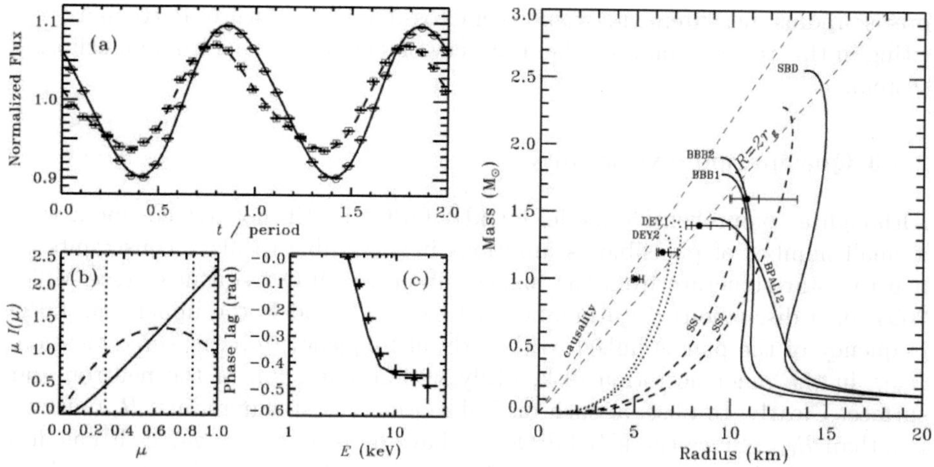

Fig. 2.6. Pulse profiles and neutron star mass–radius constraints from modelling of pulse profiles from the accreting millisecond pulsar SAX J1808.4-3658 (from Poutanen & Gierlinski 2003). The left panel shows pulse profiles (top) in the 3–4 keV (circles) and 12–18 keV (squares) bands, as well as the angular distribution, $\mu I(\mu)$, of radiation from the thermal (solid) and comptonized components (dashed), and the phase lags as a function of energy (bottom right). The right panel shows model constraints on the stellar radius for several assumed values of the mass.

then requires additional model parameters which may not be well constrained *a priori*, and thus could lead to systematic uncertainties in the derived constraints. Nevertheless, Poutanen & Gierlinski (2003) have carried out detailed modelling of the pulse profiles observed with RXTE from the first discovered accreting millsecond pulsar SAX J1808.4–3658. Contrary to what is found for most burst oscillation sources, the pulse profile seen from SAX J1808.4–3658 is strongly energy dependent. Figure 2.6 (upper left) shows the pulse profile in two different energy bands. The profile in the higher energy band (12–18 keV) has a much higher harmonic content than the lower energy profile. Poutanen & Gierlinski (2003) argue that this, as well as the fact that the soft photons lag the hard photons, requires an additional emission component (in addition to the thermal emission from a hot spot), which they model as a Comptonizing source with an optical depth of $0.3 - 1$, and a temperature of ≈ 50 keV. Using these two emission components they can adequately model the spectrum. They also include general and special relativistic effects on the photon propagation from the neutron star surface and are able to obtain constraints on the stellar compactness. They find from goodness of fit arguments that the stellar mass is likely in the range from 1.2 to 1.6 $M_{\odot}$ (not too surprising), and for a mass near 1.6 $M_{\odot}$ they find a radius consistent with several modern EOSs, but that EOS with radii larger than ≈ 13 km are ruled out. Figure 2.6 (right) shows there derived constraints on the radius for several assumed masses. If an independent mass constraint could be obtained for this (or other) accreting millisecond pulsars, then precise radius measurements would seem possible in the future. Particularly exciting in this regard would be the discovery of such pulsations from an eclipsing system.

2.4.3 Quasiperiodic variability

Although a comprehensive model for kHz QPOs is still not agreed upon, with a small number of plausible assumptions it is possible to place constraints on neutron star structure from the highest observed QPO frequencies (see Miller 2004 for a discussion). In a number of kHz QPO models the upper kHz QPO frequency of the pair is linked to the orbital frequency, $\nu_{\rm orb}$, at some location, $R_{\rm orb}$, in the inner accretion disk, likely within a few km of the neutron star surface. Clearly, to observe such orbital motion the stellar radius, R, must be less than $R_{\rm orb}$. Since the kHz QPOs can have high coherence values it has also been argued that they cannot be produced within the region of unstable orbits, which suggests an additional constraint, $R_{\rm orb} > R_{\rm ISCO}$ (Miller, Lamb & Psaltis 1998). Since $\nu_{\rm orb} = (GM/4\pi^2 R_{\rm orb}^3)^{1/2}$, the constraints on $R_{\rm orb}$ imply an upper limit on the mass and radius of the neutron star. Because, in general, the mass–radius relation and the location of the ISCO depend on both the spin frequency and the EOS, the derived constraints on mass and radius also depend on these quantities (see Miller, Lamb & Cook 1998). Miller (2004) summarizes these as;

$$M < 2.2 M_\odot \, (1000 \text{ Hz}/\nu_{\rm orb}) \, (1 + 0.75j) \ , \tag{2.3}$$

and

$$R < 19.5 \text{km} \, (1000 \text{ Hz}/\nu_{\rm orb}) \, (1 + 0.2j) \ , \tag{2.4}$$

where $j \equiv cJ/GM^2$ is a dimensionless spin frequency, and J is the stellar angular momentum.

The highest observed kHz QPO frequency to date is 1330 Hz from the LMXB 4U 0614+091 (van Straaten et al. 2000). Unfortunately, the spin frequency of this object has not yet been detected. Nevertheless, assuming j is not much larger than 0.1, this frequency implies $M < 1.8 M_\odot$, and $R < 15$ km, if the various model assumptions are reasonable. Figure 2.7 shows two examples of how the assumptions connecting the observed upper QPO frequency and orbital motion lead to allowed regions in the neutron star mass–radius plane. The left panel summarizes the constraints assuming that rotation has no influence on the mass–radius relation or the radius of the ISCO (Miller, Lamb & Psaltis 1998), while the right panel incorporates fully the effects of spin for a particular EOS (Miller, Lamb & Cook 1998). Although it is still difficult to draw firm conclusions based on observed kHz QPO frequencies, the method appears quite promising and could bear rich fruit when a clearer theoretical understanding of the QPOs emerges, and/or when higher frequency features are discovered.

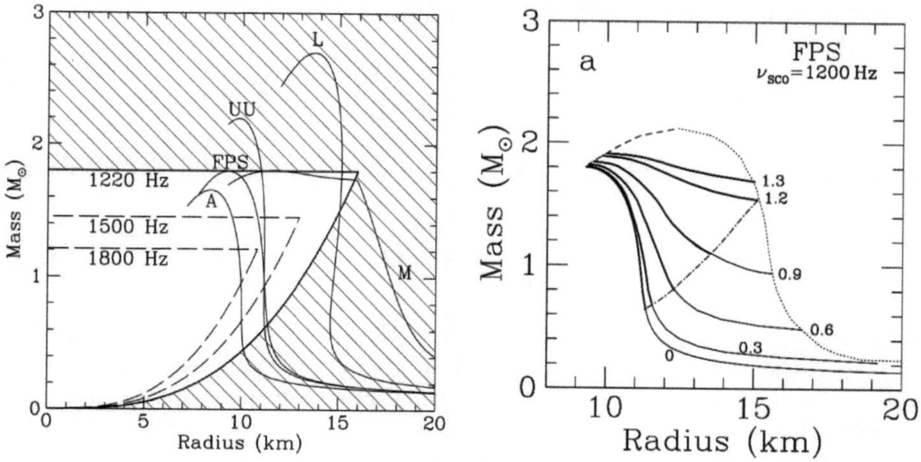

Fig. 2.7. Constraints in the neutron star mass–radius plane from measurements of kHz oscillations in accreting neutron star systems. The left panel shows allowed regions for several different values of the upper kHz QPO frequency assuming non-rotating stars and that the frequency is orbital in nature (from Miller, Lamb & Psaltis 1998). The right panel shows how rapid stellar rotation effects the constraints for a particular EOS (from Miller, Lamb & Cook 1998). See the text and the relevant papers for additional details.

2.4.4 Magnetar flares: fast X-ray variability

As discussed briefly above, recent studies of the fast X-ray timing properties of two magnetar flares have found high frequency QPOs during portions of their pulsating tails (see Israel et al. 2005; Strohmayer & Watts 2005). An exciting possibility is that these oscillation signals are caused by global torsional (toroidal) vibrations of the neutron star crust. This idea appears plausible for several reasons; the observed frequencies (≈30–150 Hz) are within the range expected theoretically for such oscillations (see Mcdermott, van Horn & Hansen 1988; Duncan 1998), and the purely horizontal motions associated with these modes could "shake" the magnetic field lines frozen into the crust, providing a possible mechanism to modulate the X-ray emission. Moreover, the giant flares themselves are theoretically linked to large scale fracturing of the neutron star crust, and, in analogy with terrestrial eartquakes, will almost certainly excite crustal modes (see Duncan 1988; Park et al. 2005).

The oscillation periods of the crustal torsional modes, (denoted ${}_lt_n$), where l is the spherical harmonic index for the mode, and n is the number of radial nodes in the eigenfunction, depend on the shear wave speed in the crust, and its size (see, for example, Strohmayer et al. 1991; Duncan 1998). These quantities depend on the global structure of the star and hence the EOS of matter in the deep interior. Note also that the magnetic field can boost the "tension" in the crust and thus modify the mode periods, perhaps non-isotropically (see Duncan 1998; Messios, Papadopolous & Stergioulas 2001). The period of the fundamental

($l = 2,\ n = 0$) toroidal mode is estimated to be

$$P({}_2t_0) = 33.6R_{10}\frac{0.87 + 0.13M_{1.4}R_{10}^{-2}}{(1.71 - 0.71M_{1.4}R_{10}^{-1})^{1/2}}\ \text{ms}\ , \qquad (2.5)$$

where $R_{10} = R/10\text{km}$ and $M_{1.4} = M/1.4M_{\odot}$ (Duncan 1998), and higher l modes scale like

$$P({}_lt_0) = P({}_2t_0)\left(\frac{6}{l(l+1)}\right)^{1/2}\left[1 + \left(\frac{B}{B_\mu}\right)^2\right]^{-1/2}\ , \qquad (2.6)$$

where the factor in square brackets is a magnetic correction, $B_\mu \approx 4\times10^{15}\rho_{14}^{0.4}$G, and ρ_{14} is the crustal density in units of 10^{14} g cm^{-3}. In all cases, modes with $n > 0$ have frequencies substantially higher than those so far observed during the SGR flares.

Israel et al. (2005) suggested that the 92.5 Hz feature detected in SGR 1806-20 could be the $l = 7$ mode. With this identification the fundamental mode would have a period of 30.4 Hz. Strohmayer & Watts (2005) found a sequence of QPOs with frequencies of 53.5, 84, and 155.1 Hz (see Figure 2.8, left). These frequencies can reasonably be associated with $l = 4$, 7, and 13 modes, and the corresponding fundamental mode would have a frequency of $\approx$ 28 Hz, consistent with the frequency of the weakest feature seen in SGR 1900+14 (Strohmayer & Watts 2005). With these identifications the fundamental frequencies in the two stars are different, suggesting that the stellar parameters differ. The above equations specify the relationship between mass and magnetic field required to give a desired mode frequency. Since the relationship depends on both the stellar mass and radius it is clearly EOS dependent, and thus secure mode identifications could provide constraints on the EOS. Strohmayer & Watts (2005) have tabulated the stellar parameters that give ${}_2t_0$ oscillations at 28 Hz (SGR 1900+14) and 30.4 Hz (SGR 1806-20). Figure 2.8 (right) shows the results for four different EOS discussed in Lattimer & Prakash (2001). The results suggest that if the stars have similar magnetic field strengths, their masses must differ by more than $0.2M_{\odot}$. Since the masses of radio pulsars (a young neutron star population) have been found to be consistent with a narrow Gaussian distribution, $M = 1.35 \pm 0.04M_{\odot}$, a perhaps more likely scenario is that the stars have similar masses but different magnetic field strengths. If both stars have masses $\approx 1.35M_{\odot}$, then it is difficult to account for both frequencies with the softest EOS (WFF1); and the stiffest EOS (MS0) predicts magnetic fields for both systems that are far larger than those inferred from timing studies (Woods et al. 2002). The moderately stiff EOSs AP3 and AP4 can account for the observed frequencies, and give magnetic fields strengths that agree reasonably well with those derived from timing measurements of both stars (Woods & Thompson 2005). This recent evidence for signatures of toroidal modes during SGR giant flares is an exciting development, and could eventually lead to important constraints on neutron star structure. An unfortunate observational drawback, however, is that SGR giant flares are rare events.

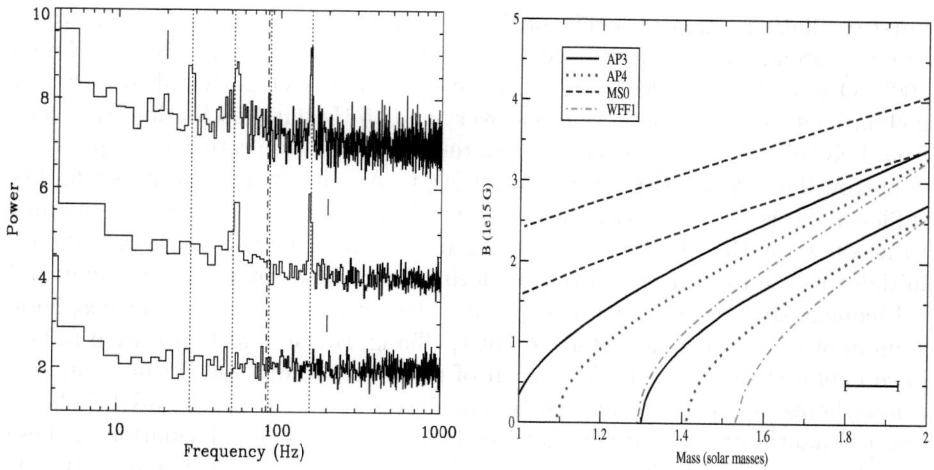

Fig. 2.8. High frequency QPOs recently discovered in the pulsating X-ray tail of the August 27, 1998 giant flare from SGR 1900+14 (left), and neutron star mass and magnetic field strengths required to give a specific $l = 2$ toroidal mode eigenfrequency (right). The left panel shows a power spectrum of X-ray intensity variations from the pulsating tail of the SGR 1900+14 flair, revealing a sequence of QPOs at ≈ 28, 53, and 155 Hz. An additional QPO at 84 Hz was also detected during a different time interval. The frequencies of these QPOs are qualitatively consistent with a sequence of $l = 2$, 4, 7, and 13 toroidal modes. The right panel shows neutron star masses and magnetic fields which have an $l = 2$ toroidal mode frequency of 30.4 (upper curves), and 28 (lower curves) Hz, the mode frequencies suggested for SGR 1806-20 (Israel et al. 2005) and SGR 1900+14 (Strohmayer & Watts 2005), respectively. The legend indicates the particular EOS used to determine the mode frequencies. See Lattimer & Prakash (2001) for a discussion of the EOSs (from Strohmayer & Watts 2005).

2.5 Constraints from High Resolution X-ray Spectroscopy

Cottam, Paerels & Mendez (2002) have recently reported evidence for redshifted absorption lines in X-ray bursts from EXO 0748–676. They co-added data from 28 bursts in order to increase the sensitivity to narrow lines. After modelling the continuum and line features thought to be produced in the highly ionized gas around the neutron star (see Cottam et al. 2001), they found a number of remaining features which were unaccounted for by their best model. Figure 2.3 shows their RGS spectra and best model. They interpreted the unmodelled dips near 13 and 14 Angstroms as redshifted absorption lines of the $n = 2$–3 transitions of hydrogen-like Fe (Fe XXVI) in the early (hotter) part of the bursts, and the same transitions in the helium-like ion in the later (cooler) portions of the bursts. The inferred neutron star surface redshift in each case was $z = 0.35$. This value is consistent with a number of modern neutron star EOS with reasonable masses (Lattimer & Prakash 2001).

When these results were first presented the narrowness of the lines was thought to be a significant problem, because if the spin frequency was like other burst oscillation sources ($\sim$ 100s of Hz), then rapid neutron star rotation would produce broad, shallow lines (see for example, Bhattacharyya, Miller & Lamb 2005; Özel & Psaltis 2003), unless the systems were being viewed at very low inclinations. However, the recent discovery of a 45 Hz burst oscillation frequency from EXO 0748–676 is consistent with a relatively slow spin rate and the presence of narrow lines (Villarreal & Strohmayer 2004). An important concern is whether sufficient Fe is present in the atmosphere at the relevant mass accretion rates to account for the observed line strengths. Bildsten, Chang & Paerels (2003) initially examined this question, exploring how sedimentation, accretion and subsequent spallation (which can produce heavy metals) sets the atmospheric column of heavy elements. More recently, Chang, Bildsten & Wasserman (2005) have explored in detail the formation of resonant atomic lines in neutron star atmospheres, including rotational broadening. They show that non-LTE effects and resonant scattering are crucial to understanding the line formation in these conditions. Their calculations imply an Fe column of $1\text{–}3 \times 10^{20}$ cm^{-2}, that is 3–10 times larger than expected based on the accretion/spallation scenario of Bildsten, Chang & Paerels (2003), but is within a factor of a few of a uniform, solar metallicity atmosphere. They also show that fine structure splitting will be important in constraining the stellar radius from modelling of the line widths, and confirm that a spin rate of 45 Hz is consistent with the presence of narrow lines. These results suggest that accurate measurements of both the mass and radius of the neutron star in EXO 0748–676 are obtainable if the proposed absorption lines can be confirmed and observed with higher statistical significance.

2.6 Future Prospects

Clearly, new observational capabilities have provided us with some powerful new ways to probe neutron star structure, but we still have not quite reached the goal of precisely measuring the neutron star mass–radius relation. Fortunately, there is a relatively clear path along which these methods can be refined, and that is primarily to increase the collecting area of future observatories. Several planned mission concepts should be able to make substantial improvements in this area, including NASA's Constellation-X, ESA's XEUS, and a larger area X-ray timing mission to further capitalize on the discoveries made with RXTE.

In the context of burst oscillation constraints, larger area allows one to obtain more precise measurements of the pulse profiles, and therefore more statistically precise constraints. Indeed, a future mission with about 10 times the collecting area of RXTE, and with similar timing capabilities should be able to provide data of sufficient statistical quality to allow tight constraints in the mass–radius plane. As an example, Figure 2.9 summarizes the results of some simulations of what burst oscillations might look like to such a "super-RXTE" observatory (see Strohmayer 2004). With this large area accurate pulse profiles can be measured during the rising phase of bursts, when the brightness pattern is likely to be well approximated by a localized hot spot, reducing the effects of some systematic uncertainties. Model fits to these simulated data suggest that statistically

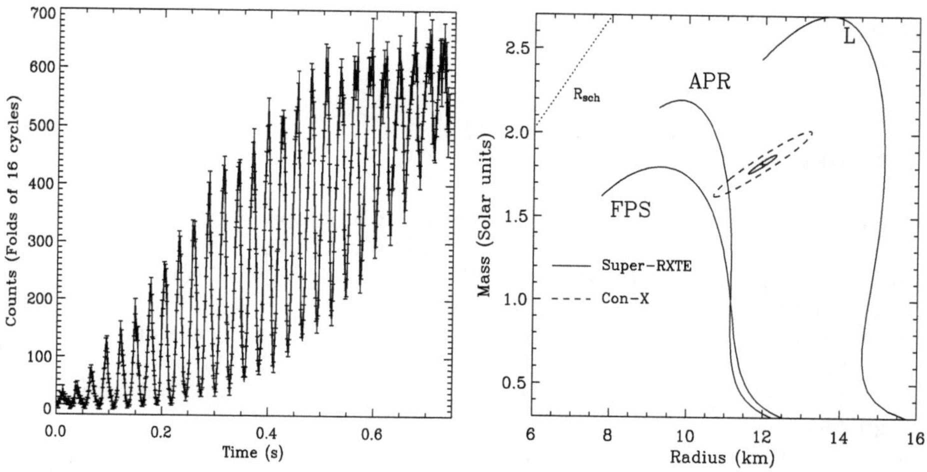

Fig. 2.9. Simulated pulse profiles and mass–radius constraints for X-ray burst observations with a detector collecting area 10 times that of RXTE/PCA. The left panel shows what burst oscillation pulses would look like to such an instrument. The right panel shows 1σ confidence contours on mass and radius for detector areas appropriate for the Constellation-X mission and a super-RXTE with 10x the area of RXTE (from Strohmayer 2004).

tight constraints are, in principle, achievable. There has also been substantial advancement in the theoretical tools needed for modelling the phenomenon and the pace of these developments suggests that it is not unrealistic to expect that by the time future missions fly, the impact of theoretical uncertainties can be substantially reduced. Moreover, new data almost always provide fresh insights which cannot be anticipated beforehand. An added strength of attacking the EOS problem using burst oscillations is that the signals are known to exist, and it is virtually guaranteed that by studying these oscillations with a factor of 10 better sensitivity we will learn something new.

If redshifted spectral lines are relatively common in X-ray bursters, then sensitive, high resolution X-ray spectroscopy may be the key that unlocks the secrets of the neutron star interior. Current observations with XMM-Newton and *Chandra* are not quite sensitive enough to establish whether or not surface absorption lines are present in enough systems and at sufficient statistical quality to enable accurate neutron star constraints, but the potential seems undeniable. We may have to wait for the next generation of large area, high spectral resolution observatories to realize the full potential of this technique.

Acknowledgements

I sincerely thank Jim Lattimer, Jean Cottam, Divas Sanwal, Marten van Kerkwijk, Sudip Bhattacharyya, Juri Poutanen, Anna Watts, and Cole Miller for either providing graphics or allowing the use of previously published figures. Additional thanks to Anna Watts and Sudip Bhattacharyya for comments on the initial draft. I thank NASA for its support of high energy astrophysics research.

References

Alford, M., Jotwani, P., Kouvaris, C., Kundu, J., & Rajagopal, K. 2005, A Hot Water Bottle for Aging Neutron Stars, Phys. Rev. D, **71**, 114011

Alford, M., Rajagopal, K., Reddy, S., & Wilczek, F. 2001, Minimal color-flavor-locked nuclear interface, Phys. Rev. D, **64**, 074017

Alford, M., Rajagopal, K., & Wilczek, F. 1999, Color-flavor locking and chiral symmetry breaking in high density QCD, Nuclear Physics B, **537**, 443

Baade, W. & Zwicky, F. 1934, Cosmic Rays from Super-novae, Proc. Nat. Acad. Sci., **20**, 259

Baym, G. & Pethick, C. J. 1975, Neutron Stars, Ann. Rev. Nucl. Part. Sci., **25**, 75

Bhattacharyya, S., Strohmayer, T. E., Miller, M. C., & Markwardt, C. B. 2005, Constraints on Neutron Star Parameters from Burst Oscillation Light Curves of the Accreting Millisecond Pulsar XTE J1814-338, ApJ, **619**, 483

Bhattacharyya, S., Miller, M. C. & Lamb, F. K. 2004, The Shapes of Atomic Lines from the Surfaces of Weakly Magnetic Rotating Neutron Stars and Their Implications, ApJ, submitted, astro-ph/0412107

Bildsten, L., Chang, P. & Paerels, F. 2003, Atomic Spectral Features during Thermonuclear Flashes on Neutron Stars, ApJ, **591**, L29

Braje, T. M., Romani, R. W., & Rauch, K. P. 2000, Light Curves of Rapidly Rotating Neutron Stars, ApJ, **531**, 447

Burwitz, V., Zavlin, V. E., Neuhauser, R., Predehl, P., Trumper, J., & Brinkman, A. C. 2001, The Chandra LETGS high resolution X-ray spectrum of the isolated neutron star RX J1856.5-3754, A&A, **379**, L35

Chakrabarty, D., Morgan, E. H., Muno, M. P., Galloway, D. K., Wijnands, R., van der Klis, M. & Markwardt, C. B. 2003, Nuclear-powered millisecond pulsars and the maximum spin frequency of neutron stars, Nature, **424**, 42

Chang, P., Bildsten, L. & Wasserman, I. 2005, Formation of Resonant Atomic Lines during Thermonuclear Flashes on Neutron Stars, ApJ, **629**, 998

Cline, T. L. et al. 1980, Detection of a fast, intense and unusual gamma-ray transient, ApJ, **237**, L1

Cottam, J., Paerels, F. & Mendez, M. 2002, Gravitationally redshifted absorption lines in the X-ray burst spectra of a neutron star, Nature, **420**, 51

Cottam, J., Kahn, S. M., Brinkman, A. C., den Herder, J. W., & Erd, C. 2001, High-resolution spectroscopy of the low-mass X-ray binary EXO 0748-67, A&A, **365**, L277

Cumming, A. 2005, Latitudinal Shear Instabilities during Type I X-Ray Bursts, ApJ, **630**, 441

Drake, J. J. et al. 2002, Is RX J1856.5-3754 a Quark Star?, ApJ, **572**, 996

Duncan, R.C., 1998, Global Seismic Oscillations in Soft Gamma Repeaters, ApJ, **498**, L45

Duncan, R. C., & Thompson, C. 1992, Formation of very strongly magnetized neutron stars - Implications for gamma-ray bursts, ApJ, **392**, L9

Fryer, C. L. & Warren, M. S. 2004, The Collapse of Rotating Massive Stars in Three Dimensions, ApJ, **601**, 391

Glendenning, N. K. 1992, First-order phase transitions with more than one conserved charge: Consequences for neutron stars, Phys. Rev. D, **46**, 1274

Haberl, F. 2004, AXPs and X-ray-dim isolated neutron stars: recent XMM-Newton and Chandra results, Advances in Space Research, **33**, 638

Hailey, C. J. & Mori, K. 2002, Evidence of a Mid-Atomic Number Atmosphere in the Neutron Star 1E 1207.4-5209, ApJ, **578**, L133

Hewish, A., Bell, S. J., Pilkington, J. D., Scott, P. F., & Collins, R. A. 1968, Observation of a Rapidly Pulsating Radio Source, Nature, **217**, 709

Heyl, J. S. 2005, R-modes on rapidly rotating, relativistic stars - II. Blackbody emission, MNRAS, **361**, 504

Heyl, J. S., 2004, r-Modes on Rapidly Rotating, Relativistic Stars. I. Do Type I Bursts Excite Modes in the Neutron Star Ocean?, ApJ, **600**, 939

Ho, W. C. G., & Lai, D. 2004, Spectral Features in the Thermal Emission from Isolated Neutron Stars: Dependence on Magnetic Field Strengths, ApJ, **607**, 420

Israel, G., et al., 2005, The Discovery of Rapid X-Ray Oscillations in the Tail of the SGR 1806-20 Hyperflare, ApJ, **628**, L53

Lattimer, J. M. & Prakash, M. 2001, Neutron Star Structure and the Equation of State, ApJ, **550**, 426

Lee, U., & Strohmayer, T. E. 2005, Light curves of oscillating neutron stars, MNRAS, **361**, 659

Lee, U. 2004, Surface r-Modes and Burst Oscillations of Neutron Stars, ApJ, **600**, 914

Lewin, W. H. G., van Paradijs, J. & Taam, R. E. 1993, X-Ray Bursts, Space Sci. Rev., **62**, 223

Link, B., Epstein, R. I., & Lattimer, J. M. 1999, Pulsar Constraints on Neutron Star Structure and Equation of State, Physical Review Letters, **83**, 3362

London, R. A., Howard, W. M. & Taam, R. E. 1986, Model atmospheres for X-ray bursting neutron stars , ApJ, **306**, 170

Mazets, E. P. et al. 1979, Observations of a flaring X-ray pulsar in Dorado, Nature, **282**, 587

McDermott, P.N., van Horn, H.M., Hansen, C.J., 1988, Nonradial oscillations of neutron stars, ApJ, **325**, 725

Messios, N., Papadopoulos, D.B., Stergioulas, N., 2001, Torsional oscillations of magnetized relativistic stars, MNRAS, **328**, 1161

Miller, M. C. 2004, Interpreting QPOs from Accreting Neutron Stars, in *X-ray Timing 2003: Rossi and Beyond*, AIP Conf. Proc. Vol. 714, Eds. P. Kaaret, F. Lamb, & J. H. Swank, 365

Miller, M. C. & Lamb, F. K. 1998, Bounds on the Compactness of Neutron Stars from Brightness Oscillations during X-Ray Bursts, ApJ, **499**, L37

Miller, M. C., Lamb, F. K. & Psaltis, D. 1998, Sonic-Point Model of Kilohertz Quasi-periodic Brightness Oscillations in Low-Mass X-Ray Binaries, ApJ, **508**, 791

Miller, M. C., Lamb, F. K., & Cook, G. B. 1998, Effects of Rapid Stellar Rotation on Equation-of-State Constraints Derived from Quasi-periodic Brightness Oscillations, ApJ, **509**, 793

Muno, M. P., Fox, D. W., Morgan, E. H. & Bildsten, L. 2000, ApJ, Nearly Coherent Oscillations in Type I X-Ray Bursts from KS 1731-260, **542**, 1016

Muno, M. P., Özel, F. & Chakrabarty, D. 2002, The Amplitude Evolution and Harmonic Content of Millisecond Oscillations in Thermonuclear X-Ray Bursts, ApJ, **581**, 550

Nath, N. R., Strohmayer, T. E., & Swank, J. H. 2002, Bounds on Compactness for Low-Mass X-Ray Binary Neutron Stars from X-Ray Burst Oscillations, ApJ, **564**, 353

Nice, D. J., Splaver, E. M., & Stairs, I. H. 2004, Heavy Neutron Stars? A Status Report on Arecibo Timing of Four Pulsar - White Dwarf Systems, IAU Symposium, **218**, 49

Oppenheimer, J. R., & Volkoff, G. M. 1939, On Massive Neutron Cores , Physical Review , **55**, 374

Özel, F., & Psaltis, D. 2003, Spectral Lines from Rotating Neutron Stars, ApJ, **582**, L31

Paerels, F. 1997, Pressure Broadening of Absorption Lines in Neutron Star Atmospheres and Prospects for Measuring Neutron Star Masses and Radii, ApJ, **476**, L47

Page, D., Lattimer, J. M., Prakash, M., & Steiner, A. W. 2004, Minimal Cooling of Neutron Stars: A New Paradigm, ApJS, **155**, 623

Palmer, D.M., et al., 2005, A giant ?-ray flare from the magnetar SGR 1806 - 20, Nature, **434**, 1107

Pavlov, G. G. et al. 2001, The X-Ray Spectrum of the Vela Pulsar Resolved with the Chandra X-Ray Observatory, ApJ, **552**, L129

Pechenick, K. R., Ftaclas, C., & Cohen, J. M. 1983, Hot spots on neutron stars - The near-field gravitational lens, ApJ, **274**, 846

Piro, A. L., & Bildsten, L. 2005, Surface Modes on Bursting Neutron Stars and X-Ray Burst Oscillations, ApJ, **629**, 438

Poutanen, J., & Gierliński, M. 2003, On the nature of the X-ray emission from the accreting millisecond pulsar SAX J1808.4-3658, MNRAS, **343**, 1301

Prakash, M., Lattimer, J. M., Pons, J. A., Steiner, A. W., & Reddy, S. 2001, Evolution of a Neutron Star from Its Birth to Old Age, LNP Vol. 578: Physics of Neutron Star Interiors, **578**, 364

Psaltis, D., & Chakrabarty, D. 1999, The Disk-Magnetosphere Interaction in the Accretion-powered Millisecond Pulsar SAX J1808.4-3658, ApJ, **521**, 332

Rajagopal, K., & Wilczek, F. 2001, Enforced Electrical Neutrality of the Color-Flavor Locked Phase, Physical Review Letters, **86**, 3492

Rutledge, R. E., Bildsten, L., Brown, E. F., Pavlov, G. G., & Zavlin, V. E. 2001, Quiescent Thermal Emission from the Neutron Star in Aquila X-1, ApJ, **559**, 1054

Rutledge, R. E., Bildsten, L., Brown, E. F., Pavlov, G. G., & Zavlin, V. E. 2002, A Possible Transient Neutron Star in Quiescence in the Globular Cluster NGC 5139, ApJ, **578**, 405

Sanwal, D., Pavlov, G. G., Zavlin, V. E. & Teter, M. A. 2002, Discovery of Absorption Features in the X-Ray Spectrum of an Isolated Neutron Star, ApJ, **574**, L61

Schatz, H., et al. 2001, End Point of the rp Process on Accreting Neutron Stars, Physical Review Letters, **86**, 3471

Shaposhnikov, N., & Titarchuk, L. 2004, On the Nature of the Flux Variability during an Expansion Stage of a Type I X-Ray Burst: Constraints on Neutron Star Parameters for 4U 1820-30, ApJ, **606**, L57

Shaposhnikov, N., Titarchuk, L. & Haberl, F. 2003, The Bursting Behavior of 4U 1728-34: The Parameters of a Neutron Star and the Geometry of a Neutron Star-Disk System, ApJ, **593**, L35

Smith, D., Morgan, E. H. & Bradt, H. V. 1997, Rossi X-Ray Timing Explorer Discovery of Coherent Millisecond Pulsations during an X-Ray Burst from KS 1731-260, ApJ, **479**, L137

Spitkovsky, A., Levin, Y. & Ushomirsky, G. 2002, Propagation of Thermonuclear Flames on Rapidly Rotating Neutron Stars: Extreme Weather during Type I X-Ray Bursts, ApJ, **566**, 1018

Steiner, A. W., Prakash, M., Lattimer, J. M., & Ellis, P. J. 2005, Isospin asymmetry in nuclei and neutron stars, Phys. Reps., **411**, 325.

Stella, L., & Vietri, M. 1998, Lense-Thirring Precession and Quasi-periodic Oscillations in Low-Mass X-Ray Binaries, ApJ, **492**, L59

Strohmayer, T. E. & Watts, A. L. 2005, Discovery of Fast X-Ray Oscillations during the 1998 Giant Flare from SGR 1900+14, ApJ, **632**, L111

Strohmayer, T. E. & Bildsten, L. 2004, New Views of Thermonuclear Bursts, in *Compact Stellar X-ray Sources*, Eds. W. H. G. Lewin and M. van der Klis, (Cambridge University Press: Cambridge), astro-ph/0301544

Strohmayer, T. E. 2004, Future Probes of the Neutron Star Equation of State Using X-ray Bursts, in *X-ray Timing 2003: Rossi and Beyond*, AIP Conf. Proc. Vol. 714, Eds. P. Kaaret, F. Lamb, & J. H. Swank, 245

Strohmayer, T. E., Markwardt, C. B., Swank, J. H., & in't Zand, J. 2003, X-Ray Bursts from the Accreting Millisecond Pulsar XTE J1814-338, ApJ, **596**, L67

Strohmayer, T. E. & Markwardt, C. B. 2002, Evidence for a Millisecond Pulsar in 4U 1636-53 during a Superburst, ApJ, **577**, 337

Strohmayer, T. E. & Markwardt, C. B. 1999, On the Frequency Evolution of X-Ray Brightness Oscillations during Thermonuclear X-Ray Bursts: Evidence of Coherent Oscillations, ApJ, **516**, L81

Strohmayer, T. E. et al. 1998, On the Amplitude of Burst Oscillations in 4U 1636-54: Evidence for Nuclear-powered Pulsars, ApJ, **498**, L135

Strohmayer, T. E., Zhang, W. & Swank, J. H. 1997, 363 HZ Oscillations during the Rising Phase of Bursts from 4U 1728-34: Evidence for Rotational Modulation, ApJ, **487**, L77

Strohmayer, T. E. et al. 1996, Millisecond X-Ray Variability from an Accreting Neutron Star System, ApJ, **469**, L9

Strohmayer, T., van Horn, H. M., Ogata, S., Iyetomi, H., & Ichimaru, S. 1991, The shear modulus of the neutron star crust and nonradial oscillations of neutron stars, ApJ, **375**, 679

Terasawa, T., et al., 2005, Repeated injections of energy in the first 600ms of the giant flare of SGR1806 - 20, Nature, **434**, 1110

Thompson, C., & Duncan, R. C. 2001, The Giant Flare of 1998 August 27 from SGR 1900+14. II. Radiative Mechanism and Physical Constraints on the Source, ApJ, **561**, 980

Thompson, C., & Duncan, R. C. 1995, The soft gamma repeaters as very strongly magnetized neutron stars - I. Radiative mechanism for outbursts, MNRAS, **275**, 255

Thorsett, S. E. & Chakrabarty, D. 1999, Neutron Star Mass Measurements. I. Radio Pulsars, ApJ, **512**, 288

van der Klis, M. 2004, Rapid X-ray variability in X-ray binaries, in *Compact Stellar X-ray Sources*, Eds. W. H. G. Lewin and M. van der Klis, (Cambridge University Press: Cambridge), astro-ph/0410551

van der Klis, M. et al. 1996, Discovery of Submillisecond Quasi-periodic Oscillations in the X-Ray Flux of Scorpius X-1, ApJ, **469**, L1

van Kerkwijk, M. H., Kaplan, D. L., Durant, M., Kulkarni, S. R., & Paerels, F. 2004, A Strong, Broad Absorption Feature in the X-Ray Spectrum of the Nearby Neutron Star RX J1605.3+3249, ApJ, **608**, 432

van Straaten, S. et al. 2000, Relations between Timing Features and Colors in the X-Ray Binary 4U 0614+09, ApJ, **540**, 1049

Villarreal, A. R. & Strohmayer, T. E. 2004, Discovery of the Neutron Star Spin Frequency in EXO 0748-676, ApJ, **614**, L121

Walter, F. M., Pons, J. A., Burwitz, V., Lattimer, J. M., Lloyd, D., Wolk, S. J., Prakash, M., & Neuhäuser, R. 2004, Properties of the isolated neutron star RX J185635-3754, Advances in Space Research, **33**, 513

Walter, F. M. & Lattimer, J. M. 2002, A Revised Parallax and Its Implications for RX J185635-3754, ApJ, **576**, L145

Watts, A. L., Strohmayer, T. E. & Markwardt, C. B. 2005, Analysis of variability in the burst oscillations of the accreting millisecond pulsar XTE J1814-338, ApJ, in press

Weinberg, N., Miller, M. C., & Lamb, D. Q. 2001, Oscillation Waveforms and Amplitudes from Hot Spots on Neutron Stars, ApJ, **546**, 1098

Woods, P.M., Thompson, C., 2005, Soft Gamma Repeaters and Anomalous X-ray Pulsars: Magnetar Candidates, in *Compact Stellar X-ray Sources*, eds. W.H.G. Lewin and M. van der Klis, Cambridge University Press, (astro-ph/0406133)

Woods, P.M., et al., 2002, Large Torque Variations in Two Soft Gamma Repeaters, ApJ, **576**, 381

Zavlin, V. E., & Pavlov, G. G. 2002, Modeling Neutron Star Atmospheres, in Neutron Stars, Pulsars, and Supernova Remnants, 263

Zwicky, F. 1938, On Collapsed Neutron Stars, ApJ, **88**, 522

3 Six Years of *Chandra* Observations of Supernova Remnants

M.C. Weisskopf and J.P. Hughes

Abstract: We present a review of the first six years of *Chandra* X-ray Observatory observations of supernova remnants. From the official "first-light" observation of Cassiopeia A that revealed for the first time the compact remnant of the explosion, to the recent million-second spectrally-resolved observation that revealed new details of the stellar composition and dynamics of the original explosion, *Chandra* observations have provided new insights into the supernova phenomenon. We present an admittedly biased overview of six years of these observations, highlighting new discoveries made possible by *Chandra*'s unique capabilities.

3.1 Introduction

The *Chandra* X-ray Observatory was launched on July 23, 1999 using the Space Shuttle Columbia. Placement in its highly elliptical orbit was completed 15 days after launch. The orbit has a nominal apogee of 140,000 km and a nominal perigee of 10,000 km. With this orbit, the satellite is above the radiation belts for more than about 75% of the 63.5-hour orbital period and uninterrupted observations lasting more than 2 days are possible. The observing efficiency, which also depends on solar activity, is typically about 65%.

The spacecraft has been functioning superbly since launch. The specified design life of the mission was 5 years; however, the only perishable is gas for maneuvering and is sized to allow operation for much more than 10 years. The orbit will be stable for decades.

The heart of the Observatory is the X-ray telescope made of four concentric, precision-figured, superpolished Wolter-1 telescopes, similar to those used for both the *Einstein* and *ROSAT* observatories, but of much higher quality, larger diameter, and longer focal length.

The telescope's on-axis point spread function, as measured during ground calibration, had a full-width at half-maximum less than 0.5 arcsec and a half-power diameter less than 1 arcsec. A relatively mild dependence on energy, resulting from diffractive scattering by surface microroughness, attests to the better than 3-angstroms-rms surface roughness measured with optical metrology during fabrication and confirmed by the ground X-ray testing. The on-orbit performance met expectations.

There are two focal plane cameras on the Observatory. The Pennsylvania State University (PSU, University Park, Pennsylvania) and the Massachusetts Institute of Technology (MIT, Cambridge, Massachusetts) designed and fabricated the Advanced CCD Imaging Spectrometer (ACIS) with CCDs produced by MIT's Lincoln Laboratory. Made of a 2-by-2 array of front-illuminated (FI), 2.5-cm-square CCDs, ACIS-I provides high-resolution spectrometric imaging over a 17-arcmin-square field of view. ACIS-S, a 6-by-1 array of 4 FI CCDs and two back-illuminated (BI) CCDs mounted along the *Chandra* transmission grating dispersion direction, serves both as the primary read-out detector for the High Energy Transmission Grating (HETG), and, using the one BI CCD which can be placed at the aim point of the telescope, also provides high-resolution spectrometric imaging extending to lower energies but over a smaller (8-arcmin-square) field than ACIS-I.

The Smithsonian Astrophysical Observatory (SAO, Cambridge, Massachusetts), designed and fabricated the other focal plane camera the High Resolution Imager (HRC) (Murray et al. 2000). Made of a single 10-cm-square microchannel plate (MCP), the HRC-I provides high-resolution imaging over a 30-arcmin-square field of view. A second detector made of 3 rectangular MCP segments (3-cm $\times$ 10-cm each) mounted end-to-end along the grating dispersion direction, the HRC-S, serves as the primary read-out detector for the Low Energy Transmission Grating (LETG).

More details as to the technical performance of the Observatory may be found in Weisskopf et al. (2003).

We note that the names of SNR are often given in papers using their galactic coordinates and a "G" prefix. This prefix is, however, used by name resolvers such as Simbad for high proper motion stars appearing in the Giclas Catalog and it has been recommended [1] that one use the SNR designation instead. We shall do so in this review.

Our review is representative of *Chandra* observations and is not intended to be exhaustive. Thus, if we have omitted a particular observation no slight to the work was intended. Finally, we emphasize that there are many excellent reviews and conference proceedings concerning the astrophysics associated with supernova remnants and, where relevant, the compact objects within them. These reviews and proceedings include: Becker and Pavlov (2001), Aschenbach (2002) Pavlov, Zavlin, and Sanwal (2002), Slane (2002), Canizares (2004), Decourchelle (2004), Slane (2005), and Rakowski (2005).

Our purpose here is to review the outstanding contributions that observations using *Chandra* have made to the discipline. We urge interested readers to include a careful reading of these reviews in addition to reading this paper.

[1]see http://heasarc.gsfc.nasa.gov/docs/faq.html#nameresolvers

3.2 The Point Sources in Supernova Remnants

An excellent introduction to the topic of compact central sources in supernova remnants may be found in Kaplan et al. (2004).

3.2.1 The compact central objects

Although not originally discovered with *Chandra*, observations using the observatory have both clarified, and drawn further attention to, the existence of a group of compact objects associated with SNR that are radio quiet, unpulsed in X-rays, and have characteristic X-ray spectra described by a blackbody with characteristic temperatures of order 0.4 keV, without indication of a non-thermal component. The associated blackbody radii are smaller than the radii of canonical neutron stars. These objects have been aptly termed Compact Central Objects (CCOs) by Pavlov et al. (2002a). (Sources of this ilk have previously been referred to as "Radio Quiet Neutron Stars" by Caraveo, Bignami, & Trümper 1996.) Although several sources exhibit these properties, it is not clear that they form a class in the sense that there is an universally accepted explanation for their characteristics.

The prevailing, perhaps contradictory, interpretation is that the spectral properties of the CCOs are the result of hot spots on the surfaces of neutron stars formed during the supernova. The inability then to find any evidence for pulsations in the X-ray flux from the majority of these sources must be confronted. Many discussions leave one with the impression that new observations and more sensitive searches for pulsations will ultimately discover periodic behavior and thus validate the explanation. In general, detection of pulsations has not been forthcoming, and so perhaps deeper looks at other potential explanations might be in order. One possible explanation involves very long periods requiring some mechanism to carry off the angular momentum. One might also begin to question the interpretation of the spectral data. The tie between spectral fits and physical interpretations can be overdone. Thus, simply because a blackbody spectrum fits the data, this does not necessarily imply that the source is a blackbody emitter. Another spectrum, more representative of the true physical situation, may fit the data equally well. It is a challenge to the theorists to tell us just what this alternative spectrum might be.

Cas A

Cassiopeia A (Cas A) is often referred to as the "first light" observation made with *Chandra*. It is historically interesting that this is not quite accurate. The first X-rays that were observed with *Chandra* when the last door was opened took place on August 12, 1999 and led to the discovery of the $z = 0.32$ AGN nicknamed "Leon X-1" (Weisskopf et al. 2005). The nickname honors the *Chandra* Telescope Scientist, Leon Van Speybroeck. The Observatory was next pointed at the radio-loud quasar PKS 0637−752 (Schwartz et al. 2000) chosen as the target for which on-orbit optimization of the imaging properties of the

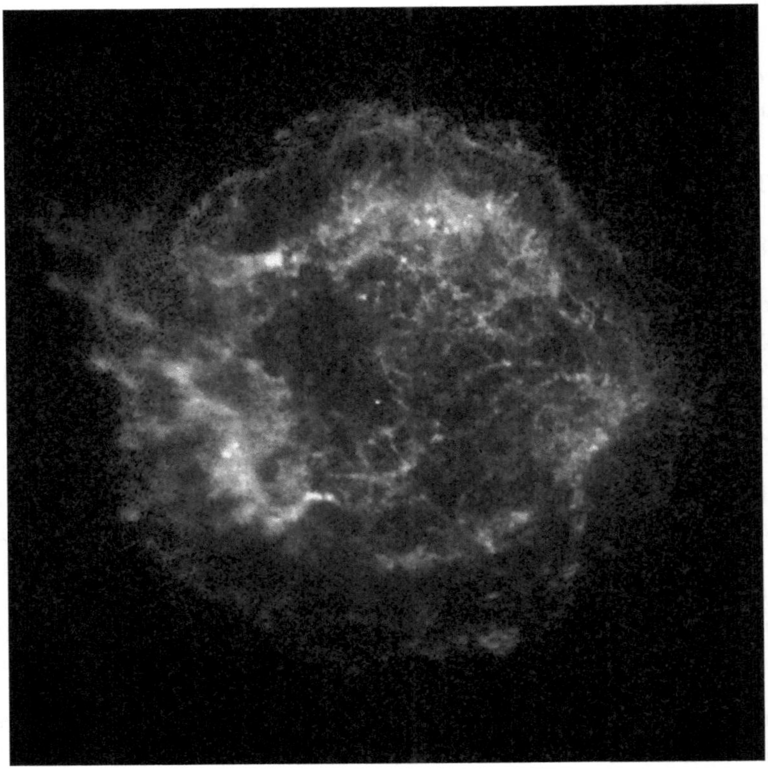

Fig. 3.1. Early *Chandra* ACIS-S3 image of Cas A. The image is 6' x 6'. Courtesy NASA/CXC/SAO.

Observatory were performed. After bore-sighting and focusing the telescope, only then was the official "first-light" image of Cas A obtained. One of the principal features of the first Cas A image, shown in Figure 3.1, was the clear presence of a point source (Tananbaum 1999) near the center of the remnant and apparent after only a few minutes of observing.

Pavlov et al. (2000) argued, convincingly, that the source was the compact object associated with the remnant. Pavlov et al. (2000) also established the nature of the spectral parameters, noting that no unique spectral fits were forthcoming from the data in hand at the time. Fits to a power law yielded high photon indices (of order 3 to 4) whereas fits to a blackbody led to temperatures of about 0.5 keV. As noted above, these authors coined the very apt description for this source – Compact Central Object (CCO) – and favored the interpretation that the CCO is a neutron star whose X-ray emission is primarily from a restricted region of the surface.

Subsequently, Chakrabarty et al. (2001) provided additional analyses of both spectral and timing data, setting a 3-σ upper limit to the pulsed fraction of less than 35% for periods longer than 20-ms. Additional period searches were performed by Murray et al. (2002a) utilizing HRC-S data in order to extend the period searches to shorter periods. Using a 50 ksec observation they also failed to detect any significant evidence for pulsations. More recently, and discussed in detail in, Sect. 3.4.1, a 10^6 s ACIS-S3 observation has been performed. To our knowledge, no pulsations have thus far been detected in these data, although the search would be limited to periods longer than several seconds due to the time resolution in this mode.

SNR 266.2−1.2 (RX J09852.0−4622)

The SNR 266.2−1.2 (aka RX J09852.0−4622 and sometimes referred to as "Vela Junior") was discovered with *ROSAT* by Aschenbach (1998). Prior to observations with *Chandra*, a number of X-ray observations had been performed with at least two, if not more, possible X-ray point sources being associated with this remnant – see the introduction to, and references in, Pavlov et al. (2001). Pavlov et al. (2001) used *Chandra* and the ACIS imaging array to pinpoint the location of a bright X-ray source (CXOU J085201.4−461753) 4′ north of the SNR center. This was the only bright X-ray source they found near the center of the remnant. Using the *Chandra*-based position to refine and guide a search for optical counterparts, they found none and established stringent upper limits to the optical flux in both B and R. These authors then used the corresponding lower limit to the X-ray to optical flux ratio to argue that the source is the compact remnant of the SN. The quality of the X-ray spectra determined from this particular observation was limited by the short observing time (3 ksec) and by pulse pileup due to the brightness of the source coupled with the readout time (frame time) of the CCDs.

To refine the spectral parameters, and to provide moderately high time resolution data useful for searching for pulsations, the source was again observed with *Chandra* by Kargaltsev et al. (2002) using ACIS-I in the continuous-clocking mode. This mode provides only one-dimensional images but yields higher time resolution (2.85 ms in this case) especially useful for mitigating pulse pileup. This observation was also ten times longer than the previous *Chandra* observation. Spectral fits to a power law were found to be statistically unacceptable, whereas a blackbody model, with a temperature of 0.4 keV, provided a statistically acceptable fit. As with Cas A and the other CCOs, the radius of the emitting sphere was found to be much smaller than a canonical neutron star radius.

Kargaltsev et al. (2002) also attempted to constrain the temperature of the entire (putative) neutron star surface by fitting the spectral data with a two-component blackbody model where one of the two components was constrained to a radius of 10 km. They found an upper limit to the surface temperature (at infinity) of 89 eV (99% confidence) which would imply accelerated cooling compared to standard neutron star cooling models assuming an age of a few thousand years. (Inferring accelerated cooling is not exceptional, see, e.g. the discussion of 3C58 in Sect. 3.2.3.) Kargaltsev et al. (2002) also searched for narrow spectral features and discussed the hint of a feature at 1.68 keV. It would be interesting to follow up on the possible spectral feature with *XMM-Newton* to see if this is similar to the feature(s) discovered using *Chandra* in 1E 1207.4−5209

(Sect. 3.2.3). A search for pulsations found no statistically significant periods with pulsed fractions in excess of 13% in the frequency range from 0.001 to 100 Hz.

We urge other observers to follow the example set by Kargaltsev et al. (2002) in their data analysis. In particular, use the data to establish an upper limit to a possible full-surface blackbody in order to constrain a neutron star temperature, and search for possible spectral features that appear as residuals to the continuum models. The results of these analyses are very useful in addressing some of the potentially interesting astrophysical questions concerning compact objects and in particular the CCOs.

Puppis A

The bright point source at about 6′ from the dynamical center of Puppis A was discovered by Petre et al. (1982) using the *Einstein* Observatory. The object has been studied with numerous X-ray satellites and in other wavelength bands (see e.g. the introduction to Hui and Becker (2005) and references therein). The early *Chandra* ACIS observation discussed by Pavlov et al. (2002a) found a spectrum both consistent with previous X-ray observations and the CCO classification – namely that the data were described by a blackbody with an associated radius much less than that of a neutron star. In this case, spectral fitting with a neutron star hydrogen atmosphere yielded a radius more like that of a neutron star. The ACIS observations were followed by observations with the HRC designed to search both for pulsations and a pulsar wind nebula (PWN). The HRC image is shown in Figure 3.2 and clearly there is no obvious evidence for a PWN, however Pavlov et al. (2002a) gave no quantitative upper limits. Note that Gaensler, Bock and Stappers (2000) did provide quantitative upper limits as to the presence of a radio PWN (on scales greater than about 30" x 30" up to about 30'). Pavlov et al. (2002a) used the HRC-data to search for pulsations and found no evidence for pulsations for pulses with a pulsed fraction of greater than 10% in the period range from 0.003 to 300 seconds. Pavlov et al. (2002a) do not state the confidence level to be associated with this upper limit.

RCW 103 (SNR 332.4−0.4)

The source 1E 161348−5055 was discovered using the *Einstein* Observatory by Tuohy and Garmire (1980) and is associated with the SNR RCW 103 (aka SNR 332.4−0.4). The central point source was considered to be the prototype of a cooling neutron star (Becker and Aschenbach 2002) prior to observations with *Chandra*. Interestingly, the RCW 103 source was also the first radio-quiet X-ray source found in a young SNR and, as such, may be considered to be the first detected CCO. The source's detailed characteristics (see below) are, however, somewhat different from the majority of CCOs such as Cas A and, as has also been noted by others (e.g. Pavlov, Zavlin, and Sanwal 2002), it is not clear that all CCOs form a single class of objects.

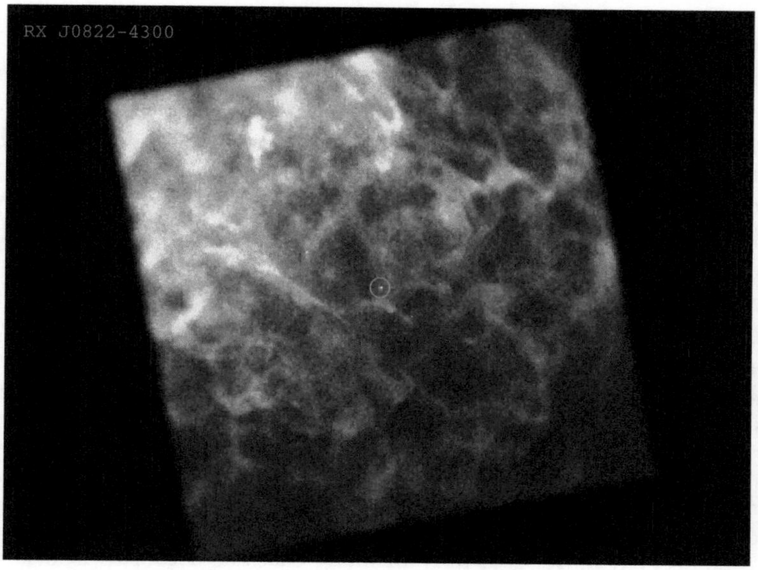

Fig. 3.2. *Chandra* HRC-I image of Puppis A. The image is 30′ × 30′. Courtesy G. Pavlov.

Garmire et al. (2000) using *Chandra* reported a precise position and strong evidence for the detection of a significant portion of a potentially periodic light curve corresponding to a ≈ 6-hr period based on the 5.5-hr *Chandra* observation. These same authors reported finding a similar period in archival *ASCA* data, and speculated as to the possibility that a binary period had been detected. We note that accretion was also listed as one of many possible explanations for the long term variability observed prior to the *Chandra* observations (see e.g., Gotthelf, Petre, and Vasisht 1999).

Binarity may not be the only explanation for the periodicity. Heyl and Hernquist (2002) presented the case for a freely-precessing, short-period neutron star with hot spots as the possible explanation for the light curve. Their model predicted that the spin period should be detectable at minimum flux, but, to date, no detection has been reported. Subsequently Becker and Aschenbach (2002) reported the results of a 5.5-hr *XMM-Newton* observation which showed what appears to be clear evidence for an eclipse. If verified, this would the first clear case of an accreting binary in a SNR. Further observations of this interesting system are clearly called for.

Hui and Becker (2005) analyzed these *Chandra* data together with the data from *XMM-Newton*, fitting the data to two blackbodies and reporting on a "promising candidate", albeit statistically weak, pulse period.

SNR 347.3−0.5

Prior to observations of SNR 347.3−0.5 with *Chandra*, an X-ray point source (1WGA J1713.4−3949) with no nearby optical counterpart (Slane et al. 1999) had been discovered with *ROSAT* by Pfeffermann and Aschenbach (1996). More interestingly, a radio source (PSR J1713−3945; Crawford et al. 2002) with a pulse period of 392 ms was also detected in the region. *Chandra* observations (Lazendic et al. 2003) played a key role as they were used to establish that the X-ray source was *not* at the same location as the radio pulsar.

Lazendic et al. (2003) found that the central X-ray source shares several characteristics with the CCO in Cas A, including an X-ray spectrum whose principal component is a blackbody with a temperature of about 0.4 keV and the absence of pulsations over the frequency ranges searched (0.01–128 Hz) with upper limits to the pulsed fraction of better than 25%.

3.2.2 Pulsars with pulsar wind nebulae (PWN)

Chandra has served more than ably in many cases as a tool to isolate and resolve the various spatial components that comprise the structure of pulsar wind nebulae. We begin our discussion of the *Chandra* contributions to this particular type of study with the remarkable images that have revealed more observational detail of the complex interaction between the central pulsar and the surrounding medium than ever before. All of the *Chandra* PWN images appear to be consistent with a similar structure – one that is approximately axially symmetric and elongated along the symmetry or jet axis. The most common explanation identifies the jets with collimated outflows of relativistic particles along the rotation axis of the underlying pulsar. The ubiquitous presence of jets in astrophysical contexts including these associated with young pulsars and those found with active galactic nuclei is always worth comment and provides an over-riding motivation for gathering detailed information to help one learn about jet formation mechanisms. Observations with *Chandra* have provided not only spectacular images but also valuable and unique insights. One of the most important of these insights is the dynamic nature of the phenomena that take place in the PWN. This is particularly well illustrated in the case of the observations of the Vela pulsar (Sect. 3.2.2) described by Pavlov et al. (2003) that discovered the variability of the northeast outer jet.

The Pulsar in the Crab Nebula

The *Chandra* X-Ray Observatory first observed the Crab Nebula and its pulsar during orbital calibration in 1999. Weisskopf et al. (2000) published the zeroth-order image obtained with the HETG and read out by ACIS-S that showed a striking richness of X-ray spatial structures. Figure 3.3 shows the original HETG–ACIS-S zeroth-order image. A number of features were observed for the first time: an X-ray inner ring within the X-ray torus; the suggestion of a hollow-tube structure for the torus; X-ray knots along the inner ring and (perhaps) along

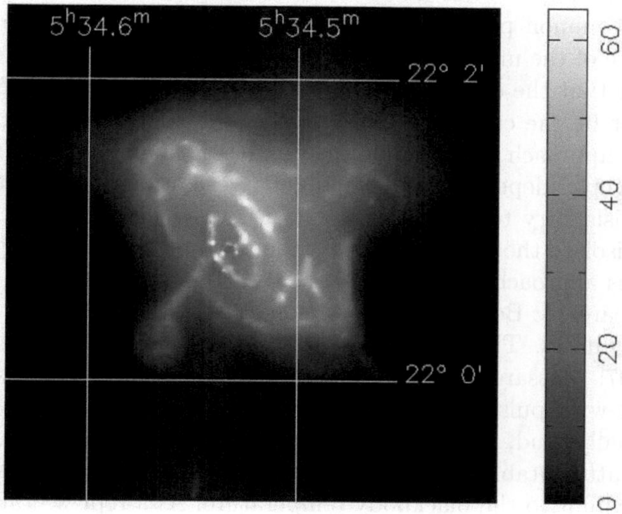

Fig. 3.3. *Chandra* HETG–ACIS-S3 adaptively smoothed first *Chandra* image of the Crab Nebula.

the inward extension of the X-ray jet. The *Chandra* image also clearly resolved the X-ray torus (Aschenbach and Brinkmann 1975) and jet and counterjet which are all features that had been previously observed (Brinkmann, Aschenbach, & Langmeier 1985; Hester et al. 1995; Greiveldinger & Aschenbach 1999) but never with such clarity. On somewhat larger scales, the image showed a sharply bounded notch (WSW of the Pulsar) into the X-ray nebular emission, earlier associated with the "west bay" of the Nebula (Hester et al. 1995). Visible-light polarization maps of the Crab Nebula (Schmidt, Angel, & Beaver 1979; Hickson & van den Bergh 1990) demonstrate that the magnetic field is parallel to the boundary of this notch, thus serving to exclude the X-ray-emitting relativistic electrons from the west bay.

The most striking feature of the X-ray image, of course, is the inner elliptical ring, lying between the pulsar and the torus. The existence of such a ring had been predicted, and corresponds to a shock in the pulsar wind (Rees and Gunn 1974; Kennel and Coroniti 1984). On the ring reside a few compact knots, and one can be seen in Figure 3.3 lying SE of the pulsar along the projected inward extension of the jet. The surface brightness of this knot is too high to be simply explained as the superposition of the ring's and jet's surface brightnesses. Ultimately the nature of these knots needs to be probed by means of high-resolution spectroscopy.

Subsequently, Tennant et al. (2001) observed the Crab Nebula and its pulsar with *Chandra* using the Low Energy Transmission Grating Spectrometer (LETGS). Time-resolved zeroth-order images were used to perform a most sensitive search for X-ray emission from the pulser as a function of pulse phase, including pulse phases that had been traditionally referred to as "unpulsed". It has been common practice to denote any minimum in a pulse profile to be representative of the unpulsed flux; subtract these data from the remainder and then purport that the difference, represents the "pulsed" component. Of course this need not be the case. One can see from the *Chandra* image, for example, that such an approach may include the flux from the inner ring if not the entire PWN as a whole, depending on the angular resolution of the instrument. It is far more satisfactory to use high-resolution spatial imaging as is possible with *Chandra* to isolate the pulsar from any background that may be present.

Using this approach, Tennant et al. (2001) discovered that, as in the visible (Golden, Shearer, & Beskin 2000; Peterson et al. 1978), the pulsar emits X-rays at *all* pulse phases. They confirmed prior observations (Pravdo, Angellini, & Harding 1997; Massaro et al. 2000) that showed that the power law spectral index varied with pulse phase and extended the measurements into the pulse minima. Finally, and, assuming that all of the flux from the pulsar at pulse minimum is attributable to thermal emission, the authors used these data to set a new upper limit to the blackbody temperature. As a representative case, they took $\theta_\infty = 2.1 \times 10^{-16}$ rad – for $R_\infty = 13$ km (e.g., $R_s = 10$ km and $M = 1.4 M_\odot$) at $D = 2$ kpc – and $N_H = 3 \times 10^{21}$ cm^{-2}. With these parameters, the blackbody temperature that would account for all the flux observed at the pulse minimum was $T_\infty = 2.12$ MK $= 0.183$ keV, which bounds the actual temperature ($L_\infty \approx 2.4 \times 10^{34}$ erg s^{-1}). Subsequent *Chandra*-LETG observations and analyses of the spectrum as a function of pulse phase (Weisskopf et al. 2004) slightly improved this upper limit to $T_\infty < (1.76(2\sigma); 2.01(3\sigma))$ MK.[2]

Weisskopf et al. (2004) also performed a detailed analysis of the pulse-averaged spectrum. They were able to study the interstellar X-ray extinction due primarily to photoelectric absorption and secondarily to scattering by dust grains in the direction of the Crab Nebula. They confirmed the findings of Willingale et al. (2001) that the line-of-sight to the Crab is under abundant in oxygen. Using the abundances and cross sections from Wilms, Allen and McCray (2000) they found $[\mathrm{O/H}] = (3.33 \pm 0.25) \times 10^{-4}$. Spectral studies such as this, where the abundances are allowed to vary, are important as it is unlikely that standard abundances apply equally well to all lines of sight, especially those that intersect large quantities of SN debris (for more on this point see the discussion in Serafimovich et al. 2004 and our comments in Sect. 3.2.2).

In 2002, Hester et al. (2002) completed one phase of a set of coordinated observations of the Crab's PWN using *Chandra* (ACIS-S in sub-array mode) and the Hubble Space telescope. These spectacular observations revealed numerous dynamical features including wisps moving outward from the inner equatorial X-ray ring at about 0.5 c. The eight *Chandra* observations are shown in Figure 3.4.

Finally, *Chandra* (and *XMM-Newton*, see Willingale et al. 2001) has been used to study spectral variations as a function of position in the nebula. Weis-

[2]These upper limits appear weaker than previous *ROSAT*-established upper limits set by Becker and Aschenbach (1995). The *ROSAT* limits were, however, "optimistic" as discussed in Tennant et al. (2001).

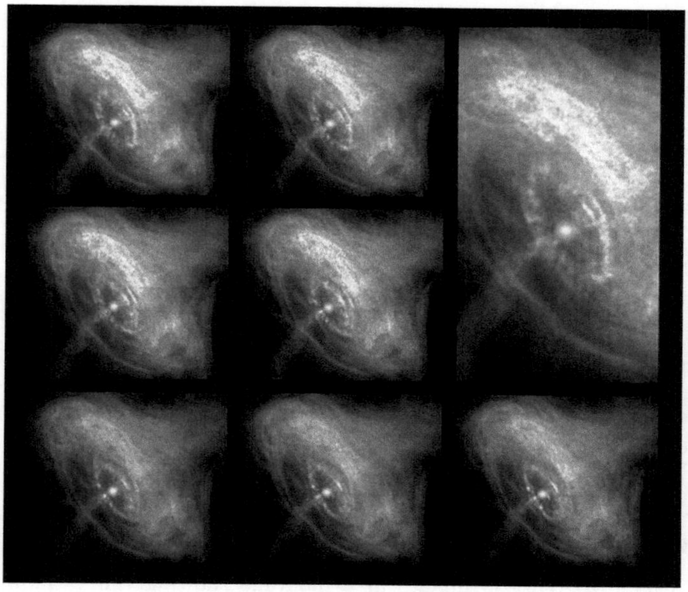

Fig. 3.4. The *Chandra* images in this collage were made over a span of several months, are $1.6' \times 1.6'$ and are time-ordered left to right and top to bottom except for the larger close-up. Image courtesy NASA/CXC/ASU/ J. Hester et al. (2002).

skopf et al. (2000) first presented the variation of a hardness ratio (the ratio of flux in two energy bands) as a function of position as seen with *Chandra* using $5'' \times 5''$ pixels. Mori et al. (2004) followed this work with studies of the variation of the power law spectral index as a function of position using $2.5'' \times 2.5''$ pixels and the same data discussed in Hester et al. (2002). Despite the particular mode (subframe) used to obtain these data, the effects of pulse pileup plagued the data analysis and required the application of corrections. These corrections, at best, did not adequately correct for pileup when bright spatial structure was present within an analysis pixel – dealing with that particular situation was noted by the authors to be beyond the scope of the paper. One hopes this problem will be addressed by some enterprising expert in pulse pileup in the future, since it is at the smallest spatial scales that the *Chandra* observations are most unique. Performing spectroscopy (perhaps even time-resolved spectroscopy) of the bright, compact features is necessary in order to understand their origin.

The Vela Pulsar and its remarkable external jet

Chandra observations of the 89-ms period Vela pulsar (Helfand, Gotthelf, & Halpern 2001; Pavlov et al. 2003) and its surroundings have been most revealing. In addition to showing the complex and time variable spatial structure of the region immediately surrounding the pulsar itself – a structure that includes two sets of arcs, a jet in the direction of the pulsar's proper motion and a counterjet – the *Chandra* images taken by Pavlov et al. (2003) also discovered

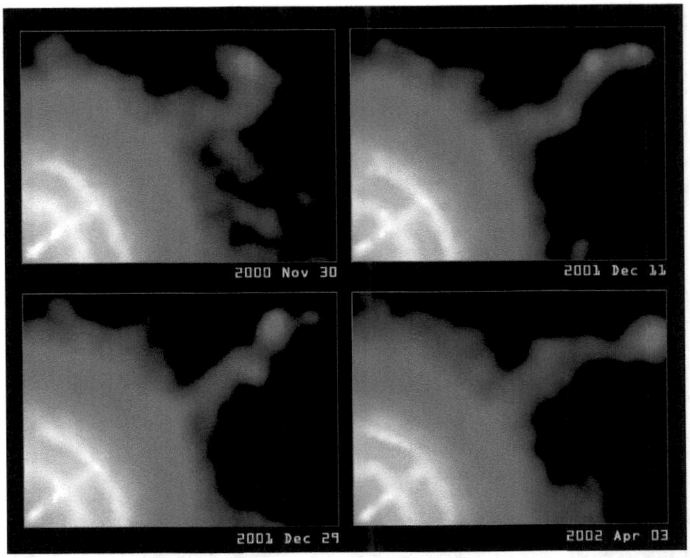

Fig. 3.5. *Chandra* ACIS-S3 images in a montage showing the variability in both intensity and position of the jet associated with the Vela pulsar. The pulsar is located at the lower left-hand corner of each image. These four images are part of a series of 13 images made over a period of two and a half years. Each image is $1.6' \times 1.2'$. Image courtesy NASA/CXC/PSU/G. Pavlov et al. (2003).

that the continuation of the jet that extends to the NW is time variable in both intensity and position on scales of days to weeks as illustrated in Figure 3.5. The brightness variations are different along the extent of the jet, and Pavlov et al. (2003) inferred flow velocities of 0.3–0.7c. Finally, the apparent width of the outer jet appears to be constant, despite large variations in appearance, indicating confinement. The analogy to a fire hose being held at its base appears very appropriate.

The *Chandra* Observatory should be used to both to search for, and to study, such behavior in all cases where it is clear that jets from neutron stars are present.

PSR1509−58

The *Chandra* image of the young PWN powered by B1509−58 in SNR 230.4−1.2 (Gaensler et al. 2002) is shown in Figure 3.6. The pulsar is the bright source at the center of the nebula. A thin jet can be seen in the image to extend to the southeast. Just above the pulsar there is a small arc of X-ray emission, which seems to mark the location of the shock wave produced by the particles flowing away from the pulsar's equator. The cloud near the top of the image may to be due to high temperature gas. This gas, possibly a remnant of the explosion associated with the creation of the pulsar, may have been heated by collisions with high-energy particles produced by the pulsar. See Yatsu et al. (2005) for a discussion of the interaction of the pulsar's jet with this material.

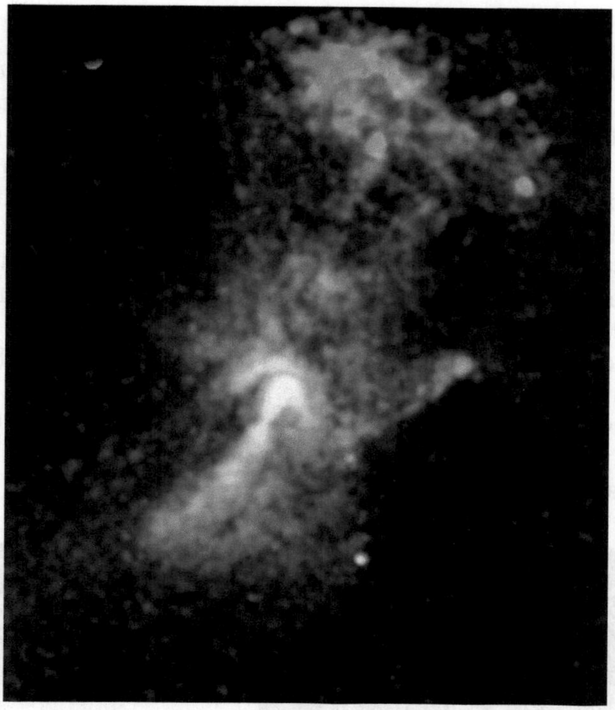

Fig. 3.6. *Chandra* ACIS-I image of pulsar B1509−58 in SNR SNR 230.4−1.2. The image is 10' by 14'. The pulsar is the bright white source at the center of the nebula. Courtesy NASA/MIT Gaensler et al. (2003).

SNR 292.2−0.5

SNR 292.2−0.5 contains the 407-ms radio pulsar J1119−6127 whose discovery (Camilo et al. 2000) led to the radio detection of the SNR by Crawford et al. (2001). The first of two *Chandra* observations with ACIS-S3 (Gonzalez and Safi-Harb 2003, 2005) provided unambiguous detection of the X-ray counterpart to the radio pulsar and strong evidence for the detection of a faint PWN. Gonzalez and Safi-Harb (2003, 2005) found that the combined X-ray emission from the pulsar and its associated nebula is described by an absorbed power law model with a photon index 2.2 (+0.6, −0.3) and an unabsorbed X-ray luminosity (0.5–10.0 keV) of 5.5(+10,-3.3) $\times 10^{32}$ ergs s^{-1} assuming a 6 kpc distance. An interesting attribute is the source's derived inefficiency in converting rotational energy into X-rays using the standard assumptions. Deeper observations are need to better establish the detailed attributes of the PWN, such as spectral variations as a function of position.

SNR 54.1+0.3

Lu et al. (2002) observed SNR 54.1+0.3 using ACIS-S3. The *Chandra* image is shown in Figure 3.7 and one sees a central bright pointlike source, a surrounding ring, jet-like elongations, and low surface brightness diffuse emission. All of these features emphasize the similarity to the Crab's PWN. Lu et al. (2002) determined that the spectra of these components are all well described by power-law models (as with the Crab PWN); the spectral index steepens (softens) with increasing distance from the point source. The similarity of SNR 54.1+0.3 to the Crab nebula and its pulsar was further strengthened by the subsequent discovery (Camilo et al. 2002a) of a 136-ms radio pulsar at the location of the *Chandra* source. (These authors then also detected the pulsations in archival X-ray observations performed with the *ASCA* satellite.) The radio pulsar, PSR J1930+1852, is very weak with a period-averaged flux density at 1180 MHz of 60 μJy. For a distance of 5 kpc, the corresponding luminosity is among the lowest for known young pulsars.

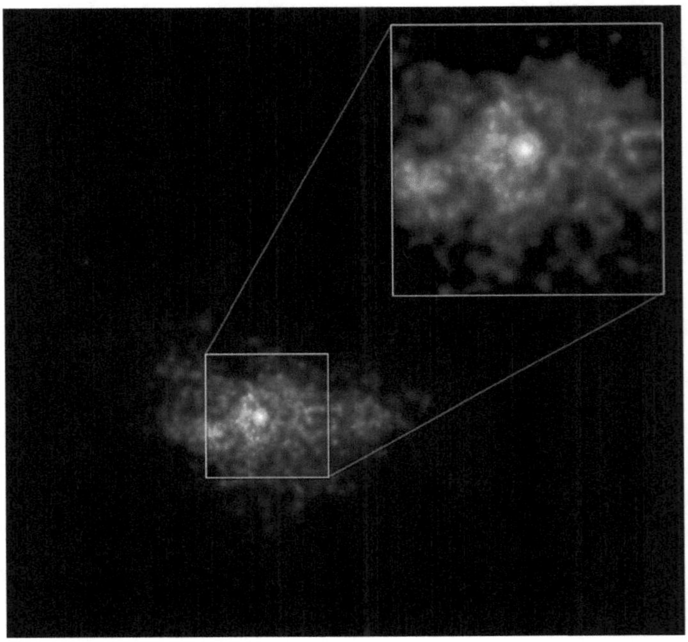

Fig. 3.7. *Chandra* ACIS-S3 image of SNR 54.1+0.3. The large image is 2.7' x 2'. Courtesy NASA/CXC/U.Mass/F. Lu et al.

SNR 39.2−0.3 (3C 396)

Olbert et al. (2003) observed SNR 39.2−0.3 (3C 396) using ACIS-S3. The *Chandra* image (Figure 3.8) resolved an extended ($55'' \times 20''$) X-ray nebula with a nonthermal energy spectrum ($\Gamma = 1.5 \pm 0.3$ at 90%-confidence) and detected

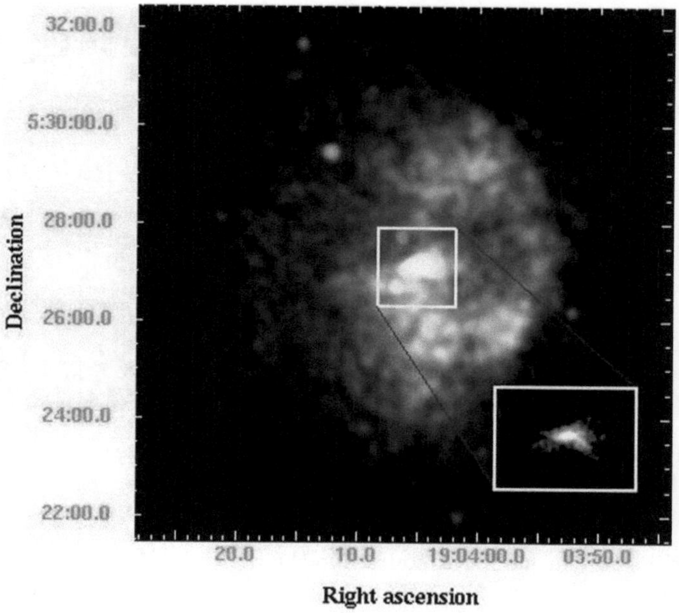

Fig. 3.8. *Chandra* ACIS-S3 image of 3C 396. Courtesy M. J. Rutkowski and J. Keohane.

what the authors refer to as a "pointlike" source at the center of the nebulosity. There is also diffuse radio emission in the same region. This discovery provided convincing evidence for the presence of a PWN, surely harboring an X-ray pulsar with a to-be-detected pulse period. The *Chandra* observations confirmed the conclusions as to the nature of this source – the existence of both thermal and non-thermal components and the possible presence of a rotating NS powering a synchrotron nebula – reached by Harrus and Slane (1999) based on observations with *ASCA*. The pulse period, if observable in our line of sight, is yet to be detected and Olbert et al. (2003) do not appear to have set upper limits over the admittedly long periods accessible using ACIS in its normal mode. A recent search (Zavlin 2005) of these data and covering the period range from 10 to 10^4 s uncovered no evidence for pulsations for pulses with a sinusoidal amplitude of greater than 34% (95%-confidence).

SNR 293.8+0.6

SNR 293.8+0.6 was observed with *Chandra* ACIS-S3 for 40 ksec by Olbert, Keohane, and Gotthelf (2003). There is no published reference to this observation other than the abstract referred to here. The abstract mentions the presence of a "soft" point source near the center of the remnant and the absence of a bright synchrotron nebula. The authors note that these results seem to be in contrast to the presence of a PWN that one might expect from the radio image. A quick glance at the *Chandra* and radio images together (shown in Figure 3.9)

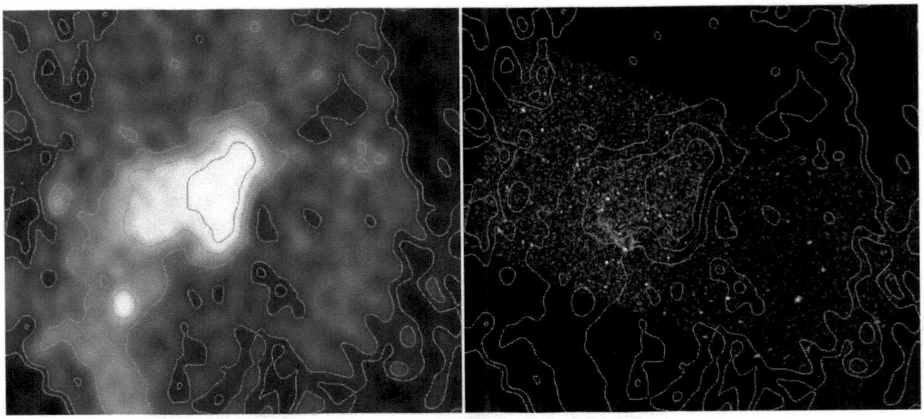

Fig. 3.9. Radio (left) and *Chandra* (right) with radio contours superimposed of SNR 293.8+0.6. Image courtesy S. Patel.

however, shows numerous X-ray sources in the region containing the SNR, any of which might be candidates for an associated compact object. The X-ray image in Figure 3.9 is rich in structure and clearly more work needs to be done before one adds SNR 293.8+0.6 to the list of SNR with established and identified compact objects.

N158A (SNR B0540−69 in the LMC)

The 50-ms pulsar PSR B0540−69 is very "Crab-like" possessing a similar pulse period, spin-down age, and spin-down power. The pulsar was discovered by Seward, Harnden, and Helfand (1984) using the *Einstein* Observatory. A *Chandra* HRC observation, performed by Gotthelf and Wang (2000), was motivated by a desire to search for an X-ray plerion, expected based on the similarity to the Crab and previous observations in the optical, and radio (see e.g. the introduction in Gotthelf and Wang (2000) and references therein). Using the HRC allowed for precision timing and subsequent detection of the pulsed emission from the central, barely-resolved, extended emission. Defining the data at pulse minimum, which lasted about 0.5 in pulse phase, as "off-pulse" these authors separated the central image into that of the "pulsed" component and that for the "off-pulse" component. The former appeared point-like whereas the latter appeared extended as shown in Figure 3.10 revealing the presence of a PWN and even possibly a jet.

Kaaret et al. (2001), in addition to reanalyzing the HRC data and applying an improved aspect solution and improved HRC processing to remove spatial artifacts, also performed ACIS-I measurements in continuous clocking mode. As noted previously, this mode allows for higher time resolution albeit at the price of one spatial dimension in the image. Thus, the data available for spectral analysis had to contend with a significant contribution, not so much from the PWN but from the larger SNR. The spectrum of the pulsed component was found to be

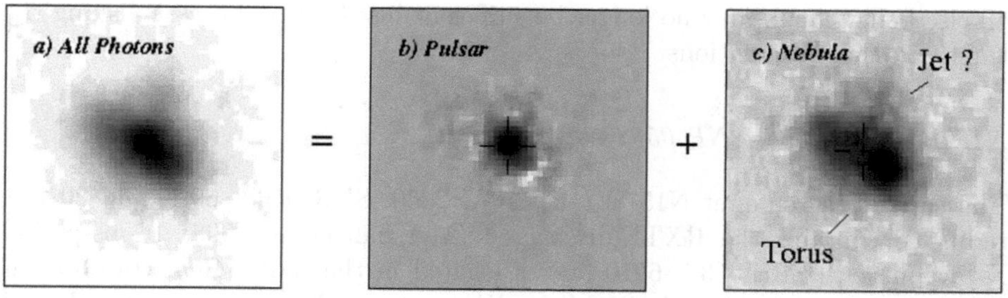

Fig. 3.10. *Chandra* HRC-I images of the pulsar wind nebula of B0540−69 in the LMC. These images are approximately 12.5″ × 12.5″. Images courtesy E. Gotthelf.

consistent with a power law with a photon index of 1.83 (±0.13) fixing N_H at 4.6×10^{21} cm^{-2}. Spectral analyses of the nebula by Kaaret et al. (2001), again fixing N_H, showed it to be softer than the pulsar with spectral indices varying from 1.85 to 2.26 although it is not totally clear that the variations are that statistically significant. The variation is in contrast to the lack of a spectral variation over the PWN measured in the visible (Serafimovich et al. 2004).

Serafimovich et al. (2004) also made important contributions to the study of this system analyzing both new (VLT) and archival (HST) observations. Moreover, these authors raised the important point, which to our knowledge has been neglected when analyzing spectra (except in the case of the Crab pulsar – as discussed in Sect. 3.2.2) that the use of standard abundances may not be (is not) justified, perhaps even more so for an extragalactic source. Of general interest (although not necessarily directly applicable to B0540−69 as they noted) is their discussion of the potential impact of the absorption produced by the SN ejecta on the X-ray spectrum which can enhance the X-ray absorption at energies above the oxygen K-edge depending on the state of evolution of the SNR – the less evolved, the more possible that there is a significant contribution to the absorption from the ejecta. *We strongly urge readers interested in understanding the spectra of X-ray sources in SNR to carefully examine the discussion in Sect. 2.7.1 of Serafimovich et al. (2004).*

Serafimovich et al. (2004) did note that the pulsar could well have a non-power-law (thermal?) spectrum at energies below 1 to 2 keV, although this conclusion must be tempered by the knowledge that the understanding of the ACIS response function decreases with decreasing energy. Serafimovich et al. (2004) concluded that a reanalysis of the X-ray spectrum of B0540−69 is called for, but noted that this was beyond the scope of their paper. The reanalysis should be done, and similar considerations such as allowing for absorbing effects of ejecta, allowing abundances to vary, etc. need to be applied in a systematic way to all spectral analyses performed in the *Chandra* era where not prevented by poor statistics. Indeed, it might be argued that spectral observations should not be performed unless there are sufficient statistics to pursue such studies – simply being able to differentiate between a power law and say a blackbody spectrum, all else being equal, may no longer be sufficient justification for establishing the length of such observations.

N157B (NGC 2060; SNR 0538−691)

The supernova remnant N157B (aka NGC 2060, SNR 0538−69.1, and 20 Doradus B) contains the RXTE-discovered (Marshall et al. 1998) 16-ms-period X-ray pulsar, PSR J0537−6910 and is located in the LMC. *Chandra* observations using the HRC-I and HRC-S by Wang et al. (2001) served a number of purposes including obtaining a precise location for the pulsar. In addition they were able to spatially resolve the pulsar, a surrounding compact yet elongated (0.6 × 1.7 pc) nebula, and an even larger-scale feature of diffuse emission trailing from the pulsar and oriented nearly perpendicular to the major axis of the nebula. These features, shown in Figure 3.11, indicate interesting interactions between the pulsar-powered nebula as it is moving through the surrounding medium. A subsequent ACIS-S observation (ObsID 2783) seems never to have been formally analyzed.

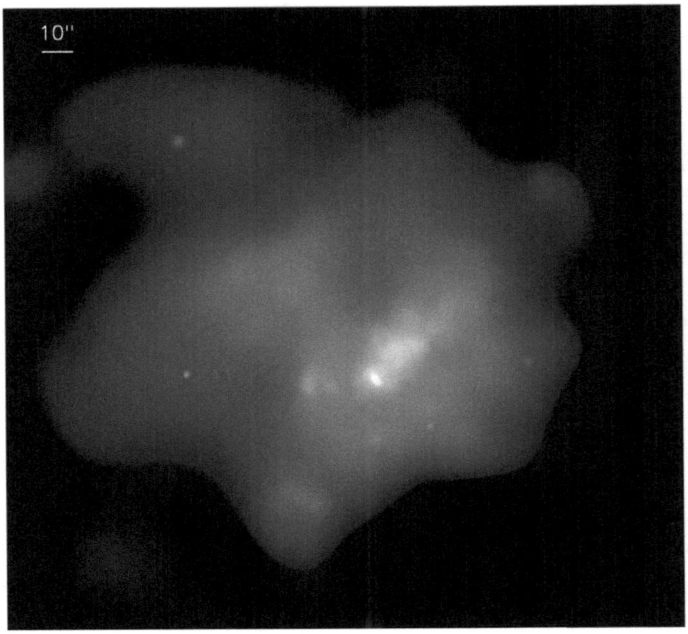

Fig. 3.11. *Chandra* image of SNR N157B. The image combines HRC-S and -I observations and is smoothed to bring out the extended structure near the pulsar. Courtesy Q.D.Wang.

B0453–685

Gaensler et al. (2003) performed both radio and *Chandra* observations of the supernova remnant B0453–685 in the Large Magellanic Cloud (LMC) and discovered a new PWN. Gaensler et al. (2003) detected a strongly linearly polarized (8% and 6% at 2.4 GHz and 1.4 GHz respectively) and elongated ($20'' \times 30''$) central radio core, similar in morphology to the X-ray core ($14'' \times 7''$) seen as part of a 40 ksec ACIS-S3 observation (Figure 3.12). They also found that the X-ray spectrum of the central core is fit by a power-law with number index of -1.9 ± 0.4. Unfortunately no mention is made as to the quality of the fit. This is important as the statistical uncertainties (measured by extremes on the $\chi^2 + \epsilon$ contours) get *underestimated* if the fit is poor. Based on the quoted uncertainties and the limited number of counts, we suspect that the fit to the power-law spectrum may not have been compelling in comparison to other models for the continuum. However, the radio and X-ray images, and their similarity leave little doubt that a PWN has been detected. The search for the underlying pulsar proved unrewarding, albeit not surprisingly so, as these authors found a scant 58-count upper limit to the contribution of a putative point source to the central image. The authors noted that the corresponding limits to the X-ray luminosity (0.5–10.0 keV) of $< 6D^2 \times 10^{33}$ ergs s^{-1} (assuming $N_H = 1.3 \times 10^{21}$ cm^{-2} and a power law index $\Gamma = 1.5$) were hardly restrictive. It is worth emphasizing that Gaensler et al. (2003) propose an interesting approach to determining the properties of the system in contrast to the usual one where the equations for a

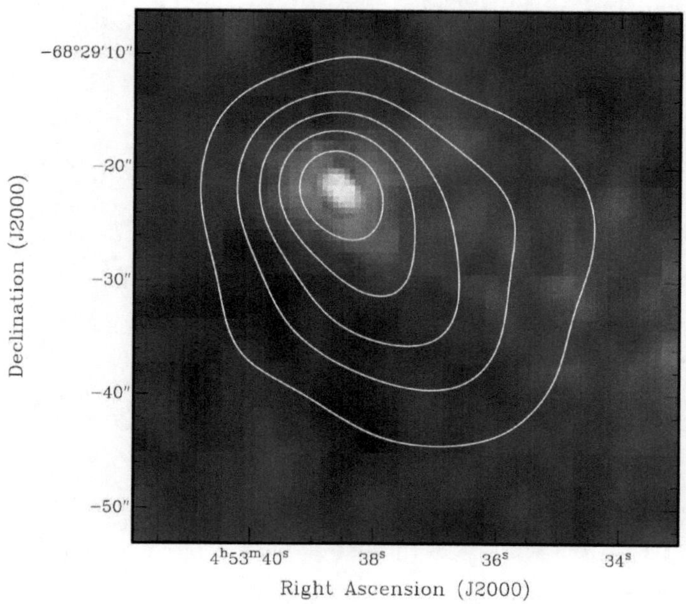

Fig. 3.12. *Chandra* ACIS-S3 image of the pulsar wind nebula and corresponding radio contours of B0453–685 in the LMC. Image courtesy B. Gaensler.

pulsar's age, spin-down luminosity, and surface magnetic field are written down in the dipole approximation and one assumes a braking index of 3, an initial period of 0, and a luminosity that is a carefully selected fraction of the spin-down luminosity in order to determine the pulse period, period derivative, and the neutron star's surface magnetic field. We won't repeat their discussion here, but refer interested readers to Sect. 3.1 of their paper.

Deeper observations of this system, which would allow one to perform spatially resolved spectroscopy in addition to facilitating sensitive searches for pulsations, are clearly called for.

CTA 1 (SNR 119.5+10.2)

CTA 1 (aka SNR 119.5+10.2) is a radio shell SNR with emission from the center that appeared harder and brighter than the emission from the limb as seen with *ROSAT* (Seward, Schmidt, & Slane 1995). Of special interest is the fact that the position of 3EG J0010+7309, one of the brighter of the EGRET unidentified sources (Hartman et al. 1999), lies inside the boundary of CTA 1.

Halpern et al. (2004) used *Chandra* ACIS-S3 to image the central portion of the remnant. The *Chandra* image (Figure 3.13) reveals a point source, a compact nebula, and a (curved) "jet". These characteristics, together with upper limits

Fig. 3.13. *Chandra* ACIS-S3 image of CTA 1. The image is $2' \times 2'$. Courtesy J. Halpern.

to the optical flux at the *Chandra* location of the point source which gives an X-ray-to-optical flux ratio in excess of 100, clearly establish the central object as a rotation-powered pulsar, although pulsations are yet to be detected.

Slane et al. (2004a) using *XMM-Newton* set a rather unrestrictive upper limit of 61% pulsed fraction for periods between 1.2 ms and tens of ksec. Halpern et al. (2004) established a restrictive upper limit to a possible radio counterpart at both 1425 and 820 MHz (implying less than 0.02 mJy kpc^2 at 1400 MHz). Halpern et al. (2004) also found that the X-ray spectrum of the point source is best fit by a power law plus blackbody model with $\Gamma = 1.6$, $kT_\infty = 0.13$ and $R_\infty = 0.4$ km. We wish to emphasize that the number of counts from the point source (and the PWN and the jet) were quite limited being only 187 (136 and 45) implying large uncertainties in the derived parameters, and indicting that a deeper *Chandra* observation is called for since the angular scale of the observed features is far to small to be adequately resolved with *XMM-Newton*.

Finally, Halpern et al. (2004) derived an upper limit to the surface temperature of the underlying neutron star. The upper limit was perhaps conservative in that it only made use of the data from the lowest energy bin rather than examining the full spectrum. On the other hand, the column was fixed and allowed variations of the column might possibly increase the upper limit of 6.6×10^5 K.

In general there is a real need for a systematic presentation (e.g. was $N_{\rm H}$ fixed or not, what is the confidence level associated with the limit? etc.) of upper limits to neutron star surface temperatures when confronting theory.

SNR 69.0+2.7 (CTB 80)

Moon et al. (2004) observed the SNR CTB 80 (aka SNR 69.0+2.7) with *Chandra* ACIS-S3 (see Figure 3.14) as part of a multi-wavelength study. The remnant contains the 40-ms radio pulsar PSR B1951+32 discovered by Kulkarni et al. (1988). The *Chandra* observations clarified the morphology of the X-ray emission and showed what appears to be a cometary PWN elongated (about 40″) along the direction of the pulsar's proper motion and seemingly confined by a bow shock produced by that proper motion and thus confirming prior speculations (see e.g. Moon et al. 2004 and references therein).

More recently Li, Lu, and Li (2005) have further analyzed these same data with emphasis on providing spatially-resolved spectra. Li, Lu and Li (2005) find that a power-law plus blackbody model fits the spectrum of the pulsar better than a pure power-law model. In this case the blackbody comprises about 10% of the total flux (with large uncertainties) and a temperature of 0.13 keV. The accompanying power law index was about 1.6. The corresponding blackbody radius was small, about 2 km at the assumed distance of 2 kpc, perhaps implying emission from hot spots. These authors also fit the data to a blackbody plus power law, while fixing the blackbody radius to a value more appropriate to the entire star in an effort to set an upper limit to the surface thermal emission. This approach led to a temperature upper limit (3σ) of 7.8×10^5 K, much below the predictions of standard neutron star cooling models (see Figures 8 and 9 of

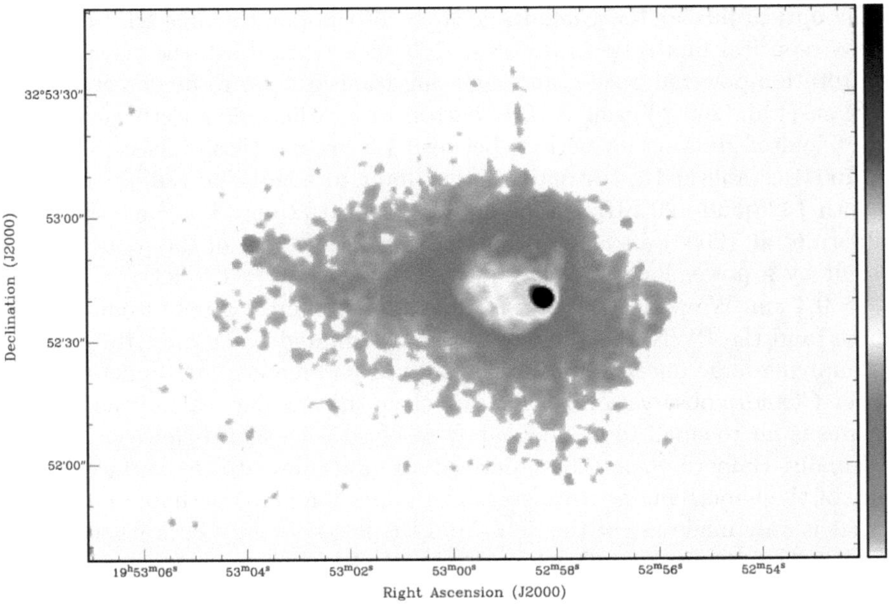

Fig. 3.14. *Chandra* ACIS-S3 image of CTB 80. Courtesy D.-S Moon.

Li, Lu, and Li 2005) as with 3C58 (Sect. 3.2.3). The time resolution of the data did not permit an analysis for 40-ms pulsations.

SNR 359.23–0.82 (The Mouse)

Gaensler et al. (2004) observed the unusually shaped radio source SNR 359.23–0.82, sometimes referred to as the "mouse", using ACIS-S3. For a history of observations of this object in all wavelength bands see the introduction to Gaensler et al. (2004) and references therein. Observations with *ROSAT* by Predehl and Kulkarni (1995) had already detected X-ray emission and they correctly proposed that this source was a bow shock PWN. The *Chandra* observations confirmed this conclusion and, as shown in Figure 3.15, provided spectacular details resolving numerous components referred to as the "halo, head, tongue and tail" by Gaensler et al. (2004). We note that Gaensler et al. (2004) goes well beyond the presentation of new observational results and they combine theory and hydrodynamic simulations of bow shocks to unravel the implications of the data. The paper is essential reading for those interested in the interaction of a PWN with the environment, especially when the pulsar has a moderately large velocity.

Geminga

Geminga was observed using ACIS-S3 by Sanwal, Pavlov and Zavlin (2004). The resulting image is shown in Figure 3.16 where one sees what Sanwal, Pavlov, and Zavlin (2004; 2005) term a "wake", about $10''$ – $15''$ in projected length to the

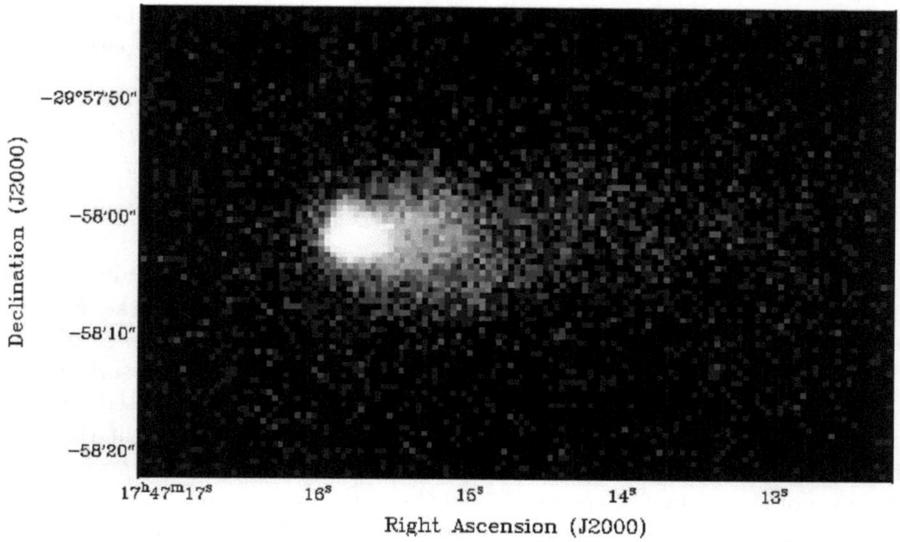

Fig. 3.15. *Chandra* ACIS-S3 image of the "mouse". Courtesy B. Gaensler.

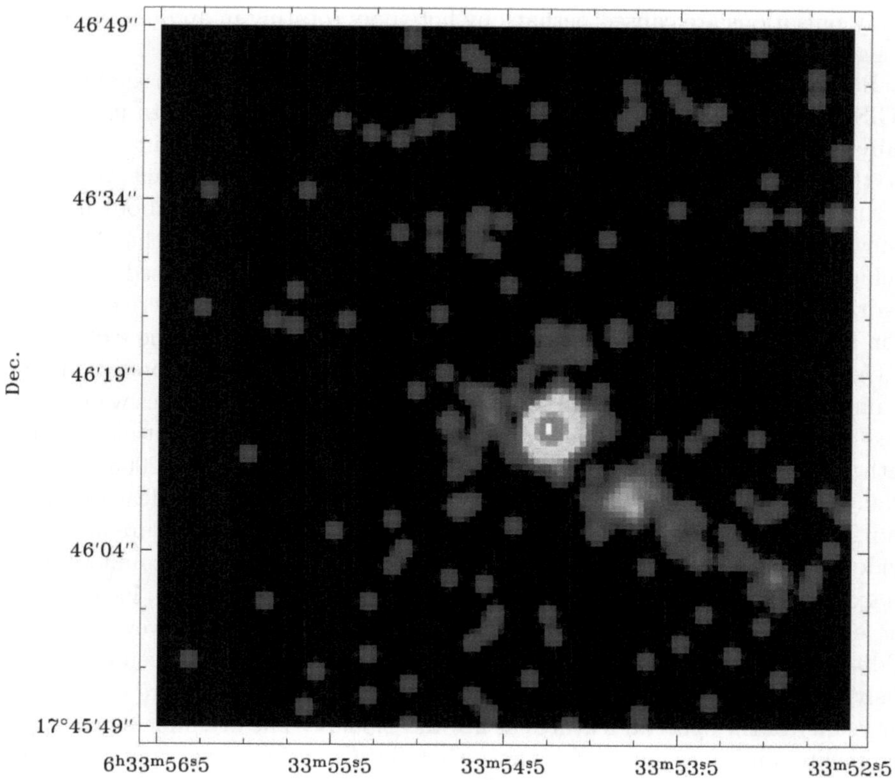

Fig. 3.16. *Chandra* ACIS-S3 image of Geminga. Courtesy D. Sanwal & V. Zavlin.

southwest of the pulsar. If one looks carefully, one can also see extended emission a few arc seconds away from the pulsar in the opposite direction, perhaps the head of a bow-shock.

3.2.3 Finding the pulsars

SNR 296.5+10.0 and 1E 1207.4−5209

Observations with *Chandra* have contributed at least two significant new insights into the source 1E 1207.4−5209 initially discovered with the *Einstein* Observatory (Helfand & Becker 1984) and located 6' from the center of SNR PKS 1209−51/52 (aka SNR 296.5+10.0). The source was first observed with *Chandra* by Zavlin et al. (2000) using ACIS-S3. These observers used ACIS in continuous clocking mode which allows time resolution of 2.85 ms at the price of one dimension of spatial information and discovered a 424-ms period. The detection of the period, of course, provided compelling evidence that the source is a neutron star. Since the source appears to be radio-quiet (Mereghetti, Bignami, & Caraveo 1996; Kaspi et al. 1996), it may be either an active pulsar beamed out of our line of sight or a truly radio-quiet neutron star, where the X-ray pulsations are caused perhaps by hot spots rotating in and out of our line of sight.

Subsequent to the detection of pulsations, Sanwal et al. (2002) analyzed two ACIS-S3 continuous-clocking-mode observations including the data used to initially detect the pulse period. In addition to establishing a preliminary estimate for the period derivative, these authors also found two significant absorption features centered at 0.7 and 1.4 keV with equivalent widths of about 0.1 keV. Sanwal et al. (2002) discussed several possible interpretations for the absorption including cyclotron resonances and atomic features. They presented arguments favoring atomic transitions of once-ionized helium in the atmosphere of the neutron star assumed to be very strongly ($\approx 10^{14}$G) magnetized. The exact cause of the *Chandra*-discovered features is not without different interpretations. For example Hailey and Mori (2002) argued that the absorption features were associated with He-like oxygen or neon in a field of $\approx 10^{12}$G. More recent observations with *XMM-Newton* (e.g., Mereghetti et al. 2002; Bignami et al. 2003; De Luca et al. 2004) not only confirmed the *Chandra*-detected absorption features at 0.7 and 1.4 keV, but also seemed to have uncovered an additional feature at 2.1 and evidence for a fourth 2.8 keV. Taking all these latter data into account supports an explanation involving the fundamental and two, possibly three, harmonics of the electron cyclotron absorption in a field of order 10^{11}G. However, the two additional spectral features in the *XMM-Newton* data have not been unambiguously accepted. Mori, Chonko, and Hailey (2005) have cast severe doubt as to the reality of the spectral features at 2.1 and 2.8 keV. The arguments given seem compelling and it is thus unfortunate that the *Chandra* response is insufficient to weigh in on this question without expending significant amounts of observing time.

Zavlin et al. (2004) have continued to observe this target using both *Chandra* and *XMM-Newton*. They have detected significant variations in the spin period, which they interpreted in light of three hypotheses: a glitching pulsar; variations in an accretion rate from a fallback disc; and variations in accretion produced by being in a wide binary.

Thus the sequence of *Chandra* observations have provided important discoveries, especially the detection of the pulse period and firm detection of two absorption features. An important and unanswered question is what are the limits as to the presence of such spectral features for the other NSs in SNR. A systematic comparison, if not already in progress, should be performed.

Finally we note that 1E 1207.4−5209 is a source that, in some critical respects, is similar to Cas A in that it is in a SNR, is radio-quiet, and has a low-energy spectrum that may be fit by a blackbody with a temperature falling in the range from 0.2–0.6 keV, however it pulses. Thus, on the one hand, the source therefore can be used to give us confidence that all CCOs will ultimately be found to pulse. On the other, this source's *Chandra*-revealed characteristics may be used to separate it from the CCO-group of objects.

SNR 292.0+1.8

SNR 292.0+1.8 is, along with Cas A and Puppis, one of three known oxygen-rich supernova remnants in the Galaxy. Hughes et al. (2001) performed an observation with ACIS-S3 (Figure 3.17), detecting a bright, spectrally hard, point source within an apparently extended region. This detection suggested the presence of a pulsar and its pulsar wind nebula. Radio observations (Camilo et al. 2002b) then found a 135-ms pulsar in SNR 292.0+1.8 localized to within the Parkes beam (14 arcmin FHWM). The detection by Hughes et al. (2003b) of X-ray pulses at the expected period from the compact X-ray star secured its identification. The X-ray spectrum is modeled with a simple power law, although, as with Vela, (and many other sources) the fit to the data is not unique. From the motions of oxygen-rich optical knots and the size of the remnant, Ghavamian, Hughes, and Williams (2005) recently estimated a kinematic age for SNR 292.0+1.8 of 3000–3400 years assuming a distance of 6 kpc. This value is in good agreement with the pulsar spin-down age of 2900 years.

3C58

The *Chandra* observations of 3C58 (aka SNR 130.7+3.1) were first performed by Murray et al. (2002b) using the High Resolution Camera (HRC) which offers excellent time resolution but no spectral information. These data imaged the previously detected X-ray point source (Becker, Helfand, and Szymkowiak 1982) which is associated with the historical SN 1181 (see Stephenson & Green 2002 and references therein). The early *Chandra* data also revealed the extended PWN and the presence of 66 ms pulsations from the central point source (J0205+6449). Deeper *Chandra* observations using ACIS-S3 by Slane et al. (2004b) produced

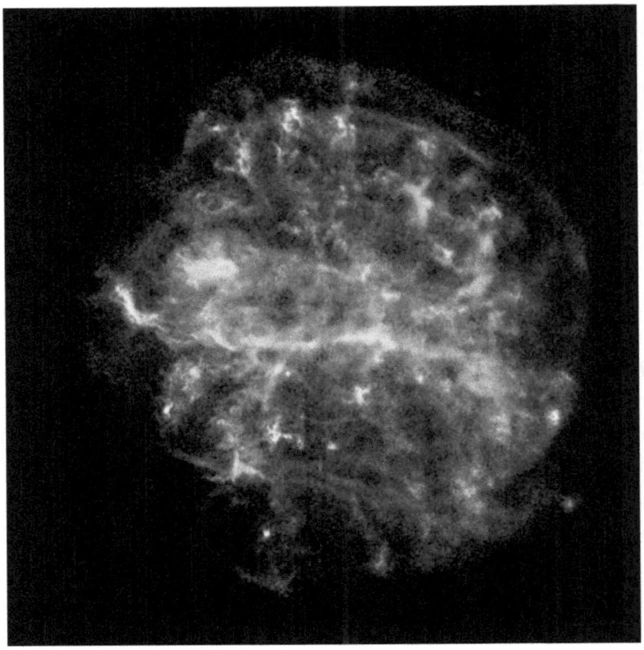

Fig. 3.17. *Chandra* ACIS-S3 image of SNR 292.0+1.8. The image is $9' \times 9'$. The point source identified with the 135 ms PSR J1124−5916 is just southeast of the center of the remnant. Courtesy NASA/CXC/Rutgers/J. Hughes et al.

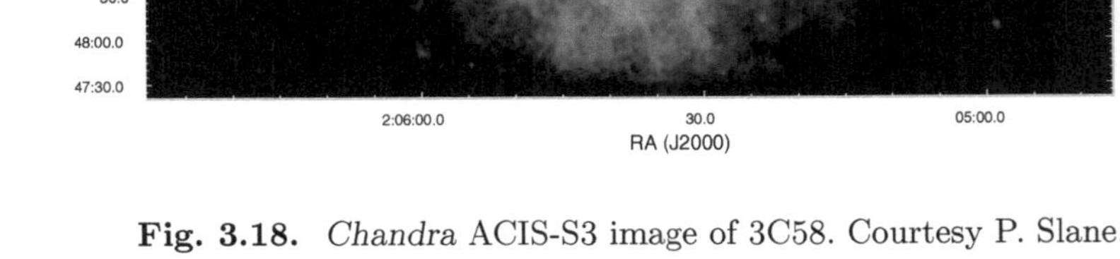

Fig. 3.18. *Chandra* ACIS-S3 image of 3C58. Courtesy P. Slane.

images such as that shown in Figure 3.18 showing the similarity of this PWN with the Crab and Vela.

One aspect of the *Chandra*-based research of 3C58 of special importance were the limits as to any thermal emission from the surface of this young cooling neutron star. The search for thermal emission was presented by Slane, Helfand,

and Murray (2002) and then refined by Slane et al. (2004b) who found that as with SNR 266.2−1.2 (Sect. 3.2.1) their upper limit ($T_\infty < 1.02 \times 10^6$ K) falls well below predictions of standard neutron star cooling. Yakovlev et al. (2002) discuss calculations of neutron star cooling in the context of 3C58 and concluded that the observations can be explained by the cooling of a superfluid neutron star where the direct Urca process is forbidden.

We note that it is far easier to derive a stringent upper limit to any thermal component for 3C58, in contrast for example to the Crab pulsar, because the flux of 3C58 is much lower. Of course neutron stars may be different, so that limits to the thermal components of both sources, indeed all the young neutron stars, are relevant to compare with theoretical predictions of neutron star cooling. In general, such analyses are not simple, requiring enhanced sensitivity for the detection of the putative thermal component often in the presence of a much stronger non-thermal flux from the magnetosphere of the pulsar, if one wants to measure the temperature – as opposed to setting an upper limit. *Chandra* is uniquely poised to provide the raw data for such studies due to its ability to maximally separate the pulsar from the surrounding nebulosity, yet often long observations are required.

IC443

Historically, the *Chandra* ACIS-I3 image of IC443 shown in Figure 3.19 was a publicity tour-de-force for the *Chandra* project as the first three authors of Olbert et al. (2001) were high school students at the time. This remnant had been previously well studied as there is a large variety of shocked molecules present

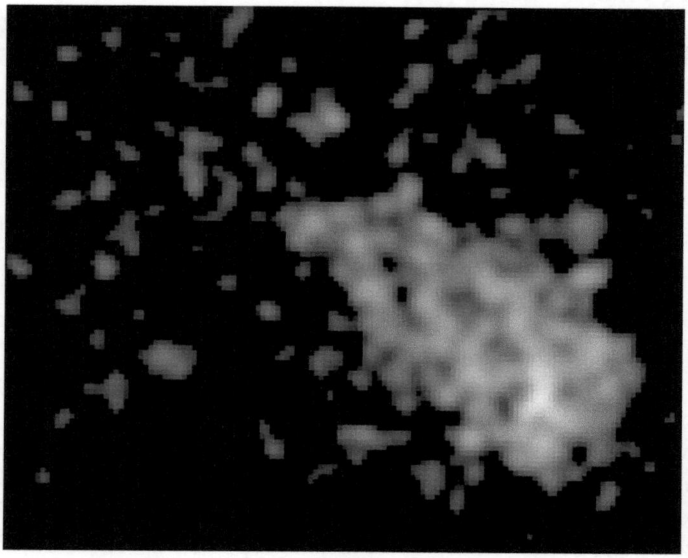

Fig. 3.19. *Chandra* ACIS-I image of IC443. The image is 1.0′ × 0.8′. Courtesy NASA/NCSSM/C. Olbert et al.

due to the interaction with surrounding molecular clouds (see e.g., references in Olbert et al. 2001 and Bykov, Bocchino, & Pavlov 2005). IC 443 is also a candidate counterpart to the EGRET source, 3EG J0617+2238. The image (Figure 3.19) shows what appears to be a point source behind a bow shock and surrounded by a nebulosity that looks somewhat like a cometary tail. Olbert et al. (2001) also reported accompanying VLA observations which confirmed and complimented the X-ray spatial structure and exhibited varying degrees of polarization as strong as 25%. No pulsations were reported either from the X-ray or the flat spectrum radio observations. Subsequent observations with both *Chandra* and *XMM-Newton* (Bocchino and Bykov 2001; Bykov, Bocchino, Pavlov 2005) have also not detected pulsations.

3.2.4 Not finding the compact objects

In a number of cases, high sensitivity searches with *Chandra* have been unable to specifically identify a compact object associated with a SNR, although often numerous candidate objects have been detected.

SN 1987A

Since launch, SN 1987A has been the focus of a series of repeated observations with *Chandra* (see Park et al. 2005 and references therein). One goal of these observations is to detect the emergence of the X-ray flux from a newly born compact object. To date, no such object has been detected and Park et al. (2005) assume a spectral form – a power law of photon index 1.7 – and use an absorbing column derived from their fit to the entire remnant to set a 90% confidence upper limit to the 3–10 keV luminosity of 1.3×10^{34} ergs s^{-1}. As noted by the authors, the uncertainty as to the correct column to apply to this calculation could easily increase this upper limit.

γ-Cygni (SNR 78.2+2.1)

Becker et al. (2004) used *Chandra* to search for the X-ray counterpart to 3EG J2020+4017 (2CG078+2). In particular these authors were following up on the possibility (Brazier et al. 1996) that RX J2020.2+4026 was the counterpart. These observations, thanks to the precision with which X-ray sources in the field could be located, demonstrated conclusively that RX J2020.2 + 4026 is associated with a K field star and therefore an unlikely counterpart of the bright EGRET source.

This observation also demonstrated the difficulties one sometimes encounters in searching for compact objects associated with a SNR. Thus, 37 additional X-ray sources were detected in the field searched (which was only a fraction of the full size of the SNR). Radio observations reported by these authors, which covered the complete 99% EGRET likelihood contour of 3EG J2020+4017 with a sensitivity limit of $L_{820} = 0.09$ mJy kpc^2, were unable to find a pulsar. The absence of radio pulsations suggests that if there is a pulsar operating in γ-Cygni, the pulsar's emission geometry is such that the radio beam does not intersect with the line of sight. Alternatively, the pulsar is perhaps a CCO-like object which does not produce significant amounts of radio emission.

Without high-precision X-ray spectra of each of the candidate X-ray sources, and detailed follow up in other wavelength bands, there is essentially no satisfactory way in which to eliminate most of the candidates from consideration. In such cases, the principal and important *Chandra* contribution is to provide target lists with accurate positions as a basis for future studies.

SNRs 315.4−2.30, 093.3+6.9, 084.2+0.8, and 127.1+0.5

Gvaramadze and Vikhlinin (2003) analyzed archival ACIS-I observations of SNR 315.4−2.30 (aka MSH 14−63, RCW 86) a bright, radio shell-like SNR (see the introduction to Gvaramadze & Vikhlinin 2003 and references therein; see also Sect. 5.2 of Kaplan et al. 2004 and references therein). They concentrated their study to a protrusion in the southwest of the remnant based on the hypothesis that the SNR resulted from an off-centered explosion of a moving and massive star. Two X-ray sources were detected in this region, one of which they identified with a foreground star. The second source they identified as a candidate for the compact remnant, in part because of the positional coincidence in support of their hypothesis, and in part because of the absence of an optical counterpart. The location implies a transverse velocity of over 1500 km s^{-1}.

SNR 315.4−2.30 is also one of the four thoroughly-studied SNR by Kaplan et al. (2004). The additional targets are listed in the title of this section above. Kaplan et al. have embarked on a program to perform systematic studies to search for compact central objects in a distance-limited sample of 23 SNR that lie within 5 kpc. These authors include with their *Chandra* (and *XMM-Newton*) observations an accompanying optical/IR identification program taking account of the fact that any such counterparts are expected to be very faint. Using the brightness of the *Chandra*-discovered CCO in Cas A as a reference, they find no compact central objects associated with these four SNR to a limit of 0.1 Cas A or $L_x < 10^{31}$ ergs s^{-1}. We eagerly await the subsequent papers covering the remainder of the observations.

SNR 41.1−0.3 (3C 397)

Safi-Harb et al. (2005) used a 66 ksec *Chandra* ACIS-S3 exposure to study 3C 397 (SNR 41.1−0.3). One goal of this study was to search the central X-ray hot spot for a compact remnant left by the original supernova. No viable counterpart was found, and these authors placed an upper limit to the 0.5–10.0 keV flux of 6×10^{-13} ergs cm^{-2} s^{-1} ($L_x(0.5 - 10.0$ keV$) = 7 \times 10^{33} D^2$ ergs s^{-1}). Unfortunately there are some ambiguities as to how these numbers were obtained as the spectrum used and the confidence levels associated with the uncertainties are not mentioned.

N63A in the LMC

Warren, Hughes, and Slane (2003) observed the supernova remnant N63A in the LMC using ACIS-S3. No hard X-ray point-source was apparent in these data and these authors could rule out a young, energetic, Crab-like pulsar. They set a 3σ upper limit to the flux of a point source assuming a power law spectrum with a number index of -2. The 2.0–8.0 keV flux upper limit was 2.5×10^{-14} ergs cm^{-2} s^{-1}, or a luminosity of 7×10^{33} ergs s^{-1} (assuming a distance to the LMC of 50 kpc) and applied to the region anywhere in the interior of the SNR. The luminosity limit for a source extended over a diameter of 2 pc (appropriate to a pulsar wind nebula) was higher: 4×10^{34} ergs s^{-1}.

1E 0102.2−72.2 in the SMC

This remnant was discovered during the *Einstein Observatory* survey of the Small Magellanic Cloud (SMC) (Seward and Mitchell 1981). Shortly after its discovery, Dopita, Tuohy, and Mathewson (1981) found oxygen-rich optical emission from the remnant extended over a diameter of $\sim 24''$. Subsequent optical spectroscopy (Tuohy and Dopita 1983) revealed that this material was moving rapidly ($\sim$6500 km s^{-1} FWHM), identifying 1E 0102.2−72.2 as the first O-rich SNR in the SMC. Amy and Ball (1993) suggested that a compact feature near the remnant's projected geometric center in their high resolution ($3''$) radio image might be "plerionic" (i.e., emission from a pulsar or pulsar wind nebula). Using a 9 ks ACIS-S3 observation taken during *Chandra*'s orbital activation and checkout period, Gaetz et al. (2000) set a 3σ upper limit to the luminosity of a hard power law component (with number index of -2.05) of $\sim 9 \times 10^{33}$ erg s^{-1} (in the energy band above $\sim$3 keV).

1E 0102.2−72.2 is used as a *Chandra* calibration target and therefore a large number of observations are available in the archive. From a merged data set of on-axis ACIS-S3 observations with a total exposure time of 125 ks, we set a 3σ count rate limit of 2×10^{-4} s^{-1} (3-8 keV band) on the X-ray emission near the center of the SNR at the location of the central radio feature. For a Crab-like spectrum this rate corresponds to an unabsorbed flux of 7×10^{-15} erg cm^{-2} s^{-1} (2–8 keV) or an X-ray luminosity of 3×10^{33} erg s^{-1} for a distance of 60 kpc.

3.2.5 Miscellaneous

Kes 73 and 1E 1841−045

The anomalous X-ray pulsar 1E 1841−045 is associated with the SNR Kes 73. See the introduction to Morii et al. (2003) and references therein for an overview of previous observations. The *Chandra* observations by Morii et al. (2003) using ACIS-S3 in both timed-exposure (30 ksec) and continuous-clocking (10 ksec) modes were the first for which the pulsar could be spatially separated from the surrounding SNR – totally for the timed-exposure observation and in one dimension for the continuous-clocking observation. The spatial advantage was partially mitigated by these authors, who chose to perform all analyses with the continuous-clocking mode data, no doubt in order to exploit the high time resolution (2.85 ms) for this 11.8 s pulsar. The spectral parameters they found from fitting the data to a power law plus black body were $\Gamma = 2.0 \pm 0.3$, $kT = 0.44 \pm 0.02$ keV, and $N_{\rm H} = 2.54(+0.15, -0.13)$ cm^2. We note that the *Chandra* response function provided for the timed-exposure mode is not precisely transferable to continuous clocking mode data, especially the gain. Sophisticated users tend to let the gain be an additional free parameter when spectrally fitting continuous-clocking mode data. A further indication that an incorrect response function may been applied to these data may be indicated by a rather dramatic and large residual in the spectral fitting at about 1.6 keV, which the authors attributed to the aluminum in the ACIS filters. A similar feature, and at the same energy, was also seen in continuous-clocking mode data by Patel et al. (2003; their Figure 4) who also applied the timed-exposure mode response to continuous-clocking mode data. Since such a feature should *not* be present, given a proper response function, this coincidence may well indicate that the incorrect response function was used in both cases. If so, it is not clear, however, as to what impact (if any) this might have on the derived spectral parameters. Further work is needed to clarify this issue. It is also possible that the timed-exposure mode response functions used at the time were simply incorrect, leading to spurious features near the aluminum edge. A good check would have been (and is) to compare the phase-averaged pulsar spectrum determined from the data in both modes. Such a comparison would be meaningful if pileup had not been a problem – unfortunately in this case it was (Wachter et al. 2004). Even so, comparing timed-exposure mode and continuous-clocking mode spectra from sufficiently large regions, and well away from the pulsar, might have sufficed and would have been informative.

Wachter et al. (2004) used the timed-exposure mode data to achieve a precise position for 1E1841−045. The location enabled them to accomplish a refined, and successful search for an infrared counterpart. The archival data from this observation should be also be analyzed both to establish the validity of the spectrum deduced by Morii et al. (2003) and for any insights that they might provide concerning the spectrum of the extended emission.

SNR 109.1−1.0 and 1E 2259+586

We include this *Chandra* observation of 1E 2259+586 and SNR 109.1−1.0 (aka CTB 109) as the anomalous X-ray pulsar 1E 2259+586 lies along the line of sight to the SNR and may well be associated with it. For a discussion of the AXP-SNR connection see Gaensler et al. (2001). Patel et al. (2001) used *Chandra* ACIS-S3 to, amongst other things, determine the most precise X-ray position of 1E 2259+586. The *Chandra* image also shows clear evidence for extended emission, extending from about 4″ to more than 100″, which Patel et al. (2001) attributed to the SNR. The precise position enabled Hulleman et al. (2001) to perform deep optical and near-infrared observations with Keck and they found a faint ($K_s = 21.7 \pm 0.2$ mag) candidate counterpart.

Deeper observations, which might serve to establish a possible PWN, would seem called for.

N49

Park et al. (2003) analyzed two ACIS S-3 observations of SNR N49 (aka SNR 0525−66.0) in the LMC. These authors concentrate their discussion as to the X-ray properties of the gaseous remnant (see Sect. 3.4.1 below), but do present a brief discussion of the point source located in the north-eastern portion of the remnant. This supernova is best known in the high-energy community as the probable site (Cline et al. 1982 – published in all seriousness on April 1) of the famous transient event of March 5, 1979, which has become known as the soft gamma-ray repeater (SGR 0526−66). The tie to N49 is as follows. First, Cline et al. (1982) used data from a network of satellites, including the *Einstein* Observatory, to provide a precise position described by a narrow rectangle which lay in the northern portion of N49. The association with N49 was compelling. Next, twelve years later, Rothschild, Kulkarni, and Lingenfelter (1994) using *ROSAT* discovered an X-ray source whose positional error circle intersected the error region associated with the gamma-ray repeater as indicated in their Figure 1. The *Chandra* image of the region containing the *ROSAT* source is shown in Figure 3.20. The *Chandra* point source lies within the *ROSAT* error circle and appears to be within the earlier positional uncertainty of the SGR.

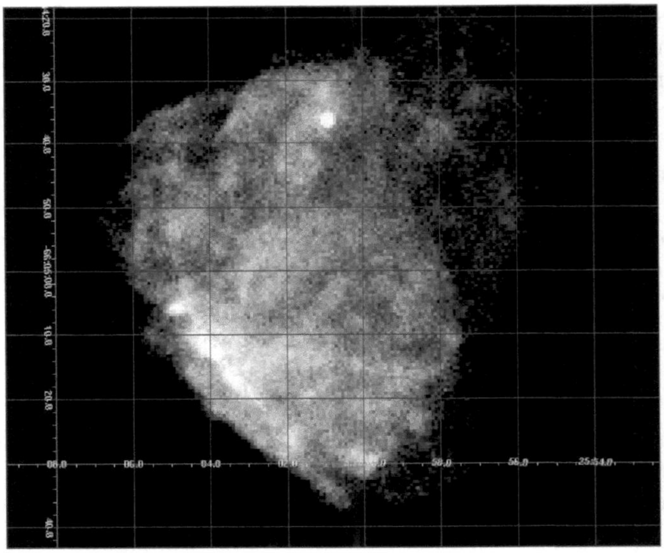

Fig. 3.20. *Chandra* ACIS-S3 image of N49. The bright point source in the northern part of the remnant is clearly visible. Note that southern portions of the remnant are cut-off, since the data were taken in 1/8 subarray mode in order to search for pulsations from the point source. Courtesy S. Patel.

The X-ray source would therefore seem to be the SGR in quiescence, and using *Chandra* ACIS-S3 data Kulkarni et al. (2003) claim a detection of 8-s periodic pulsations, as were observed during the original transient gamma-ray event. The statistical significance of the pulsed X-ray signal is only modest (~99.98% confidence level) and suggests a spin-down rate of 6.5×10^{-11} s s^{-1} based on two measurements separated by approximately a year and a half. The pulse fraction is low ~10%. The characteristic age of the pulsar is ~2000 yr, somewhat less than the estimated remnant age of 5000 yr (Vancura et al. 1992).

Kulkarni et al. (2003) and Park et al. (2003) also analyzed the spectral data for the point source. The spectrum was relatively featureless, in contrast to the line-dominated spectra from the rest of the remnant. Kulkarni et al. (2003) fit the spectral data with several models and found that a powerlaw or powerlaw with a blackbody produced acceptable fits. The significance associated with inclusion of the blackbody component, which required a temperature (at infinity) of 0.5–0.6 keV, was only 90%. The photon index of the powerlaw was ~3.1 and the absorbing column to the point source, $\sim 5.5 \times 10^{21}$ cm^{-2} is consistent with the absorption to the SNR. Park et al. (2003) fit the spectral data using power law models with $\Gamma = 2.8$ to 2.9 which provided statistically acceptable fits. They also noted that a BB model with a temperature of about 0.5 keV is not ruled out for the low energy data, especially if one also invokes a hard tail. The 0.5 keV BB is characteristic of what one finds for a CCO and it is tempting to speculate as to a possible tie in.

The inability to distinguish simple spectral models seems to be a characteristic of the *Chandra* and *XMM-Newton* era of CCD-resolution spectroscopy. In general, the observations are typically much too short to accomplish such an objective. It is also not clear that more sophisticated (admittedly unidentified) mathematical tools are needed to tackle this problem. For example, can one take advantage of the fact that the χ^2 statistic used as a measure of goodness-of-fit is *not* distributed as χ^2 when the model spectral distribution is not representative of the true, underlying, spectrum?

NGC 6822 (Ho 12)

Kong, Sjouwerman, and Williams (2004) analyzed archival *Chandra* ACIS-I observations that included SNR Ho 12 in the nearby dwarf irregular galaxy NGC 6822. Ho 12 was known to be a SNR based on optical imaging and spectroscopy and an X-ray source had been associated with the SNR based on *Einstein* imaging – see the introduction to Kong, Sjouwerman, and Williams (2004) and references therein. The *Chandra* observation (Figure 3.21) resolves Ho 12 into what appears to be a shell-shaped object about 10″ (about 24 pc at 500 kpc) in diameter. The extended image, together with a comparison to optical and radio observations, unambiguously confirms the identification with the SNR. No evidence for, nor an upper limit to, the presence of a point source was presented. This object is an excellent candidate for further study and deeper observations.

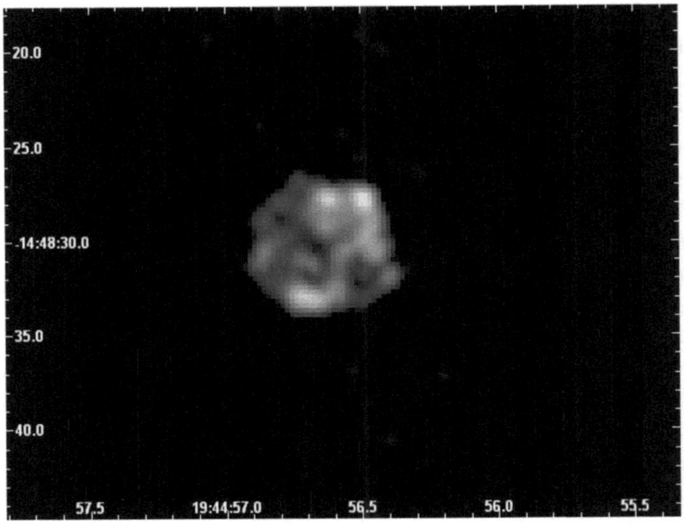

Fig. 3.21. *Chandra* ACIS-I image of Ho 12. Image courtesy D.S. Moon.

3.3 Basic View of the Debris and Gaseous Parts of Remnants

Here we summarize the basic picture of the origin, development, and evolution of the gaseous remains of supernovae in order to provide some context for the succeeding discussion. We will see that *Chandra* observations have done much to confirm the basic scenario and also have revealed new complexities.

As the stellar ejecta stream from the site of the SN explosion, they expand and interact with the ambient medium (AM), ultimately evolving into a visible supernova remnant (SNR). A blast wave, which precedes the ejecta, forms in the AM, while a reverse shock propagates back though the ejecta. SNRs gradually become strong X-ray sources over the course of typically hundreds of years as progressively greater amounts of ejecta and AM are shock-heated to X-ray temperatures, a consequence of the high shock velocities (several thousand km s^{-1}) during the early phases of evolution. Typically the X-ray emission is dominated by the SN ejecta at these early stages, a result of both the higher density and metal-rich composition of the reverse shocked ejecta. As the amount of swept-up AM grows, the ejecta tend to decrease in importance both in terms of producing X-ray emission and influencing the remnant's dynamical evolution. According to theory in the adiabatic phase of evolution, the dynamics depend only on the initial SN explosion energy and the density of the AM (Taylor 1950; Sedov 1959). As a remnant ages further its blast wave velocity drops and radiative cooling at the shock becomes important. A dense cool shell of swept up AM forms as all newly shocked material loses its thermal energy to radiation. This shell surrounds a hot, low density cavity of previously shocked ejecta and AM.

The typically low densities and short evolutionary timescales introduce an important ingredient to the X-ray emission from SNRs, namely the effect of time-dependent or nonequilibrium ionization. A plasma with an electron density of 1 cm^{-3} and temperature $kT \sim 1$ keV takes of order 10^5 yr to reach collisional ionization equilibrium. Thus the shocked plasma in remnants, which are almost all significantly younger than this, will be in a lower state of ionization than expected based on the plasma temperature. Departures from equilibrium ionization are characterized by the ionization timescale, $n_e t$, which is the product of the plasma electron density and the time since the material was heated.

The Rankine-Hugoniot (RH) relations (e.g., Shu 1992, p. 214ff) tell us about the properties of shocks in general, such as how the temperature and density jump across the shock front depend on the shock velocity, ratio of specific heats, and other properties of the shocked medium. However these relations fail to tell us about a number of astrophysically interesting questions, such as whether individual species in the fluid (electrons, protons, heavy elements) are heated to the same temperatures, or how much of the shock energy might be diverted into a population of relativistic particles (i.e., cosmic rays). These complications remain open to investigation, even though the basic properties of SNR shocks are well established. In the case of an ideal gas with a ratio of specific heats equal to 5/3, the RH relations say that a strong shock moving at speed v_s will heat the gas to a characteristic temperature $kT \sim 5\,\mathrm{keV}(v_s/1000\,\mathrm{kms}^{-1})^2$. This clearly corresponds to emission in the X-ray band, where *Chandra* has significant sensitivity.

3.4 Studying Ejecta in Supernova Remnants with *Chandra*

3.4.1 Remnants of Young Core Collapse SNe

The most secure way to identify the remnant of a core collapse SNe is through the presence of an associated compact object. In lieu of this we consider oxygen-rich ejecta to be the likely indicator of a core collapse SN. In the following we discuss a selection of remnants that satisfy these requirements. Objects are ordered by approximate chronological age.

SN 1987A

During the six years covered by this review SN1987A has been monitored at least yearly by *Chandra*. From Oct 1999 until Jan 2004 the 0.5–2 keV X-ray flux of the SNR has increased by a factor of 5 and appears to be growing brighter exponentially (Park et al. 2005). Even from the earliest *Chandra* observations, deconvolved ACIS images have shown a ring-like geometry. That ring is expanding radially at ~4000 km s^{-1} (Park et al. 2004b). The ACIS-S3 spectra are consistent with circumstellar matter; there is no evidence yet for X-ray emitting ejecta in SN1987A.

Cas A

Cas A (Figure 3.22) was the subject of the first refereed publication on *Chandra* observations (Hughes et al. 2000a) and has remained an intense focus of study through the subsequent years. This article laid out evidence for three clearly distinct spectral types in Cas A: Si-rich thermal emission, Fe-rich thermal emission, and featureless continuum emission. The Si-rich spectra were dominated by emission from Si and S with little Fe, with a composition similar to that expected from explosive oxygen burning. The Fe-rich spectra showed strong Fe-L and Fe-K shell emission with much weaker Si and S lines, consistent with explosive silicon burning. A further and perhaps more remarkable finding was that the Fe-rich material along the eastern edge of Cas A lay outside the Si-rich material, an inversion of the relative ordering in which these species should have been produced during the SN explosion. This argues for an energetic process of mixing or overturning of the ejecta possibly from neutrino-driven convection. The spectrum of the northeast jet was shown to be less Fe-rich than the knots toward the east. The filaments showing featureless continuum emission are possibly due to synchrotron radiation from relativistic electrons (see Sect. 3.5.2 below).

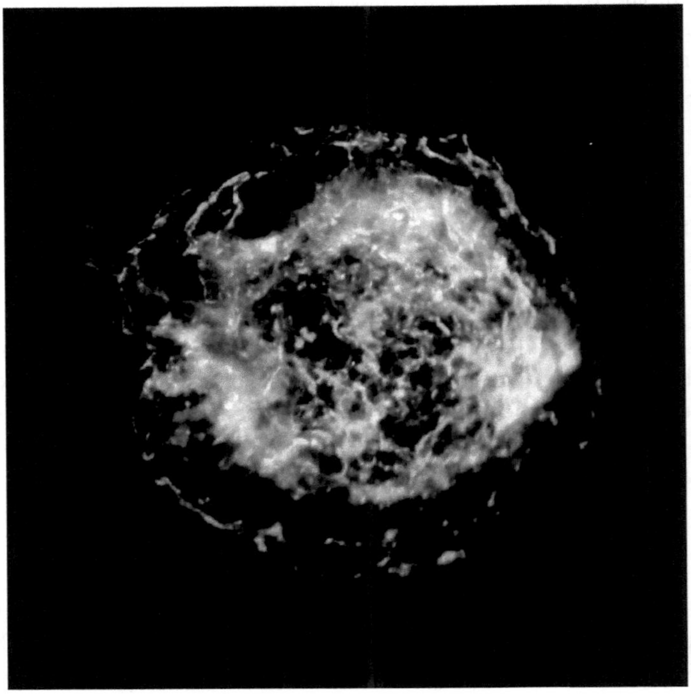

Fig. 3.22. Color coded energy image of Cas A using *Chandra* ACIS-S3 data. The image is $6' \times 6'$. The red band contains the Si XIII Heα line blend, green band the 4.2–6.4 keV continuum, and the blue band contains the Fe Kα line. Image courtesy NASA/CXC/GSFC/U. Hwang et al.

Willingale et al. (2002) used *XMM-Newton* data to measure the radial velocities of the Si and Fe emission in Cas A. The southeastern knots are generally blueshifted while the northern ones are redshifted and their total velocity range covers ∼5000 km s^{-1}. The Si velocity results are consistent with earlier measurements (Markert et al. 1983; Hwang et al. 2001) while the results on the velocity of Fe are new. The velocities of the southeastern Si and Fe knots are consistent with each other and confirm that the inversion of the relative ordering between the Si and Fe emission in this region seen in the *Chandra* images is not a result of some strange velocity structure in the ejecta. In addition Willingale et al. (2002) show that the Fe emission in the northern region is separated from the Si in terms of velocity, with the Fe showing higher speeds. This suggests that the Si and Fe ejecta experienced a spatial inversion of their original locations in the north as well.

Laming & Hwang (2003) and Hwang & Laming (2003) present a sophisticated analysis of the X-ray ejecta in Cas A. Temperature and ionization timescales determined from spectral fits to individual X-ray knots were compared to self-similar hydrodynamic models that incorporate effects of time-dependent ionization, radiative losses, and the exchange of energy between electrons and ions through Coulomb collisions. Several sequences of O-rich knots selected from different parts of the remnant are analyzed in Laming & Hwang (2003) to infer variations in the amount of explosion energy directed into different azimuthal directions. The asymmetry they determine is as much as a factor of two with more energy going into the polar regions (near the jet) than into equatorial regions. This level of asymmetry is, however, less than that expected from asymmetric core collapse explosion models. Hwang and Laming (2003) study the Fe-rich knots in Cas A to obtain important constraints on the extent of mixing in the ejecta. Several Fe-rich knots along the eastern edge that lie beyond the Si-rich ejecta are at a mass coordinate of approximately 2 $M_{\odot}$ measured from the center. An extremely Fe-rich feature in the same general vicinity is identified as a possible site of α rich freeze-out (when complete Si-burning occurs at low density). This material is produced closest to the center of the exploding star and is therefore most sensitive to the explosion mechanism and the position of the mass cut between the ejecta and compact remnant.

Gotthelf et al. (2001) locate the forward and reverse shocks in Cas A. The position of the forward shock was determined by the set of thin tangential wisps of X-ray continuum emission that extend to a radius of $153'' \pm 12''$. An increase in radio intensity as well as a large jump in radio polarization angle are coincident with the (local) peak in X-ray continuum emission at this radius. This appears to be the location of the forward shock in Cas A. These authors deproject the Si-line image and radio continuum to estimate the location of the reverse shock, which they determine to be at a radius of $95'' \pm 10''$. Under the assumption of standard adiabatic shock models (e.g., Truelove and McKee 1999), the relative positions of the forward shock and reverse shock were used to infer that the forward shock in Cas A has swept-up roughly as much mass as was ejected. The location of the contact discontinuity should be determined in order to assess whether the assumption of standard adiabatic shock models should be revised by, for example, including the dynamical effects of cosmic ray acceleration (see Sect. 3.5.2 below).

Delaney and Rudnick (2003) determine the average expansion rate of the forward shock in Cas A to be 4900 km s^{-1} with a range from place to place that covers ~4000–6000 km s^{-1}. In a subsequent article Delaney et al. (2004) carry out an extended study of the kinematics of Cas A as a function of the spectral character of the X-ray emission. They identify four spectrally distinct classes of emission that they designate as Si-dominated, Fe-dominated, low-energy–enhanced, and continuum-dominated. The first two "ejecta-dominated" classes show a mean X-ray expansion rate of 0.2% yr^{-1}, less than the 0.3% yr^{-1} of the optical ejecta. This discrepancy is posited to be due to a greater deceleration of the lower density X-ray knots compared to the higher density optical knots. The low-energy–enhanced component has a low mean expansion rate 0.05% yr^{-1} and likely corresponds to the clumpy circumstellar medium. Continuum-dominated filaments in the interior of the remnant also show a low mean expansion rate, although they are spread over a wide range of expansion rates that include some large inward motions. The continuum-dominated filaments around the rim of Cas A are expanding at the same rate as the Si- and Fe-rich ejecta.

Hwang et al. (2004) introduce the Megasec observation of Cas A taken in 2004 (Figure 3.22). These data reveal clearly a bipolar jet-like structure extending from the northeast to the southwest. The jet was mostly likely produced in the SN explosion process itself rather than being the result of the interaction of the ejecta with the projenitor's axisymmetric stellar wind. We expect to learn much over the next years from this rich data set.

1E 0102.2–72.2

In the *Chandra* ACIS-S3 data of 1E 0102.2–72.2 (see Figure 3.23) Gaetz et al. (2000) identify a bright, clumpy, incomplete ring of emission prominent in X-ray lines from He- and H-like oxygen and neon that correlates well with the optically-emitting ejecta. Beyond the ejecta, the *Chandra* data show a faint shelf of emission, identified as the blast wave interacting with the AM and extending to a diameter of ~44″. The rim of the remnant shows a very sharp, smooth edge, while the outer edge of the ejecta are "scalloped" likely as a result of Rayleigh-Taylor instabilities acting at the contact discontinuity.

Although there is a hint in the ACIS data of spectral variation as a function of radius through the shock-heated ejecta, this effect is most clearly seen in the *Chandra* High Energy Transmission Grating (HETG) Spectrometer observation presented by Flanagan et al. (2004). The size of 1E 0102.2–72.2 is well matched to the dispersion scale of the grating, so that images of the remnant in individual lines (e.g., O VIII and Ne X Ly α) are cleanly resolved. Furthermore the prominence of the remnant's O, Ne and Mg line emission, especially as compared to Fe, significantly reduces the amount of line-blending in the 0.7–1 keV band and allows for the derivation of quantitative results. Constraints on the global plasma properties were obtained from measured O and Ne line fluxes (principally the H-like Ly α and He-like resonance and forbidden lines). The global temperature

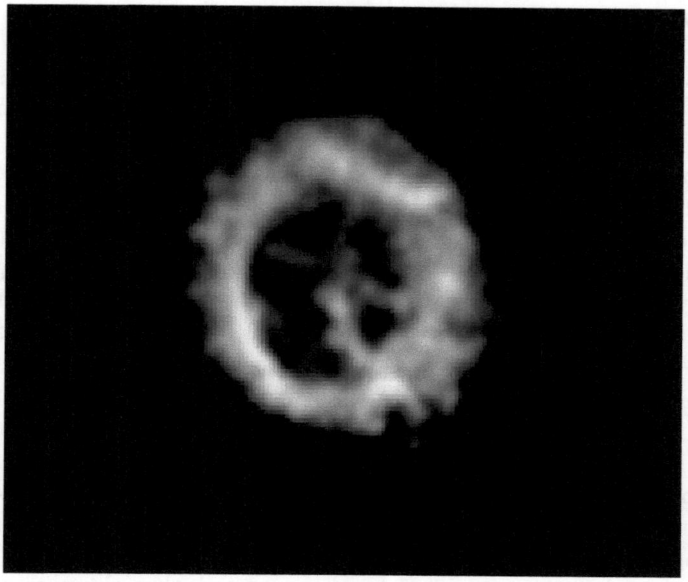

Fig. 3.23. Color coded energy image of 1E 0102−72.2 using *Chandra* ACIS-S3 data. The SNR is roughly 45″ in diameter. The red band contains the O VII Heα line blend, green band the Ne IX Heα line blend, and the blue band contains the Ne X Lyα line as well as Mg and Si lines.

values obtained ($kT \sim 0.34$ keV for O and $kT \sim 0.58$ keV for Ne) are consistent with those from the *XMM-Newton* RGS observation Rasmussen et al. (2001).

However, as has been know since the time of *ASCA* (Hayashi et al. 1994), the global spectrum of this remnant cannot be adequately described by a single component plasma model. The HETG data provide keen insight into why this so. There are azimuthal variations in the relative brightnesses of HETG spectral line images that suggest large scale variations in the plasma properties (for example, the ionization state toward the north is more advanced than the south). But perhaps of greater import is the clear evidence for differences in size of the various spectral line images. The sense of these differences is that lines preferentially produced at lower ionization timescales have images that are smaller, while higher ionization-timescale lines have larger images. This trend is fully compatible with a reverse shock origin, where the innermost visible ejecta are the most recently shocked, while the outermost has been shocked for the longest time. These observational results provide important constraints on the profiles of temperature and ionization timescale throughout the shocked ejecta.

There are systematic distortions in the spectral line images between the + and − arms of the HETG spectrum due to Doppler shifts that indicate internal velocity differences within the SNR. The inferred bulk matter flows are of order 1000 km s^{-1}. Additionally the ejecta appear to be distributed in a nonspherical manner (toroidal or cylindrical geometry). This type of study appears to be one that only *Chandra* is capable of pursuing.

SNR 292.0+1.8

The ejecta and shocked ambient medium of SNR 292.0+1.8 were first studied by Park et al. (2002) using an ACIS-S3 observation (Sect. 3.2.2). The remnant is bright in thermal emission from O, Ne, and Mg, but weak in Fe, in strong contrast to the other young Galactic O-rich SNR Cas A. The *Chandra* image (Figure 3.17) shows a complex set of knots and filaments, as well as more diffuse, extended emission. Equivalent width maps were used to identify metal-rich ejecta, which are preferentially distributed around the periphery. A prominent feature in the broadband image is a narrow, twisted belt of emission running approximately east-west across the center of the SNR. It is composed of solar-abundance material and hence plausibly is shocked circumstellar gas. Park et al. (2002) argue that it is the relic of the axisymmetric dense stellar wind lost by the progenitor star during its red supergiant phase of evolution. Another feature noted by these authors was a set of narrow filaments with normal composition located around nearly the entire circumference of the SNR. It was suggested that these filaments might be a relic structure in the circumstellar (perhaps related to the progenitor's stellar wind) being overrun by the blast wave. The more conventional explanation presented by these authors that the narrow, thermal rims represent the outer blast wave may be supported by the presence of similar features at the outer edge of 1E 0102.2−72.2 (Figure 3.23).

In a subsequent study Park et al. (2004a) examine in more detail the ACIS-S3 spectra of a number of small-scale emission features that cover a representative range of spectra from across the SNR. A spectrum from the central belt is shown to have abundances that, within a factor of two, are consistent with solar. Other regions, chosen for their prominent line emission, are clearly from SN ejecta and display greatly enhanced abundances of O, Ne, and Mg compared to Si, S and Fe. These abundance patterns are similar to those in 1E 0102−72.2, while they are quite unlike those in Cas A, which is dominated by Si- and Fe-rich ejecta. Additionally SNR 292.0+1.8 and 1E 0102−72.2 did not appear to undergo the same sort of extensive overturning of their ejecta that Cas A experienced. The origin for these differences among the class of O-rich SNRs in not understood.

N49

Toward the southwest corner of the SNR N49 Park et al. (2003a) found a spectrally hard X-ray knot beyond the main boundary of the remnant. This feature shows a metal-rich spectrum with apparently enhanced abundances of O, Mg, Si, S, and perhaps Fe. Park et al. (2003a) argue that it is a fragment of ejecta. Since this remnant hosts the famous soft γ-ray repeater SGR 0526−66, insights into the nature of the progenitor star based on the properties of the ejecta would be important to obtain. Unfortunately no other metal rich ejecta fragments have been reported and the statistical quality of the knot just described was barely adequate to reveal its metal-rich nature. Deeper insights into the nature of the progenitor will require a considerably longer observation of N49.

N63A and the Vela Bullets

A dramatic example of an LMC remnant where ejecta fragments, bullets, or shrapnel may extend beyond the main shock boundary is the case of N63A (Warren, Hughes, & Slane 2003). The crescent-shaped features seen around the rim here (Figure 3.24) are very reminiscent of similar features discovered by *ROSAT* surrounding the Vela SNR (Aschenbach et al. 1995). In N63A none of the outlying regions shows strongly enhanced abundances, so if the features are driven by high speed ejecta clumps, at this point they must have been mixed with considerable amounts of ambient medium. Support that the Vela fragments are in fact SN ejecta have come from *Chandra* observations by Miyata et al. (2001) of the feature called shrapnel A. Their ACIS-S3 spectrum from the head of the fragment required twice the solar abundance of Si. Shrapnel D is a more definitive case: Katsuda & Tsunemi (2005) find factors of 5-10 enhancements in the O, Ne, and Mg abundances from their *XMM-Newton* observations. Confidently identifying the site (or sites) in the exploding star where these fragments originated based on the observed nucleosynthetic products is a very important next step.

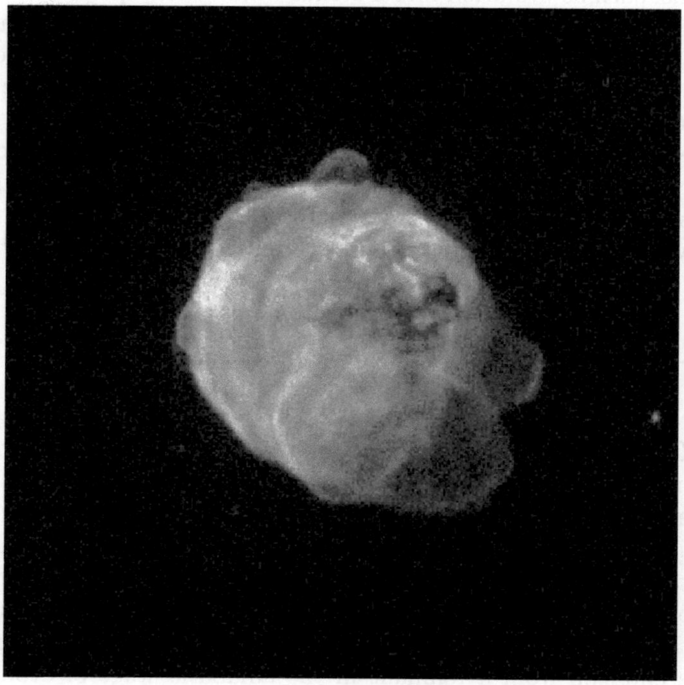

Fig. 3.24. *Chandra* ACIS-S3 image of N63A. The region displayed is 150″ × 150″.

SNR 0049−73.6

This object and the one we discuss next are surprisingly similar in many respects and both appear to be mature versions of 1E 0102.2−72. SNR 0049−73.6 is 72″ in diameter (∼21 pc at the distance of the SMC, 60 kpc). The *Chandra* ACIS-S3 data show a central ring-like zone of relatively bright emission within a faint, circular shell (Hendrick, Reynolds, and Borkowski 2005). The outer limb-brightened emission is consistent with an adiabatic shock model with equal electron and ion temperatures ($kT \sim 0.39$ keV) and a preshock hydrogen density of $n_0 \sim 0.14$ cm^{-3}. The remnant's inferred age and explosion energy are 14000 yr and 0.8×10^{51} ergs, respectively. The central bright emission is enhanced in O and Ne and contains of order 0.35 $M_\odot$ of O, which suggests a progenitor with a mass of at least 13 $M_\odot$. The authors claim that the observed radius of the ejecta shell (∼6 pc) is inconsistent with the expected location based on 1D hydrodynamical simulations of a 15 $M_\odot$ progenitor evolving into a uniform AM. To explain the apparent discrepancy Hendrick, Reynolds, and Borkowski (2005) propose the "Ni bubble" effect (i.e., the dynamical effect of the energy deposited within the center of the SN by the radioactive decay of^{56} Ni). This is an intriguing idea, but needs to be developed more fully, in addition to alternative explanations, such as evolution into a stellar wind density profile.

No mention is made of a compact remnant. We inspected the *Chandra* data and note the presence of a compact source in the hard band (> 3 keV) about two-thirds of the way out from the center to the outer limb of the remnant. Although this is likely to be an unrelated extragalactic background source we use it to set a limit on the presence of an associated X-ray pulsar. The 3σ count rate upper limit for this source is 1×10^{-3} s^{-1} (3–8 keV band), which for a Crab-like spectrum corresponds to an unabsorbed flux of 3.4×10^{-14} erg cm^{-2} s^{-1}. This yields an X-ray luminosity upper limit of 1.5×10^{34} erg s^{-1} (2–8 keV band).

SNR 0103−72.6

Park et al. (2003b) observed SNR 0103−72.6, the second brightest X-ray remnant in the SMC, with ACIS-S3. This observation resolved the remnant into a nearly complete, remarkably circular shell of emission 85″ in diameter (∼25 pc at the SMC distance) surrounding a smaller region with bright clumpy O- and Ne-rich emission. The fitted O and Ne abundances of the central ejecta emission are significantly enhanced above the SMC values, while Fe is rather low. The relative abundances of O, Ne, Mg, and Si are reasonably consistent with the nucleosynthetic yields from an 18 $M_\odot$ star. The outer zone is consistent with blast wave emission from a remnant in the adiabatic stage of evolution. The preshock hydrogen density of the AM is $n_0 \sim 0.16$ cm^{-3}, the SN explosion energy is 1.2×10^{51} ergs, and the age of the remnant is ∼18000 yrs. This is the oldest known example of an O-rich SNR and was the first one discovered by its X-ray emission.

Park et al. (2003b) state that all the remnant's emission lies below 3 keV, but give no quantitative upper limit on the flux of a compact source with a hard power-law spectrum. The brightest point source within the extent of the SNR (according to our own inspection of the *Chandra* data) provides a 3σ count rate upper limit of 4×10^{-4} s^{-1} (3-8 keV band), which for a Crab-like spectrum corresponds to an unabsorbed flux of 1.4×10^{-14} erg cm^{-2} s^{-1} or a luminosity of 6×10^{33} erg s^{-1} (2-8 keV band).

3.4.2 Remnants of Type Ia SNe

Other than with an optical spectrum from near the time of optical maximum, there is no definitive way to identify the remnant of a Type Ia SN (SN Ia). (In a remarkable advance it may become possible to use light echoes to obtain optical spectra from ancient SN; see Rest et al. 2005) In some cases, historical light curves have been used, although this can only be suggestive. Several other methods have been proposed based on the expected properties of the progenitor or its environment, such as (1) a Chandrasekhar mass of ejecta, (2) partially neutral surrounding ambient medium (Tuohy et al. 1982), or (3) the elemental composition of the ejecta (Hughes et al. 1995). An additional factor, although not definitive, is the absence of a compact remnant. One final introductory note is that, because the ejecta in SN Ia remnants are essentially invisible in the optical band, the best way to study it is in the X-ray band.

Tycho

It was Hamilton, Sarazin, and Szymkowiak (1986b), who presented the first detailed argument that this bright remnant, observed as a SN by Tycho Brahe and others in 1572, had a SN Ia origin. They demonstrated consistency between the global X-ray spectra available at the time and a model of 1.4 solar masses of layered SN Ia ejecta interacting with a uniform ambient medium. Much of the Fe in their model was interior to the reverse shock and hence unshocked. Since this study, our understanding of Tycho's SNR has advanced in several areas. The ejecta are clumpy (see Figure 3.25), although nowhere near as inhomogenous as the core collapse remnants Cas A and SNR 292.0+1.8, and reach to the very edge of the rim in places (Hwang et al. 2002). Spectral inhomogeneities throughout the ejecta are generally modest, with two important exceptions. There is a radial variation in the line intensities: Si, S, and Fe-L shell line emission reach their peak surface brightness at larger radii than does the Fe Kα line emission (Hwang and Gotthelf 1997; Hwang et al. 2002). Toward the eastern limb are a few knots that bulge out beyond the rim; the spectral differences between these knots (the northern one is Si-rich, like the bulk of the ejecta, while the southern one is Fe-rich) were first noted by Vancura, Gorenstein, and Hughes (1995). The origin of these small-scale spectral inhomogenities is not yet explained, although an interesting idea is that they may be related to the SN Ia ignition process (Hughes et al. 2005a). Based on the analysis of a 150 ksec ACIS-I observation, Warren et al. (2005) found the contact discontinuity to be considerably more

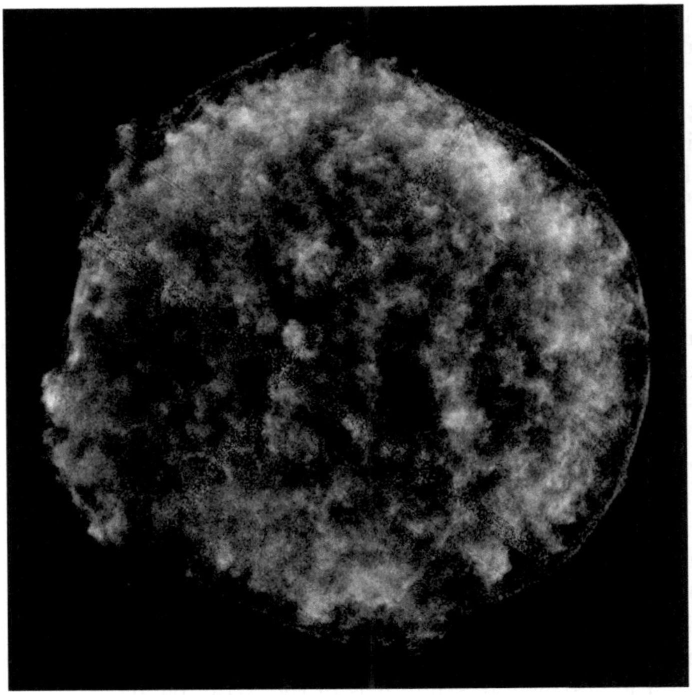

Fig. 3.25. Color coded energy image of the Tycho SNR using *Chandra* ACIS-S3 data. The image is $9.5' \times 9.5'$. The red band contains the Fe L line blend, the green band the Si Kα lines, and the blue band contains the 4.1–6.1 keV continuum emission. Image courtesy Warren et al. (2005)

structured than the blast wave, indicating the action of the Rayleigh-Taylor instability there.

Badenes et al. (2003), following firmly in the footsteps of Hamilton, Sarazin, and Szymkowiak (1986b), have developed a detailed model for the X-ray emission from remnants of SN Ia. Starting from realistic SN models that span the range of proposed explosion models, the ejecta are evolved to the remnant stage with attention to relevant heating, ionization and thermal effects. Comparison of their model to an *XMM-Newton* spectrum of the Tycho SNR (Decouchelle et al. 2001) suggests the delayed detonation model as the best choice. Badenes, Borkowski, & Bravo (2005) further utilize their model to explain the radial variation in the X-ray spectrum of Tycho spectra as being due to the temperature profile through the ejecta. They require a region of higher temperature and lower ionization timescale in the Fe-rich zone near the reverse shock, which can result from a modest amount of collisionless electron heating at the reverse shock. They also argue that SN Ia ejecta need to be stratified to some extent which puts serious constraints on current three-dimensional SN Ia explosion models which tend to produce well-mixed ejecta.

SNR0509−67.5

Like the Tycho SNR this remnant (Figure 3.26) has provided valuable clues to the SN Ia explosion process. Warren & Hughes (2004) show that the integrated ACIS-S spectrum is dominated by emission from ejecta. The integrated abundances of O, Ne, Mg, Si, S Ar and Ca from their fits are consistent with yields from SN Ia models, with some preference for a delayed detonation model (Iwamoto et al. 1999). There is considerably less hot Fe than expected, arguing that most of the Fe in the ejecta is cold and lies interior to the reverse shock. The mass of Si inferred from the spectral analysis is $\sim 0.2\ M_{\odot}$ when clumping is taken into account, which is in line with SN Ia models.

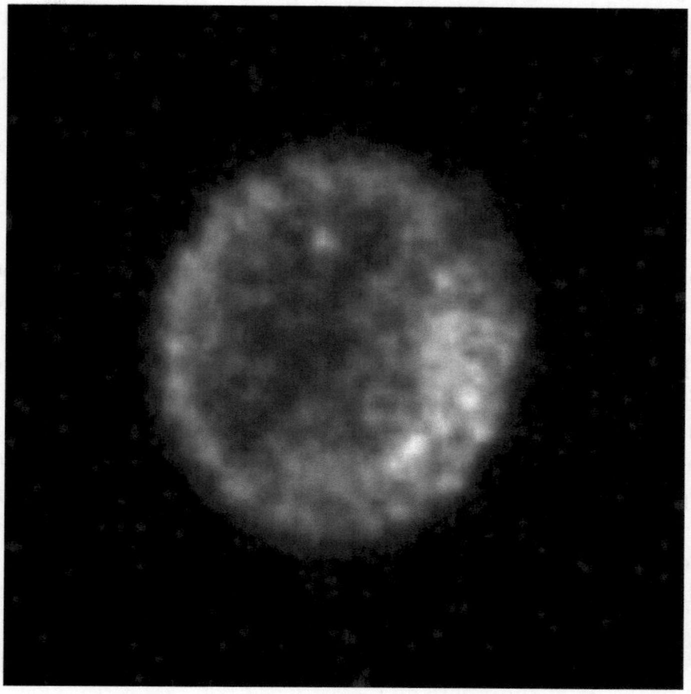

Fig. 3.26. Color coded energy image of SNR E0509−67.5 using *Chandra* ACIS-S3 data. The image is $50'' \times 50''$. The red band contains the 0.2–0.69 keV band, the green band the 0.69–1.5 keV and, and the blue band contains the 1.5–7 keV band. Image courtesy Warren and Hughes (2004).

N103B

N103B was first suggested to have a SN Ia origin by Hughes et al. (1995) based on qualitative inspection of its *ASCA* X-ray spectrum. This was somewhat of a surprise given the remnant's relative proximity to the young star cluster NGC 1850, only about 40 pc away in projection (Chu & Kennicutt 1988). Optically,

N103B consists of several small bright knots that show the usual set of emission lines seen in radiative shocks: [O III] λ 5007, [S II] $\lambda\lambda$ 6716, 6731, Hα, and so on (Danziger & Leibowitz 1985). The abundances of N103B inferred from optical spectroscopy are consistent with the swept-up interstellar medium of the LMC (Russell & Dopita 1990). These characteristics suggested a massive star progenitor and a dense ambient medium.

In their analysis of the N103B *XMM-Newton* data van der Heyden et al. (2002) argue for a massive star progenitor. They base this conclusion on abundances derived from RGS data integrated over the entire SNR. They find more total O compared to Si and Fe than expected from a SN Ia and suggest a core collapse SN as a viable alternative. Unfortunately in their comparison to the model yields they do not include a swept-up component (which would contribute a significant mass of O) in addition to the ejecta and so their conclusion on the originating SN type cannot be considered definitive.

According to Lewis et al. (2003), who studied the *Chandra* ACIS-S3 data, N103B consists of several spectrally distinct spatial regions. These include a centrally located region of hot Fe with low ionization timescale surrounded by a shell-like region dominated by the Si-group elements, features that resemble those of the Tycho SNR. On the other hand the O, Ne, Mg, and continuum emission, which are clearly required by the spectra, do not show a definitive radial trend and, in particular, are not distributed in a shell-like geometry. It was suggested that this emission could be clumpy foreground or background ambient material overrun by the forward shock. Based on the estimated masses of the Si-group elements and Fe, Lewis et al. (2003) argued that a SN Ia interpretation was preferred.

N103B bears similarities to the Kepler SNR, which also shows an Fe-rich X-ray spectrum (Cassam-Chenaï et al. 2004), a large asymmetry in brightness from one hemisphere to the other, more centrally located Fe emission, and evidence for interaction with a dense ambient medium. If, at some point, these remnants are confidently identified with SN Ia events, they might be able to shed light on the still poorly understood properties of the environment and/or companion star.

SN1006

Hamilton, Sarazin, and Szymkowiak (1986a) attempted to model the global spectrum of SN1006 in terms of an exploded white dwarf as they suceeded in doing for the Tycho SNR. However, unbeknownst to them, the global spectrum of SN1006 is dominated by nonthermal X-ray emission coming from the northeast and southwest rims as first demonstrated by Koyama et al. (1995) using *ASCA*. These appear as the whitish rims in Figure 3.27, which is a complete *Chandra* image of SN1006 from a mosaic of eleven individual 22 ksec long ACIS-I pointings (Hughes et al. 2005b).

The thermal emission from SN1006 is dominated by O emission as shown by Long et al. (2003) in their analysis of two deep *Chandra* observations (O emission appears reddish in Figure 3.27). Toward the northwest there is a long

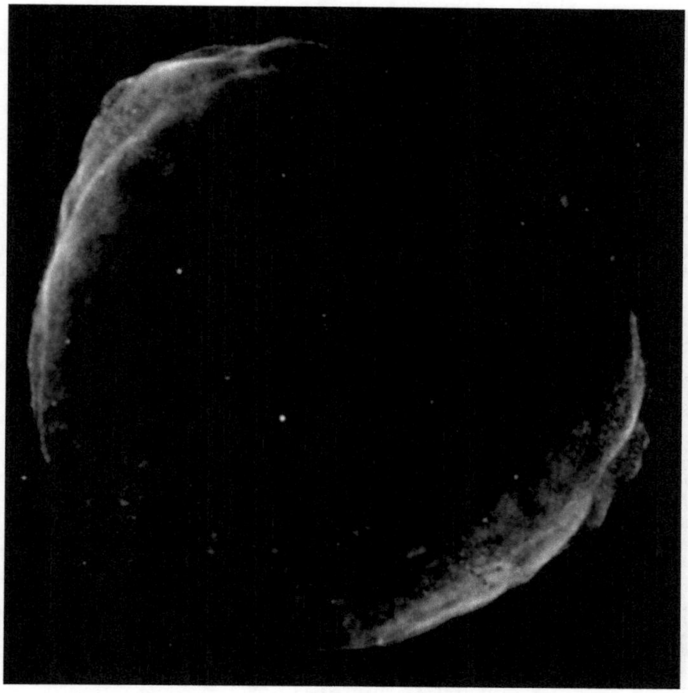

Fig. 3.27. Color coded energy image of SN1006 using a mosaic of 11 separate *Chandra* ACIS-I observations. The image is $33.5' \times 33.5'$. The three bands (red, green, and blue) correspond to energy bands of 0.50–0.91 keV, 0.91–1.34 keV, and 1.34–3.00 keV.

thin feature, coincident with an Hα optical filament, whose spectrum is consistent with shocked ambient medium. Elsewhere the ejecta dominate. SN1006's ejecta display a fluffy structure that is similar in appearance and physical scale to the fluff seen in Tycho. Ejecta clumps in SN1006 also appear to extend to the forward shock. ACIS-S3 spectral fits to the ejecta emission near the northwest filament require enhanced abundances of O, Mg, Si, and Fe according to Long et al. (2003). In an analysis of *XMM-Newton* RGS grating data Vink et al. (2003) find no evidence for Fe XVII emission and can fit the EPIC data with emission only from the species N, O, Ne, Mg, and Si, as well as a hydrogen continuum.

It has been known for some time that SN1006 contains signficant amounts of cold Fe in its interior based on UV absorption lines in the continuum emission of a background star (Wu et al. 1983). This has been one of the strongest arguments in support of its SN Ia origin. The presence of Fe in the shocked ejecta is therefore of some importance, since it would tell us about the extent of radial mixing in these explosions.

DEM L71 and other older SN Ia SNRs

The LMC SNR DEM L71 is some 4000 yr old (Ghavamian et al. 2003). The *Chandra* ACIS-S3 data (Figure 3.28) show a clear double shock morphology: an

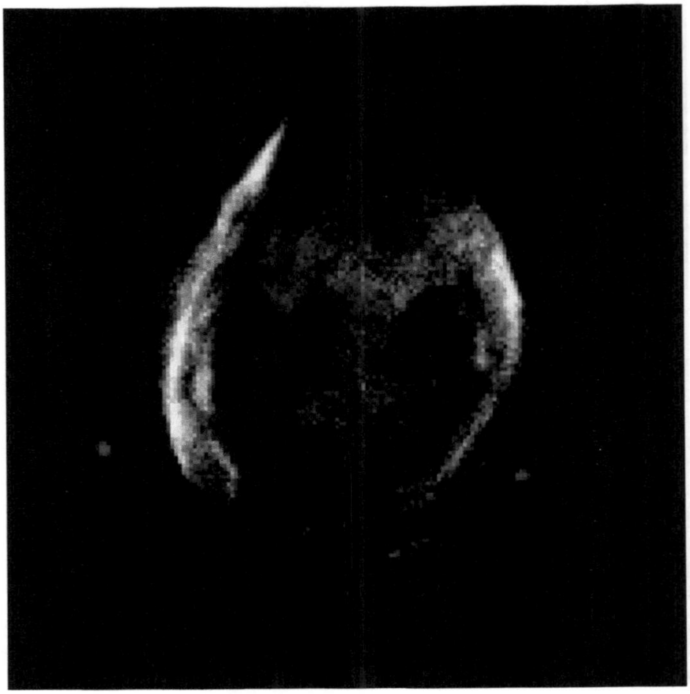

Fig. 3.28. Composite image of DEM L71. The image is $2' \times 2'$. The red band contains Hα optical emission while the green and blue bands include two energy bands from the *Chandra* ACIS-S3 data (0.2–0.7 keV and 0.7–2.6 keV, respectively). Image courtesy Hughes et al. (2003)

outer rim present in both the broadband X-ray and Hα optical emission and a central excess only seen in the X-ray band, predominantly above 0.7 keV (Hughes et al. 2003). The outer rim is the blast wave in the ambient medium, while the central excess corresponds to Si and Fe-rich ejecta. The ratio of Fe to O in the ejecta is at least 5 times the solar ratio. From the relative locations of the contact discontinuity and the blast wave shock an estimate of the total ejecta mass of $\sim$1.5 $M_{\odot}$ was obtained. Masses estimated from the X-ray emission are generally consistent with this value and further show that the mass of Fe is some 6 times that of Si. These properties are fully consistent with the picture that DEM L71 is a middle-aged SN Ia remnant.

It is possible that even older SN Ia remnants may have been identified. Hendrick, Borkowski, & Reynolds (2003) report on the discovery of metal-rich ejecta in the LMC SNRs 0548−70.4 and 0534−69.9, which are estimated to be $\sim$10,000 yr old. Their spectral analysis yields a value for the ratio of O/Fe in the ejecta that is intermediate between the typical SN Ia and core collapse SN. On the other hand the mass limits they derive are of order 1 $M_{\odot}$ or less, supportive of the SN Ia hypothesis.

3.5 Shock Properties

3.5.1 Electron-ion temperature equilibration

For a current review of this topic see Rakowski (2005). In the following we highlight studies that utilize X-ray emission to investigate the heating of electrons, protons, and heavier ions at high Mach number shocks.

In a pair of papers on the LMC SNR DEM L71, Rakowski, Ghavamian, and Hughes (2003) and Ghavamian et al. (2003) use optical and X-ray observations to separately measure the post-shock proton and electron temperatures. The proton temperature comes from the width of the broad Hα line component (for details see Ghavamian et al. 2001 and references therein) and corresponds to a very thin region ($<10^{-3}$ pc) right behind the shock front. The X-ray emission constrains the electron temperature. Since even *Chandra*'s spatial resolution is insufficient to resolve the post-shock region, modeling was required in order to infer the immediate post-shock electron temperature. Model effects included the temporal evolution in temperature and ionization timescale as well as equilibration due to Coulomb collisions between the electron and protons. Using this technique it was possible to constrain the initial post-shock ratio of electron to proton temperatures over a factor of two in shock velocity (500–1000 km s^{-1}). Lower velocity shocks were found to have nearly equal post-shock electron and proton temperatures, while the higher velocity ones showed significantly lower levels of temperature equilibration. These results are in good agreement with the published estimates of the electron to proton temperatures (done with a different method using only the optical data) from four other SNRs: SN1006, Tycho, RCW 86, and the Cygnus Loop (Ghavamian et al. 2001).

In their *Chandra* HETG study of SN1987A, Michael et al. (2002) investigated electron-ion temperature equilibration using X-ray line widths from a combined line profile including emission from N, O, Ne, Mg, and Si. The widths are due to thermal and turbulent broadening as well as the bulk motion expansion of the remnant. The first two effects should produce symmetric, Gaussian profiles, while the latter profile will depend on the detailed geometry of the expanding source. Michael et al. (2002) tried several different geometries for this and arrived at a shock velocity, $v_s = 3400 \pm 700\,\mathrm{km\,s^{-1}}$, that is consistent with the radio and X-ray expansion rates. Their measured electron temperature, $kT \sim 2.6$ keV, is much lower than the post-shock temperature, $\sim$17 keV inferred from the line widths. After accounting for Coulomb equilibration between the electron and ions, the immediate post-shock ratio of electron to shock temperature can only be $\sim$10%.

Vink et al. (2003) also use line widths to estimate ion temperatures. They observed a bright knot along the northwestern limb of SN1006 with *XMM-Newton* and extracted a spectrum using the RGS. They found that the O lines were broadened by 3.4 ± 0.5 eV, which, if interpreted as a thermal broadening, indicated an O temperature of $kT = 528 \pm 150$ keV. The observed electron temperature from this knot was only 1.5 keV, which argues for incomplete electron-ion heating.

Hwang et al. (2002) observed the forward shock region in the Tycho SNR with *Chandra* ACIS-S3 and set a limit of <2 keV on the electron temperature there. This was shown to be significantly less than the shock temperature inferred from the radio (Reynoso et al. 1997) and X-ray (Hughes 2000) expansion rates. A limit of 0.03–0.12 was set on the amount of collisionless electron heating at the forward shock of Tycho. We also note that Badenes, Borkowski & Bravo (2005) require roughly the same amount of collisionless electron heating (<10%) at the reverse shock to explain the X-ray properties of the Fe-rich ejecta in the Tycho SNR.

3.5.2 Cosmic ray acceleration

One of the most important and possibly paradigm-shifting results from *Chandra* observations has come in demonstrating that SNRs are the sites for the shock acceleration of high energy (TeV range) cosmic rays. Our discussion of this is divided in two broad categories: evidence for the electron component and evidence for the hadronic component. Again we focus on results from the 0.2–10 keV X-ray band.

Evidence for the electron component

Prior to *Chandra* there were three shell-like remnants whose emission in the 0.5-10 keV band was dominated by featureless powerlaw continuum: SN1006 (Koyama et al. 1995), SNR 347.3−0.5 (aka RX J1713.7−3946) (Koyama et al. 1997, Slane et al. 1999), and SNR 266.2−1.2 (Slane et al. 2001). Other young SNRs (Cas A, Tycho, RCW 86, and Kepler) showed evidence for hard power-law emission at energies beyond 10 keV (Allen, Gotthelf, & Petre 1999). These featureless power-law spectra, widely believed to arise from synchrotron radiation, suggested the presence of highly energetic (TeV range) electrons in these remnants.

Chandra has contributed to this area in several important ways. First was the discovery of geometrically thin, spectrally featureless filaments in remnants with predominantly thermal emission in the *Chandra* X-ray energy band. Examples include Cas A, where these filaments are visible as the green-colored network in Figure 3.22 (Hughes et al. 2000, Gotthelf et al. 2001), and Tycho, where the thin bluish rim around the rim in Figure 3.25 displays a featureless X-ray spectrum (Hwang et al. 2002; Warren et al. 2005). Similar thin, spectrally featureless filaments are present in RCW 86 (Rho et al. 2002) and Kepler (Bamba et al. 2005b). In these cases as well as SN1006 (Long et al. 2003; Bamba et al. 2003), SNR 347.3−0.5 (Uchiyama et al. 2003; Lazendic et al. 2004) and SNR 266.2−1.2 (Bamba et al. 2005a), the structure of the featureless rims are consistent with thin sheets of X-ray emission, with widths in most cases significantly less than a parsec.

Synchrotron cooling of the high energy electrons accelerated at the shock front is one widely discussed interpretation for the origin of the thin featureless X-ray filaments (Vink & Laming 2003; Berezhko, Ksenofontov, & Völk 2003;

Berezhko & Völk 2004; Völk, Berezhko, & Ksenofontov 2005; Warren et al. 2005). These studies provide estimates for the magnetic field within the filaments of order 100 μG, far above the Galactic value and evidently requiring magnetic field amplification at the shock front. An alternate view (Pohl, Yan, & Lazarian 2005) posits that the X-ray filaments are actually magnetic filaments. In this case the X-ray emission falls off rapidly behind the shock because the magnetic field, initially enhanced, is efficiently damped over a post-shock distance of some 10^{16}–10^{17} cm. Careful comparison of radio and X-ray observations of young remnants might be able to discriminate between the two interpretations presented here.

Chandra has also discovered new cases of shell-like SNRs with powerlaw continuum. In the case of the ejecta-dominated SNR 0509−67.5, Warren & Hughes (2004), based on dynamical arguments, propose that the continuum emission is likely to have a nonthermal origin. Bamba et al. (2004) report the discovery of nonthermal X-ray emission from a portion of the shell of the LMC superbubble 30 Dor C. The spectral index of the emission is consistent with the other cases mentioned here. These objects, both in the LMC, are too distant for even *Chandra* to provide useful constraints on the thickness of the nonthermal emission regions.

Evidence for the hadronic component

Direct detection of the hadronic component of cosmic rays in SNRs is a critical apect of the overall scenario that remains to be demonstrated. A prime method for doing this relies on observing TeV γ-rays from the decay of pions produced when cosmic ray protons accelerated at the SNR shock interact with local interstellar gas. Such emission has not yet been confidently detected.

If the acceleration process is efficient, a significant fraction of the SNR shock energy can be diverted from the thermal gas and end up in the relativistic component. In a study of the SMC SNR 1E 0102.2−72.2, Hughes, Rakowski, & Decourchelle (2000b) measured the expansion rate (proper motion) of the SNR to determine a blast wave velocity of ∼6000 km s^{-1}. From this value they determined a range of electron temperatures 2.5–45 keV, dependent on the degree of collisionless electron heating. Their measured electron temperature from *Chandra* ACIS-S3 spectroscopy ($kT \sim 0.4 - 1$ keV) was significantly lower than the range expected from the shock velocity. They argued that the only plausible way to reconcile the different electron temperature values was if a significant fraction of the shock energy, rather than contributing to the heating of the post-shock electrons and ions, had gone into generating cosmic rays.

It should also be possible to rely on the dynamical effects that cosmic ray protons have on the evolution of SNRs (Decourchelle, Ellison, & Ballet 2000). The softer equation of state of a relativistic fluid and the possibility that cosmic ray particles may escape from the shock result in a compression factor that is larger than the typical value of 4. This results in a shrinking of the gap between the contact discontinuity and forward shock. Warren et al. (2005) measure the relative locations of the forward shock, contact discontinuity, and reverse shock in the Tycho SNR using *Chandra* observations. The observed distance between the contact discontinuity and forward shock is too small to be explained by hydrodynamic models that ignore cosmic ray acceleration. The authors show that the energy density in relativistic electrons alone is insufficient to alter the dynamics of the forward shock, which points to the requirement for a signficant component of relativistic ions. The energy density of the cosmic ray ions needs to be 50 times or more that of the electrons, close to the well-established ratio of 100 for Galactic cosmic rays.

3.6 Acknowledgements

Images labeled "courtesy NASA/..." are publicly available at http://chandra.harvard.edu/. Other contributors of figures are acknowledged in the figure captions. MCW appreciates a thorough reading of, and comments to, a draft of this manuscript by V. Zavlin. JPH acknowledges C. Badenes, G. Cassam-Chenaï, and J. Warren for helpful discussions related to this article.

References

Allen, G., Gotthelf, E. V., & Petre, R. 1999, in Proc. of the 26th International Cosmic Ray Conference, edited by D. Kieda, M. Salamon, and B. Dingus, vol. 3, p. 480

Amy, S. W., & Ball, L. 1993, ApJ, **411**, 761

Aschenbach, B., & Brinkmann, W. 1975, A&A, **41**, 147

Aschenbach, B. 1998, Nature, **396**, 141

Aschenbach, B. 2002, in Proceedings of the 270. WE-Heraeus Seminar on Neutron Stars, Pulsars, and Supernova Remnants. MPE Report 278. Edited by W. Becker, H. Lesch, and J. Trümper. Garching bei München: Max-Plank-Institut für Extraterrestrial Physik, 13

Aschenbach, B., Egger, R., & Trumper, J. 1995, Nature, **373**, 587

Badenes, C., Borkowski, K. J., & Bravo, E. 2005, ApJ, **624**, 198

Badenes, C., Bravo, E., Borkowski, K. J., & Domínguez, I. 2003, ApJ, **593**, 358

Bamba, A., Ueno, M., Nakajima, H., & Koyama, K. 2004, ApJ, **602**, 257

Bamba, A., Yamazaki, R., & Hiraga, J. S. 2005a, ApJ, **632**, 294

Bamba, A., Yamazaki, R., Ueno, M., & Koyama, K. 2003, ApJ, **589**, 827

Bamba, A., Yamazaki, R., Yoshida, T., Terasawa, T., & Koyama, K. 2005b, ApJ, **621**, 793

Becker, R. H., Helfand, D. J., & Szymkowiak, A. E. 1982, ApJ, **255**, 557

Becker, W., & Aschenbach, B. 1995, in *The Lives of Neutron Stars*, Kluwer Academic Publishers, 47

Becker, W., & Aschenbach, B. 2002, in Proceedings of the 270. WE-Heraeus Seminar on Neutron Stars, Pulsars, and Supernova Remnants. MPE Report 278. Edited by W. Becker, H. Lesch, and J. Trümper. Garching bei München: Max-Plank-Institut für extraterrestrische Physik, 64

Becker, W., & Pavlov, G.G. 2001, in *The Century of Space Science*, eds. J. Bleeker, J. Geiss & M. Huber, Kluwer Academic Publishers, 721 (also astro-ph/0208356).

Becker, W,, Weisskopf, M, C., Arzoumanian, Z., Lorimer, D., Camilo, F., Elsner, R. F., Kanbach, G., Reimer, O., Swartz, D. A., Tennant, A. F., & O'Dell, S. L. 2004, ApJ, **615**, 897

Berezhko, E. G., Ksenofontov, L. T., & Völk, H. J. 2003, A&A (Letters), **412**, L11
Berezhko, E. G., & Völk, H. J. 2004, A&A, **419**, L27
Bignami, G. F., Caraveo, P. A., De Luca, A., & Mereghetti, S. 2003, Nature, **423**, 725-727
Bocchino, F., & Bykov, A. M. 2001, A&A, **376**, 248
Brazier, K.T.S., Kanbach, G., Carraminana, A., Guichard, J., & Merck, M. 1996, MNRAS, **281**, 1033
Brinkmann, W., Aschenbach, B., & Langmeier, A. 1985, Nature, **313**, 662
Bykov, A. M., Bocchino, F., & Pavlov, G. G. 2005, ApJ (Letters), **624**, L41
Camilo, F., Kaspi, V. M., Lyne, A. G., Manchester, R. N., Bell, J. F., D'Amico, N., McKay, N. P. F., & Crawford, F. 2000, ApJ, **541**, 367
Camilo, F., Lorimer, D. R., Bhat, N. D. R., Gotthelf, E. V., Halpern, J. P., Wang, Q. D., Lu, F. J., & Mirabal, N. 2002a, ApJ (Letters), **574**, L71
Camilo, F. Manchester, R.N., Gaensler, B.M., Lorimer, D.R. & Sarkissian, J. 2002b, ApJ (Letters), **576**, L71
Canizares, C. R. 2004, In Frontiers of X-ray astronomy. Edited by A. C. Fabian, K. A. Pounds, R. D. Blandford. Cambridge (UK): Cambridge University Press, 107
Caraveo, P. A., Bignami, G. F., & Trümper, J. E. 1966, The Astronomy and Astrophysics Review, **7**, 209
Cassam-Chenaï, G., Decourchelle, A., Ballet, J., Hwang, U., Hughes, J. P., & Petre, R. 2004, A&A, **414**, 545
Chakrabarty, D., Pivovaroff, M. J., Hernquist, L. E., Heyl, J. S., Narayan, R. 2001, ApJ, **548**, 800
Chu, Y.-H., & Kennicutt, R. C., Jr. 1988, AJ, **96**, 1874
Cline, T. L., Desai, U. D., Teegarden, B.J., Evans, W.D., Klebesadel, R.W., Laros, J. G., Barat, C., Hurley, K., Niel, M., Vedrenne, G., Estuline, I.V., Kurt, V.G., Mersov, G.A., Zenchenko, V.M., Weisskopf, M.C., & Grindlay. J. 1982, ApJ (Letters), **255**, L45
Crawford, F., Gaensler, B. M., Kaspi, V. M., Manchester, R. N., Camilo, F., Lyne, A. G., & Pivovaroff, M. J. 2001, ApJ, **554**, 152
Crawford, F., Pivovaroff, M. J., Kaspi, V. M., & Manchester, R. N. 2002, in Neutron Stars in Supernova Remnants, ASP Conference Series, Vol. 271, held in Boston, MA, USA, 14-17 August 2001. Ed. P. O. Slane and B. M. Gaensler (San Francisco), 37
Danziger, I. J., & Leibowitz, E. M. 1985, MNRAS, **216**, 365
Decourchelle, A. 2004, New Astronomy Review, **48**, 75
Decourchelle, A., Ellison, D. C., & Ballet, J. 2000, ApJ (Letters), **543**, L57
Decourchelle, A., Sauvageot, J. L., Audard, M., Aschenbach, B., Sembay, S., Rothenflug, R., Ballet, J., Stadlbauer, T., & West, R. G. 2001, A&A, **365**, L218
De Luca, A., Mereghetti, S., Caraveo, P. A., Moroni, M., Mignani, R. P., & Bignami, G. F. 2004, A&A, **418**, 625
DeLaney, T., & Rudnick, L. 2003, ApJ, **589**, 818
DeLaney, T., Rudnick, L., Fesen, R. A., Jones, T. W., Petre, R., & Morse, J. A. 2004, ApJ, **613**, 343
Dopita, M. A., Tuohy, I. R., & Mathewson, D. S. 1981, ApJ, **248**, L105
Flanagan, K. A., Canizares, C. R., Dewey, D., Houck, J. C., Fredericks, A. C., Schattenburg, M. L., Markert, T. H., & Davis, D. S. 2004, ApJ, **605**, 230
Gaensler, B. M., Bock, D. C.-J., & Stappers, B. W. 2000, ApJ (Letters), **537**, L35
Gaensler, B. M., Slane, P. O., Gotthelf, E. V., & Vasisht, G. 2001, ApJ, **59**, 963

Gaensler, B. M., Arons, J., Kaspi, V. M., Pivovaroff, M. J., Kawai, N., & Tamura, K. 2002, ApJ, **569**, 878

Gaensler, B. M., Hendrick, S. P., Reynolds, S. P., & Borkowski, K. J. 2003, ApJ (Letters), **594**, L111

Gaensler, B. M., van der Swaluw, E., Camilo, F., Kaspi, V. M., Baganoff, F. K., Yusef-Zadeh, F., & Manchester, R. N. 2004, ApJ, **616**, 383

Gaetz, T. J., Butt, Y. M., Edgar, R. J., Eriksen, K. A., Plucinsky, P. P., Schlegel, E. M., & Smith R. K. 2000, ApJ, **534**, L47

Garmire, G. P., Pavlov, G. G., Garmire, A. B., & Zavlin, V. E. 2000, IAU circ., **7350**, 2

Ghavamian, P., Hughes, J. P., and Williams, T. B. 2005, ApJ, in press (astro-ph/0509309)

Ghavamian, P., Rakowski, C. E., Hughes, J. P., and Williams, T. B. 2003, ApJ, **590**, 833

Ghavamian, P., Raymond, J., Smith, R. C., & Hartigan, P. 2001, ApJ, **547**, 995

Golden, A., Shearer, A. & Beskin, G.M. 2000, ApJ, **535**, 373

Gonzalez, M., & Safi-Harb, S. 2003, ApJ (Letters), **591**, 143

Gonzalez, M. & Safi-Harb, S. 2005, ApJ, **619**, 856

Gotthelf, E. V., Petre, R., & Vasisht, G. 1999, ApJ (Letters), **514**, L107

Gotthelf, E. V., & Wang, Q. D. 2000, ApJ (Letters), **532**, L117

Gotthelf, E. V., Koralesky, B., Rudnick, L., Jones, T. W., Hwang, U., & Petre, R. 2001, ApJ, **552**, L39

Greiveldinger, C., & Aschenbach, B. 1999, ApJ, **510**, 305

Gvaramadze, V. V., & Vikhlinin, A. A. 2003, A&A, **401**, 625

Hailey, C. J. & Mori, K. 2002, ApJ (Letters), **578**, L133

Hamilton, A. J. S., Sarazin, C. L., & Szymkowiak, A. E. 1986a, ApJ, **300**, 698

Hamilton, A. J. S., Sarazin, C. L., & Szymkowiak, A. E. 1986b, ApJ, **300**, 713

Halpern, J. P., Gotthelf, E. V., Camilo, F., Helfand, D. J., & Ransom, S. M. 2004, ApJ, **612**, 398

Harrus, I. M. & Slane, P. O. 1999, ApJ, **516**, 811

Hayashi, I., Koyama, K., Masanobu, O., Miyata, E., Tsunemi, H., Hughes, J. P., & Petre, R. 1994, PASJ, **46**, L121

Helfand, D. J. & Becker, R. H. 1984, Nature, **307**, 215

Helfand, D. J., Gotthelf, E. V., & Halpern, J. P. 2001, ApJ, **556**, 380

Hendrick, S. P., Borkowski, K. J., & Reynolds, S. P. 2003, ApJ, **593**, 370

Hendrick, S. P., Reynolds, S. P., & Borkowski, K. J. 2005, ApJ (Letters), **622**, L117

Hester, J.J., Scowen, P. A., Sankrit, R., Burrows, C.J., Gallagher, J.S., III, Holtzman, J.A., Watson, A., Trauger, J.T., Ballester, G.E., Casertano, S., Clarke, J.T., Crisp, D., Evans, R.W., Griffiths, R.E., Hoessel, J.G., Krist, J., Lynds, R., Mould, J.R., O'Neil, E.J.,Jr., Stapelfeldt, K.R., & Westphal, J.A. 1995, ApJ, **448**, 240

Hester, J. J., Mori, K., Burrows, D., Gallagher, J. S., Graham, J. R., Halverson, M., Kader, A., Michel, F. C., & Scowen, P. 2002, ApJ (Letters), **577**, L49

Heyl, J. S., & Hernquist, L. 2002, ApJ, **567**, 510

Hickson, P. & van den Bergh, S. 1990, ApJ, **365**, 224

Hughes, J. P. 2000, ApJ (Letters), **545**, L53

Hughes, J. P., Hayashi, I., Helfand, D., Hwang, U., Itoh, M., Kirshner, R., Koyama, K., Markert, T., Tsunemi, H., & Woo, J. 1995, ApJ (Letters), **444**, L81

Hughes, J. P., Rakowski, C. E., & Decourchelle, A. 2000b, ApJ (Letters), **543**, L61

Hughes, J. P., Rakowski, C. E., Burrows, D. N., & Slane, P. O. 2000a, ApJ (Letters), **528**, L109

Hughes, J. P., Slane, P. O., Burrows, D. N., Garmire, G., Nousek, J. A., Olbert, C. M., & Keohane, J. W. 2001, ApJ (Letters), **559**, L153

Hughes, J. P., Ghavamian, P., Rakowski, C. E., & Slane, P. O. 2003a, ApJ (Letters), **582**, L95

Hughes, J. P., Slane, P. O., Park, S., Roming, P. W. A., & Burrows, D. N. 2003b, ApJ (Letters), **591**, L139

Hughes, J. P., et al. 2005b, ApJ, in preparation

Hughes, J. P., Warren, J. S., & Badenes, C. 2005a, ApJ, in preparation

Hui, C.Y. & Becker, W. 2005, submitted to A&A, astro-ph/0508655

Hulleman, F., Tennant, A. F., van Kerkwijk, M. H., Kulkarni, S. R., Kouveliotou, C., & Patel, S. K. 2001, ApJ (Letters), **563**, L49

Hwang, U., & Gotthelf, E. V. 1997, ApJ, **475**, 665

Hwang, U., et al. 2004, ApJ (Letters), **615**, L117

Hwang, U., Holt, S., & Petre, R. 2000, ApJ (Letters), **537**, L119

Hwang, U., Szymkowiak, A. E., Petre, R., & Holt, S. S. 2001, ApJ (Letters), **560**, L175

Hwang, U., Decourchelle, A., Holt, S. S., & Petre, R. 2002, ApJ, **581**, 1101

Hwang, U., & Laming, J. M. 2003, ApJ, **597**, 362

Iwamoto, K., Brachwitz, F., Nomoto, K., Kishimoto, N., Umeda, H., Hix, W. R., & Thielemann, F.-K. 1999, ApJ (Supplement), **125**, 439

Kaaret, P., Marshall, H. L., Aldcroft, T. L., Graessle, D. E., Karovska, M., Murray, S. S., Rots, A. H., Schulz, N. S., & Seward, F. D. 2001, ApJ, **546**, 1159

Kaplan, D. L., Frail, D. A., Gaensler, B. M., Gotthelf, E. V., Kulkarni, S. R., Slane, P. O., & Nechita, A. 2004, ApJ (Supplement), **153**, 269

Kargaltsev, O., Pavlov, G. G., Sanwal, D., & Garmire, G. P. 2002, ApJ, **580**, 1060

Kaspi, V. M., Manchester, R. N., Johnston, S., Lyne, A. G., & D'Amico, N. 1996, AJ, **111**, 2028

Katsuda, S., & Tsunem, H. 2005, PASJ, **57**, 621

Kennel, C.F., & Coroniti, F.V. 1984, ApJ, **283**, 694

Kong, A. K. H., Sjouwerman, L. O., & Williams, B. F. 2004, AJ, **128**, 2783

Koyama, K., Petre, R., Gotthelf, E. V., Hwang, U., Matsuura, M., Ozaki, M., & Holt, S. S. 1995, Nature, **378**, 255

Koyama, K., Kinugasa, K., Matsuzaki, K., Nishiuchi, M., Sugizaki, M., Torii, K., Yamauchi, S., & Aschenbach, B. 1997, PASJ, **49**, L7

Kulkarni, S. R., Clifton, T. C., Backer, D. C., Foster, R. S., & Fruchter, A. S. 1988, Nature, **331**, 50

Kulkarni, S. R., Kaplan, D. L., Marshall, H. L., Frail, D. A., Murakami, T., & Yonetoku, D. 2003, ApJ, **585**, 948

Laming, J. M., & Hwang, U. 2003, ApJ, **597**, 347

Lazendic, J. S., Slane, P. O., Gaensler, B. M., Plucinsky, P. P., Hughes, J. P., Galloway, D. K., & Crawford, F. 2003, ApJ (Letters), **593**, L27

Lazendic, J. S., Slane, P. O., Gaensler, B. M., Reynolds, S. P., Plucinsky, P. P., & Hughes, J. P. 2004, ApJ, **602**, 271

Lewis, K. T., Burrows, D. N., Hughes, J. P., Slane, P. O., Garmire, G. P.,& Nousek, J. A. 2003, ApJ, **582**, 770

Li, X. H., Lu, F. J., & Li, T. P. 2005, ApJ, **628**, 931

Long, K. S., Reynolds, S. P., Raymond, J. C., Winkler, P. F., Dyer, K. K., & Petre, R. 2003, ApJ, **586**, 1162

Lu, F. J., Wang, Q. D., Aschenbach, B., Durouchoux, P., & Song, L. M. 2002, ApJ (Letters), **568**, L49

Markert, T. H., Canizares, C. R., Clark, G. W., & Winkler, P. F. 1983, ApJ, **268**, 13

Marshall, F. E., Gotthelf, E. V., Zhang, W., Middleditch, J., & Wang, Q. D. 1998, ApJ (Letters), **499**, L179
Massaro, E., Cusumano, G., Litterio, M., & Mineo, T. 2000, A&A, **361**, 695
Mereghetti, S., Bignami, G. F., & Caraveo, P. A. 1996, ApJ, **464**, 842
Mereghetti, S., De Luca, A., Caraveo, P. A., Becker, W., Mignani, R., & Bignami, G. F. 2002, ApJ, **581**, 1280
Miyata, E., Tsunemi, H., Aschenbach, B., & Mori, K. 2001, ApJ (Letters), **559**, L45
Michael, E., Zhekov, S., McCray, R., Hwang, U., Burrows, D. N., Park, S., Garmire, G. P., Holt, S. S., & Hasinger, G. 2002, ApJ, **574**, 166
Moon, D.-S., Lee, J.-J., Eikenberry, S. S., Koo, B.-C., Chatterjee, S., Kaplan, D. L., Hester, J. J., Cordes, J. M., Gallant, Y. A., & Koch-Miramond, L. 2004, ApJ (Letters), **610**, L33
Mori, K., Burrows, D, N., Hester, J. J., Pavlov, G. G., Shibata, S., & Tsunemi, H. 2004, ApJ, **609**, 186
Mori, K., Chonko, J. C., & Hailey, C. J. 2005, ApJ, **631**, 1082
Morii, M., Sato, R., Kataoka, J., & Kawai, N. 2003, Publications of the Astronomical Society of Japan, **55**, L45
Murray, S. S., Ransom, S. M., Juda, M., Hwang, U., & Holt, Stephen S. 2002a, ApJ, **566**, 1039
Murray, S. S., Slane, P. O., Seward, F. D., Ransom, S. M., & Gaensler, B. M. 2002b, ApJ, **568**, 226
Olbert, C. M., Clearfield, C. R., Williams, N. E., Keohane, J. W., & Frail, D. A. 2001, ApJ (Letters), **554**, L205
Olbert, C. M., Keohane, J. W., Arnaud, K. A., Dyer, K. K., Reynolds, S. P., & Safi-Harb, S. 2003, ApJ (Letters), **592**, L45
Olbert, C. M., Keohane, J. W., & Gotthelf, E. V. 2003, BAAS, **203**, 39.07
Park, S., Roming, P. W. A., Hughes, J. P., Slane, P. O., Burrows, D. N., Garmire, G. P., & Nousek, J. A. 2002, ApJ (Letters), **564**, L39
Park, S., Burrows, D. N., Garmire, G. P., Nousek, J. A., Hughes, J. P., & Williams, R. M. 2003a, ApJ, **586**, 210
Park, S., Hughes, J. P., Burrows, D. N., Slane, P. O., Nousek, J. A., & Garmire, G. P. 2003b, ApJ (Letters), **598**, L95
Park, S., Hughes, J. P., Slane, P. O., Burrows, D. N., Roming, P. W. A., Nousek, J. A., & Garmire, G. P. 2004a, ApJ (Letters), **602**, L33
Park, S., Zhekov, S. A., Burrows, D. N., Garmire, G. P., & McCray, R. 2004b, ApJ, **610**, 275
Park, S., Zhekov, S. A., Burrows, D. N., Garmire, G. P., & McCray, D. 2005, Advances in Space Research, **35**, 991
Patel, S. K., Kouveliotou, C., Woods, P. M., Tennant, A. F., Weisskopf, M. C., Finger, M. H., Göğüş, E., van der Klis, M., Belloni, T. 2001, ApJ (Letters), **563**, L45
Patel, S. K., Kouveliotou, C., Woods, P. M., Tennant, A. F., Weisskopf, M. C., Finger, M. H., Wilson, C. A., Göğüş, E., van der Klis, M., & Belloni, T. 2003, ApJ, **587**, 367
Pavlov, G. G., Zavlin, V. E., Aschenbach, B., Trümper, J., & Sanwal, D. 2000, ApJ (Letters), **531**, L53
Pavlov, G. G., Sanwal, D., Kızıltan, B., & Garmire, G. P. 2001, ApJ (Letters), **559**, L131
Pavlov, G. G., Zavlin, V. E., Sanwal, D., & Trümper, J. 2002, ApJ (Letters), **569**, L95
Pavlov, G. G., Sanwal, D., Garmire, G. P., & Zavlin, V. E. 2002a, in Neutron Stars in Supernova Remnants, ASP Conference Series, 271, held in Boston, MA, USA,

14-17 August 2001. Eds. P. O. Slane & B. M. Gaensler. San Francisco: ASP, 247
Pavlov, G. G., Zavlin, V. E., & Sanwal, D. 2002, Proceedings of the 270. WE-Heraeus Seminar on Neutron Stars, Pulsars, and Supernova Remnants. MPE Report 278. Ed. W. Becker, H. Lesch, and J. Trümper. Garching bei Mnchen: Max-Plank-Institut fr extraterrestrische Physik, 273
Pavlov, G. G.,Teter, M. A., Kargalstev, O., & Sanwal, D., 2003, ApJ, **591**, 1171
Peterson, B.A., Murdin, P., Wallace, P., Manchester, R.N., Penny, A.J., Jorden, A., Hartley, K.F. & King, D. 1978, Nature, **276**, 475.
Petre, R., Kriss, G. A., Winkler, P. F., & Canizares, C. R. 1982, ApJ, **258**, 22
Pfeffermann, E. & Aschenbach, B. 1996, in Rntgenstrahlung from the Universe, eds. H.U. Zimmermann, J. Trümper, & and H. Yorke (MPE Report 263, Garching: MPE) 267
Pohl, M., Yan, H., & Lazarian, A. 2005, ApJ (Letters), **626**, L101
Predehl, P., & Kulkarni, S. R. 1995, A&A, **294**, L29
Pravdo, S. H., Angelini, L., & Harding, A. K. 1997, ApJ, **491**, 808
Rasmussen, A. P., Behar, E., Kahn, S. M., den Herder, J. W., & van der Heyden, K. 2001, A&A, **365**, L231
Rakowski, C. E., Ghavamian, P., & Hughes, J. P. 2003, ApJ, **590**, 846
Rakowski, C. E. 2005, Advances in Space Research, **35**, 1017
Rees, M. J., & Gunn, J. E. 1974, MNRAS, **167**, 1
Rest, A., Suntzeff, N. B., Olsen, K., Prieto, J. L., Smith, R. C., Welch, D. L., Becker, A., Bergmann, M., Clocchiatti, A., Cook, K., Garg, A., Huber, M., Miknaitis, G., Minniti, D., Nikolaev, S., & Stubbs, C. 2005, Nature, in press (astro-ph/0510738)
Reynoso, E. M., Moffett, D. A., Goss, W. M., Dubner, G. M., Dickel, J. R., Reynolds, S. P., & Giacani, E. B. 1997, ApJ, **491**, 816
Rho, J., Dyer, K. K., Borkowski, K. J., & Reynolds, S. P. 2002, ApJ, **581**, 1116
Rothschild, R. E., Kulkarni, S. R., & Lingenfelter, R. E. 1994, Nature, **368**, 432
Russell, S. C., & Dopita, M. A. 1990, ApJ (Supplement), **74**, 93
Sanwal, D., Pavlov, G. G., Zavlin, V. E., & Teter, M. A. 2002, ApJ, **574**, L61
Sanwal, D., Pavlov, G. G., & Zavlin, V. E. 2004, BAAS, HEAD meeting #8, 11.06
Sanwal, D., Pavlov, G. G., & Zavlin, V. E. 2005, in preparation
Safi-Harb, S., Dubner, G., Petre, R., Holt, S. S., & Durouchoux, P. 2005, ApJ, **618**, 321
Schmidt, G. D., Angel, J. R. P., & Beaver, E. A. 1979, ApJ, **227**, 113
Schwartz, D. A., Marshall, H. L., Lovell, J. E. J., Piner, B. G., Tingay, S. J., Birkinshaw, M., Chartas, G., Elvis, M., Feigelson, E. D., Ghosh, K. K., Harris, D. E., Hirabayashi, H., Hooper, E. J., Jauncey, D. L., Lanzetta, K. M., Mathur, S., Preston, R. A., Tucker, W. H., Virani, S., Wilkes, B. & Worrall, D. M. 2000, ApJ, **540**, 69
Sedov, L. I. 1959, Similarity and Dimensional Methods in Mechanics, (New York: Academic Press)
Serafimovich, N. I., Shibanov, Yu. A., Lundqvist, P., & Sollerman, J. 2004, A&A, **425**, 1041
Seward, F. D., Harnden, F. R., Jr., & Helfand, D. J. 1984, ApJ (Letters), **287**, L19
Seward, F. D., Schmidt, B., & Slane, P. 1995, ApJ, **453**, 284
Seward, F. D., & Mitchell, M. 1981, ApJ, **243**, 736
Shu, F. H. 1992, The Physics of Astrophysics, Volume II, Gas Dynamics, (Mill Valley: University Science Books)
Slane, P., Gaensler, B. M., Dame, T. M., Hughes, J. P., Plucinsky, P. P., & Green, A. 1999, ApJ, **525**, 357

Slane, P., Hughes, J. P., Edgar, R. J., Plucinsky, P. P., Miyata, E., Tsunemi, H., & Aschenbach, B. 2001, ApJ, **548**, 814
Slane, P. 2002, In Neutron Stars in Supernova Remnants, ASP Conference Series, Vol. 271, held in Boston, MA, USA, 14-17 August 2001. Edited by Patrick O. Slane and Bryan M. Gaensler. San Francisco: ASP, 165
Slane, P. O., Helfand, D. J., & Murray, S. S. 2002, ApJ, **571**, L45
Slane, P., Helfand, D. J., van der Swaluw, E., & Murray, S. S. 2004b, ApJ, **616**, 403
Slane, P., Zimmerman, E. R., Hughes, J. P., Seward, F. D., Gaensler, B. M., & Clarke, M. J. 2004a, ApJ, **601**, 1045
Slane, P. 2005, Advances in Space Research, 35, 1092
Stephenson, F.R., & Green, D.A. 2002, Historical Supernovae and their Remnants (Oxford: Clarendon)
Tananbaum, H. 1999, IAU Circ. 7246
Taylor, G. 1950, Royal Society of London Proceedings Series A, **201**, 159.
Tennant, A. F., Becker, W., Juda, M., Elsner, R. F., Kolodziejczak, J. J., Murray, S. S., O'Dell, S. L., Paerels, F., Swartz, D. A., Shibazaki, N., & Weisskopf, M. C. 2001, ApJ (letters), **554**, L173
Truelove, J. K., & McKee, C. F. 1999, ApJ (Supplement), **120**, 299
Tuohy, I. R., & Dopita, M. A. 1983, ApJ (Letters), **268**, L11
Tuohy, I. & Garmire, G. 1980, ApJ (Letters), L107
Tuohy, I. R., Dopita, M. A., Mathewson, D. S., Long, K. S., & Helfand, D. J. 1982, ApJ, **261**, 473
Uchiyama, Y., Aharonian, F. A., & Takahashi, T. 2003, A&A, **400**, 567
Vancura, O., Blair, W. P., Long, K. S., & Raymond, J. C. 1992, ApJ, **394**, 158
Vancura, O., Gorenstein, P., & Hughes, J. P. 1995, ApJ, **441**, 680
van der Heyden, K. J., Behar, E., Vink, J., Rasmussen, A. P., Kaastra, J. S., Bleeker, J. A. M., Kahn, S. M., & Mewe, R. 2002, A&A, **392**, 955
Vink, J. & Laming, J. M. 2003, ApJ, **584**, 758
Vink, J., Laming, J. M., Gu, M. F., Rasmussen, A., Kaastra, J. S. 2003, ApJ Letters, **587**, L31
Völk, H. J., Berezhko, E. G., & Ksenofontov, L. T. 2005, A&A, **433**, 229
Wachter, S., Patel, S. K., Kouveliotou, C., Bouchet, P., zel, F., Tennant, A. F., Woods, P. M., Hurley, K., Becker, W., & Slane, P. 2004, ApJ, **615**, 887
Wang, Q. D., Gotthelf, E. V., Chu, Y.-H., & Dickel, J. R. 2001, ApJ, **559**, 275
Warren, J. S., & Hughes, J. P. 2004, ApJ, **608**, 261
Warren, J. S., Hughes, J. P., & Slane, P. O. 2003, ApJ, **583**, 260
Warren, J. S., Hughes, J. P., Badenes, C., Ghavamian, P., McKee, C. F., Moffett, D., Plucinsky, P. P., Rakowski, C., Reynoso, E., and Slane, P. 2005, ApJ, **634**, 376 (astro-ph/0507478)
Weisskopf, M.C., Hester, J. J., Tennant, A. F., Elsner, R. F., Schulz, N. S., Marshall, H. L., Karovska, M., Nichols, J. S., Swartz, D. A., Kolodziejczak, J. J., O'Dell, S. L. 2000, ApJ, **536**, 81
Weisskopf, M. C., Aldcroft, T. L., Bautz, M., Cameron, R. A., Dewey, D., Drake, J. J., Grant, C. E., Marshall, H. L., Murray, S. S. 2003, Experimental Astronomy, **16**, 1
Weisskopf, M.C., O'Dell, S. L., Paerels, F., Elsner, R. F., Becker, W., Tennant, A. F., Swartz, D. A. 2004, ApJ, **601**, 1050
Weisskopf, M.,C., Aldcroft, T.L., Cameron, R.A., Gandhi, P., Foellmi, C., Elsner, R.F., Paterl, S.K., & O'Dell, S.L. 2005, ApJ, accepted.
Willingale, R., Aschenbach, B., Griffiths, R. G., Sembay, S., Warwick, R. S., Becker, W., Abbey, A. F., & Bonnet-Bidaud, J.-M. 2001, A&A, **365**, L212

Willingale, R., Bleeker, J. A. M., van der Heyden, K. J., Kaastra, J. S., & Vink, J. 2002, A&A, **381**, 1039
Wilms, J., Allen, A., & McCray, R. 2000, ApJ, **542**, 914
Wu, C.-C., Leventhal, M., Sarazin, C. L., & Gull, T. R. 1983, ApJ (Letters), **269**, 5
Yakovlev, D. G., Kaminker, A. D., Haensel, P., & Gnedin, O. Y. 2002, A&A, **389**, L24
Yatsu, Y., Kataoka, J., Kawai, N., Tamura, K., & Brinkmann, W. 2005, Advances in Space Research, **35**, 1066
Zavlin, V. E., Pavlov, G. G., Sanwal, D., & Trümper, J. 2000, ApJ (Letters), **540**, L25
Zavlin, V. E., Pavlov, G. G., Sanwal, D. 2004, ApJ, **606**, 444
Zavlin, V. 2005, private communication

4 Galaxy Collisions – Dawn of a New Era

C. Struck

Abstract: The study of colliding galaxies has progressed rapidly in the last few years, driven by observations with powerful new ground- and space-based instruments. These instruments have been used for detailed studies of specific nearby systems, statistical studies of large samples of relatively nearby systems, and increasingly large samples of high redshift systems. Following a brief summary of the historical context, this review attempts to integrate these studies to address the following key issues. What role do collisions play in galaxy evolution, and how can recently discovered processes like downsizing resolve some apparently contradictory results of high redshift studies? What is the role of environment in galaxy collisions? How is star formation and nuclear activity orchestrated by the large scale dynamics, before and during merger? Are novel modes of star formation involved? What are we to make of the association of ultraluminous X-ray sources with colliding galaxies? To what degree do mergers and feedback trigger long-term secular effects? How far can we push the archaeology of individual systems to determine the nature of precursor systems and the precise effect of the interaction? Tentative answers to many of these questions have been suggested, and the prospects for answering most of them in the next few decades are good.

4.1 Introduction: Some Past Highlights and Current Issues

4.1.1 Early days

The study of galaxy collisions is not an ancient one; Erik Holmberg, Fritz Zwicky, and a few others did quite a bit of work relevant to colliding galaxies before the 1950s, but that decade opened and closed with two landmark papers. Thus, we can justify taking it as the first decade of general interest in the subject, and view the earlier work as pioneering. The first of the two papers, by Spitzer & Baade (1951), revived Zwicky's suggestion that collisions would be frequent within dense galaxy clusters, and considered what would happen in direct collisions between two galaxies. Specifically, they correctly argued that the stellar distribution might be only moderately disturbed, while strong shock waves could push the interstellar gas out of the galaxies. Their primary conclusion was that this process could account for the scarcity of late-type spiral galaxies with substantial ongoing star formation in clusters. For the first time galaxy collisions were seen to have an important role in galaxy evolution.

The second landmark paper was Zwicky's review of his extensive imagery of morphologically peculiar galaxies, together with arguments that many of these peculiarities were caused by tidal forces in collisions (Zwicky, 1959). At the time, Zwicky's theory seemed doubtful, since collisions between the widely separated "island universes" were deemed improbable. Moreover, his arguments were generally semi-quantitative, and so, not compelling.

In the middle of this first decade Baade & Minkowski (1954) suggested that one of the most prominent members of the newly discovered class of radio galaxies, Cygnus A, was in collision. Thus, we already have the first hints of many of the most important themes in this field, including: the generation of unique tidal morphologies, induced nuclear activity, induced star formation, the important role of collisions in galaxy evolution, and the dependence of these effects on the clustering environment. In this review I will focus on the last few items – induced star formation, galaxy evolution, and environmental effects – and say relatively little about the first two and many other related topics.

In the second decade, much of the work was of a more detailed, and sometimes indirect nature, which I cannot review here (see e.g., Struck 1999). The major exception to this generalization was publication of Arp's pictoral atlas of more than 300 peculiar galaxies (Arp, 1966). Arp derived the atlas objects from the Palomar sky survey. In the following decades this atlas became the standard 'field guide' for workers in this field. The Arp galaxies were arranged in categories somewhat like Zwicky's, but with many more examples, and excellent photographic images. One psychological effect of so many images may well have been to make the peculiar galaxies seem less like freakish rarities, and more like zoological families in need of explanation.

4.1.2 The 1970s

Toomre & Toomre took a giant step toward these explanations. Their numerical models were not the first, and were simple by modern standards, but they were more extensive than previous efforts. They were able to account for many of the Arp atlas forms in detail, thereby making a strong case for the collisional origins theory. They also made a number of important predictions for observation, such as that strong tidal waves would lead to enhanced star formation and gas transfer to nuclear regions, which could fuel nuclear activity. These would become dominant themes in subsequent work.

However, Toomre and Toomre's models did not directly account for the 'messiest' objects in the Arp atlas (see examples in Fig. 4.1). Alar Toomre returned to these objects in his contribution to the seminal Yale galaxies conference (Toomre, 1977). He pointed out that the earlier models had not included the effect of Chandrasekhar's dynamical friction (Chandrasekhar, 1943), and showed that the effect would draw the colliding galaxies into a merger. He further considered how merger remnants would evolve, and how they would appear

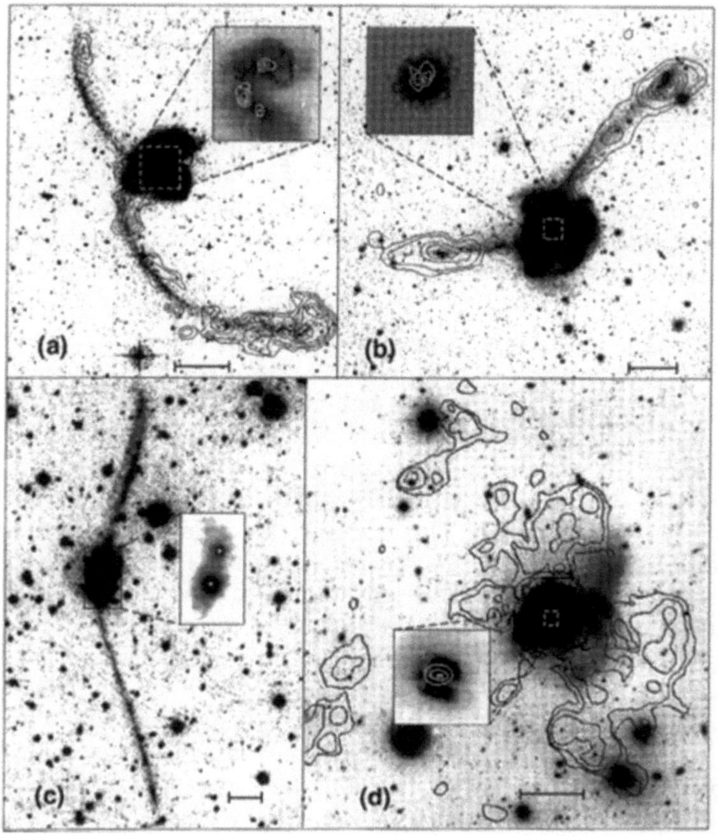

Fig. 4.1. Multiwaveband images of several well-known merger remnants (courtesy D. B. Sanders and I. F. Mirabel): a) NGC 4038/39 (Arp 244, "The Antennae"), b) NGC 7252 (Arp 226, "Atoms for Peace"), c) IRAS 19254-7245 ("The Super-Antennae"), d) IC 4553/54 (Arp 220). An optical image shown in greyscale, HI (21 cm) surface intensity shown by contours, and K band (2.2 mm) shown in insets. Scale-bar represents 20 kpc in each case; see Sanders & Mirabel (1996) for details.

observationally. This lead him to some radical conclusions for that time, that mergers between comparable large spiral galaxies could lead to the formation of elliptical galaxies, and that reasonable extrapolation of the statistics of such collisions suggested that a large fraction of ellipticals could be formed this way. The debate still continues on many aspects of this scenario, but it immediately had an important effect. Toomre had opened the door to the possibility that collisions were the dominant factor in the evolution of an important class of galaxies. Collisions were more than just a means of accounting for rare freaks, or a specialized process peculiar to the environment of dense clusters.

Other important developments in the 1970s included the work of Larson & Tinsley (1978), who suggested that Arp atlas galaxies had a wider range of optical colors and star formation rates (SFRs) than the more normal Hubble atlas (Sandage, 1961) galaxies. Extension of that work suggested that infrared colors would provide even more sensitive indications of varying SFRs (Struck-Marcell & Tinsley, 1978). Ever increasing evidence that galaxies (and groups and clusters) possessed massive dark halos (see Sofue & Rubin (2001) for a history of rotation curve studies) completely changed our understanding of what a galaxy is. The ten-fold increase of galaxy masses and sizes in the new picture provided an explanation of why collisions could be common, despite the great separations of the visible parts of galaxies. Their cross section were much larger than previously thought, and collision partners were born bound together in larger entities.

4.1.3 The 1980s and early 1990s

This period saw expansion of the field into many new directions, with a number of major developments that defined the current epoch. One of the highest points was the discovery of ultraluminous far-infrared galaxies (ULIRGs) with the observations of the IRAS satellite (see reviews of Soifer, Houck & Neugebauer 1987 and Sanders & Mirabel 1996). This discovery set off a gold rush of studies of these objects, as illustrated by the papers of the 1986 Pasadena meeting (Lonsdale Persson, 1986) and the 1989 Alabama meeting (Sulentic, Keel & Telesco, 1989), and which continues to some degree up to the present. A primary focus of most ULIRG papers has been the relative role of nuclear starbursts versus active nuclei in generating the huge emissions. This is a difficult question to answer because both are usually buried deeply in the gas and dust of the merger remnant; most observational techniques give only indirect clues. While elucidating the connection between starburst and nuclear activity is very important, the ULIRGs and their somewhat less luminous cousins, the LIRGs, offer a wealth of other information on questions of galaxy evolution. See also the review by Lonsdale, Farrah and Smith in this volume.

A second focus of ULIRG studies was the determination of what sort of remnant would ultimately emerge from a major merger. ULIRGs could be seen as the missing link in Toomre's theory of elliptical formation from major mergers. They are recent mergers with prodigious amounts of star formation, which might eventually either consume or heat and disperse the gas, as required by the theory. The fact that the old star surface brightness profile approximated the de Vaucouleurs profile characteristic of ellipticals in the inner regions of some ULIRGs, despite the presence of tidal distortions in the outer parts, gave further support to the theory.

This was generally a period of rapid development of numerical models. It began with the publication of the first fully self-consistent three-dimensional models of galaxy collisions followed through the merger (see review of Barnes & Hernquist 1992). In these models the galaxies were of comparable size and consisted of a single spheroidal component, i.e. like two elliptical galaxies without dark haloes. They showed that mergers occurred much more quickly than expected, as orbital energy was efficiently channeled into internal collective modes. They also revealed the rapid appearance of a de Vaucouleurs surface density profile in some major merger remnants. This profile can be viewed as a kind of metastable state, resulting from the prompt relaxation of collective modes. Its appearance in ULIRGs indicated agreement between observations and models, and provided more support for the ellipticals from mergers theory.

By the end of this period the state of the numerical art had advanced to self-consistent merger models of galaxies with stellar disk, gas disk, and dark halo components (Barnes & Hernquist, 1992). These models showed that different galaxy components behaved somewhat differently during the (major) merger process, with dynamically hot halo components generally merging more quickly than the disk components. Even more exciting from the point of view of ULIRG studies, the models showed that a fraction of the gas carried much of the angular momentum out into extended tidal structures, while the rest of the gas fell into a small volume in the remnant center. This mass of highly compressed gas could readily fuel ULIRG superstarbursts.

This period did not see many models of mini or micro mergers, in part because ULIRGs and major mergers were the focus, but also because adequate numerical resolution of small companions was difficult. Another lacuna of modeling in this period was realistic gas dynamics; most models used either an isothermal equation of state for the gas or 'sticky particle' algorithms with phenomenological collision rules between particles representing gas clouds. Cooling, heating and stellar feedback processes were not generally included, (but see e.g., Appleton & Struck-Marcell 1987b).

Alongside the major thrusts of merger studies several quiet revolutions occurred in this period. One of these was based on the sensitive mapping of atomic hydrogen in galaxies generally, as well as collisional systems, by many observers using the Westerbork array, and later the VLA (Very Large Array of the National Radio Astronomy Observatory). These observations first made clear that the gas disks of typical disk galaxies were much larger than the stellar, and then as one might have expected, that these extended gas disks were more strongly affected by collisional encounters than the inner stellar disks. It soon became clear that such observations were essential for determing the full extent of tidal tails and bridges. HI mapping also provides a map of the line-of-sight velocities of the gas. Kinematic maps provide us a view in a third dimension of the six dimensional position-velocity space, and this information is usually crucial to the success of models of individual systems, thereby to detailed tests of collision theory. The accomplishments of the VLA were summarized at a recent symposium (Hibbard, Rupen & van Gorkom, 2001), and a valuable legacy of that meeting was the creation of the HI Rogue's Gallery website of colliding galaxy HI maps by J. Hibbard (www.nrao.edu/astrores/Hirogues/).

Another discovery that can be described as revolutionary is that tidal interactions can induce the formation of a bar component out of disk material. This was shown by the numerical models of Noguchi (1987), and studied in detail by Athanassoula (see review of Athanassoula (2004) and references therein). This result is important because bars transfer angular momentum outward in the disk, and so can drive gas into the central regions before merger. The bar can also drive spiral density waves. Both the increased central gas concentration and the bar/spiral waves can induce star formation.

We will examine the question of SF induced before merging in more detail below. However, we should note here that Keel, Kennicutt and collaborators carried out an extensive program of Hα imagery and spectra of both collisional systems and of a control sample (Keel et al. 1985, Kennicutt et al. 1987). They found indications of enhanced SF in the collision sample, and particularly of SF enhancements in galaxy cores which were kinematically disturbed. On larger scales, Schombert, Wallin & Struck-Marcell (1990) observed the broad band colors of a sample of tidal bridges, plumes and tails, and found that while SF in these structures was not especially strong, it did continue after their formation. This is somewhat surprising given the great extent of many of these structures, which would seem to imply diminished gas densities and SF.

In his continuing studies of putative merger remnant-to-elliptical systems, Schweizer also discovered large, young star clusters or dwarf galaxies formed in tidal tails, most notably in the "Antennae" system (Schweizer, 1983). These discoveries would inspire a great deal of new work in the 1990s and the present decade. More generally, Schweizer's detailed, multi-waveband studies of specific merger remnants, whose appearance suggested that they were on the road to becoming ellipticals, advanced Toomre's merger theory (see Schweizer (1998) and references therein).

As a final example of quiet revolutions of the 1980s, I would include the extensions to dynamical friction theory by Tremaine & Weinberg (1984), and the application of the new theory to the evolution of galactic bars (Weinberg, 1985). The classical Chandrasekhar (1943) theory was too idealized to account for the frictional effects in major mergers, and even more so in the case of a "sinking satellite" orbiting outside of, but interacting with the disk of the primary galaxy. The Tremaine and Weinberg theory included the collective effects not accounted for in the classical theory, and is able to account for the rapidity of major mergers seen in numerical models.

Even beyond these revolutionary examples the tapestry of colliding galaxy studies also grew with the addition of more new threads in this period. These included studies of many specific types of collisional system, such as: colliding ring galaxies (see review of Appleton & Struck-Marcell 1987a), polar rings (see review of Sparke 2002), ocular ovals (Elmegreen et al., 1991), and shell galaxies (e.g., Hernquist & Quinn 1988). Numerical modeling demonstrated how these distinctive morphologies could be produced in collisions, and thus confirmed earlier conjectures on the broad scope of collision theory. In addition, distinctive morphologies were generally found to be the result of a relatively narrow set of collision parameters. Examples in each class can be viewed as a set of related natural experiments, seen at different times and with slightly different initial parameter values, which have the potential to provide much insight into difficult or obscure collision processes (e.g., hydrodynamic or SF processes).

4.1.4 Key issues up to the present

The 1990s saw continued rapid expansion of the field, driven in part by new ground and satellite-based instrumentation, and by rapidly increasing computer power. It is very difficult to summarize the accomplishments of that decade briefly. Queries to NASA's Astrophysical Data System show that the number of literature papers with abstracts containing the words "galaxy" and "collision" grew very rapidly with each decade: 27 (1950s), 75 (1960s), 326 (1970s), 826 (1980s), 1413 (1990s). Similar increases in the number of studies in the related fields of galaxy formation and galaxy evolution at high redshift make the task even more difficult. In this review we will focus our attention on key issues relating to star formation and galaxy evolution.

It is clear that over the second half of the 20th century this field has gone from bare beginnings to a considerable maturity, providing answers to some of its most important questions and early paradoxes. Yet many questions remain, including some that have been common threads through the whole history of the subject, and which are connected to the deepest questions in astrophysics. For reference in the rest of this review, I list here some of the most important ones.

1. How do collisions and interactions affect galaxy evolution overall? More precisely, what are the relative roles of major and minor mergers in building galaxies? This question is related to that of how galaxies form, since major mergers are very important in hierarchical build-up models, and negligible in monolithic collapse models of galaxy formation.

2. How does the answer to the previous question depend on environment? How do collisions differ in cluster, group, or nearly isolated environments? Some partial answers to these questions have been known for a long time. For example, collisions between field galaxies are very different from those between cluster galaxies because the latter have typical relative velocities of thousands of km/s versus velocities of hundreds of km/s in the former case. High velocity collisions can remove interstellar gas and produce moderate tidal distortions, but are unlikely to result in merger, while mergers are generally inevitable in the lower velocity collisions in groups. Research over the last few decades has provided a great deal of information on these questions, and it has become clear that environment plays a very large role in determining the nature of collisions that can occur, and the relative importance of galaxy collisions versus other evolutionary processes (like gas sweeping in dense clusters).

3. How do the large-scale dynamics of collisions and interactions orchestrate star formation (SF) and nuclear activity, which are inherently small scale processes? The clear answer from the 1980s is that activity is induced by dumping a great deal of gas into the central regions of major merger remnants. Major mergers may be the way to make most of the stars in a significant fraction of early type galaxies, but they are a rare event in the world of galaxy collisions, and the question remains for other types of collision. Related questions include: when do galactic winds and fountains result from interaction induced SF, and what feedback role do they play in the subsequent SF?

4. To what degree do mergers trigger long-term (more than 1.0 Gyr) secular processes? Examples include the long-term effects of collisionally induced bar components, and the fallback of large scale tidal structures.

5. How far can we push the archaeology of individual systems? Do enough clues remain to determine the morphology of the precursor galaxies, and decipher the details of the interaction up to the present?

In the remainder of this review we will consider how developments in the last decade and the near future help to answer these questions. The first three sets of questions include the key questions of this review. The last two push beyond its scope, and I will not treat them in any detail, despite their intrinsic interest.

4.2 Induced Star Formation and Winds

4.2.1 Star formation processes in interactions

Star formation induced by galaxy collisions appears similar to SF in isolated galaxies in several ways. Before merging it is often concentrated in spiral waves or bars, and tidal tails often look like extensions of the spirals. Both before and after merger it is often concentrated in nuclear starbursts. These can be orders of magnitude stronger than core bursts in isolated galaxies, but they can appear qualitatively similar.

However, there are theoretical reasons to think that the nature of collisionally induced SF is very different from that in isolated disks. In isolated star-forming disks there is evidence that SF, and gas disk structure, are self-regulated by energy and momentum feedbacks from young SF regions (e.g., Kennicutt 1989). The self-regulation processes work to maintain a gas surface density close to the threshold for local gravitational instability throughout the disk. SF is usually concentrated in grand design or flocculent phase waves, which compress the gas, pushing its density over the stability threshold. Thus, isolated gas galaxy disks are likely examples of self-regulated, non-equilibrium steady states, at least in regions where the rotation curve is essentially flat (Note that the details of the self-regulatory processes are not well understood. See Struck & Smith (1999) for a self-consistent model in the case of strong global SF. See Zhang (2003) for a discussion of how spiral waves may be maintained for relatively long periods.)

Collisions upset steady state disks, even if they don't tear regions in them apart, as occurs in the case of direct collisions between two gas disks (e.g., Struck 1997), or major mergers (e.g., Barnes & Hernquist 1996, Mihos & Hernquist 1996). The waves in these disturbed disks are of a different nature than those in steady disks. For example, tidal tails are material, rather than phase waves, and in most cases induced spirals and bars have mixed material and phase aspects. Induced waves can have a very different combinations of Fourier modes than steady waves. For example, odd numbered, asymmetric modes are evidently more common.

Compressions in steady spirals can push the gas above instability thresholds and drive SF, but the degree of compression is limited by the passage time through the wave (e.g., half the epicyclic period). In material waves, compressed gas elements can move together, and maintain their compression for longer periods. Beyond this, gas clouds can be partially separated from their original surroundings, and launched like collisionless stars over substantial distances, to interact with other clouds from very different radial positions in the initial disk. This tidal mixing can sometimes involve substantial relative velocities, and may play a great role in induced SF. This point has not been studied in any detail, probably because of the difficulty in obtaining observational evidence of the mixing.

Tidal mixing is similar to collisional splash effects, where direct collisions between gas disks drive gas out of both disks, and both disks experience later fallback. Both effects are analogous to splash and mixing effects in water waves. Tidal tails are breaking waves in galaxies.

These examples highlight how detailed studies of SF in colliding galaxies can advance of understanding of SF processes in general, as well as allow us to study modes that simply don't occur in quasi-steady isolated disks. These modes are likely to be very important in the early stages of galaxy buildup. We will return to the subject of high redshift galaxies below.

4.2.2 Observational samples of star formation before merger

Given these theoretical motivations, let us consider observational results. In Sect. 4.1 we discussed the discovery of ULIRG super-starbursts in gas-rich, major mergers in the 1980s. Generally, no such strong signal of enhanced SF has been found in pre-merger interactions. Since in most interactions there is no wholesale gas compression like that found in major merger remnants this is not surprising. The questions remain, however, do interaction induced disturbances lead to substantial SF enhancements, and if so, where, when and how? These questions were raised by Toomre & Toomre (1972) and Larson & Tinsley (1978). They have been the focus of much interest in observational studies in many wavebands of both individual systems and samples of systems (see Sect. 7 in the review of Struck 1999).

The common conclusion was that there is an average SF enhancement in interacting systems, and that this could be result of a modest starburst in most cases. However, SF is not obviously enhanced in all interactions, and may be suppressed in some. The galaxy samples studied in the 1980s and 1990s were not generally large enough to provide strong enough statistical results to be definitive, let alone to tease out details of the relevant processes. The larger samples tended to contain systems from a wide range of pre-merger or merger stages (like the the Larson and Tinsley Arp Atlas sample), and so, could be dominated by the merger-burst effect. On the other hand, samples of specific types of interaction (e.g., the ring galaxy sample of Appleton & Struck-Marcell 1987a) or specific stages (e.g., the Bushouse 1987 violently interacting sample) tended to be small. Interactions are rare in the present day, and specific types are therefore doubly rare!

Nonetheless, interesting clues came out of many of these studies. An important example is the Keel (1993) spectroscopic study of SF correlations in a sample of 75 Karachentsev spiral pairs (see Fig. 4.2 for an example system).

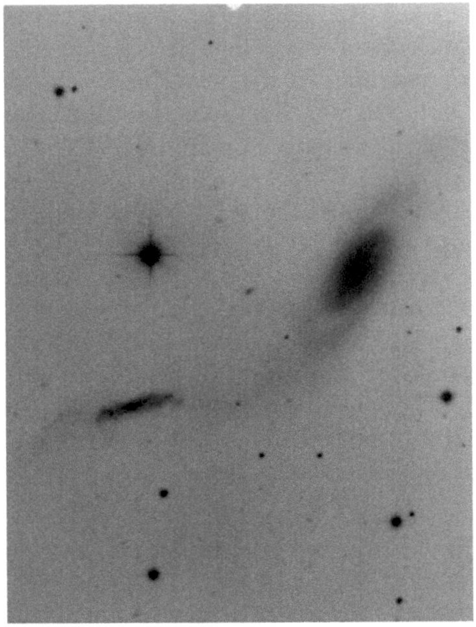

Fig. 4.2. Image of the Arp 89 system (NGC 2648, from Arp (1966). It is an example of systems studied by Keel (1993). The companion has one of the strongest nuclear SFRs in the sample.

This work built on a decade of earlier work by Keel, Kennicutt and collaborators (Keel et al. 1985, Kennicutt et al. 1987). Keel found that the current SFR (as measured by Hα equivalent width) did not depend much on the projected separation of the two galaxies, nor on whether a galaxy experienced the collision as prograde or retrograde. These results seem to defy the intuitive notion that strong perturbations at closest approach should drive strong responses, which could result in enhanced SF (but see Keel & Borne 2003). In prograde encounters the companion orbits in the same sense as the galaxy's spin, and so the encounter is prolonged, undoubtedly resulting in more disturbance, e.g., tidal tails. Thus, it was surprising that Keel did not find a spin/orbit effect in the SF.

What Keel did find was SF enhancement in systems with disturbed kinematics or in galaxies with large regions of solid body rotation. Disturbed kinematics was measured by the largest difference between the measured velocity and that of a mean symmetric rotation curve. Such kinematic disturbances can be seen in barred galaxies. However, Keel's sample did not include many barred galaxies. Keel also found that both disk and nuclear SF enhancements were linked to kinematic disturbance, which at first sight seems to be another mysterious result.

Keel considered some of the theoretical mechanisms proposed to account for induced SF in light of his observational results. He found contradictions between several of the observational results and the predictions of models on the enhancement of collisions between massive gas clouds. The correlation of enhanced SF with the size of solid-body rotation regions lead him to favor gravitational instability processes, since such regions are very susceptible to these instabilities.

Recently, Barton and collaborators (Barton, Geller & Kenyon 2000, Barton Gillespie, Geller & Kenyon 2003) re-examined these questions with a larger sample of 502 galaxy pairs and groups drawn from Harvard redshift surveys. In contrast to Keel they found a significant anti-correlation between SF (again measured by Hα equivalent width) and separation of the galaxies. The two samples have comparable ranges of separation and equivalent width. Although Barton et al.'s SF-separation anticorrelation is statistically strong, it does appear to be strongly influenced by the approximately two dozen sample galaxies with equivalent widths greater than or about equal to 50. Given the relative sample size, we would expect to find only about 3-4 such systems in Keel's sample. Indeed, there are 4. Thus, it appears that the effect is too weak to have been easily detected in a sample much smaller than Barton et al.'s. Barton et al. speculate that the cause of this anticorrelation is driven gas inflow before merger in some systems.

Barton and collaborators find a second anticorrelation between SF and line of sight velocity separation between the two galaxies in each pair. This is in accord with the intuitive notion that slower passages induce stronger collisional responses. Among their pairs and groups they find a very strong anticorrelation between SF and galaxy density, which they interpret as a symptom of the well-known density-morphology relation in groups and clusters. And finally, Barton Gillespie, Geller & Kenyon (2003) have compared their observations and models of Hα equivalent width and B-R broad band color, taking careful account of reddening effects. They find a significant correlation between burst population age and separation. They attribute this correlation and post-starburst spectral indicators in some systems to starburst triggering at closest approach, and subsequent aging as the galaxies move to apogalacticon.

On the other hand, Bergvall, Laurikainen & Aalto (2003) have recently questioned the whole notion that there are statistically significant SF enhancements before merger. They examined the UBV broad band colors of a sample of 59 interacting or merging systems, and compared them to a control sample of 38 isolated galaxies. They find no significantly greater scatter in the colors of Arp atlas galaxies relative to controls, in contrast to the Larson and Tinsley result, and no evidence for a significant enhancement in global SF in their interacting sample relative to the control. They do find evidence for a modest enhancement, by a factor of 2-3, in the central SF of their interacting sample. Given the previous result this implies a diminution of the average extra-nuclear disk SF in the interacting sample. Keel (1993) found no such distinction between net and nuclear SF enhancement in his sample.

On the face of it, Bergvall et al.'s primary result about the lack of SF enhancement in interactions seems to contradict many previous studies. However, these studies also find that the effect is weak if we exclude merger remnants, and the Barton et al. papers in particular suggest that we may need a sample of at least several hundred galaxies to find it. Given Bergvall et al.'s sample size, their work may not provide strong evidence for the complete absence of an effect, and they may even be a bit pessimistic in their estimate that the frequency of strong, triggered starbursts in interacting systems is of order 0.1%. Recent very large surveys of galaxy properties, like the Sloan Digital Sky Survey (SDSS) and the Two Degree Field (2dF) survey, could provide the answers, and indeed, a couple of analyses based on these surveys have been published recently.

Nikolic, Cullen & Alexander (2004) selected nearly 12,500 pair systems with companions within 300 kpc of the primary from the SDSS. This is a volume-limited, low redshift sample with SFRs determined from SDSS (extinction and aperture corrected) Hα data, supplemented by IRAS data. They also reject very close pairs, i.e., most merger remnants. They find that "the mean projected star formation rate is significantly enhanced for projected separations less than 30 kpc." (see Fig. 4.3). Like Barton et al. they find an anticorrelation between SF and the pair velocity difference. Despite its statistical significance they also found the the SF-separation anticorrelation is relatively weak, in accord with previous studies.

With such a large sample, they were able to look at subsamples, for example, subsamples consisting of two late-type disks, two early-type disks, or mixtures. The anticorrelation is present in all three subsamples, with some indication that it extends to larger radii in the late-type subsample. Nikolic, Cullen and Alexander also presented SFRs normalized by galaxy mass, and show that the magnitude of the normalized SF-separation relation depends on how the normalization is performed.

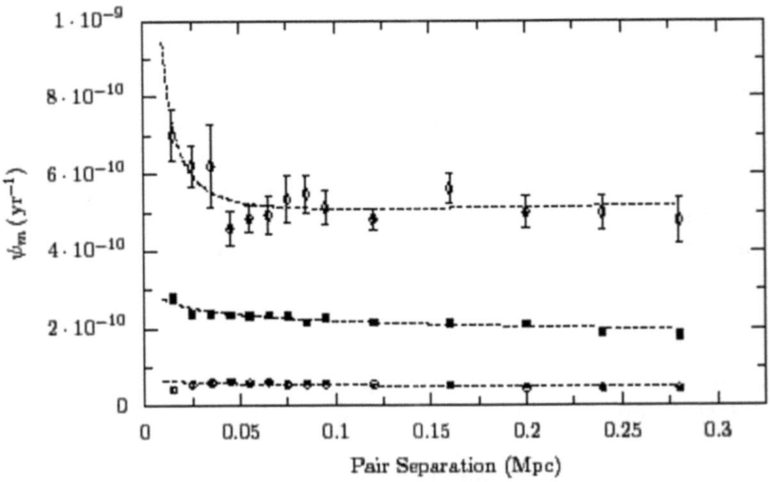

Fig. 4.3. Specific star formation rates for 3 subsamples of Sloan Digital Sky Survey galaxies, selected according to absolute SFR in the ranges: 0-3, 3-10, and > 10 $M_\odot$ yr^{-1} (courtesy B. Nikolic). See Nikolic, Cullen & Alexander (2004) for details.

Lambas et al. (2003) carried out a similar pair study with 1258 pairs from the 2dF survey, and found anticorrelations of SF with separation and velocity like those in the Nikolic et al. study.

Bergvall, Laurikainen, and Aalto noted that "the interacting and in particular the merging galaxies are characterized by increased far infrared luminosities and temperatures that weakly correlate with the central activity." This result, in turn, agrees with much evidence that many specific types of interacting galaxy have enhanced far-infrared emission. For example, M51 type systems (e.g., Klimanov & Reshetnikov 2001) on one hand, and the collisional ring galaxies (Appleton & Struck-Marcell, 1987a) on the other hand, both show modestly enhanced IRAS fluxes relative to the late-type disk norms.

For a broader perspective, one can turn the table and ask about the nature of galaxies with enhanced infrared emission (and usually radio continuum emission as well). We have discussed ULIRGs above, and noted that they are primarily merger remnants, and so not of interest in the present context. Luminous Infrared Galaxies (LIRGs or LIGs) and Very Luminous Infrared Galaxies (VLIRGs or VLIGs) are variously defined as galaxies with far-infrared luminosities in the approximate range 3×10^{10}–10^{12} $L_\odot$, and have not been studied as intensively (see examples in Fig. 4.4). However, it appears that a large fraction of these objects are pre-merger, collisional systems with a relatively strong starburst in the core of at least one of the galaxies (see e.g., Young et al. (1996), Gao & Solomon (1999), Corbett et al. (2003), Arribas et al. (2004) and references therein). Based on

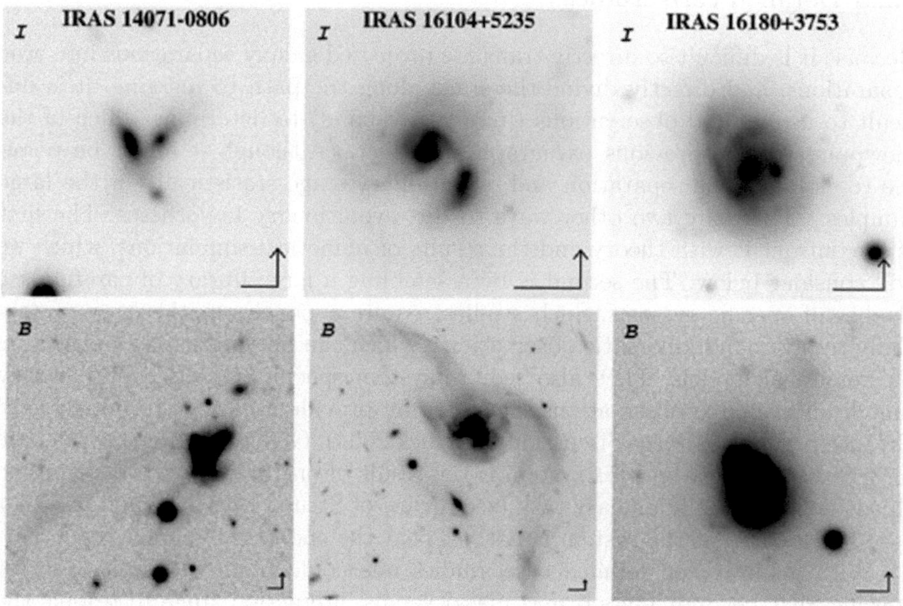

Fig. 4.4. Some examples of LIRG systems in B and I wavebands. Note the scale bars and the change of scale between rows. From Arribas et al. (2004), courtesy S. Arribas.

statements like that of Bergvall et al. in the previous paragraph, and the rarity of LIRGs (like the ULIRGs), it seems likely that they are the same as, or more obscured relatives of, the few starburst galaxies that seem to be responsible for the weak SF enhancement found in optical pair samples.

In summary, optical studies show that interactions lead to only a very small SF enhancement before merger, on average. Given that core starbursts are likely to have quite short durations (unless they have prolonged driving, e.g., Struck 2005), it is natural to interpret this as the result of random sampling of a common process with a short duty cycle. The LIRG studies suggest another possibility, that a small minority of galaxies (the LIRGs) are responsible for the general weak enhancement, and that these galaxies are near the end of the road to merger, though not yet merged. The latter clause is supported by the fact that the few starbursts in pair samples generally have small separations and velocity separations, and this is also true for many LIRGs. In this alternative view, SF is not significantly enhanced in the early stages of interaction despite strong morphological disturbances. Also there is a more continuous increase in SF as merger is approached, an idea suggested in some of the LIRG studies. Gao and Solomon, in particular, have suggested that the phase structure of galactic ISM changes through the merger process, with an increase in the molecular phase in the final pre-merger stage (Gao & Solomon 1999, also see Gao & Solomon 2004 for similar results concerning molecular abundance changes through core starburst evolution). We will return to this discussion in Sect. 4.2.5.

4.2.3 Detailed case studies

Because it is difficult to directly translate projected galaxy separations into true separations, and directly divine the stage along the path to merging, it is difficult to use limited observations (in any waveband) to determine which of the viewpoints of the previous paragraph is correct. (Although it might be possible to estimate the separation and evolutionary stage statistically in the large samples.) There are two other ways to test evolutionary hypotheses. The first is to confront it with theory and the results of numerical simulations, which we will consider below. The second is by assembling a large library of careful case studies of specific systems. Such studies require a panchromatic array of spatially resolved and kinematic observations, which can provide strong constraints on numerical models. They also require system specific models, which closely match all available observations, and thereby provide a clear determination of the interaction stage (see discussion in Struck 2004). Given the prolonged debate on whether the nearby M51 system is the result of one or two close encounters, this is not necessarily an easy task (see review of Struck 1999), though in either case it is clear that the system is not yet near the end of the merger road.

With a library of detailed case studies one could hope to graph SFR (or specific SFR per unit mass or gas mass) versus interaction stage to resolve the issues above. The "interaction stage" would require careful definition, however.

Detailed color and spectral synthesis modeling can in fact yield constraints, if not yet unique solutions, for the SF history of some nearby well-studied systems, e.g., the Magellanic Clouds (Zaritsky & Harris 2004, Javiel, Santiago & Kerber 2005), M51 (Bastian et al. 2005, Bianchi et al. 2005), M82 (De Grijs 2001, De Grijs, O'Connell & Gallagher 2002) and Arp 284 (Lançon et al. 2001). From such cases, one can add a few points to the hypothetical SFR-interaction stage plot.

4.2.4 Modes of star formation

Detailed case studies are also a primary tool for studying a number of specialized modes of SF, some of which have received a good deal of attention in recent years. These include: the formation of super star clusters (SSCs), SF in tidal bridges and tails, and SF in induced disk waves. Except perhaps in the last case, these modes do not usually dominate the SF in interacting systems, but they may involve physical processes unique to collisional environments, and produce especially interesting products like dwarf galaxies and halo globular clusters.

Tidal dwarf galaxy formation

We have already mentioned early studies of SF in tidal tails in Sect. 4.1, but there has been a great deal of recent work. Work in this area has been energized by the possibility that, not only massive star clusters, but also actual dwarf galaxies might be formed out of material in tidal tails (Fig. 4.5). If so, this could be a means of forming dwarf galaxies at the present time, and in observable environments. In the introduction to a recent paper Duc, Bournaud & Masset (2004) review much of the literature of the last decade, and additionally a section of the proceedings of a recent IAU symposium is also dedicated to the topic (Duc, Braine & Brinks 2004). These two sources provide good entry points to the literature.

We should begin by noting that the tidal dwarf formation has been controversial, and difficult to prove (or disprove). Most tailed galaxies do not have an obvious luminous SF region at the end of their tails. To date, only a few examples of dwarfs forming in tails have been studied in detail, so it is not clear how rare is that circumstance, nor what is the general nature of SF in tails. In fact, there are a number of difficulties in finding these objects, and confirming that they are dwarf galaxies in formation. Sometimes the tail is viewed edge-on, and if it is curved in the vicinity of the candidate dwarf a good deal of material that is not physically connected can be superposed along the line-of-sight, including multiple SF regions (e.g., Duc et al. 2000). This can lead to large overestimates of the mass and extent of the tidal dwarf candidate, leading, in turn, to a bias for such systems in the candidate list. Determining whether a tidal dwarf candidate is truly a gravitationally bound object, and will persist as a distinct entity is also challenging, if only because of the resolution limits of observations of HI and molecular gas in these small objects.

Discussions of this latter question have been entwined with those concerning two early theories for the formation of tidal dwarfs. Barnes & Hernquist (1992) suggested that they could form as a result of gravitational instabilities in tails consisting of collisionless stars, while Elmegreen et al. (1991) argued for the

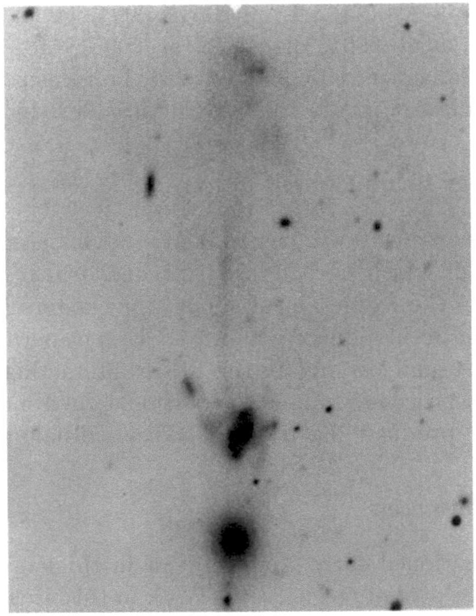

Fig. 4.5. Arp atlas image (Arp 1966) of the Arp 105 system. This system contains a probable tidal dwarf galaxy at the end of the long tidal tail in the north. See Duc et al. (1997) for details.

dominance of gas dynamical processes in regions of enhanced turbulence (i.e., enhanced velocity dispersion). One of the difficulties in the modeling is that some density concentrations may not be persistent, and the models are not generally able to follow their evolution for very long times, or with sufficient particle resolution (but see the high resolution model of Hibbard & Barnes 2004). Another problem in confronting these theories to observation is that since gas disks are more extensive than stellar disks, all tidal dwarf candidates are likely to contain a large fraction of gas, so it is not possible to find a case of assembly by gravitational means alone. Based on new simulations, Duc, Bournard & Masset (2004, and references therein) argue that only if the parent galaxy has an extensive dark halo is it likely that large amounts of gas will accumulate at the end of a tidal tail, and that this is the most efficient route to forming true tidal dwarfs with masses in excess of $10^9 M_\odot$. These authors also find that the gas accumulation process is primarily kinematic, with self-gravity playing only a minor role. It will be interesting to see how these new ideas develop in the next few years.

Massive and super star cluster formation

One of the greatest contributions of the Hubble Space Telescope to extragalactic astronomy was to resolve individual star clusters in relatively nearby galaxies and allow us to take the census of the cluster populations in them. As a result of such studies it has become clear that a large fraction of new stars in colliding

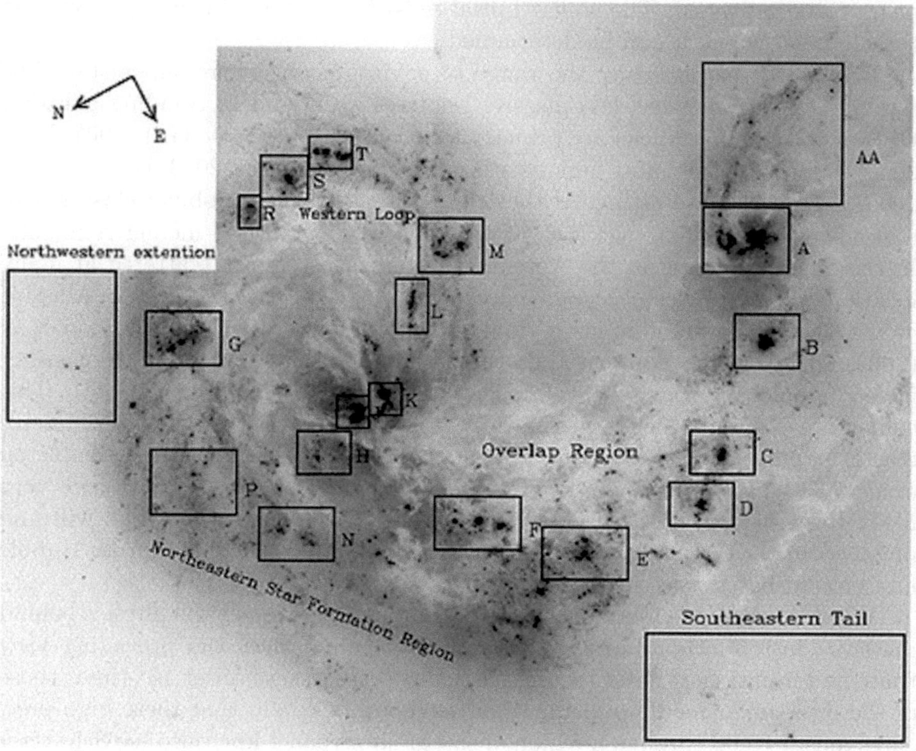

Fig. 4.6. HST image of young star clusters in the merging Antennae galaxies, from Whitmore et al. (1999, courtesy of B. Whitmore).

galaxies are formed in clusters (see Fig. 4.6). This is difficult to quantify, but has been estimated at 50–100%. The characteristics of the most massive of these clusters, the super star clusters with estimated masses in the range 10^5–$10^8 M_\odot$, are just what we would expect from young clusters, so it appears that we are now able to study the formation and development of globular clusters at a variety of stages by direct observation. These studies have given rise to a considerable literature, which extends far beyond the topic of this review, so we only describe a few of the relevant highlights.

In a summary of a recent conference on this topic, O'Connell (2004) emphasized the universality of the properties of young cluster populations, despite a huge range of formation environments and scales. These properties include a nearly universal power law mass function, which evolves naturally with time to the exponential function of old globulars. The number of clusters and the maximum cluster luminosity in a star-forming region both scale with total SFR. Most cluster populations have a very small range of formation ages. This is especially true of populations in galactic nuclei, but in colliding galaxies with widely spread SF regions there can be distinct populations, each with small age spreads (e.g., Alonso-Herrero et al. 2002). The spatial structure of super star clusters also appears to be universal. The stellar initial mass function is universal, at least at the high end where it can be determined.

It is worth emphasizing the range of environments where super star clusters and their somewhat less massive relatives are found in colliding galaxies. Of course, starburst nuclei are primary locations and M82 (De Grijs 2001, Melo 2005) is probably the most famous example. M51 (Bik et al. 2003, Bastian et al. 2005) is also very interesting. At the other extreme we have globular cluster populations of intermediate age (i.e., of order a few Gyr) around merger remnants. In the ongoing merger in the Antennae system, clusters are scattered at many locations in the bodies of the galaxies (Whitmore & Schweizer, 1995). Massive young clusters are found in many tidal tails, though interestingly Knierman et al. (2003) make a suggestion, based on their study of 6 tail regions, that they either have a population of massive young clusters or a tidal dwarf, but not both. This conjecture certainly merits further observational and modeling study. Massive cluster populations are also found in ring galaxies like the Cartwheel (Appleton et al., 1996) and ocular waves like IC 2163 (Elmegreen et al., 2000). It seems very likely that the mid-infrared detectors on the Spitzer Space Telescope will find massive cluster populations in more environments that are hidden from Hubble and ground-based telescopes.

What do these environments have in common and what's the physics behind massive cluster formation? O'Connell (2004) summarizes the prevailing view that the formation of SSCs requires high gas pressures, of order 10^4 times those of the interstellar medium in the solar neighborhood, and that these high pressures must extend through a region of size greater than 1 kpc (also see Schweizer 1998). Strong turbulence also pervades the formation region. O'Connell emphasize that the energetic environment inside a forming massive cluster must be truly extraordinary.

It is likely that all of the colliding galaxy environments noted above are able to achieve the high pressures and turbulence that the theory says are necessary to form the super star clusters. This is not entirely clear in the case of disk waves and tidal tails. However, in the former case the process may be aided by feedback effects from the first stars to form. In the case of tails it may simply be that some achieve the requisite conditions and form massive clusters, and others do not. We have much to learn yet about these processes.

Finally, O'Connell notes a couple of examples of nuclear starbursts where the super star clusters are much more massive than the other clusters, and so, the mass function is discontinuous. He speculates there may be a special formation mode for these cases, though the nature of that mode is not clear. As in the case of tidal dwarf formation there are competing mechanisms, and one of these may dominate only in the exceptional cases. These mechanisms again include the formation of massive progenitor clouds triggered directly via gravitational instability, or indirectly in dense environments assembled by large scale gravitational instability. They may also include hydrodynamic effects like cloud crushing that occurs when giant clouds experience an abrupt pressure increase after impacting large-scale shocks or other high pressure regions (Jog & Solomon 1992, Braine et al. 2004, Bekki et al. 2004). The combination of these processes could probably generate discontinuous cluster mass functions, but at present, this is only speculation.

A few more exotic ideas have also been discussed recently. Scannapieco, Weisheit & Harlow (2004) have suggested that strong winds from young galaxies could have shocked their dwarf companions, stripping gas and compressing it to form globular clusters. Burstein et al. (2004) also suggest that globulars might form in dwarf companions. Their argument is based on a hierarchical clustering model of galaxy formation as applied to the cluster populations in the Milky Way and M31. On the other hand, Hibbard, Vacca & Yun (2000) have found several examples of systems where winds seem to have swept the gas out of parts of tidal tails, without the production of massive star clusters.

ULXs

Ultraluminous X-ray sources (ULXs) are defined as having X-ray luminosities of order 10^{39}–10^{41} ergs per s^{-1}, which extends beyond the luminosities of the well-studied high mass X-ray binaries, but is still much less than a typical active galactic nucleus. X-ray sources of this luminosity have been detected in galactic nuclei for two decades, but the arcsecond resolution of the Chandra Observatory has facilitated their discovery and definition as a class of objects (see reviews of Mushotsky 2004a, van der Marel 2004, Ward 2003).

Estimates indicate that they may be a quite common constituent of the nuclei of normal galaxies. However, their nature remains somewhat mysterious. There are two leading theories. The first is that they are indeed an extension of the high mass ($\simeq 10 M_{\odot}$) binary phenomenon, but with highly super-Eddington accretion rates and beamed emission (e.g., Begelman 2002, King 2004, King & Dehnen 2005, Liu, Bregman & Irwin 2005). The second is that these are in fact black hole accretion systems of intermediate mass between stellar black holes and active nuclei, e.g., masses $\gg$ 100 $M_{\odot}$ (e.g., Hopman et al. 2004, Krolik 2004, Miller, Fabian & Miller 2004, Miller 2005). There are strong arguments for both models. For the most luminous ULXs the stellar mass explanation is strained. For the most rapidly variable ones the source size and mass is limited from above. Since the luminosity bounds of the class are ad hoc, it is certainly possible that the class includes both kinds of source. Gutiérrez & López (2005) also present a cautionary tale of how apparent ultraluminous sources can in fact be background sources.

A number of galaxies are now known to contain enough ULXs to allow the construction of luminosity functions, and these luminosity functions are found to scale with total SFR or a combination of SFR and galaxy mass (e.g., Swartz et al. 2004). Over the last 5 years ULX populations have been found in several colliding galaxies, including M82 (Fuiruti & Titarchuk 2004, Matsumoto et al. 2004, Portegies Zwart, Dewi & Maccarone 2004 and references therein), the Antennae, the Cartwheel ring galaxy, and the ring-tailed Arp 284 system (Fig. 4.7). The ULXs are widely spread across the Antennae (Fabbiano et al. 2004, Miller et al. 2004. They are found in the outer ring and an X-ray bridge of the Cartwheel (Gao et al. 2003, Wolter & Trinchieri 2004). In Arp 284, most are associated with

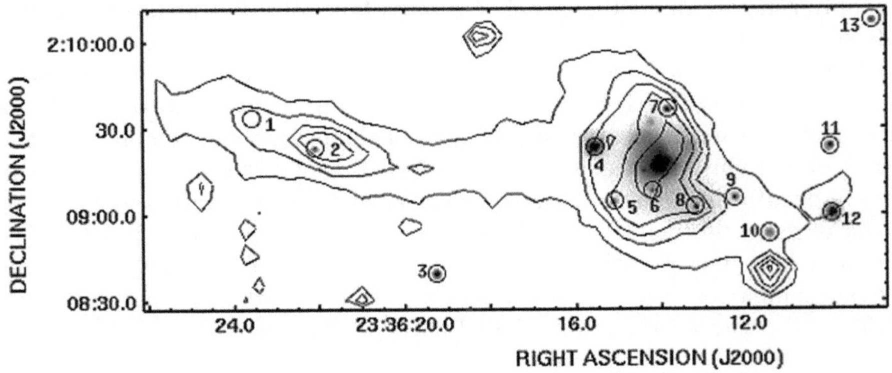

Fig. 4.7. Smoothed greyscale X-ray image of the Arp 284 system from Chandra Observatory data, with luminous point sources numbered. Contours showing the optical outline are derived from the Digital Sky Survey image. See Smith, Struck & Nowak (2005) for details.

tidal features, especially a prominent tail, but with one of the brightest known contained in the ring (Smith, Struck & Nowak 2005). Some of these ULXs are associated with star-forming regions, and so they provide another locator and probe of such regions that is not easily obscured.

The environment of these "tidal ULXs" can be less confused than that of many nuclear ULXs. A few are associated with super star clusters. Models can often provide a good deal of information about the dynamical history of specific tidal structures, and thus, about the formation environment of the ULXs. The discovery of any specific phenomenology in these environments could provide useful information on the nature of ULXs. However, at present the number of cases studied is too small to allow any firm conclusions beyond the association between collision morphologies and the occurence of ULXs.

4.2.5 Summary and theory

Clearly, there has been a vast amount of work on interaction induced SF in the last few decades, and we have only been able to skim through it above. Is all of this work leading to a comprehensive understanding of the phenomenon? Perhaps not quite yet, but we may be getting close to that goal. The question can be broken down into several separate questions. First of all, do we understand the general physical processes, and do we understand enough about how they work to account for the general observational systematics? Secondly, do we understand these processes well enough to reproduce their effects in numerical models, both comprehensive models of specific systems, and models of SF in particular dynamical processes?

Numerical models

We'll begin by considering recent numerical models and some aspects of the last question. Star and star cluster formation take place on scales that are orders of magnitude smaller than those typically resolved in simulations of galaxy collisions. However, separate models of the process on those small scales are beginning to advance our understanding greatly. Because of this, and the fact that much of the dynamics on the intermediate scales is essentially scale-free turbulence, we may be able to develop reasonably accurate SF formulations, without needing to resolve the scales on which it occurs. However, to date, relatively simple SF prescriptions have been used in galaxy collision models. Moreover, these prescriptions have been based on several different ideas about the dominant SF triggering process. Three of the most popular are: 1) a simple density-dependent SFR, 2) triggering by strong compressions in cloud-cloud collisions or large-scale shock waves, or 3) triggering by gravitational instability above a threshold density (or surface density or pressure).

With regard to the first of these, the Schmidt Law, in which the SFR is proportional to a low power of the gas density, surface density, or the gas density divided by a local dynamical time has been surprisingly resilient. Mihos, Bothun & Richstone (1993) used it (and isothermal particle hydrodynamics) to model collisions between two disk galaxies, and Mihos & Hernquist (1994) used it to model the Cartwheel ring galaxy. In both cases they found that the models gave about the "relative intensity and morphology of induced star formation." Later, Mihos & Hernquist (1996) found that with this formulation mergers between disk galaxies with bulges could produce burst SFRs a hundred times larger than those of isolated galaxies. They also explored how the presence of a bulge component affected the merger SFR. Given the simplicity of the prescription the results are impressive. However, Cox et al. (2005) have recently shown that the amount of SF in mergers may have been overestimated in earlier models, because this quantity depends on how conservation conditions are implemented in the SPH algorithm.

Phenomenological cloud collision models for gaseous dissipation in galaxies go back to the 1970s. The obvious disadvantage of such models is that interstellar gas clouds are transient, ever-changing structures, and not the coherent entities implicitly assumed when equating them with the 'sticky' (i.e., dissipative) particles of a numerical model. On the other hand, it is a straightforward way to model the cloud collisions and shock encounters which occur in many types of collision. In models of polar ring galaxies (Bekki, 1997), models of starbursts in multiple mergers (Bekki, 2001), and in other applications, Bekki has used a hybrid particle model. In his models there is dissipation from cloud collisions, but a probabilistic Schmidt Law is used to convert selected gas particles to stars. The local gas density is computed for each gas particle and used in the Schmidt Law.

Recently, Barnes (2004) has proposed a rather sophisticated phenomenological model, in which SF depends on the amount of energy dissipation in shocks. He argues that with this prescription he is able to produce a much better model of the Mice system (NGC 4676) than with a Schmidt Law. This is one of the few significant comparisons of different formulations in models of the same system. Barnes also notes that Schmidt Law models are quite insensitive to details of the interaction, while shock induced SF is very sensitive, and could be checked observationally.

Threshold instability models have been used frequently in the areas of galaxy formation and multiphase models of galaxy disks in recent years. This author has used such a model with feedback and gas with a continuous range of thermal phases in studies of direct collisions between two gas disks and their reformation (Struck, 1997). More recent work on disk collisions with many more particles has been carried out by Springer & Hernquist (2005). Cox et al. (2005) have recently presented an efficient effective equation of state approach to handling the thermal physics.

I have also used this type of SF formulation in detailed N-body hydrodynamic models of a couple of specific systems with extensive observational data (Struck & Smith 2003, Struck 2005). Both the spatial distribution of SF and the history of net SF fit the observational constraints, though the constraints on the SF history are not stringent. At low threshold densities this type of formulation is probably much like the Schmidt Law, since the SF will occur in regions with the most particles (i.e., high density). With a high threshold density, a violent process like shock compression and subsequent cooling will be needed in many cases to drive SF, more like the Barnes model.

In the end we see that many different numerical treatments can simulate induced SF reasonably well, and so none are immediately falsifiable. The answer to the question posed at the beginning of this discussion is yes, we can reproduce observations, but not because the models represent the underlying physical processes especially faithfully. The universality of those processes, and their highly interconnected properties, allow modelers to use simple formulations on large scales. Stringent tests of feedback prescriptions may eventually come by fitting the mass fraction and distribution of warm-to-hot phases in the interstellar gas. However, this will take much more realistic modeling of the thermal physics, and the stellar initial mass function, than is currently the norm.

Detailed observational studies on kiloparsec scales in various environments may provide insight into how sensitive SF is to compression and dynamical timescales. Spitzer Space Telescope observations in the mid-infrared have the ability to see through obscuring dust and provide a complete SF census on these scales in nearby galaxies, so the prospects are exciting in the next few years.

Squeezing out stars

All of the models described above form stars by compressing gas (albeit in more or less finely tuned ways). This recalls the Kennicutt (1989) observational result on the universality of the Schmidt Law over a range from isolated galaxies with modest SFRs to ULIRGs. Apparently, the first law of induced SF is – it's just the (large-scale) compression. More precisely, it appears that large-scale compression drives a turbulent cascade, which enhances star-forming compressions on the small scales (e.g., Krumholz & McKee 2005, and references therein). Because of the universality of the cascade, this process doesn't necessarily depend much on the details at the large and small scales.

In the case of ULIRGs the spectacular response is the result of spectacular angular momentum transport and compression in the major merger. For rapid or distant encounters the most that can be achieved are relatively small compressions in bars and waves. It is worth recalling that basic tidal forces stretch along the line of galaxy centers and compress in the perpendicular directions. For an approximately two-dimensional disk this means (very roughly!) stretching in one dimension and compressing in one dimension. Alternately, in terms of a simple impulsive torque, it means that angular momentum is added to one side of the disk (stars are pulled ahead in their orbits), and subtracted from the other side (stars are pulled back). Thus, net compression across the disk is roughly balanced by stretching or torque-induced rarefaction. In either case, the global effects are modest for small amplitude disturbances, implying little induced SF, as observed in such cases. On the other hand, in strong disturbances, the torque-induced decelerations of the gas orbital motion, and subsequent compression of a significant fraction of the gas, may be enough to induce a strong starburst, regardless of the fate of the rest of the gas.

LIRGs seem to be a heterogeneous class, but they include interacting galaxies that are separated by about 1–2 diameters of the larger. In such cases the tidal effects are nonlinear. In addition, the gravitational forces within each disk will be augmented by dark matter from the other galaxy's halo, which is coextensive with the disk. (In fact, the importance of this effect must be estimated quantitatively, but generally it will become important at the separations cited.) The resulting global compressions can account for the SF enhancement. It appears that such cases play an important part in creating the observed anti-correlation between SFR and the separation of the two galaxies.

In summary, it appears that the general systematics of induced SF can indeed be accounted for, to first order, as the direct result of compression. The consequence of this, that we can learn little more about SF physics from large scale studies, is disappointing. On the other hand, it means that colliding galaxy model results are not sensitive to many details of the SF/feedback formulation, and that there is little point in trying to extend numerical resolutions to very small scales. (However, we will eventually have much higher observational and modeling resolutions, which would allow the study of the cloud turbulent cascade, and the full effects of the non-equilibrium interaction environment. The point is that modest resolution improvements will not help much.)

We conclude with a brief mention of some possible exceptions or refinements to the "it's just compression" rule. The first might be found in the environment of core starbursts. If these are triggered by strong shocks, or with a sensitive threshold, then they may turn on rapidly. If, instead, they obey the Schmidt Law they will turn on more slowly if the central gas mass and density accumulate slowly. Generally, current spectral synthesis techniques are not able to provide SF histories that are accurate enough to distinguish. That is, except in a few nearby starburst cores, where the evidence seems to favor rapid turn-on of local density concentrations (because of the small age spreads within these concentrations, e.g., Harris et al. 2004). Similarly, the study of core burst turnoff might be enlightening. There seem to be regions in the core of M82 where the cloud system is disrupted, SF is turned off, but pressures and gas densities remain high (Mao et al. 2000 and references therein). The existence of such regions may necessitate at least a caveat in the Schmidt formulation.

We might be able to derive more information by studying SF within and behind density waves in disks. Spiral waves are ubiquitous, but ring waves produced in direct galaxy collisions are simpler. Asymmetric rings produced in slightly off-center collisions are the most interesting because they are still simple, but the wave amplitude varies continuously with azimuth. This is an excellent environment for confronting threshold versus continuous theories, at least if the compression in part, but not all, of the ring exceeds the threshold. To date, there is some evidence in support of thresholds, but not without complications ranging from incompletely known obscuration to unknown details of the collision parameters. The primary example is the Cartwheel ring with large variations of SFR and cluster populations around the ring, but with uncertainty about the details of the collision, and no old star component in the ring to provide independent information about wave amplitude as a function of azimuth (Appleton & Struck-Marcell 1996). With its relatively high resolution and ability to see through much obscuration, the Spitzer Space Telescope could resolve these ambiguities in carefully chosen systems.

There are many more examples of how to get beyond the simple compression law and the first-order theory of induced SF, and such work should become increasingly important in this field.

4.3 Environmental Effects

We have seen that the study of galaxy collisions is a relatively young, but rapidly maturing field. Thus, it is understandable that most progress to date has been in understanding the most spectacular collisions, ULIRG/major mergers, and the nature of some of the closest systems which can be studied in detail. Most of the latter occur in quite small groups like our own local group. However, studies of collisions in other environments date back to Spitzer & Baade (1951), and their number has been increasing recently.

4.3.1 Cluster bustle

At the opposite end of the spectrum from the local group environment is that of massive, dense clusters of galaxies. It will suffice for our purposes to briefly note the different processes in this environment relative to that of local groups. These differences include: high speed collisions, galaxy 'harassment,' ram pressure stripping, 'strangulation,' and induced slow collisions (see Fig. 4.8).

Spitzer & Baade (1951) first suggested that high velocity collisions in galaxy clusters might have little effect on the stellar components, but could blast away

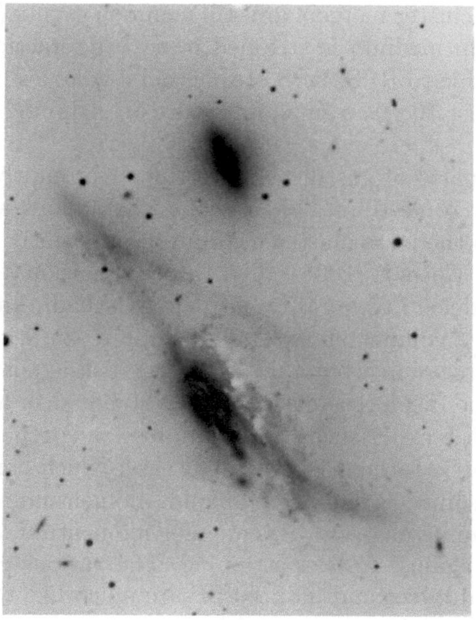

Fig. 4.8. Arp (1966) atlas image of Arp 120 (NGC 4438), a Virgo cluster system in collision, with a starburst galactic wind, and likely also experiencing environmental effects. See Boselli et al. (2005) and Kenney & Yale (2002) for details.

the overlapping parts of gas disks. This is because the typical random galaxy velocities in clusters of up to several thousand km/s, are not only highly supersonic for the intercluster gas, but are in excess of normal disk escape velocities. Generally, we expect a moderation of tidal gravitational effects, but in some cases a drastic increase in hydrodynamic effects.

High-speed collisions may have weak gravitational effects, but encounters are much more common in the cluster environment. Thus, in the aggregate, tidal effects are not negligible in clusters. The cumulative effect of many weak (high-speed or distant) galaxy-galaxy interactions in clusters, as well as perturbations from the cluster potential, and possibly from intermediate scale sub-structure is called 'harassment' (see Moore et al. 1996, Moore et al. 1999). In recent high resolution numerical studies of the growth of moderate clusters, Gnedin (2003a, b) has shown how this process can secularly erode galaxy halos, thicken moderate mass stellar disks and truncate SF, and destroy small disks.

Ram pressure stripping (RPS) by the intra-cluster medium can have somewhat similar effects on disks. RPS is an interesting subject, with a number of recent developments, and worthy of a review of its own. Thus, it is beyond the scope of this review, except for a few comments. First of all, long time residents of dense galaxy clusters were probably stripped long ago, so RPS is most relevant to gas-rich galaxies falling into the intra-cluster medium for the first time. X-ray satellites have provided much evidence that the infall of galaxy groups into clusters and cluster-cluster mergers are still common events (Mushotsky 2004b). Once the intra-group medium is stripped in such interactions, individual disk galaxies are vulnerable to RPS. RPS of spheroidal galaxies has been studied for 30 yrs., but in the last 10 yrs. a small literature on stripping of disk galaxies has blossomed.

The outermost parts of gas disks are stripped promptly, and slower interactions continue for some time. Slow viscous interactions at the edge of disks moving face-on into the intra-cluster medium have recently been studied in detail by Roediger & Hensler (2005). The three-dimensional dynamics in tilted cases have been modeled in several recent papers (Abadi, Moore & Bower 1999, Vollmer et al. 2000, Vollmer et al. 2001, Schulz & Struck 2001). Schulz and Struck, in particular, pointed out that if the gas disk is not promptly stripped, it can nonetheless be displaced relative to the stellar disk and the halo center. The displaced gas disk experiences tidal compression (perpendicular to the disk plane) and asymmetric torques in the tilted case, which generate spiral waves. The waves transfer some gas and much angular momentum outwards, where it is stripped. The remaining gas, with less angular momentum, compresses radially, which "anneals" it against further stripping. The various compressions should induce SF. The tidal forces and induced SF are much like those in galaxy collisions. Thus, after the stripped material is gone, it could be difficult to discern whether excess SF is the result of RPS, a minor merger, or harassment. In fact, since these processes could work simultaneously, the question may be academic.

Vollmer and Schulz & Struck emphasized another aspect of stripping: some of the removed material can later fall back onto the disk. This can occur either when the galaxy moves into regions of lower intra-cluster medium densities, where the levitating pressure is reduced, or when gas clouds move into the disk 'shadow' where the pressure is also reduced. In either case we would expect effects akin to those of mass transfer in galaxy collisions.

Strangulation is a weaker cousin of RPS. It is the process of removing the potential feedstock of disk SF, gas in the galactic halo, usually via RPS (Larson, Tinsley & Caldwell, 1980). The feedstock could include gas blown out of the disk by supernovae or stellar or galactic winds, it could include gas tidally removed from companions, or primordial gas falling into the halo for the first time. This process is likely to be most important in the young universe, when there is still much gas outside of galaxy disks, but it also hampers gas recycling from dying populations in cluster galaxies.

4.3.2 Cluster slow dance

A final process that may be very important in cluster environments is induced slow encounters in infalling groups. Recent studies of the Butchler-Oemler effect (see review of Pimbblet 2003), which is an excess of blue galaxies in clusters at redshifts of less than 1, provide evidence that many of the blue galaxies are mergers or interacting (also see Zabludoff et al. 1996, Hashimoto & Oemler 2000, Ellingson et al. 2001, Goto 2005). It seems unlikely that high speed interactions could be responsible for this effect. Mihos (2004) has emphasized that large-scale cluster formation models show that slow interactions continue to occur even in large clusters, and are quite common during cluster formation. He also notes that slow encounters can occur in groups or small clusters with modest velocity dispersions falling into large clusters (also see Poggianti et al. 2004).

This is an interesting phenomenon that has not been much investigated. Mihos notes that a number of different processes may be involved and it may be impossible to disentangle them. For example, tidal forces from the cluster potential could perturb orbits in the infalling group inducing interactions, and intracluster medium annealing could induce increased SFRs. Personally, I suspect these are secondary processes.

At least for galaxies that fall through the cluster core the primary process may well be gravitational shocking, which depends directly on the cluster potential rather than indirectly or on the derivative of the potential (tidal forces). Consider the relatively simple example of a galaxy group containing about 30-100 galaxies, falling into a large dense cluster. Cold dark matter structure formation models predict a common density profile across the range of structures from galaxy halos to the dark halos of large clusters, and observational tracers (e.g., intracluster starlight, see Feldmeier et al. 2002 and references therein) show good agreement with profile functions derived from these simulations, like the popular NFW profile (Navarro, Frenk & White, 1997). Then it is reasonable to assume our example group and cluster have similar density profiles, though not necessarily with the central cusp of the NFW profile.

The observations also suggest that the central density decreases slowly with mass in dark matter halos. We might, for example, model our large cluster after Abell 2029, whose mass profile was studied in detail by Lewis, Buote & Stocke (2003). They find a mass of about $9.2 \times 10^{13}/h_{70}\ M_{\odot}$ contained within a radius of $260/h_{70}$ kpc, yielding a central density of 0.0052 $\mathrm{M}_{\odot}/\mathrm{pc}^3$. As an example of a group, on the other hand, we can take a 'poor' group like those studied by Zabludoff & Mulchaey (1998). These groups have virial masses of about $7 \times 10^{13}/h_{70}\ \mathrm{M}_{\odot}$ within radii of about $300/h_{70}$ Mpc. The authors estimate that 80–90% of the virial mass is in the group halo, so the mean group halo density is about 0.0022 $\mathrm{M}_{\odot}/\mathrm{pc}^3$. These numbers are meant to be representative, not precise. Group parameters, e.g., in compact versus poor groups, could easily range over factors of a few. Nonetheless, the message is clear that passing through the cluster core would substantially increase the instantaneous group halo mass.

Moreover, the time to pass through the core is of the same order as the group dynamical time. We can estimate the former by dividing the core radius above by a typical cluster velocity of about 2000 km/s; the result is about 300 Myr. The free fall time at the edge of our example group is about 200 Myr. Therefore there is time for group galaxies to be pulled into a much denser and compact configuration. For roughly comparable group and cluster halo core densities, galaxies could be pulled in to roughly half their previous distance from the core, increasing the galaxy density by nearly an order of magnitude and their collisions by about a factor of 100 (density squared).

Like stars in clusters that pass through the galactic disk, when the group leaves the cluster core, and gravity is reduced, the galaxies will fly outward. (This is also like collisional ring galaxies.) However, collisions between galaxy halos are stickier than those between cluster stars. During the compression period, along near parallel but converging trajectories on the way down, encounters are likely, and there is time for dynamical friction to dissipate relative orbital energy. The result would frequently be a fairly slow interaction and eventual merger.

Considering all the various ways that clusters accelerate galaxy evolution, one can view life for average galaxies in small local groups as more or less a holding pattern, or at least a matter of slow maturation. Evolution doesn't begin in earnest until they fall into a larger group or cluster; the classic tale of youth leaving the farm for the big city.

4.4 Interactions and Galaxy Evolution

Galaxy collisions drive galaxy evolution, but how much compared to other processes like the passive conversion of gas to stars in isolated galaxy disks or compared to other dynamical processes like ram pressure stripping? Also, how does the role of collisions change with cosmological epoch?

To begin, we note that it has become quite clear that at least some massive galaxies and massive disk galaxies, in particular, formed very early, and had already attained a respectable age by redshifts of 1-2 (see review of Spinrad 2004). From a practical point of view, this means that observations must push to very high redshifts to see big evolutionary changes. We will come back to what has been seen in a moment. This fact has also been taken as evidence that at least some galaxies formed in a rapid monolithic collapse, rather than building up steadily in a prolonged sequence of mergers.

4.4.1 Models of structure buildup

What do theory and cosmological structure formation simulations lead us to expect? Currently, hierarchical build-up, ΛCDM models (i.e., models with cold dark matter plus "Λ" dark energy) in the "concordance cosmology" are the dominant paradigm. This picture suggests the occurrence of many mergers of small building blocks in the earliest stages, and continuing mergers thereafter. Moreover, recent analysis shows that it is possible to form some massive galaxies, including disk galaxies, at early times in these models (Nagamine et al., 2005).

Thus, part of the solution to the paradox of early massive galaxy formation is that fully nonlinear ΛCDM models do not yield exactly the same results as simple, analytic hierarchical structure formation models. Early massive disk galaxies may be a roughly 2σ outcome of the simulations, but that may be sufficient to account for the observations.

From another point of view, the early formation of massive disks allows for the possibility that major mergers form elliptical galaxies at early times, i.e., accounts for ellipticals containing only old stellar populations within the merger theory (see discussion in Schweizer 2005).

To return to the ΛCDM paradigm, another thing the simulations show us is that when substantial entities merge, not all their substructure is erased. In fact, the absence of hundreds of dwarf satellites around the Milky Way has been cited as a problem for this kind of model (e.g., Klypin et al. 1999). However, RPS in the hot halo, tidal disruption, and collisions with the galactic disk may have destroyed many of the leftover building blocks. Indeed, digesting the leftovers may be an important secondary evolutionary process, operating alongside the primary hierarchical merging process.

Thus, the picture of sequential buildup of galaxies via successive major mergers in the simplest hierarchical models is not the only one in which collisions and mergers are crucial. In more realistic models minor mergers and the accretion of numerous small companions play important roles. Such lesser collision events are probably very common in groups and clusters. As in solar system formation, "core accretion" may be as important as monolithic collapse and hierarchical buildup.

4.4.2 Observations of evolution

Let us return to observation. As in the case of induced SF discussed above, there are two approaches – study of the statistical properties of large samples, or study of individual objects in detail. In the last decade there have been a great many surveys to provide data for the first type of analysis (see the overview of Irion 2004). These include Hubble Space Telescope projects like the Medium Deep Survey, the Hubble Deep Fields North and South, and most recently GOODS (the Great Observatories Origins Deep Survey), carried out in collaboration with the Chandra Observatory and the Spitzer Space Telescope (see special issue of the Astrophysical Journal Letters, Jan, 10, 2004), and GEMS (Galaxy Evolution from Morphology and Spectral energy distributions, e.g., Bell et al. 2005).

The science and observing techniques of deep field survey and high redshift studies in general are well beyond the scope of this review. This is also not the place to consider the many different classes of high redshift galaxies in any detail. These topics have become the subject of wide interest and a burgeoning literature. However, specific products of these studies, such as the merger rates, the cosmic star formation rate and mean morphological statistics as a function of redshift, can provide information on the history of galaxy collisions.

Merger rate versus redshift

The differing predictions of the different models of galaxy formation, and the interest in the role of major mergers/ULIRGs, have motivated a continuing interest in the merger rate as a function of redshift. For example, the CNOC cluster galaxy redshift project has reported quite modest evolution in the merger rate at redshifts less than 1 (see Patton et al. 2002 and references therein). Specifically, Patton et al. examined a sample of 4184 galaxies, found 88 galaxies in close pairs, and derived a merger rate of $(1+z)^{2.3\pm0.7}$.

This result of low (and not rapidly changing) merger rate is confirmed by several other recent studies, including the Caltech Faint Galaxy Redshift Survey (Carlberg et al., 2000). Moreover, Lin et al. (2004) report initial results of the DEEP2 survey, in which they find a merger rate of $(1+z)^m$ with exponent of $m = 0.51 \pm 0.28$ assuming mild luminosity evolution, or $m = 1.6 \pm 0.29$ assuming no luminosity evolution, since z = 1.2. They note that this implies only 9% of L_* galaxies have undergone major mergers over this redshift interval. Using deep infrared observations from the Subaru telescope Bundy et al. (2004) found that the fraction of close pairs (which usually define the merger rate in these studies) increases "modestly" to only about $7 \pm 6\%$ at z $\simeq$ 1. This is less than that found by typical optical studies, and they note that the optical studies may be "inflated" by unrepresentative "bright star-forming regions."

Going in the other direction, Lavery et al. (2004) find a very rapid increase in the number of colliding ring galaxies with redshift. Head-on ring galaxy collisions generally result in merger, so if the rings represent a small randomly chosen fraction of all mergers, this would imply a very rapidly evolving merger rate. On the other hand, if the number of ring galaxies increases much more rapidly with redshift than other types of merger, one wonders why?

For reference, we note that Xu, Sun & He (2004) recently used data from the 2MASS near infrared survey to estimate the local merger rate; they found the fraction of close major merger pairs to be $1.70 \pm 0.32\%$. For completeness, we note that at the time of writing, mergers rates based on the Sloan Digital Sky Survey or the 2dF survey have not been published, though an initial atlas of SDSS merger pairs has (Allam et al., 2004). In the coming years it will be very interesting to see statistically significant estimates of the merger rate extended to well beyond z = 1.

Cosmic star formation

Estimates of the mean SFR as a function of redshift are usually based on color or emission line indicators (as opposed to the morphology used to estimate merger rates in the pair surveys). In recent years there have been surveys in a variety of wavebands. Cram (1998) carried out a novel radio continuum survey and found a local SFR of about twice the optical Hα value – 0.025 $M_\odot$ yr^{-1} Mpc^{-3}. He found a value about 12 times greater at z $\simeq$ 1. An analysis based on the 2dF survey also found an strong increase ($\propto (1+z)^b$, with $b < 5$) back to z $\simeq$ 1, and a more moderate increase at redshifts of 1–5 (Baldry et al., 2002). Analyses based on SDSS data come to similar conclusions (Glazebrook et al., 2003; Brinchman et al., 2004). The HST STIS Parallel Survey found an SFR at z $\simeq$ 1 of 0.043 ± 0.014 $M_\odot$ yr^{-1} Mpc^{-3}, based on [OII] emission (Teplitz et al., 2003), which is lower than most of the previous results.

The CADIS survey found SFR decreased by about a factor of 20 between redshift 1.2 and the present, and the authors note the agreement of their extinction corrected results with far infrared results (Hippelein et al. 2003, also see the Herschel Telescope survey of Glazebrook et al. 2004). Results from the Gemini Deep Deep Survey indicate that the SFR was about 6 times higher at z = 2 than at present (Juneau et al., 2005). One of the most dramatic changes in SFR was the factor of 30 found in a GALEX (Galaxy Evolution Explorer satellite) ultraviolet survey between $z = 1$ and the present (Schiminovich et al., 2005). Ultraviolet luminous galaxies may not very representative of the cosmic SFR, but they could be related to colliding galaxies.

The newest and deepest surveys indicate that SFR declines from a peak at moderate redshifts to lower values at the highest redshifts. Bundy et al. (2004) identify and study 54 galaxies in the Hubble Ultra Deep field and conclude that the SFR at $z = 6$ is about 6 times less than at $z \simeq 3$. Heavens et al. (2004) come to similar conclusions about the general history of SF on the basis of an analysis of SDSS and other surveys. Juneau et al. (2005) describe the situation as a cosmic starburst at $z \simeq 2$.

Very recently, survey results have revealed the phenomenon of "downsizing," wherein the most massive galaxies form first, and most of the SF takes place in progressively smaller galaxies as time goes on (e.g., Poggianti et al. 2004, Bouche & Lowenthal 2005, Juneau et al. 2005, Le Borgne et al. 2005, Shapley et al. 2005). Bundy, Ellis & Conselice (2005) argue (based in part on GOODS data) that downsizing also proceeds from early to late Hubble types, and that merging plays a key role. The implication is that there is a mass dependence in the merger rate at any given epoch.

These new cosmic SFR results provide very interesting inputs to the story of galaxy evolution. However, the relation of these results to interactions and mergers remains to be clarified. Actually, the same is true of the merger rate results, which are not sensitive to many minor mergers or other interaction phenomena.

Morphology versus redshift

For an outsider the phenomenology of high redshift galaxies, which is much constrained by detection techniques, is a daunting jungle of jargon. Moreover, the relation between increasingly elaborate simulations of structure formation and the observations is complex. With rapid advances on both fronts, and increased efforts in analysis and synthesis, we can expect much more clarity in the coming decade (see review of Spinrad 2004 for a lucid current discussion). For the present we focus on a few simple questions. Are we directly observing galaxy evolution, i.e., the changing appearance (build-up) of galaxies with redshift? Are collisions and mergers an important part of this evolution?

There is much new evidence in favor of an affirmative answer to both questions (see commentary of Conselice 2004, with further details on GOODS data in Conselice et al. 2004). To say it a bit more emphatically, these papers and those referenced within them suggest that we may be beginning to acquire the observations that directly show the buildup of typical Hubble sequence galaxies (see Fig. 4.9).

There is much information to be found in the literature on the properties of individual high redshift objects (individual galaxies and clusters). We cannot review this literature here, and refer the reader to the review of Spinrad (2004).

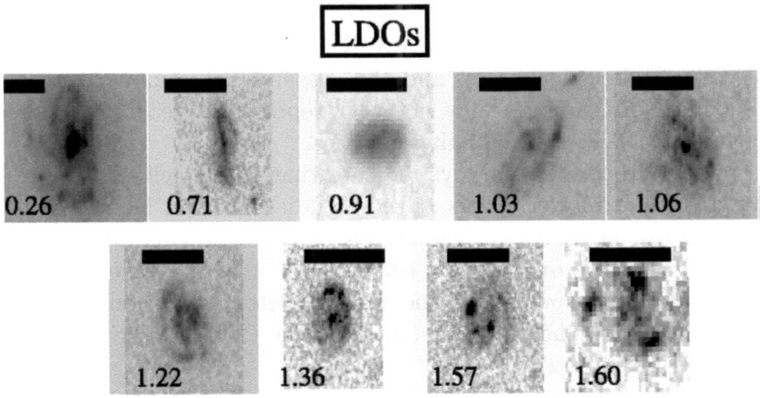

Fig. 4.9. Conselice et al. (2004) sequence of "low density objects" at varying redshifts, illustrating the development of Hubble type galaxies. (Courtesy C. Conselice)

Instead, let us return to the subject of massive, or at least luminous, galaxies at high redshift, and in particular, the interesting classes of extremely red galaxies and submillimeter galaxies (or SCUBA galaxies, after the detector on the James C. Maxwell Telescope). The latter are very infrared luminous, high redshift objects (e.g., Conselice, Chapman & Windhorst 2003, Genzel et al. 2004, Swinbank et al. 2004, Pope et al. 2005). Until recently, only a few were known, but recent deep searches are beginning to detect a substantial number (Greve et al. 2004, Smail et al. 2004, Wang, Cowie & Barger 2004). It appears that most of these objects are either dust obscured quasars or high redshift LIRGs or ULIRGs, with perhaps about 2/3 being the latter (Conselice, Chapman & Windhorst 2003, Neri et al. 2003, Smail et al. 2004, Swinbank et al. 2004). As with their local counterparts, the LIRGs and ULIRGs are generally mergers or mergers-in-progress (see Georgakakis et al. 2005).

It appears that the submillimeter LIRGs are much more common than present day LIRGS, and that they generate a substantial fraction of the IR background (e.g., Genzel et al. 2004, Wang, Cowie & Barger 2004). They have typical redshifts of 2–3, and thus, coincide with the peak of the cosmic SFR. They may be well represented among the most luminous galaxies in that peak, but the evidence is preliminary. These results are beginning to provide direct evidence that major mergers, if not hierarchical buildup, were major contributors to galaxy evolution and the cosmic star formation rate at these redshifts.

The submillimeter galaxies may be related to another high redshift class, the Lyman Break Galaxies (Shu, Mao & Mo, 2001). However, few of the latter are detected as the former (Chapman et al. 2000, but note the outstanding exception Westphal-MMD 11 discussed in Chapman et al. 2002). The deep ISO (Infrared Space Observatory) ELAIS survey also found a number of ULIRGs at $z < 1$ (Rowan-Robinson et al., 2004). These objects may bridge the gap between local ULIRGs and SCUBA galaxies. Recent observations with the Spitzer Space Telescope show that SCUBA galaxies are generally detectable at 24 microns, but with a wide range of mid-infrared colors (Frayer et al., 2004). Spitzer observations also promise to delineate active nuclei from starburst powered submillimeter galaxies (Ivison et al., 2004). All of this work should contribute substantially to our understanding of the "ULIRG rate" as a function of redshift (see the review of the cosmic evolution of luminous infrared galaxies by Sanders 2004).

Before submillimeter galaxies were discovered, observers were already very interested in "extremely red objects" at high redshift. Naively, one might expect to find more and more blue galaxies at high redshift, and as described above, this is generally the case. In this context, finding very red galaxies is surprising. On the other hand, with a knowledge of dust-enshrouded starbursts in ULIRGs, maybe this is not so surprising, but are EROs ULIRGs? Recent studies suggest not, but rather many of them may be the already (at typical redshifts of > 1–2) old, red progenitors of present-day early type galaxies (e.g., Franx et al. 2003, Förster Schreiber et al. 2004, Yan & Thompson 2003, Yan et al. 2004, Yan, Thompson & Soifer 2004, Bell et al. 2005).

Redness and age are relative terms. The typical age of the stellar populations in these galaxies may be about 1 Gyr, which locally would not be described as an old, red population. However, at the high redshifts where these galaxies are found, the age of the universe when the light was emitted was only a few Gyr or less.

Nonetheless, a fraction of the extremely red objects may be ULIRGs. Thompson (2003) in an HST study of the morphology of a sample at redshifts of about 1–2, estimate that about $17 \pm 4\%$ of the objects are mergers or interactions. However, for the majority dominated by older stellar populations we will have to seek merging and interacting progenitors at still higher redshifts.

In conclusion, the above paragraphs describe the great advances that have been made in recent years in studies of galaxy evolution at high redshift. This work is impressive, but it is still hampered by resolution, sensitivity and statistical limitations. There are hints that mergers and interactions are important at all stages, but there is a great deal more work to do before we understand the details.

4.5 Archaeology

As discussed in several places above, models of particular classes of collision have become quite sophisticated at this stage in the development of the field. However, in the case of specific systems this generalization is true only for systems that have experienced only one close encounter, or where the time between encounters is so long that the signatures of the first encounter have been largely erased. The one notable exception to this caveat is the M51 system, which may well be the result of two close encounters (see discussion in Struck 1999).

There are wave and tidal morphologies characteristic of cases with two close passes separated by a time interval of order the mean internal dynamical (e.g., rotational) time in the primary disk. Stuck-Marcell & Lotan (1990) demonstrated this explicitly in the case of colliding ring galaxies, and it is quite clear in a number of merger models. It also seems likely that a number of objects in the colliding galaxy atlases require two close encounters to explain their morphology (e.g., the M51 types).

Another kind of double encounter that has so far received only exploratory attention is the case when a galaxy falls into a group and has close encounters with more than one group member.

The study, and ultimately the classification, of double encounter morphologies and merger remnant systematics is one that could advance a long way in the next decade. There are no insurmountable technical difficulties preventing advancement, though a great deal of numerical effort will be required. Not only would a large number of simulations have to be run, but they would have to be fully self-consistent models. The ability to decipher two stages of development in colliding systems would represent a substantial advance in galaxy archaeology.

4.6 Coming Attractions

For the patient reader, it should be clear from the above that this field has had a very exciting first fifty years or so. It must be admitted that this is in large measure due to outside influences. Like all other parts of astronomy, the study of galaxy collisions has ridden the breaking waves of the vast technological advances in detectors, satellite engineering, and computational resources. The subject has received further boosts from the enormous interest in parallel areas of study within the fields of galaxy evolution and star formation, and has contributed back to those areas. Never again will so many new, information-rich wavebands be opened. On the other hand, wide scientific frontiers remain to be explored with the aid of continuing increases in observational sensitivity, resolution, computational power, and synergistic interactions with allied fields.

In the last few sections I have attempted to clarify where we stand on the key questions posed at the end of Sect. 4.1. The first group of questions concerned the role of galaxy collisions within the overall picture of galaxy formation and evolution. Toomre's work in the 1970s held out the possibility that collisions and interactions were a dominant process, and that possibility has energized the field for most of the time since. However, there have always been counter-arguments. One of the strongest in the present era is the modest increase in the merger rate with redshift found in deep surveys. On the other hand, the relations between cosmic SFR or ULIRG numbers and redshift tell a different story. Downsizing may be an important part of the resolution between the different stories. Presently, we only see a hazy outline of the full portrait of the relation between galaxy morphology and redshift. Progress has been rapid in these areas, and we can expect a great deal more in the next decade or two. For optimistic theorists the answers are already available (if not yet fully extracted) from large-scale numerical models of structure formation.

The second group of key questions concerned the role of environment on collision dynamics and evolutionary processes. Although the study of galaxy collisions in groups and clusters has been around since Spitzer and Baade's work, it is being reborn in the present era. There is currently a great deal of interest in groups and clusters among observers, with new tools to facilitate that work. The theory and modeling side of this area is more complicated than that of binary collisions and mergers because of the interaction between several strong dynamical processes (e.g., ram pressure stripping, group/cluster direct or tidal effects, etc.). Nonetheless, it is also reasonable to expect significant advances in this area on decadal timescales.

The third group of questions concerned the orchestration of SF and nuclear activity by large-scale interaction dynamics. In the recent past it seemed likely that progress in this area would be hindered by the interplay of a number of complex dynamical processes. It now appears that this view was overly pessimistic. Due to the universal properties of turbulent interstellar gas, it now seems that wherever you compress cool gas you will enhance SF (in quantifiable ways). Thus, large-scale orchestration is mostly about gathering and compressing gas; feedback effects are mostly about heating and dispersing gas. There are more complexities than this, but the big picture does not appear impossibly complicated. Work in the coming decades should provide a much firmer foundation for this scenario, and a much better understanding of the exceptional cases.

So the reviewer's crystal ball conveys a bright outlook for answers to the first three groups of key questions. The glass gets more murky when we ask the last couple of questions. The fourth group of questions concerned secular effects and the fifth was about the archaelogy of individual systems. Of course we can model long-term processes on the computer with ever more precision. However, as discussed in several contexts above, it is hard to compare to observation either statistically or in individual cases.

That said, I would expect more progress from statistical studies, even though that will require the relatively slow accumulation of good datasets on numerous systems, for example acquiring large libraries of faint tidal structures in numerous galaxies. That slow work is not likely to be taken up by professional astronomers, but with the increasing availability of moderate sized telescopes and sensitive CCD detectors, it could become the realm of serious amateurs or robot astronomers.

It seems possible that the majority of key questions discussed above will be resolved within the next 50 years. However, new phenomena will be discovered, and more detailed understandings will be demanded. Recent, and possible near-future, examples support the point. As an example, consider the exotic forms or products of SF, like the ULX sources, and the possibility that some of these X-ray sources are intermediate mass black holes formed in dense, young star clusters. It will take a lot more observational work to explicate this phenomenon, and probably new theoretical insights to explain it. We could use several more Chandra observatories!

There are also still a few wavebands that remain largely unexplored, including low-frequency radio waves, very high-energy gamma rays (new more sensitive Cherenkov telescope arrays are presently coming on line), and gravitational waves of many frequencies. Equally exciting is the prospect that within a few decades our understanding of galaxy disk hydrodynamics may advance to point that we understand both the small scale, relatively short time, weather changes occurring in isolated disks, and the longer term climate changes wrought by various types of collisions and interactions. However, there is a great deal of work to be done before that goal is achieved.

4.7 Acknowledgements

I am very grateful to my research collaborators for teaching me much about colliding galaxies and related topics. I want to thank Bev Smith, in particular, for making a number of helpful suggestions on this manuscript. I'd also like to acknowledge support from a NASA Spitzer GO Cycle 1 grant.

References

Abadi, M. G., Moore, B., & Bower, R. G. (1999) Ram Pressure Stripping of Spiral Galaxies in Clusters, MNRAS, **308**, 947

Allam, S. S., et al. (2004) Merging Galaxies in the Sloan Digital Sky Survey Early Data Release, AJ, **127**, 1883

Alonso-Herrero, A., et al. (2002) Massive Star Formation in Luminous Infrared Galaxies: Giant HII Regions and Their Relation to Super-Star Clusters, AJ, **124**, 166

Appleton, P. N., & Struck-Marcell, C. (1987a) Star Formation Rates in Ring Galaxies from IRAS Observations, ApJ, **312**, 566

Appleton, P. N., & Struck-Marcell, C. (1987b) Models of Ring Galaxies: II. Extended Starbursts, ApJ, **323**, 480

Appleton, P. N., & Struck-Marcell, C. (1996) Collisional Ring Galaxies, Fun. Cos. Phys., **16**, 111

Appleton, P. N, et al. (1996) Mapping Stellar Evolution in the Wake of Density Waves in Ring Galaxies. In: *New Light on Galaxy Evolution, I.A.U. Symp. 171*, ed by R. Bender & R. L. Davies, (Kluwer, Dordrecht) p. 337

Arp, H. (1966) Atlas of Peculiar Galaxies, ApJS, **123**, 1

Arribas, S., et al. (2004) Optical Imaging of Very Luminous Infrared Galaxy Systems: Photometric Properties and Late Evolution, AJ, **127**, 2522

Athanassoula, E. (2004) Dynamical Evolution of Barred Galaxies, Amer. Astr. Soc. (DDA mtg.), **35**, #0305 (astroph 0501196)

Baade, W., & Minkowski, R. (1954) Identification of the Radio Sources in Cassiopeia A, Cygnus A, and Puppis A, ApJ, **119**, 206

Baldry, I. K., et al. (2002) The 2dF Galaxy Redshift Survey: Constraints on Cosmic Star Formation History from the Cosmic Spectrum, ApJ, **569**, 582

Barnes, J. E. (2004) Shock-induced Star Formation in a Model of the Mice, MNRAS, **350**, 798

Barnes, J. E., & Hernquist, L. (1992) Dynamics of Interacting Galaxies, ARAA, **30**, 705

Barnes, J. E., & Hernquist, L. (1996) Transformations of Galaxies. II. Gasdynamics in Merging Disk Galaxies, ApJ, **471**, 115

Barton, E. J., Geller, M. J., & Kenyon, S. J. (2000) Tidally Triggered Star Formation in Close Pairs of Galaxies, ApJ, **530**, 660

Barton Gillespie, E. J., Geller, M. J., & Kenyon, S. J. (2003) Tidally Triggered Star Formation in Close Pairs of Galaxies. II. Constraints on Burst Strengths and Ages, ApJ, **582**, 668

Bastian, N., et al. (2005) The Star Cluster Population of M51: II. Age Distribution and Relations Among the Derived Parameters, A&A, **431**, 905

Begelman, M. C. (2002) Super-Eddington Fluxes from Thin Accretion Disks?, ApJ, **568**, L97

Bekki, K. (1997) Formation of Polar Ring S0 Galaxies in Dissipative Galaxy Mergers, ApJ, **490**, L37

Bekki, K. (2001) Starbursts in Multiple Galaxy Mergers, ApJ, **546**, 189

Bekki, K., et al. (2004) Formation of Star Clusters in the Large Magellanic Cloud and Small Magellanic Cloud: I. Preliminary Results on Cluster Formation From Colliding Gas Clouds, ApJ, **602**, 730

Bell, E. F., et al. (2005) Towards an Understanding of the Rapid Decline of the Cosmic Star Formation Rate, ApJ, **625**, 23

Bergvall, N., Laurikainen, E., & Aalto, S. (2003) Galaxy Interactions - Poor Starburst Triggers . III. A Study of a Complete Sample of Interacting Galaxies, A&A, **405**, 31

Bianchi, L., et al. (2005) Recent Star Formation in Nearby Galaxies from GALEX Imaging: M101 and M51, ApJ, **619**, L71

Bik, A., et al. (2003) Clusters in the Inner Spiral Arms of M51: The Cluster IMF and the Formation History, A&A, **397**, 473

Boselli, A., et al. (2005) GALEX Ultraviolet Observations of the Interacting Galaxy NGC 4438 in the Virgo Cluster, ApJ, **623**, L13

Bouche, N., & Lowenthal, J. D. (2005) The Star Formation Rate-Density Relationship at Redshift Three, ApJL, **623**, L75

Braine, J., et al. (2004) Colliding Molecular Clouds in Head-on Galaxy Collisions, A&A, **418**, 419

Brinchman, J., et al. (2004) The Physical Properties of Star Forming Galaxies in the Low Redshift Universe, MNRAS, **351**, 1151

Bundy, K., Ellis, R. S., & Conselice, C. J. (2005) The Mass Assembly Histories of Galaxies of Various Morphologies in the GOODS Fields, ApJ, **625**, 621

Bundy, K., et al. (2004) A Slow Merger History of Field Galaxies since $z \simeq 1$, ApJ, **601**, L123

Bunker, A. J., et al. (2004) The Star Formation Rate of the Universe at $z \simeq 6$ from the Hubble Ultra Deep Field, MNRAS, **355**, 374

Burstein, D., et al. (2004) Globular Cluster and Galaxy Formation: M31, The Milky Way, and Implications for Globular Cluster Systems of Spiral Galaxies, ApJ, **614**, 158

Bushouse, H. A. (1987) Global Properties of Interacting Disk-type Galaxies, ApJ, **320**, 49

Carlberg, R., et al. (2000) Caltech Faint Galaxy Redshift Survey. XI. The Merger Rate to Redshift 1 from Kinematic Pairs, ApJ, **532**, L1

Chandrasekhar, S. (1943) Dynamical Friction. I. General Considerations: the Coefficient of Dynamical Friction, ApJ, **97**, 255

Chapman, S. C., et al. (2000) A Search for the Submillimetre Counterparts to Lyman Break Galaxies, MNRAS, **319**, 318

Chapman, S. C., et al. (2002) Westphal-MMD 11: An Interacting, Submillimeter Luminous, Lyman Break Galaxy, ApJ, **572**, L1

Conselice, C. J. (2004) Unveiling the Formation of Massive Galaxies, Science, **304**, 399

Conselice, C. J., Chapman, S. C., & Windhorst, R. A. (2003) Evidence for a Major Merger Origin of High-Redshift Submillimeter Galaxies, ApJ, **596**, L5
Conselice, C. J., et al. (2004) Observing the Formation of the Hubble Sequence in the Great Observatories Origins Deep Survey, ApJ, **600**, L139
Corbett, E. A., et al. (2003), COLA. II. Radio and Spectroscopic Diagnostics of Nuclear Activity in Galaxies, ApJ, **583**, 670
Cox, T. J., et al. (2005) The Effects of Feedback in Simulations of Disk-galaxy Major Mergers, MNRAS, submitted (astro-ph 0503201)
Cram, L. E. (1998) The Global Star Formation Rate from the 1.4 GHz Luminosity Funcition, ApJ, **506**, 85
De Grijs, R. (2001) Star Formation Timescales in M82, Astr. & Geophys., **42**, 12
De Grijs, R., O'Connell, R. W., & Gallagher III, J. S. (2002) Tidally-induced Super Star Clusters in M82, in *Extragalactic Star Clusters*, I.A.U. Symp. **207**, eds. D. Geisler, E. K. Grebel, and D. Minniti, (ASP, San Francisco) p. 477
Duc, P.-A., et al. (2000) Formation of a Tidal Dwarf Galaxy in the Interacting System Arp 245 (NGC 2992(93), AJ, **120**, 1238
Duc, P.-A., et al. (1997) Gas Segregation in the Interacting System Arp 105, A&A, **326**, 537
Duc, P.-A., Bournaud, F., & Masset, F. (2004) A Top-down Scenario for the Formation of Massive Tidal Dwarf Galaxies, A&A, **427**, 803
Duc, P.-A., Braine, J., & Brinks, E. (2004) *Recycling Intergalactic and Interstellar Matter*, I.A.U. Symp. **217**, (A.S.P., San Francisco)
Ellingson, et al.(2001) The Evolution of Population Gradients in Galaxy Clusters: The Butcher-Oemler Effect and Cluster Infall, ApJ, **547**, 609
Elmegreen, B. G., et al. (2000) Hubble Space Telescope Observations of the Interacting Galaxies NGC 2207 and IC 2163, AJ, **120**, 630
Elmegreen, D. M., et al. (1991) Properties and Simulations of Interacting Spiral Galaxies with Transient "Ocular" Shapes, A&A, **244**, 52
Fabbiano, G., et al. (2004) X-raying Chemical Evolution and Galaxy Formation in the Antennae, ApJ, **605**, L21
Feldmeier, J. J., et al. (2002) Deep CCD Photometry of Galaxy Clusters. I. Methods and Initial Studies of Intracluster Starlight, ApJ, **575**, 779
Fiorito, R., & Titarchuk, L. (2004) Is M82 X-1 Really an Intermediate-Mass Black Hole? X-ray Spectral and Timing Evidence, ApJ, **614**, L113
Förster Schreiber, N. M., et al. (2004) A Substantial Population of Red Galaxies at z¿2: Modeling of the Spectral Energy Distributions of an Extended Sample, ApJ, **616**, 40
Franx, M., et al. (2003) A Significant Population of Red, Near-infrared-selected High-redshift Galaxies, ApJ, **587**, L79
Frayer, D. T., et al. (2004) Infrared Properties of Radio-selected Submillimeter Galaxies in the Spitzer First Look Survey Verification Field, ApJS, **154**, 137
Gao, Y., & Solomon, P. M. (1999) Molecular Gas Depletion and Starbursts in Luminous Infrared Galaxy Mergers, ApJ, **512**, L99
Gao, Y., & Solomon, P. M. (2004) The Star Formation Rate and Dense Molecular Gas in Galaxies, ApJ, **606**, 271
Gao, Y., et al. (2003) Nonnuclear Hyper/Ultraluminous X-ray Sources in the Starbursting Cartwheel Ring Galaxy, ApJ, **596**, L171
Genzel, R., et al. (2004) Submillimeter Galaxies as Tracers of Mass Assembly at Large M, astro-ph 0403183

Georgakakis, A., et al. (2005) Dust in a Merging Galaxy Sequence: the SCUBA View, in *The Spectral Energy Distribution of Gas-rich Galaxies: Confronting Models with Data*, eds. C. C. Popescu and R. J. Tuffs, (A.I.P. Conf. Series **761**, New York), 441

Glazebrook, K., et al. (2003) The Sloan Digital Sky Survey: The Cosmic Spectrum and Star Formation History, ApJ, **587**, 55

Glazebrook, K., et al. (2004) Cosmic Star Formation History to z=1 from a Narrow Emission Line-selected Tunable-filter Survey, AJ, **128**, 2652

Gnedin, O. Y. (2003a) Tidal Effects in Clusters of Galaxies, ApJ, **582**, 141

Gnedin, O. Y. (2003b) Dynamical Evolution of Galaxies in Clusters, ApJ, **589**, 752

Goto, T. (2005) 266 E+A Galaxies Selected from the Sloan Digital Sky Survey Data Release 2: The Origin of E+A Galaxies, MNRAS, **357**, 937

Greve, T. R., et al. (2004) A 1200-mu MAMBO Survey of ELAISN2 and the Lockman Hole - I. Maps, Sources and Number Counts, MNRAS, **354**, 779

Gútierrez, C. M., & López-Corredoira, M. (2005) The Nature of Ultra Luminous X-ray Sources, ApJ Letters, **622**, L89

Harris, J., et al. (2004) The Recent Cluster Formation Histories of NGC 5253 and NGC 3077: Environmental Impact on Star Formation, ApJ, **603**, 503

Hashimoto, Y., & Oemler Jr., A. (2000) The Effect of Environment on Galaxy Interactions, ApJ, **530**, 652

Heavens, A., et al. (2004) The Complete Star Formation History of the Universe, Nature, **428**, 625

Hernquist, L., & Quinn, P. J. (1988) Formation of Shell Galaxies: I. Spherical Potentials, ApJ, **331**, 682

Hibbard, J. E., & Barnes, J. E. (2004) The Dynamical Masses of Tidal Dwarf Galaxies, in *Recycling Intergalactic and Interstellar Matter*, I.A.U. Symp. **217**, eds. P.-A. Duc, J. Braine, and E. Brinks, ASP, p. 510

Hibbard, J. E., Rupen, M. & van Gorkom, J. H., eds. (2001) *Gas and Galaxy Evolution*, ASP Conf. Proceedings **240**, Astronomical Society of the Pacific

Hibbard, J. E., Vacca, W. D., & Yun, M. S. (2000) The Neutral Hydrogen Distribution in Merging Galaxies: Differences Between Stellar and Gaseous Tidal Morphologies, AJ, **119**, 1130

Hippelein, H., et al. (2003) Star Forming Rates between z = 0.25 and z = 1.2 from the CADIS Emission Line Survey, A&A, **402**, 65

Hopman, C., et al. (2004) Ultraluminous X-ray Sources as Intermediate-Mass Black Holes Fed by Tidally Captured Stars, ApJ, **604**, L101

Irion, R. (2004) Surveys Scour Cosmic Deep, Science (News Note), **303**, 1750

Ivison, R. J., et al. (2004) Spitzer Observations of MAMBO Galaxies: Weeding Out Active Nuclei from Starbursting Protoellipticals, ApJS, **154**, 124

Javiel, S. C., Santiago, B. X., & Kerber, L. O. (2005) Constraints on the Star Formation History of the Large Magellanic Cloud, A&A, **431**, 73

Jog, C. J., and Solomon, P. M. (1992) A Triggering Mechanism for Enhanced Star Formation in Colliding Galaxies, ApJ, **387**, 152

Juneau, S., et al. (2005) Cosmic Star Formation History and Its Dependence on Galaxy Stellar Mass, ApJ, **619**, L135

Keel, W. C. (1993), Kinematic Regulation of Star Formation, AJ, **106**, 1771

Keel, W. C., & Borne, K. D. (2003) Massive Star Clusters in Ongoing Galaxy Interactions: Clues to Cluster Formation, AJ, **126**, 1257

Keel, W. C., et al. (1985), The Effects of Interactions on Spiral Galaxies. I. Nuclear Activity and Star Formation, AJ, **90**, 708

Kenney, J. D. P., & Yale, E. E. (2002) Hubble Space Telescope Imaging of Bipolar Nuclear Shells in the Disturbed Virgo Cluster Galaxy NGC 4438, ApJ, **567**, 865
Kennicutt Jr., R. C. (1989) The Star Formation Law in Galactic Disks, ApJ, **344**, 685
Kennicutt Jr., R. C. (1998) The Global Schmidt Law in Star-forming Galaxies, ApJ, **498**, 541
Kennicutt Jr., R. C., et al. (1987) The Effects of Interactions on Spiral Galaxies. II. Disk Star-Formation Rates, AJ, **93**, 1011
King, A. R. (2004) Ultraluminous X-ray Sources and Star Formation, MNRAS, **347**, L18
King, A. R., & Dehnen, W. (2005) Hierarchical Merging, Ultraluminous and Hyperluminous X-ray Sources, MNRAS, **357**, 275
Klimanov, S. A., & Reshetnikov, V. P. (2001) A Statistical Study of M51-type Galaxies, A&A, **378**, 428
Klypin, A., et al. (1999) Where are the Missing Galactic Satellites?, ApJ, **522**, 82
Knierman, K. A., et al. (2003) From Globular Clusters to Tidal Dwarfs: Structure Formation in the Tidal Tails of Merging Galaxies, AJ, **126**, 1227
Krolik, J. H. (2004) Are Ultraluminous X-ray Sources Intermediate-Mass Black Holes Accreting from Molecular Clouds?, ApJ, **615**, 383
Krumholz, M. R., & McKee, C. F., (2005) A General Theory of Turbulence-Regulated Star Formation, From Spirals to ULIRGs, ApJ, **630**, 250
Lambas, D. G., et al. (2003) Galaxy Pairs in the 2dF Survey - I. Effects of Interactions on Star Formation in the Field, MNRAS, **346**, 1189
Lançon, A., et al. (2001) Multiwavelength Study of the Starburst Galaxy NGC 7714. II. The Balance Between Young, Intermediate-Age, and Old Stars, ApJ, **552**, 150
Larson, R. B., & Tinsley, B. M. (1978) Star Formation Rates in Normal and Peculiar Galaxies, ApJ, **219**, 46
Larson, R. B., Tinsley, B. M., & Caldwell, C. N. (1980) The Evolution of Disk Galaxies and the Origin of SO Galaxies, ApJ, **237**, 692
Lavery, R. J., et al. (2004) Probing the Evolution of the Galaxy Interaction/Merger Rate Using Collisional Ring Galaxies, ApJ, **612**, 679
Le Borgne, D., et al. (2005) Gemini Deep Deep Survey VI: Massive Post-starburst Galaxies at z=1, ApJ, submitted (astro-ph 0503401)
Lewis, A. D., Buote, D. A., & Stocke, J. T., (2003) Chandra Observations of A2029: The Dark Matter Profile Down to below 0.01 r_{VIR} in an Unusually Relaxed Cluster, ApJ, **586**, 135
Lin, L., et al. (2004) The DEEP2 Galaxy Redshift Survey: Evolution of Close Galaxy Pairs and Major-Merger Rates up to $z \sim 1.2$, ApJ, **617**, L9
Liu, J.-F., Bregman, J. N., & Irwin, J. (2005) Ultra-luminous X-ray Sources in Nearby Galaxies from ROSAT HRI Observations, II. Statistical Properties, ApJS, **157**, 59
Lonsdale Persson, C. J., ed., (1986), *Star Formation in Galaxies*, NASA Conf. Pub. 2466
Mao, R. O., et al. (2000) Dense Gas in Nearby Galaxies. XIII. CO Submillimeter Line Emission from the Starburst Galaxy M 82, A&A, **358**, 433
Matsumoto, H., et al. (2004) Peculiar Characteristics of the Hyper-luminous X-ray Source M82 X-1, Prog. Th. Phys. Suppl., **155**, 379
Melo, V. P., (2005) Young Super Star Clusters in the Starburst of M82: The Catalog, AJ, **619**, 270
Mihos, J. C. (2004) Interactions and Mergers of Cluster Galaxies, in *Clusters of Galaxies: Probes of Cosmological Structure and Galaxy Evolution*, eds. J. S. Mulchaey, A. Dressler, & A. Oemler, (Cambridge Univ. Press, Cambridge) p. 277

Mihos, J. C., Bothun, G. D., & Richstone, D. O. (1993) Modeling the Spatial Distribution of Star Formation in Interacting Disk Galaxies, ApJ, **418**, 82

Mihos, J. C., & Hernquist, L. (1994) Star-forming Galaxy Models: Blending Star Formation into TREESPH, ApJ, **437**, 611

Mihos, J. C., & Hernquist, L. (1996) Gasdynamics and Starbursts in Major Mergers, ApJ, **464**, 641

Miller, J. M. (2005) Present Evidence for Intermediate Mass Black Holes in ULXs and Future Prospects, in *From X-ray Binaries to Quasars: Black Hole Accretion on All Mass Scales*, eds. T. J. Maccarone, R. P. Fender, & L. C. Ho, (Kluwer, Dordrecht), in press (astro-ph 0412526)

Miller, J. M., Fabian, A. C., & Miller, M. C. (2004) A Comparison of Intermediate-Mass Black Hole Candidate Ultraluminous X-ray Sources and Stellar-Mass Black Holes, ApJ, **614**, L117

Miller, J. M., et al. (2004) XMM-Newton Spectroscopy in the Antennae Galaxies (NGC 4038/4039), ApJ, **609**, 728

Moore, B., et al. (1996) Galaxy Harrassment and the Evolution of Clusters of Galaxies, Nature, **379**, 613

Moore, B., et al. (1999) On the Survival and Destruction of Spiral Galaxies in Clusters, MNRAS, **304**, 465

Mushotsky, R. (2004a) Ultra-luminous Sources in Nearby Galaxies, Prog. Th. Phys. Suppl., **155**, 27

Mushotsky, R.. (2004b) Clusters of Galaxies: an X-ray Perspective, in *Clusters of Galaxies: Probes of Cosmological Structure and Galaxy Evolution*, eds. J. S. Mulchaey, A. Dressler, & A. Oemler, (Cambridge Univ. Press, Cambridge) p. 134

Nagamine, K., et al. (2005) Massive Galaxies in Cosmological Simulations: Ultraviolet-selected Sample at Redshift z=2, ApJ, **618**, 23

Navarro, J. F., Frenk, C. S., & White, S. D. M. (1997) A Universal Density Profile from Hierarchical Clustering, ApJ, **490**, 493

Neri, R., et al. (2003) Interferometric Observations of Powerful CO Emission from Three Submillimeter Galaxies at z = 2.39, 2.51, and 3.35, ApJ, **597**, L113

Nikolic, B., Cullen, H., & Alexander, P. (2004) Star Formation in Close Pairs Selected from the Sloan Digital Sky Survey, MNRAS, **355**, 874

Noguchi, M. (1987) Close Encounter Between Galaxies: II. Tidal Deformation of a Disc Galaxy Stabilized by a Massive Halo, MNRAS, **228**, 635

O'Connell, R. W. (2004) Ten Years of Super Star Cluster Research, in *The Formation and Evolution of Massive Young Clusters*, A.S.P. Conf. **322**, eds. H. J. G. L. M. Lamers, L. J. Smith, & A. Nota, (A.S.P., San Francisco) p. 551

Patton, D. R., et al. (2002) Dynamically Close Galaxy Pairs and Merger Rate Evolution in the CNOC2 Redshift Survey, ApJ, **565**, 208

Pimbblet, K. (2003) At the Vigintennial of the Butcher-Oemler Effect, PASA, **20**, 294

Poggianti, B. M., et al. (2004) A Comparison of the Galaxy Populations in the Coma and Distant Clusters: The Evolution of the k+a Galaxies and the Role of the Intracluster Medium, ApJ, **601**, 197

Pope, A., et al. (2005) The Hubble Deep Field North SCUBA Super-map III - Optical and Near-infrared Properties of Submillimetre Galaxies, MNRAS, **358**, 149

Portegies Zwart, S., Dewi, J., & Maccarone, T. (2004) Intermediate Mass Black Holes in Accreting Binaries: Formation, Evolution and Observational Appearance, MNRAS, **355**, 413

Roediger, E., & Hensler, G. (2005) Ram Pressure Stripping of Disk Galaxies A&A, in press (astro-ph 0412518)

Rowan-Robinson, M., et al. (2004) The European Large-area ISO Survey (ELAIS): the Final Band-merged Catalogue, MNRAS, **351**, 1290
Sandage, A. (1961) *The Hubble Atlas of Galaxies*, (Carnegie Inst. of Washington, Washington) (Pub. 618)
Sanders, D. B. (2004) The Cosmic Evolution of Luminous Infrared Galaxies: from IRAS to ISO, SCUBA, and SIRTF, AdSpR, **34**, 535
Sanders, D. B., & Mirabel, I. F. (1996) Luminous Infrared Galaxies, ARAA, **34**, 749
Scannapieco, E., Weisheit, J., & Harlow, F. (2004) Triggering the Formation of Halo Globular Clusters with Galaxy Outflows, ApJ, **615**, 29
Schiminovich, D., et al. (2005) The GALEX-VVDS Measurement of the Evolution of the Far Ultraviolet Luminosity Density and the Cosmic Star Formation Rate, ApJ, **619**, L47
Schulz, S., & Struck, C. (2001) Multi Stage Three-dimensional Sweeping and Annealing of Disc Galaxies in Clusters, MNRAS, **328**, 185
Schweizer, F. (1983) Observational Evidence for Mergers, in *Internal Kinematics and Dynamics of Galaxies*, ed. E. Athanassoula, (Reidel, Dordrecht) p. 319
Schweizer, F. (1998) Observational Evidence for Interactions and Mergers, in *Galaxies: Interactions and Induced Star Formation*, Saas Fee Advanced Course **26**, eds. D. Friedli, D. Martinet, & D. Pfenniger, (Springer, Berlin) p. 105
Schweizer, F. (2005) in *Starbursts: From 30 Doradus to Lyman Break Galaxies*, eds. R. de Grijs & R. M. Gonzalez Delgado, (Kluwer, Dordrecht) p. 143 (astro-ph 0502111)
Schombert, J. M., Wallin, J. F., & Struck-Marcell, C. (1990) A Multicolor Photometric Study of the Tidal Features in Interacting Galaxies, AJ, **99**, 497
Shapley, A. E., et al. (2005) UV to Mid-IR Observations of Star-forming Galaxies at z 2: Stellar Masses and Stellar Populations, ApJ, **626**, 698
Shu, C., Mao, S., & Mo, H. J. (2001) The Host Haloes o f Lyman-break Galaxies and Submillimetre Sources, MNRAS, **327**, 895
Smail, I., et al. (2004) The Rest-Frame Optical Properties of SCUBA Galaxies, ApJ, **616**, 71
Smith, B. J., Struck, C., & Nowak, M. A. (2005) Chandra X-ray Imaging of the Interacting Starburst Galaxy NGC 7714/7715: Tidal Ultra-luminous X-ray Sources, Emergent Wind, and Resolved HII Regions, AJ, **129**, 1350
Sofue, Y., & Rubin, V. (2001) Rotation Curves of Spiral Galaxies, ARAA, **39**, 137
Soifer, B. T., Houck, J. R., & Neugebauer, G. (1987) The IRAS View of the Extragalactic Sky, ARAA, **25**, 187
Sparke, L. (2002) Off-plane Gas and Galaxy Disks, in *Disks of Galaxies: Kinematics, Dynamics and Perturbations*, ASP Conf. Series **275**, eds. E. Athanassoula, A. Bosma, & R. Mujica, (ASP, San Francisco) p. 367
Spinrad, H. (2004) The Most Distant Galaxies, in *Astrophysics Update: Topical and Timely Reviews in Astrophysics*, ed. J. W. Mason, (Springer Praxis, Berlin) p. 155
Spitzer, L., Jr., & Baade, W. (1951) Stellar Populations and Collisions of Galaxies, ApJ, **113**, 413
Springel, V., & Hernquist, L. (2005) Formation of a Spiral Galaxy in a Major Merger, ApJ, **622**, L9
Struck, C. (1997) Simulations of Collisions Between Two Gas-Rich Galaxy Disks with Heating and Cooling, ApJS, **113**, 269
Struck, C. (1999) Galaxy Collisions, Phys Rep, **321**, 1
Struck, C. (2004) Case Studies of Mass Transfer and Star Formation in Galaxy Collisions, in *Recycling Intergalactic and Interstellar Matter*, I.A.U. Symp. **#217**, eds. P.-A. Duc, J. Braine, & E. Brinks, (ASP, San Francisco) p. 400

Struck, C. (2005) The Recurrent Nature of Central Starbursts, in *Starbursts: From 30 Doradus to Lyman Break Galaxies*, eds. R. de Grijs & R. M. Gonzalez Delgado, (Kluwer, Dordrecht) p. 163

Struck, C. & Smith, B. J. (2003) Models of the Morphology, Kinematics and Star Formation History of the Prototypical Collisional Starburst System: NGC 7714/7715 = Arp 284, ApJ, **589**, 157

Struck, C., & Smith, D. C. (1999) Simple Models for Turbulent Self-Regulation in Galaxy Disks, ApJ, **527**, 673

Struck-Marcell, C., & Lotan, P. (1990) The Varieties of Symmetric Stellar Rings and Radial Caustics in Galaxy Disks, ApJ, **358**, 99.

Struck-Marcell, C., & Tinsley, B. M. (1978) Star Formation Rates and Infrared Radiation, ApJ, **221**, 562

Sulentic, J. W., Keel, W. C., & Telesco, C. M., eds., (1989) *Paired and Interacting Galaxies:* I.A.U. Colloq. No. **124**, NASA Conf. Pub. 3098

Swartz, D. A., et al. (2004) The Ultra-luminous X-ray Source Population from the Chandra Archive of Galaxies, ApJS, **154**, 519

Swinbank, A. M., et al.(2004) The Rest-frame Optical Spectra of SCUBA Galaxies, ApJ, **617**, 64

Teplitz, H. I., et al. (2003) Emission-line Galaxies in the STIS Parallel Survey. II. Star Formation Density, ApJ, **589**, 704

Toomre, A. (1977) Mergers and Some Consequences, in *The Evolution of Galaxies and Stellar Populations*, eds. B. M. Tinsley & R. B. Larson, (Yale University Observatory, New Haven) p 401

Toomre, A., & Toomre, J. (1972) Galactic Bridges and Tails, ApJ, **178**, 623

Tremaine, S., & Weinberg, M. D. (1984) Dynamical Friction in Spherical Systems, MNRAS, **209**, 729

van der Marel, R. P. (2004) Intermediate-mass Black Holes in the Universe: a Review of Formation Theories and Observational Constraints, in *Coevolution of Black Holes and Galaxies*, Carnegie Observatories Astrophysics Series, Vol. I., ed. L. C. Ho, (Cambridge University Press, Cambridge) p. 37

Vollmer, B., et al. (2000) The Consequences of Ram Pressure Stripping on the Virgo Cluster Spiral NGC 4522, A&A, **364**, 532

Vollmer, B., et al. (2001) Ram Pressure Stripping and Galaxy Orbits: The Case of the Virgo Cluster, ApJ, **561**, 708

Wang, W.-H., Cowie, L. L., & Barger, A. J. (2004) An 850 Micron SCUBA Survey of the Hubble Deep Field-North GOODS Region, ApJ, **613**, 655

Ward, M. (2003) X-ray Components in Spiral and Star-forming Galaxies, in *Frontiers of X-ray Astronomy*, eds. A. C. Fabian, K. A. Pounds, & R. D. Blandford, (Cambridge Univ. Press, Cambridge) p. 117

Weinberg, M. D. (1985) Evolution of Barred Galaxies by Dynamical Friction, MNRAS, **213**, 451

Whitmore, B. C., & Schweizer, F. (1995) Hubble Space Telescope Observations of Young Star Clusters in NGC 4038/4039, the 'Antennae' Galaxies, AJ, **109**, 960

Whitmore, B. C., et al. (1999) The Luminosity Function of Young Star Clusters in "the Antennae" Galaxies (NGC 4038-4039), AJ, **118**, 1551

Wolter, A., & Trinchieri, G. (2004) A Thorough Study of the Intriguing X-ray Emission from the Cartwheel Ring, A&A, **426**, 787

Xu, C. K., Sun, Y. C., & He, X. T. (2004) The Near Infrared Luminosity Function of Galaxies in Close Major-Merger Pairs and the Mass Dependence of the Merger Rate, ApJ, **603**, L73

Yan, L., et al. (2004) High-redshift Extremely Red Objects in the Hubble Space Telescope Ultra Deep Field Revealed by the GOODS Infrared Array Camera Observations, ApJ, **616**, 63
Yan, L., & Thompson, D. (2003) Hubble Space Telescope WFPC2 Morphologies of K-selected Extremely Red Galaxies, ApJ, **586**, 765
Yan, L., Thompson, D., & Soifer, T. (2004) Optical Spectroscopy of K-selected Extremely Red Galaxies, AJ, **127**, 1274
Young, J. S., et al. (1996) The Global Rate and Efficiency of Star Formation in Spiral Galaxies as a Function of Morphology and Environment, AJ, **112**, 1903
Zabludoff, A. I., et al. (1996) The Environments of "E+A" Galaxies, ApJ, **466**, 104
Zabludoff, A. I., & Mulchaey, J. S. (1998) Hierarchical Evolution in Poor Groups of Galaxies, ApJ, **498**, L5
Zaritsky, D., & Harris, J. (2004) Quantifying the Drivers of Star Formation on Galactic Scales. I. The Small Magellanic Cloud, ApJ, **604**, 167
Zhang, X. (2003) Secular Evolution of Spiral Galaxies, JKAS, **36**, 223
Zwicky, F. (1959) Multiple Galaxies, Handbuch der Phy, **53**, 373

5 Formation and Evolution of Supermassive Black Holes

F. Combes

Abstract: The correlation between the mass of supermassive black holes in galaxy nuclei and the mass of the galaxy spheroids or bulges (or more precisely their central velocity dispersion), suggests a common formation scenario for galaxies and their central black holes. The growth of bulges and black holes can commonly proceed through external gas accretion or hierarchical mergers, and are both related to starbursts. Internal dynamical processes control and regulate the rate of mass accretion. Self-regulation and feedback are key to the correlation. It is possible that the growth of one component, either BH or bulge, takes over, breaking the correlation, as in Narrow Line Seyfert 1 objects. The formation of supermassive black holes can begin early in the Universe, from the collapse of Population III stars, and then through gas accretion. The active black holes can then play a significant role in the re-ionization of the Universe. The nuclear activity is now frequently invoked as a feedback to star formation in galaxies, and even more spectacularly in cooling flows. The growth of SMBH is certainly self-regulated there. SMBHs perturb their local environment, and the mergers of binary SMBHs help to heat and destroy central stellar cusps. The interpretation of the X-ray background yields important constraints on the history of AGN activity and obscuration, and the census of AGN at low and at high redshifts reveals the downsizing effect, already observed for star formation. History appears quite different for bright QSO and low-luminosity AGN: the first grow rapidly at high z, and their number density decreases then sharply, while the density of low-luminosity objects peaks more recently, and then decreases smoothly.

5.1 Introduction

It is now well established that all nearby galaxies possessing a spheroidal stellar component or bulge possess a central black hole (BH), with a BH mass proportional to the bulge mass, with a proportionality factor which is now renormalised around 2×10^{-3} (Magorrian et al 1998, Gebhardt et al 2000, Merritt & Ferrarese 2001, Shields et al 2003). It soon appeared that the relation is more precise and with less scatter, between the BH mass and the central velocity dispersion (or dispersion inside the effective radius of the bulge σ_e), as shown in Fig. 5.1. The BH mass grows then close to the 4th power of the central velocity dispersion.

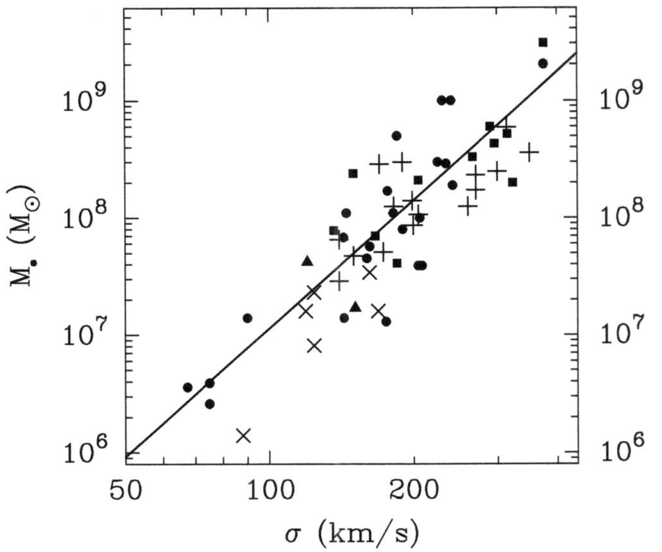

Fig. 5.1. The relation between BH mass and the velocity dispersion σ inside the effective radius of the bulge. Filled circles indicate BH mass measurement from stellar dynamics, squares from ionized gas, triangles on maser lines, crosses are from reverberation mapping, and "plus signs" from ionization models (from Kormendy & Gebhardt 2001). The relation is close to a power-law of slope 4.

The determination of this relation has been carried out by various methods:

- stellar proper motions for the Galactic center BH (Schödel et al. 2002, Ghez et al. 2003),
- stellar absorption lines, to obtain the stellar kinematics,
- ionized gas emission lines (less reliable, since affected by outflows, inflows), and also masing gas emission lines,
- reverberation mapping, exploiting time delays between variations of AGN continuum, and broad line emission, giving the size of the emitting gas region, combined with the gas Doppler velocity to give the virial mass (Peterson & Wandel, 2000)
- ionization models: method based on the correlation between quasar luminosity and the size of the Broad Line Region (BLR, Rokaki et al 1992).

The relation has recently been somewhat extended to lower masses, in dwarf Seyfert 1 nuclei, which are more difficult to measure (Barth et al 2004, 2005). Some progress has also been made in the search of intermediate mass black holes (IMBH), for example in the globular clusters M15 in our Galaxy and G1 in M31: in M15, the mass of the central object is lower than 10^3 $M_\odot$ and could be stellar remnants (van der Marel 2003), while in G1, a BH of 2×10^4 $M_\odot$ is identified, and obeys the $M_{bh} - \sigma$ relation (Gebhardt et al 2002). A 1000 $M_\odot$ IMBH has also been estimated as member of a binary, at the origin of the ULX source M82 X-1 in the starburst galaxy M82 (Portegies Zwart et al., 2004a).

The demography of SMBH, statistics on their activity frequency, and their observed mass functions, constrain the possible AGN lifetime and growth rate. It was already suspected that AGN were active during a short duty cycle of $\sim 4 \times 10^7$ yr, and that many galaxies today should host a starving black hole (Haehnelt & Rees 1993). The observed $M_{bh} - \sigma$ relation now strongly constrains the duty cycle time-scale. Also the cosmic background radiation detected at many wavelengths constrains the formation history. The volumic density of massive black holes today is derived, from the observed density of bulges, and the proportionality factor $M_{bh} = 0.002\ M_{bulge}$. And independently, the light that should have been radiated at the formation of these BHs can be computed, redshifted and compared to the observed cosmic background radiation: in the optical, we see only 10% of the expected flux, but 30% in X-rays, and 80% in the infra-red. The accretion radiation does not get out in optical light, probably due to the extinction.

5.2 BH Growth

Powerful QSOs are observed early in the Universe, at $z > 6$, with luminosities indicating very high BH masses, meaning that masses as high as 10^8–10^9 $M_\odot$ can grow in less than one Gyr. However, the time-scale to grow a black hole from a stellar mass of 10 $M_\odot$ to the Hills limit, M_c ($M_c = 3 \times 10^8$ $M_\odot$), above which stars are swallowed by the black hole without any gas radiation, is of the order of 1.6 Gyr, if the gas accretion occurs at the Eddington limit, and the efficiency is $\epsilon \sim 0.1$–0.2 (Hills 1975). The problem is therefore to accelerate the growth rate, or begin from a higher initial mass.

5.2.1 Quantifying the problem

To have an order of magnitude, and simple dimensional relations, let us assume spherical accretion, from an accretion radius $R_{acc} = 0.3\ M_6/v_2^2$ pc, where M_6 is the mass of the BH in 10^6 $M_\odot$, and v_2 the velocity in 100km/s (corresponding to the effective stellar velocity inside the galaxy nucleus, related to the bulge mass). The canonical Bondi accretion rate is then:

$$dM/dt = 4\pi R^2 v\rho = 10^{-4} M_\odot/yr M_6^2/v_2^3\rho$$

where ρ is the local density in $M_\odot/pc^3$.

Since $dM/dt \propto M^2$, then the accretion time is $\propto 1/M$, $t_{acc} \sim 10^{10}$ yr/M_6 v_2^3/ρ; for very low mass BH, this takes much larger than the Hubble time. Therefore the formation of SMBH requires a large seed, mergers of BH, or very large densities, like that characteristic of the Milky Way nucleus, 10^7 $M_\odot/pc^3$.

If these conditions are fulfilled, the growth of massive BH can then be accretion-dominated, i.e. $t_{growth} = t_{acc}$. This phase could correspond to moderate AGN, like Seyferts, and the luminosity is increasing as $L \propto dM/dt \propto M^2$. At some point, the luminosity will reach the Eddington luminosity, since $L_{edd} \propto$

M. The Eddington ratio increases as $L/L_{edd} \propto M$, the BH growth slows down when approaching L_{edd}, corresponding to a QSO phase. The time-scale of this powerful AGN phase is $t_{edd} = M/(dM/dt)_{edd} = 4.5 \times 10^7$ yr $(0.1/\epsilon)$ (where ϵ is the usual radiation efficiency). Equating $t_{acc} = t_{edd}$, this occurs for $M = 2 \times 10^8$ $M_\odot$ v_2^3/ρ $(\epsilon/0.1)$. Wang et al. (2000) propose that tidal perturbations help to grow a SMBH from a small seed, by boosting the accretion, and then lead to the $M_{bh} - \sigma$ relation.

5.2.2 Formation of the first massive black holes, in the early Universe

One solution to the growth problem could be that massive BH form very early at high redshift, as the remnants of Pop III stars. In the CDM scenario of hierarchical structure formation, it is generally thought that the first stars are expected to form in dark matter minihalos of mass $10^6 M_\odot$, at redshifts around 20. Their virial temperature is too small for atomic hydrogen cooling to be efficient, but the molecular hydrogen cooling is fast enough (Tegmark et al., 1997). Without metals and dust, the H_2 molecules form through H^- with the electrons as catalysts. The minimum halo mass at a given redshift, in which the baryons are able to cool and form stars is obtained through the condition that the cooling time is smaller than the dynamical time, and is confirmed to be $10^6 M_\odot$, at $z \approx 20$–30 (Fuller & Couchman 2000). Both semi-analytical estimations, and full 3D numerical simulations concord to find very massive first stars, with $M^* > 100 M_\odot$ (e.g. Abel et al., 2002, Bromm & Larson 2004). Fragmentation is quite inefficient for these first condensations, due to the low metallicity and negligible radiative losses. The mass spectrum of these first stars is still not well known, but according to the cosmology, it is expected that the most massive structures are significantly clustered.

Above 260 $M_\odot$, the formed objects could collapse to a BH directly (Bond et al., 1984, Madau & Rees 2001, Schneider et al., 2002). After the first subhaloes have merged in larger entities, and formed dwarf galaxies, there could exist 10^5 $M_\odot$ IMBH in each center, formed by the merging of these seeds.

The total mass in these first black holes can be quite important. If every halo corresponding to a 3 sigma peak (or higher) at $z = 24$ forms a 260 $M_\odot$ BH, then the density per comoving volume is estimated at $\rho_\bullet = 2.9 \times 10^5$ $M_\odot/\mathrm{Mpc}^3$, already half of the present SMBH density (Islam et al., 2004). It will then be sufficient to add some gas accretion to grow the BH along their lives, and to ensure the merging of all seeds. The problem at this stage is however the low efficiency of dynamical friction for objects that are still not massive enough. The consequence is that many BH will keep orbiting around subhaloes, instead of sinking to the main center. Semi-analytical merger-tree calculations have been carried out, taking into account dynamical friction, tidal disruption and encounters with the galactic disk, to determine the abundance and distribution of MBHs in present-day haloes of various masses (Islam et al., 2004): the result is that it is difficult to reproduce the observed mass distribution of SMBH with only merging of the initial seeds, and that further gas accretion is required. Also the

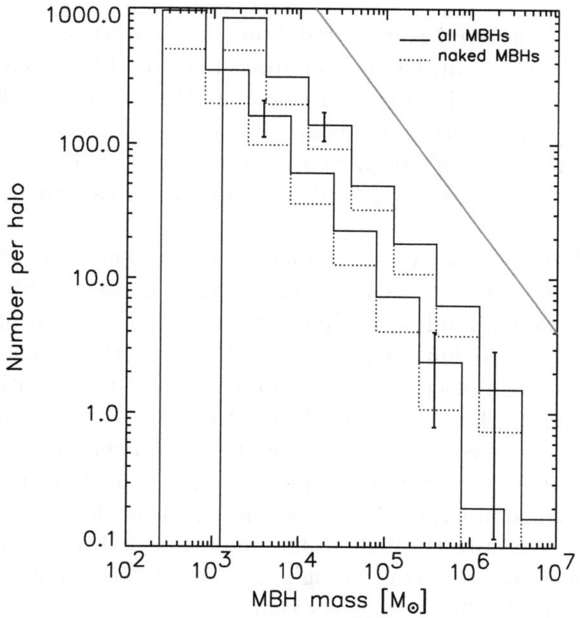

Fig. 5.2. The mass distribution of massive BH (naked, i.e. without any satellite halo, or not) predicted in a final halo of 1.2 $10^{12} M_{\odot}$, for two hypothesis on the seed mass (260 and 1300 $M_{\odot}$, for the upper and lower curves respectively), from Islam et al. (2003).

formation of binaries at the center of structures require gas accretion in order for the binary BH to merge before a triple is formed and some BH are lost in intergalactic space.

The consequence of low merging efficiency of the seed BH is the predicted large abundance of these IMBH in a given galactic halo (cf Fig. 5.2): typically a thousand or more should exist in the Milky Way. Coming from rare high density peaks, they are expected to cluster in the bulges and spheroids; when they accrete gas, they could account for ultra-luminous X-ray sources (ULX) which are offset from the galaxy centers. In particular, masses typical of large IMBH, i.e. $10^5 M_{\odot}$, should reach the number of $\sim$10 in the Galaxy. Also, it is found that hierarchical merging can only be responsible of 10% of the total mass of present SMBH, and that gas accretion should be responsible for the rest. Taking into account the progressive gas accretion along the BH growth leads to a present SMBH density comparable to what is observed (Volonteri et al., 2003a), and also to a large number of wandering IMBHs.

Numerical simulations show that the $M_{bh} - \sigma$ relation can indeed be conserved through several successive mergers, provided that gas dissipation and star formation is included at each merger (Kazantzidis et al., 2005); collisionless mergers could cause some scatter in the relation.

5.2.3 Mini-quasars and reionization

An intense UV background is expected from the first stars. These stars are so massive that they create an HII region around them, able to blow away all the gas from the mini-halo where they are seated (Whalen et al., 2004). The UV photons in the Lyman-Werner bands are able to photo-dissociate the fragile H_2 molecules in the neighbourhood, preventing the gas to cool down. Star formation will be inhibited in a large region, until accumulation of gas creates dense regions able to be shielded.

At the death of the first stars, the massive black holes created from merging of the seeds, could accrete gas and become mini-quasars, able to produce a harder radiation background, including hard and soft X-rays. It has been argued that these X-rays could have a positive feedback on the formation of the H_2 molecules in producing electrons, and compensate for the negative feedback of the UV background (Haiman et al., 2000). However, 3D detailed simulations, find that the positive feedback is barely sufficient (Machacek et al., 2003).

A fundamental question is to know precisely at which epoch the inter-galactic hydrogen has been completely re-ionized, ending the dark age, and whether this has been done essentially through stellar radiation or from mini-quasars. The discovery of the Gunn-Peterson trough in some $z > 6$ quasars of the Sloan Survey (Becker et al., 2001) suggests that reionization is occurring near $z = 6$, while the WMAP result of a high electron scattering optical depth implies that ionizing sources were present up to $z = 15$, suggesting a long reionization period, may be in two steps (very massive stars at $z = 15$, and after a feedback epoch, much less massive stars at $z = 6$). The possibility of mini-quasars as the source of reionization has been studied by Dijkstra et al., (2004), in view of the X-ray background constraints. The hard X-ray photons produced by the miniquasars would be observed today as a soft X-ray background. If the quasars were only responsible for the reionization, than they will overproduce X-rays, and be incompatible with the observed 0.5–2 keV background. The miniquasars could only be responsible to about 50% of the IGM reionization.

5.2.4 The case of IMBH

Does the $M_{bh} - \sigma$ relation extrapolates to low masses? At least below $10^6 M_\odot$, the extrapolation appears straightforward (Barth et al., 2005), although, it is difficult to bridge the gap towards the low end of IMBH (of $10^3 M_\odot$); their observation is very difficult, both by the kinematics, since their gravitational influence is small, and from their possible AGN activity, since the expected luminosity is weak. According to the extrapolation of the $M_{bh} - \sigma$ relation, these IMBH should be searched as AGN in dwarf galaxies: among the good candidates are NGC 4395 (Filippenko & Ho 2003), where the BH mass is likely to be 10^4–10^5 $M_\odot$ (radiating much below the Eddington limit), or Pox 52, with $\sim 2 \times 10^5 M_\odot$ (Barth et al., 2004). The problem of this search is that dwarf galaxies frequently host nuclear star clusters of $\sim 10^6$ $M_\odot$, hiding the weak AGN. They are best observed in the Local Group; a famous example, M33, does

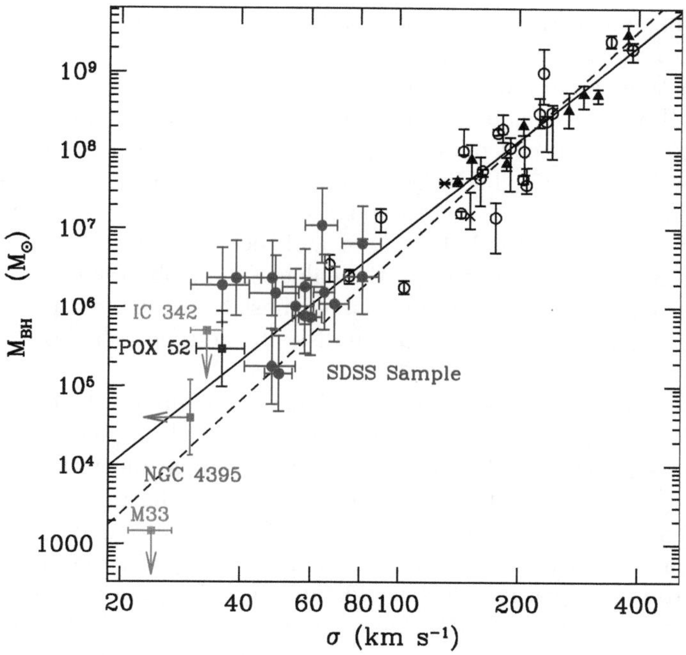

Fig. 5.3. Extension of the $M_{bh} - \sigma$ relation to IMBH (from Barth et al., 2005). Points in the upper right are black holes with dynamical mass measurements compiled by Tremaine et al., (2002). Open circles represent stellar-dynamical measurements, filled triangles are gas-dynamical detections, and crosses are from H_2O maser observations. In the middle of the diagram, filled circles are the SDSS Seyfert 1 galaxies studied by Barth et al. (2005). The low mass points are indexed by their galaxy name. Note the upper limit for any black hole mass in M33 by Gebhardt et al. (2001).

not host any BH more massive than 10^3 $M_\odot$, which is already 10 times below the value expected from the $M_{bh} - \sigma$ relation (cf Fig. 5.3).

One of the main features of lower mass IMBH is that they are no longer unique objects at the very center of the galaxies, but could be found in large numbers, spread out in galaxy haloes. The dynamical friction is no longer efficient enough, and their formation mechanism, through mergers of lower mass black holes through binary coalescence could provide them randomly large velocities (Ebisuzaki et al. 2001).

X-ray observations have revealed in nearby galaxies a class of extranuclear point sources with X-ray luminosities of 10^{39}–10^{41} ergs/s, exceeding the Eddington luminosity for stellar mass objects. These ultraluminous X-ray sources (ULXs) may be powered by IMBH of a few thousand $M_\odot$ or stellar mass black holes with special radiation processes.

Liu & Bregman (2005) find a strong association between ULXs and star formation sites, ULXs are preferentially observed in late-type galaxies, in spiral arms, and in some cases associated with supernovae. However a few ULXs are observed in old globular clusters, and Colbert et al. (2004) find them associated with Population II stars, in particular in elliptical galaxies.

These ULX sources could be due to IMBH resulting from the mergers of massive BH seeds, formed out of Population III objects in the early universe (Volonteri & Perna 2005). The large number of BH formed can merge in galaxies, through binary coalescence, but the possibility of a triple association, followed by the ejection of one of the BH, and recoil of the binary, leads to the prediction of many IMBHs wandering through the haloes. To be observed radiating at the high end of X-ray luminosity, these sources must be associated to baryons, and the most probable locations are in the disk of late-type galaxies (Volonteri & Perna 2005).

There are however observational problems in the interpretation of ULXs in terms of IMBH (Makishima et al. 2000). In most of them, the inner disk temperature is observed around 2kev, too high to be compatible with the high black hole mass, as required with the IMBH hypothesis, radiating at Eddington luminosity. They might be a mixed-bag class of objects, some could result from beamed emission during a short phase of common X-ray binaries, and they could be related to micro-quasars. This would explain their relation to star formation regions. (e.g. King et al., 2001). Some could represent the IMBH, expected in the continuity of SMBHs. An interesting case is the ULX source right at the nucleus of M33 whose nature is still debated (Dubus & Rutledge 2002). The central source corresponds to radiosources expected for micro-quasars (Trejo et al., 2004).

Evidence for an IMBH could come from the Milky Way nucleus: Hansen & Milosavljevic (2003) propose its existence to explain the observation of bright stars orbiting within 0.1pc, which are are young massive main sequence stars, in spite of an environment hostile to star-formation. Alternative solutions could be star mergers, or exotic objects (Ghez et al., 2003). In the IMBH scenario, stars were formed in a star cluster outside the central pc, and then dragged in by a BH of 10^3–10^4 $M_\odot$. The decay time-scale by dynamical friction for normal stars is too large (much longer than the massive stars lifetime), but for the IMBH, this time-scale is 1–10 Myr. Stars may be dragged inwards even after the star cluster has been disrupted.

Such a system SMBH-IMBH and a gas disk may reveal interesting dynamics; it is similar to a protosolar system, with the Sun-Jupiter couple, resonant effects like planetary migration are expected (Gould & Rix 2000).

5.2.5 NLS1 and black hole growth

The $M_{bh} - \sigma$ relation has been established locally, and it is not yet known whether the relation was already there in the primordial structures, and then was maintained during the evolution by a feedback process, or was obtained progressively, without maintaining a permanent relation.

There might be phases in the life of a galaxy, where the star formation has some advance with respect to the black hole growth, according to the various feedback and regulating mechanisms, and we should be able to recognize a subclass of AGN where the BH mass is somewhat below the standard relation. This has been proposed by Mathur (2000) for Narrow Line Seyfert 1 galaxies (NLS1). Grupe & Mathur (2004) investigated the BBR relation for a sample of broad-line Seyfert 1 galaxies (BLS1s) and narrow-line Seyfert 1 galaxies (NLS1s), and confirm that NLS1s lie below the BBR relation of BLS1s. As a consequence, black holes grow by accretion in well-formed bulges, possibly after a major merger. As they grow, they get closer to the BBR relation for normal galaxies (Mathur & Grupe 2005). The accretion is highest in the beginning and then decreases with time. There is no AGN feedback for the control of bulge growth there. Kawaguchi et al. (2004) estimate that the NLS1 phase is characterized by very efficient accretion, at a super-Eddington rate; given the high frequency of these objects (10 of all AGN), and the average duty cycle for an average AGN phase (10^7–10^8 yrs), the essential of the BH growth is occurring during this phase: the BH grows by up to 1–2 orders of magnitude, while in the BLR phase (the most frequent and common phase) at sub-Eddington rate, the BH will only multiply its mass by about 2.

When the accretion rate is much larger than Eddington, the accretion is occuring not through a "thin" but a "slim" disk, with a cooling time larger than the viscous time, so that energy is advected towards the BH before being radiated. The luminosity can then saturate, and never be larger than a few Eddington luminosities. According to the type of BH, the accretion rate can range from 60 (Schwarzschild BH) to 300 Eddington accretion rate (Kerr BH). Whatever these accretion rates, and whatever the mass of the BH, the luminosity is always no more than 10 Eddington luminosities, as shown in Fig. 5.4, and the accretion rate a few $M_\odot$/yr (Collin & Kawaguchi 2004). This is a strong indication of a mass-limited supply, with an external mechanism to regulate the accretion.

5.2.6 Downsizing, and lifetime of activity

It is now well known from the physics of baryons, that both the star formation, and gas accretion by black holes, act to compensate the hierarchical formation of dark matter haloes, which grow larger and larger with time: the most massive star-forming galaxies and the most massive SMBH are forming at high redshift, early in the universe, while only smaller masses are assembling now (Cowie et al., 1996). Semi-analytic follow up of these processes have succeeded to reproduce the downsizing, and taking into account the constraints of luminosity functions of galaxies and AGN at all redshifts can teach us more on the formation of the objects, and for instance on the duty-cycles or lifetime of activity.

Using the observed correlations between X-ray and radio luminosities of quasars and their black hole mass, Merloni (2004) has computed the past history of SMBH, assuming their growth is only due to gas accretion. The accretion rate, and radiative regime (Eddington or not) is not fixed, but derived by the

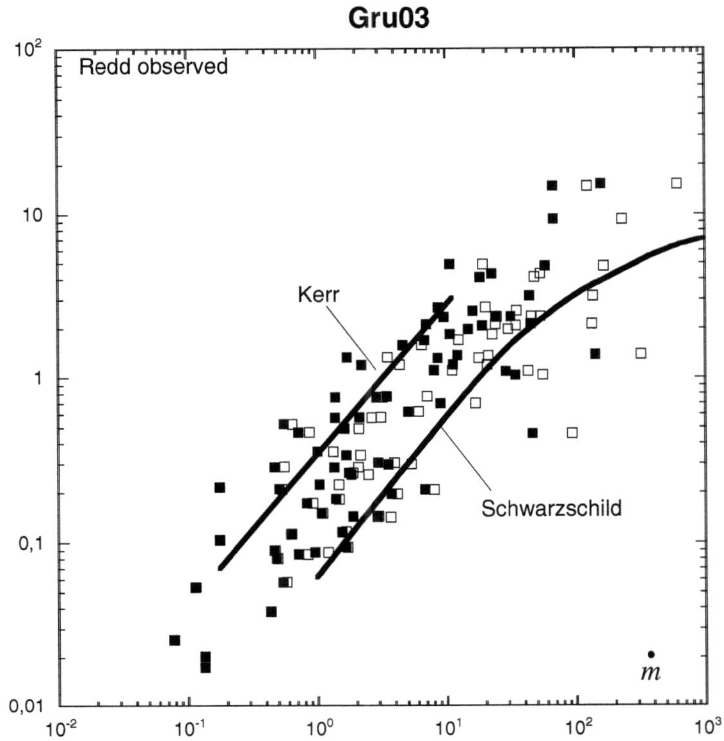

Fig. 5.4. The Eddington luminosity ratio *R*edd as a function of $\dot{m}$, the accretion rate in Eddington units, $\dot{m} = \dot{M}c^2/L_{\rm edd}$, for the NLS1 sample of Grupe & Mathur (2004). The luminosities are computed with the standard disc (open squares) and the self-gravitating disc (filled squares). The two curves correspond to the slim disc model, and respectively a Schwarzshild and a Kerr BH (from Collin & Kawaguchi 2004).

model. The results show a clear anti-hierarchical growth of the black holes, as shown in Fig. 5.5. The most massive SMBH are in place at high redshift, while at low redshift only smaller mass black holes are accreting, so that the average BH mass of observable AGN is increasing with redshift. The lifetime of activity is also varying with redshift, being shorter at early times. The mean lifetime is defined by the average over the activity of the time, weighted by the accretion rate. It is not imposed to be the doubling time of the mass at the Eddington rate, i.e. the Salpeter time $t_s = \epsilon M c^2/L_{\rm Edd} = \frac{\epsilon}{0.1} 4.5 \times 10^7$ yrs. The lifetime ranges from 10^7 yrs to assemble $10^9 M_\odot$ at z = 3, up to 10^8 yrs to assemble $10^7 M_\odot$ at z = 0 (Merloni 2004).

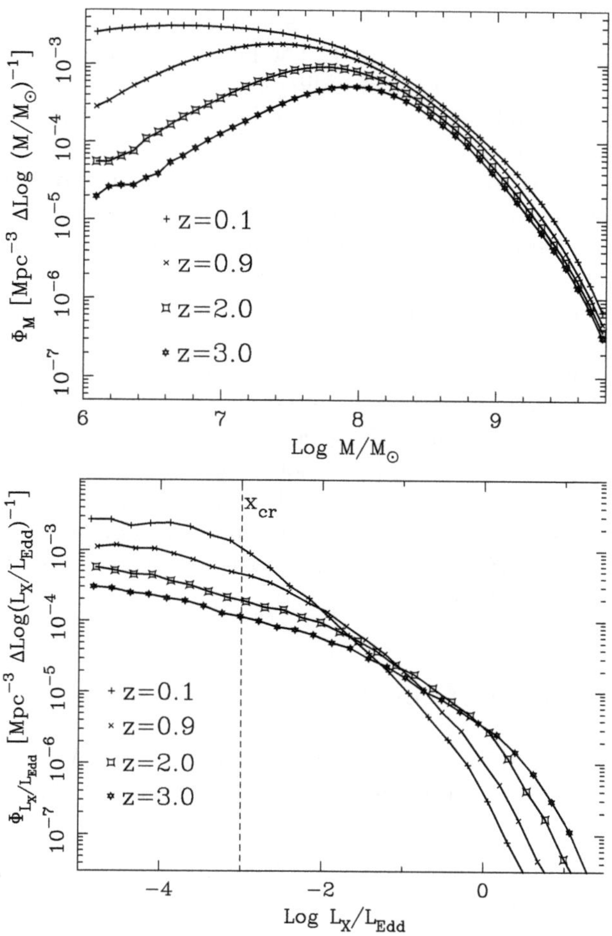

Fig. 5.5. Illustration of the downsizing effect in growing BH: **Top**: Redshift evolution of the SMBH mass function, from z = 3 (bottom) to 0.1 (top) **Bottom**: corresponding evolution of the X-ray luminosity (normalised to Eddington rate), from Merloni (2004).

5.2.7 Quasar lifetimes

The quasar lifetime can be estimated from the observations of the AGN demography, through the statistical argument that the fraction of active nuclei among the whole number of SMBH present in every early-type galaxy is a measure of the time spent in the active phase. The observation of the proximity effect (presence of an ionized region around the quasar host) puts already a lower limit to the lifetime of 10^4 yrs. The lifetime is defined as the total active time of a quasar, i.e. if the same quasar experiences episodic activity, the lifetime is the sum of the duration of the active phases. The minimum lifetime of 10^4 yrs is for one episode only. The various methods to estimate the quasar lifetime have been reviewed by Martini (2004), and results in values between 10^7 and 10^8 yrs, quite close to the Salpeter time, or mass doubling time, assuming Eddington luminosity, and with efficiency $\epsilon = 0.1$. The longest duration of an episode may be derived by the observed length of radio jets, about $t_Q \sim 10^8$ yrs. Through the measurement of the AGN-galaxy cross-correlation length, Adelberger & Steidel (2005) conclude that high and low luminosity AGN are both found in haloes of similar masses, and therefore the higher observed frequency of faint AGN must imply that their duty cycle is much longer than for bright AGN, of a few Gyr.

All these estimations are compatible with the hypothesis that the active phases of bright AGN are triggered by a major merger between two gas rich galaxies, that removes angular momentum and drives the gas towards the center (Barnes & Hernquist 1992). This hypothesis is supported by the frequent association between quasars and interactions (e.g. Hutchings & Neff 1992). However, it is possible that the observed quasar lifetime is biased in the observations, if the active phase, where the BH grows and radiation is emitted, is partially obscured by dust, as expected when a lot of gas is driven towards the galaxy centers, in the beginning of the activity. Hopkins et al. (2005) have estimated the importance of this obscuration phase in numerical simulations, and find that the quasar lifetime is then reduced from an intrinsic value of 100 Myr, to an observable value of 10–20 Myr.

5.3 Interpretation of the $M_{bh} - \sigma$ Relation

Several models have been proposed to account for the relation, all involving a simultaneous formation of bulges and SMBH, and constraining the feedback processes.

5.3.1 Radiative feedback

Although the radiative feedback is not the most efficient, it can play an important role at the end of the feeding of a giant black hole in an elliptical galaxy, which by definition does not possess much gas. Sazonov et al. (2004) have computed the equilibrium temperature T_{eq} of the gas around a quasar, heated by Compton scattering and photoionization, and cooled by continuum and line emission. When T_{eq}, which is proportional to $L/(nr^2)$ becomes larger to the virial temperature of the galaxy, proportional to the velocity dispersion σ^2, the gas is expelled, and the fueling is stopped. This occurs when the density n becomes lower than a critical density, $n_{crit} \propto L/(r^2\sigma^2)$. Assuming that the gas distribution follows the stellar distribution, which is isothermal, with an r^{-2} radial profile, then the equilibrium temperature is constant with radius, and inversely proportional to σ^2. At the critical regime, when $T_{eq} = T_{vir}$, the maximum BH-mass is then proportional to σ^4, and its growth is stopped. The radiative feedback then could explain the $M_{bh} - \sigma$ relation, for massive ellipticals, with very low gas mass content (Sazonov et al., 2005).

5.3.2 Feedback due to QSO outflows

The main cosmic formation epochs of QSOs and stars coincide (e.g. Shaver et al., 1996). Their common formation could be regulated by each other, and the QSO outflows prevent star formation (Silk & Rees 1998). The condition for the wind to be powerful enough to give escape velocity to the gas constrains the BH mass to $M_{bh} \propto \sigma^5$, which from the Faber-Jackson relation, gives $M_{bh} \propto M_{bulge}$. But the phenomenon is assumed spherical, and in reality jets are collimated, the gas is clumpy, and compressed to form stars.

The mass of the SMBH in a galaxy is quite negligible, lower than 10^{-3} in general, and the distance to which the gravitational action is significant is quite small, less than 100pc even for the more massive BH. But the energy that the SMBH can radiate or expel as jets, winds or outflows, is relatively large, in comparison to the binding energy of the gas component in the galaxy, and therefore, the energy output of the AGN could have some significant feedback action, in stopping the gas inflow favoring the accretion (Wyithe & Loeb 2002). Let us note that the binding energy of the gas, rotating at circular velocity v_c in the galaxy, is of the order of $(v_c/c)^2 \sim 10^{-6}$ of its rest mass, while the energy output of the AGN could be much larger, around 10% of the rest mass energy, so that the accretion of only a fraction 10^{-5} of the gas mass will be sufficient to release its binding energy.

This could explain the relation between the mass of the black hole and the central velocity dispersion, as a self-regulation mechanism. If the binding energy of the system of mass M is of the order $M\sigma^2$, and its dynamical time r/σ, the typical energy per dynamical time is $M\sigma^3/r$. Eliminating the mass through the virial relation $M \sim \sigma^2 r/G$, the typical energy rate or luminosity is σ^5/G. This can be considered as the maximum luminosity of the black hole before unbinding the system hosting it. Equating this to the Eddington luminosity, relates the mass of the black hole to σ^5 with an order of magnitude quite comparable to the $M_{bh} - \sigma$ relation observed (Silk & Rees 1998, Ciotti & Ostriker 2001). The proportionality factor takes into account the low coupling of the energy of the quasar (wind, outflows) to the galaxy gas. About 5–10% of the energy of the quasar must be absorbed by the galaxy to explain the self- regulation. Also the self-regulation might account for the maximum mass observed for SMBHs, which are never more massive than a few $10^9 M_\odot$.

The principle of the self-regulation is welcome to account for the very short duty cycle of nuclear activity in galaxies. The statistics of the number of AGN with respect to the quiescent SMBH in all galaxies leads to a duty cycle as short as 10^7 to 10^8 yrs, according to the strength of the AGN. The duty cycle is of the same order as the dynamical time of the gas feeding the AGN.

5.3.3 Models based on self-regulation growth

The detailed computation has been done by several groups, with different assumptions. The main lines are that the BH grows as long as the energy released in the galaxy is lower than the binding energy. If the heated gas can cool with a sufficiently short time-scale, more energy is required for the feedback, by a factor up to c/σ, and the resulting relation is then $M_{bh} \propto \sigma^4$. The mass function of quasars is obtained, assuming that the BH grows in galaxy mergers (Kauffmann & Haehnelt 2000), both by the merging of the BH, and also by gas accretion infalling during the interaction. The quasars are assumed to radiate at Eddington luminosity during their duty cycle, which is comparable to the dynamical time of the feeding system. The peak in the quasar luminosity function at $z \sim 2$ is obtained through the merging history, since it coincides to the peak of the formation of massive ellipticals, while the galaxy clusters are forming. The maximum BH masses correspond to the maximum galaxy masses, obtained at these epochs. The models then should cut off the gas infall in haloes with velocity larger than 500 km/s typically (cf. Fig. 5.6). These systems correspond to small clusters of galaxies, where the hot gas cannot cool to fuel a central SMBH. The duty cycle of quasars of 3 10^7 yrs corresponds also to the peak of quasar luminosity at $z \sim 2$, but this time scale must be shorter at high redshifts.

The main conclusions of these models is that 80–90% of the SMBH mass has been already accreted at $z \sim 1.5$ (Wyithe & Loeb 2003). The total light in galaxies can be also modelled according to the same ideas, assuming that star formation is regulated by feedback, and ceases when an energy comparable to the binding energy is released (Dekel & Woo 2003). However, the dependence in redshift of the efficiencies to accrete mass for black holes and star formation is not the same, and therefore the $M_{bh} - \sigma$ relation should be z-dependent. Only the $M_{bh} - \sigma$ relation should be constant. Indeed for a given dark halo mass, the dependence of the stellar mass is in $(1 + z)$ (constant characteristic feedback

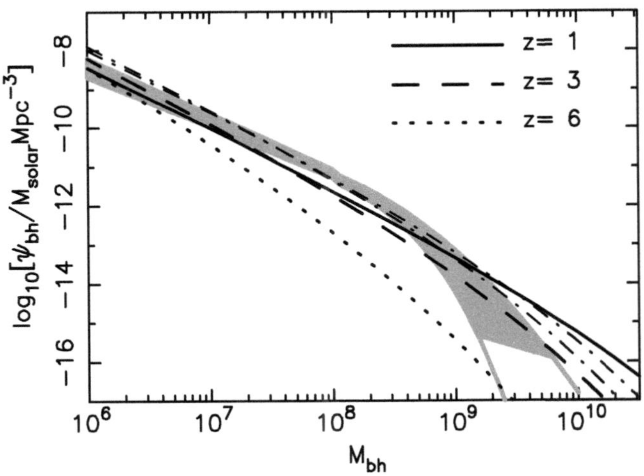

Fig. 5.6. The mass function of SMBHs. The grey region shows the mass function estimated from the velocity function of Sheth et al. (2003) and the $M_{bh} - \sigma$ relation of Merritt & Ferrarese (2001). The three lines, solid, dash, dots represent the computations at $z = 1, 3$ and 6 from the Press-Schechter halo mass function by Wyithe & Loeb (2003). The dot-dashed lines show the mass function at $z \sim 2.35$ and $z \sim 3$ implied by the observed density of quasars and a quasar lifetime $t_{\rm dyn}$ (Wyithe & Loeb 2003).

time), while that of M_{bh} is in $(1+z)^{5/2}$, and their ratio is $\sim (1+z)^{3/2}$. The BH mass is larger with respect to the stellar mass at high redshift, with the same $M_{bh} - \sigma$ relation, since stellar systems are more centrally condensed at high z.

The influence of AGN feedback due to energetic winds can be studied through numerical simulations, adopting simple recipes for the accretion of gas by the growing black holes, and energy release through winds in the interstellar medium. Di Matteo et al. (2005) have then compared major mergers between two spiral galaxies with and without the presence of BH, and shown the dramatic difference between the star formation rate, and the presence of gas in the remnant. A series of such simulations, where the star formation rate and BH accretion rate are self-regulated, can yield at the end the $M_{bh} - \sigma$ relation in the present remnants.

5.3.4 AGN feedback and cooling flows

The self-regulation between accretion and feedback appears to be at work in elliptical galaxies, where the cooling of the gas in only intermittent, and at larger scale in galaxy clusters, where huge cooling flows are impeded through re-heating by the central AGN. Ciotti et al. (1991) and Binney & Tabor (1995) developed the regulating mechanism, based on the two opposed sources: mass loss from evolving stars fuels the galaxy in gas, and the heating by Type Ia supernovae keeps it far from the cold phase, but with a faster declining efficiency. Since the heating by supernovae cannot compensate for the mass drop out, there must occur a cooling catastrophe, fueling the central black hole now known to be present in every elliptical galaxy. The energy release during the short active phase reheats efficiently the gas, which is then the source of X-ray radiation. The intermittent AGN phases are schematically shown in Fig. 5.7, revealing relaxation oscillations.

At larger scales, it has become evident in recent years, thanks to the progress in X-ray observations by Chandra and XMM-Newton, that cooling flows in galaxy clusters are completely different from the stationnary, symmetrical and abundant phenomenon expected by simple theoretical ideas. The X-ray observations have constrained the amount of cool gas observed, and the cooling rates have been reduced by at least one order of magnitude; the old view of quiet and regular, quasi-spherical cooling has given place to partial and intermittent cooling, perturbed by re-heating and feedback processes due to the central AGN. The compensation of cooling and heating could even be used to measure the power of the AGN (Churazov et al., 2002). A spectacular illustration of this perturbed cooling is the Chandra image of the cooling flow in the Perseus galaxy cluster, with bubbles, shocks, gas streaming up and down from the center, and ripples looking like emitted sound waves (Fabian et al., 2003). In the same time, cold gas in the form of CO molecules were observed in dozens of cooling flows (Edge 2001, Salomé & Combes 2003), and the amount of cold gas corresponds to the order of magnitude expected by the revised cooling rates. High spatial resolution observations show that the cold gas is associated to the dense X-ray gas, compressed by the AGN lobes, and is present around the cavity created by the lobes (cf Fig. 5.8). In these dense regions, star formation occurs, and HII regions are observed.

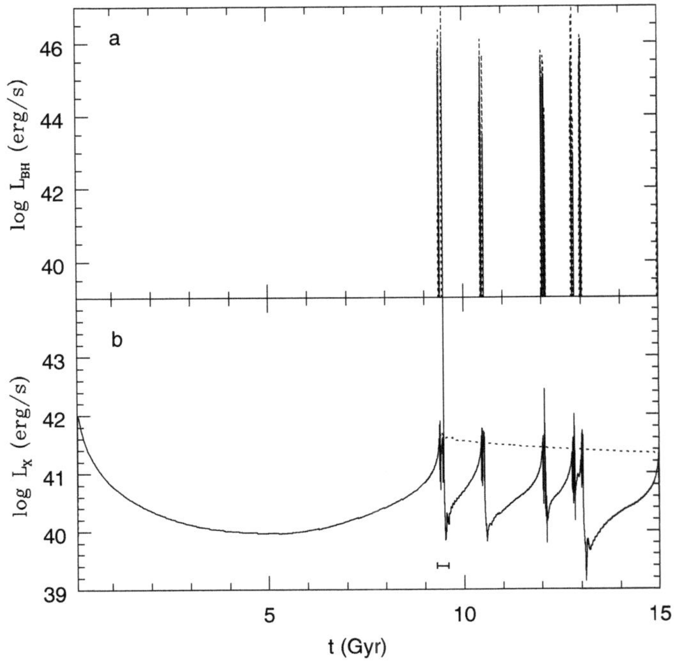

Fig. 5.7. Self-regulation due to the quasar in an elliptical galaxy: Top (a): the time evolution of L_{bh} (bolometric) emitted by the quasar at the galaxy centre. Bottom (b): the time evolution of L_x for the model with $\epsilon = 0.1$ (solid line), and that of the same model with $\epsilon = 0$ (cooling flow - dashed line), emitted by the whole galaxy, up to the truncation radius, from Ciotti & Ostriker (1997).

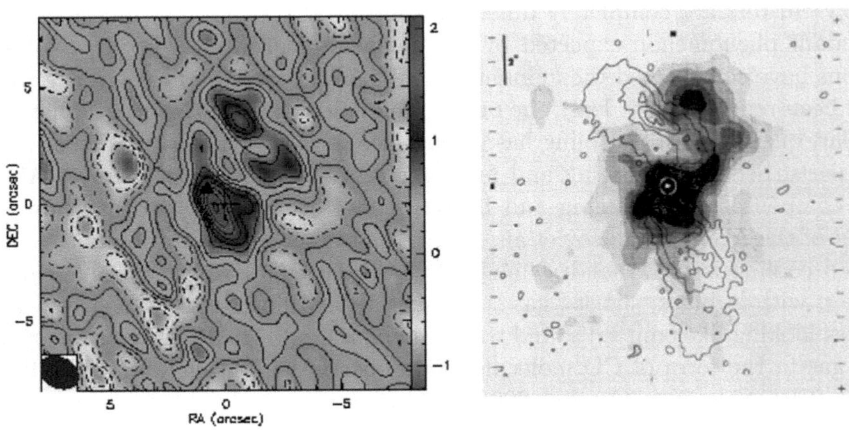

Fig. 5.8. Cold gas associated to the Abell 1795 cooling flow: **Left** CO(2-1) map obtained with the IRAM interferometer, from Salomé & Combes (2004). The AGN is indicated by the white ring. **Right** At the same spatial scale, Hα +[NII] line emission (grey scale), with 6cm contours from van Breugel et al. (1984).

All these new observations concord to draw a picture where the cooling flows are intermittent, and the AGN feedback is self-regulating both the growth of the central black hole mass, and the amount of star formation in the central galaxy.

5.3.5 Hierarchical models of galaxy formation

The mass assembly of supermassive black holes in galaxies, simultaneous to the build-up of their stars can well be integrated in the CDM scenario of hierarchical formation, and the $M_{bh} - \sigma$ relation follows (Haehnelt & Kauffmann 2000). The main hypotheses of the model are that black holes are essentially assembled in galaxy major mergers, which simultaneously form starbursts, elliptical galaxies, and fuel a QSO phase. An additional assumption is that the fraction of gas transformed into stars per dynamical time increases along the Hubble time, while the available gas fraction in galaxies decreases, as does the rate of gas accretion by galaxies. The required gas fraction accreted by the black hole grows with the mass of the halo, and the accretion time scales with the dynamical time.

In this model, the scatter in the $M_{bh} - \sigma$ relation is due to:

- M_{gas} of the bulge progenitor depends on σ, but not on the formation epoch of the bulge, while M_* depends on both;
- mergers move the galaxies on the $M_{bh} - \sigma$ relation, even at the end, when there is only BH mergers, and not enough gas left to grow the black hole.

The gas fraction in galaxies falls from 75% at $z = 3$ to 10% at $z = 0$. The gas fraction in major mergers is higher in fainter spheroids that form at high z, which are more concentrated. Elliptical/spheroids forming recently have smaller BH.

Typically a seed BH of 10^6 $M_\odot$ forms at $5 < z < 10$ and then gas is accreted. For a typical SMBH, about 30 black holes are merged. Today, black holes in big ellipticals accrete only by merging with small galaxies, but in the past gas accretion was dominant.

Both the number density of quasars as a function of redshift, and the evolution of gas abundance are found to be compatible with observations. The required duration of a QSO phase is 10^7 yrs (Haehnelt & Kauffmann 2000).

5.3.6 Feedback through bars in spiral galaxies

If quasars, which are the high luminosity end of the AGN population, are clearly associated with interactions and mergers (Hutchings & Morris 1995), it is difficult to trace evidence of dynamical triggering mechanisms for milder AGN, like Seyfert or LINERS in spiral galaxies (e.g. Combes 2003).

The accretion rates required are of course very different, of the order of a few 10 $M_\odot$/yr for quasars, and more than two orders of magnitude less for nearby Seyferts, so that the dynamical processes are much less violent, for Low Luminosity AGN (LLAGN). However, most SMBH in galaxies today have been built by gas accretion, since the successive mergers of BH from the primordial ones are far too insufficient (Islam et al., 2004), so it is of prime importance to understand the dynamical processes responsible for gas accretion in the nearby LLAGN, that can be studied in details.

Non-axisymmetries, and essentially bars, are the main providers of gravity torques, that will make the gas lose its angular momentum, and infall towards the center. This is the main mechanism both for isolated galaxies with spontaneous bar instabilities, and also during galaxy interactions, that favor bar instability: bars are then the way to propagate tidal interactions in the inner parts of galaxies (e.g. Barnes & Hernquist 1992).

The feedback mechanism due to the energy released by the AGN, such as that studied for cooling flows in elliptical galaxies (Ciotti et al., 2001) might not be efficient here, because of the low luminosity of LLAGN, and also the low coupling with the gas in a disk. Instead, other intrinsic feedback mechanisms exist, related to the dynamical mechanisms themselves that drive the gas to the center. In these cases, the $M_{bh} - \sigma$ relation could be explained only with the feedback mechanisms related to bars, that both can be responsible for bulge and BH formation (e.g. review in Combes 2001).

The demographics of nearby AGN reveals that LLAGN exist in about 40% of all galaxies, and they tend to lie in early-type galaxies (Terlevich et al., 1987, Moles et al., 1995). In an optical spectroscopic survey of 486 nearby galaxies, Ho et al. (1997) detected 420 emission-line nuclei (86% detection rate). Half of these objects can be classified as HII or star-forming nuclei, and half as some kind of AGN: Seyfert, LINERs and transition objects LINER/HII. A signature of Broad Line Region is found in 20% of the AGN, while Seyfert nuclei reside in about 10% of all galaxies. AGNs are found predominantly in luminous, early-type galaxies, while HII nuclei are in less luminous late-type objects, which is compatible with the $M_{bh} - \sigma$ relation.

Bars are present in roughly two thirds of spiral galaxies. The frequency of bars and non-axisymmetries has recently been quantified in detail from near-infrared surveys (Block et al., 2002, Laurikainen et al., 2004), and the fraction of bars does not seem to vary with redshift (e.g. Jogee et al., 2004). Since bars are observed to have a suicidal behaviour in spiral galaxies with gas (e.g. Hasan et al., 1993, Friedli & Benz 1995), bars must be reformed to explain their frequency (Bournaud & Combes 2002). The bar is destroyed by two main mechanisms: first the central mass concentration built after the gas inflow, destroys the orbital structure sustaining the bar, scatters stellar particles and pushes them into chaotic orbits (Hozumi & Hernquist 1999; Shen & Sellwood 2004). Second, the gas inflow itself weakens the bar, since the gas loses its angular momentum to the stars forming the bar (Bournaud, Combes & Semelin 2005). This increases the angular momentum of the bar wave, in decreasing the eccentricity of the orbits.

This bar destruction is reversible, and other bar episodes are driven by external gas accretion, replenishing the gas disk (cf Fig. 5.9). A typical spiral galaxy is in continuous evolution, and must accrete gas all along its life, both to maintain its star formation rate, and its spiral and bar structure. The amount of gas required is able to double the galaxy mass in about 10 Gyr. This gas cannot

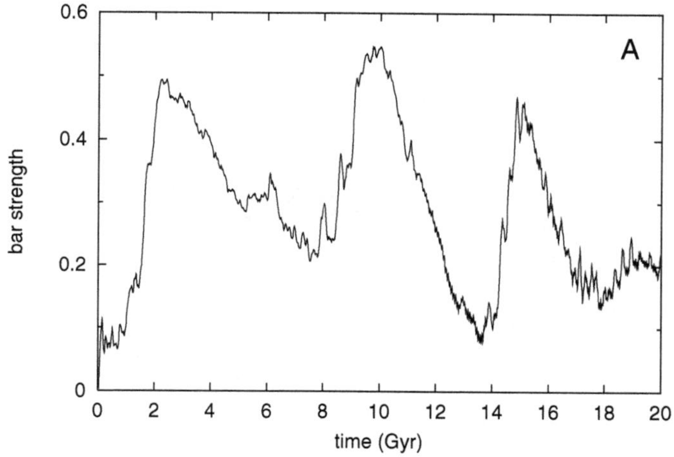

Fig. 5.9. Evolution of bar strength in a simulated spiral galaxy with gas accretion: several bar episodes provide a flow of matter towards the center, to fuel both the bulge and a central black hole (from Bournaud & Combes 2002).

be provided by accretion of gas-rich dwarf galaxies, since the interaction with companions would heat and destroy the disk. Instead, cold gas from cosmic filaments must inflow to replenish the galaxy disk; this can decrease temporarily the bulge-to-disk ratio, making the disk more self-gravitating, and triggering another bar instability. Several bar episodes can succeed each other in a Hubble time, through this dynamical feedback. At each bar episode, both bulge and BH grow in a similar manner, which explains the $M_{bh} - \sigma$ relation.

It is then more easy to understand the lack of correlation between the presence of bars and nuclear activity in spiral galaxies. The gas is driven to the very center only intermittently, through the action of a secondary nuclear bar, or even viscous torques, once the primary bar has been dissolved by the main gas flow (Garcia-Burillo et al., 2005). The first gas flow is frequently stalled at the inner Lindblad resonance, responsible for a nuclear starburst. Only when the bar has dissolved, can the gas fuel the nucleus. The activity of the nucleus can occur in short episodes, on timescales which are much shorter than the bar formation and dissolution time scales, which are of the order of 1 Gyr.

There are however some components, like resonant rings in galaxy disks, which are the remnants of the presence of bars in galaxies: once the bar has dissolved, the rings survive for some more time, from stars formed in the previously gaseous rings (e.g. Buta & Combes 1996). Unbarred galaxies observed with three resonant rings can be considered as good evidence for the bar dissolution phenomenon. The presence of outer rings has been found to be predominant in Seyfert galaxies (Hunt & Malkan, 1999, and Fig. 5.10).

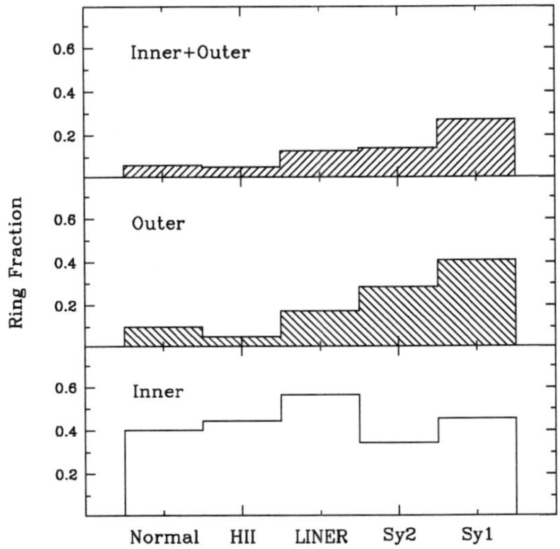

Fig. 5.10. Fraction of inner and outer rings in galaxies as a function of their activity class. Rings are thought to be formed at resonances with bars, near the corotation, for the inner rings, and the OLR (Outer Lindblad Resonance) for the outer rings. Only objects with $v < 5000 km/s$ are shown (from Hunt & Malkan 1999).

5.3.7 Other mechanisms

Since the relation between BH and bulge mass does not include the disk, ideas to involve only a spheroidal system in the required angular momentum transfer, led to the radiation drag from the bulge stars. The relativistic drag force by the radiation from bulge stars is exerted on the dusty gas distributed spheroidally in the bulge. In practice, the radiation drag saturates in case of optical thickness, but the optically thin envelope of interstellar clouds is stripped, loses angular momentum, and is accreted by the center (Umemura 2001). The M_{bh}/M_{bulge} is then a universal constant depending only on the energy conversion efficiency for nuclear fusion of hydrogen to helium. The efficiency of nuclear reactions in stars is $\epsilon = 0.007$, which would predict a too important ratio between BH and bulge mass. The efficiency falls as $1/\tau^2$, with τ the optical depth of the gas. Also the radiation drag could be strongly subject to geometrical dilution.

The global scenario relates the ultra-luminous infra-red starburst phase and the formation of a quasar. During the first phase, a large number of stars are formed, while the black hole is still growing. In this phase, the BH-to-bulge mass ratio is well below the present relation (such as has been proposed for NLS1, Mathur 2000). Then at the end of the starburst phase, the black hole has grown, and radiates at maximum luminosity, in its QSO phase, while the optical thickness of the interstellar medium decreases. Then the black hole ends its growing phase, with the well-known ratio between BH and bulge mass. Today this mechanism is no longer inefficient, since elliptical galaxies and bulges have no gas. The mechanism has been quantified, with the radiative transfer in a clumpy medium by Kawakatu & Umemura (2002). This idea has the advantage of explaining why the BH mass is not proportional also to the disk mass, where radiation drag loses its efficiency due to dilution and optical thickness (Kawakatu & Umemura 2004).

Also related to the formation of a starburst in a first phase, is the formation of Super Star Clusters (SSC) in the centers of galaxies (e.g. de Grijs et al., 2001). Sinking of these Super Star Clusters in their dark halo, due to dynamical friction, has been proposed to form cuspy stellar bulges (Fu et al., 2003); the merging of small BH associated to clusters would provide a mass ratio $M_{bh}/M_{bulge} = 10^{-4}$ only, slightly below what is observed. Some intermediate mass black holes (IMBH) of masses 800–3000 $M_\odot$ would form easily in dense and young star clusters (Portegies Zwart et al., 2004b).

5.4 Stellar Cusps and Cores and Binary Black Holes

The supermassive black hole present in every spheroid, has a gravitational influence on its stellar environment. It can form a cusp, through gravitational attraction, or a hole by swallowing the low angular momentum stars in the neighbourhood, or flatten a core, through interaction with another merging black hole. The observation of the stellar profile in galaxy centers can then teach us the formation history of the SMBH (e.g. Merritt 2004).

A well known dichotomy is observed between massive and small ellipticals (e.g. Lauer et al., 1995):

- Cusps (steep power-law in stellar central density profile) are characteristic of low-mass ellipticals, with disky isophotes and weak rotation;
- Cores (flat central density profile) are found in high-mass galaxies, with boxy isophotes and no rotation.

5.4.1 Formation of a cusp of stars around the black hole

The density profile in the stellar component around a massive black hole depends on the relative value of the 2-body relaxation time scale with respect to the Hubble time, or more precisely with respect to the formation time of the black hole.

The relaxation time can be expressed by $T_{rel} = \frac{V^3}{8\pi G^2 m\rho(r) log(\Lambda)}$ where V is the mean relative speed between the stars, m the mean stellar mass, and $\rho(r)$ the volumic density in the nucleus ($log(\Lambda)$ is the Coulomb parameter). It is well known that globally in a galaxy, the relaxation time is much longer than the Hubble time, and varies approximately as $0.2(N/logN)t_c$, if N is the total number of stars in the system, and $t_c = R/V$ is the crossing time. However, the relaxation time becomes shorter than the Hubble time in dense systems like globular clusters, and the nuclear stellar clusters may also approach these conditions (small N). For the galactic center, with a volumic density of stars of $10^7 M_\odot/pc^3$, $\langle V^2 \rangle^{1/2} = 225$ km/s, this relaxation time is 3×10^8 yr.

Young (1980) has computed the adiabatic growth of a black hole inside a nuclear stellar cluster: the growth rate is assumed to be longer than the cluster dynamical time scale but shorter than the relaxation time scale. Then a stellar cusp forms, stars being attracted by the black hole. The power-law profile has a slope larger then 2, up to 2.5. Two regimes can be distinguished, according to the initial mass of the black hole, the slope being larger for more massive black holes (Cipollina & Bertin, 1994).

If 2-body relaxation can take place among the stars, then the cusp is less pronounced, and the slope is 1.75 (7/4) (Bahcall & Wolf 1976). N-body simulations can retrieve asymptotically this result, cf Preto et al. (2004) and Fig. 5.11.

The black hole can grow by swallowing the nearby stars, that have an angular momentum lower than $J = (2GM_{bh}r_t)^{1.2}$, with r_t being the tidal radius, beyond which a star is disrupted by the black hole. After a dynamical time, if 2-body relaxation does not refill these particles in the loss-cone, the black hole will starve. In fact, the angular momentum can diffuse faster than the energy (faster than the stellar relaxation time T_{rel}), and the low angular momentum stars are replenished faster, which increases the accretion rate to $T_{rel}(1 - e^2)$, with e the

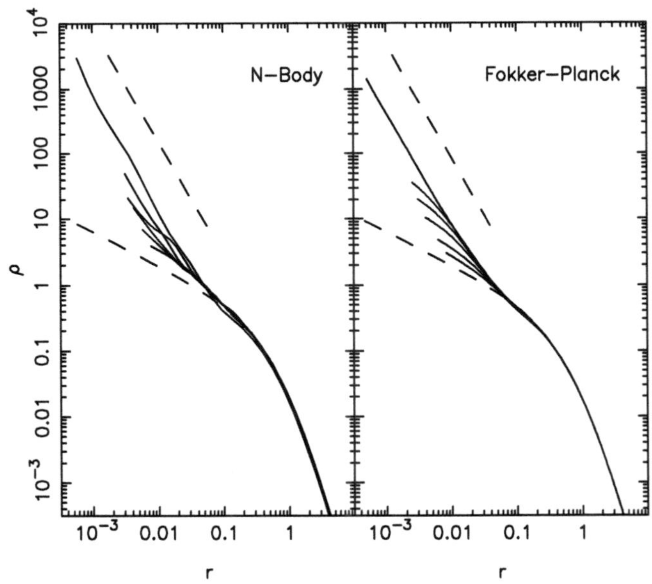

Fig. 5.11. Evolution of the mass density profile around a massive black hole: **Left**: density from N-body simulations, at times $t/T_{rel} = 0.07, 0.13, 0.2, 0.25, 0.33, 1$. **Right**: Densities predicted from the Fokker-Planck equation at the same times. The curves progressing from bottom to top, are bracketted by the lower dashed curves at $t = 0$ and the upper dashed curves showing $\rho \propto r^{-7/4}$, the asymptotic solution to the Fokker-Planck equation (from Preto et al., 2004).

excentricity of the orbits (Frank & Rees 1976, Lightman & Shapiro 1977). It is possible that black holes less massive than 10^7 $M_\odot$ can be formed only through stellar accretion.

5.4.2 Wandering of the black hole

A massive compact object, like a black hole, embedded in a dense stellar system, experiences a multitude of gravitational encounters; the total action on this body is composed of a slowly varying force, deriving from the smooth stellar system potential, and a rapidly fluctuating stochastic force due to discrete encounters with individual stars. The motion of the black hole is then similar to that of a random walk (Chatterjee et al., 2002, 2003), and this Brownian motion has been invoked to counter the effect of the empty loss-cone, and provide new stars to interact with. It is expected that equipartition of energy is reached, so that the velocity acquired by the black hole is small, even if the black hole interacts with particles with high velocity dispersion. Numerical simulations over-estimate this effect, since the number of particles is far smaller than the realistic number.

The effect of wandering might be even more interesting on a binary black hole (see next section). When interacting with a third body, the binary can eject stars at large velocity. Also the dynamical friction on a binary is less than on a single black hole. Finally, when a black hole binary merges, the gravitational waves emitted take away some momentum, producing a recoil of the merged object. This hardly ejects it out of the galaxy (except may be at high redshift, when the potential wells are not deep enough), but can produce a large wandering of the black hole, and a flattening of the cusp into a core (Merritt et al., 2004).

5.4.3 Binary black holes

The formation of binary black holes in the centers of galaxies is a natural prediction of the hierarchical scenario, given the presence of a massive black hole in nearly every galaxy. The successive physical processes able to brake the two black holes in their relative orbit have been considered by Begelman et al. (1980). Each black hole sinks first toward the merger remnant center through dynamical friction onto stars. A binary is formed; but the lifetime of such a binary can be much larger than a Hubble time, if there is not enough stars to replenish the loss cone, where stars are able to interact with the binary. Once a loss cone is created, it is replenished only through the 2-body relaxation between stars, and this can be very long (T_{rel}). If the binary lifetime is too long, another merger with another galaxy will bring a third black-hole. Since a three-body system is unstable, one of the three black-holes will be ejected by the gravitational slingshot effect (Saslaw et al., 1974).

Numerical simulations have brought more precision in the determination of the lifetime of the binary, although numerical artifacts have stimulated debates. Ebisuzaki et al. (1991) claimed that the lifetime of the binary should be much shorter if its orbit is eccentric, since then the binary can interact with more stars and overcome the loss cone problem. The first numerical simulations tended to show that orbit eccentricity should grow quickly through dynamical friction (Fukushige et al., 1992). Mikkola & Valtonen (1992) and others found that the eccentricity in fact grows only very slowly (Quinlan 1996).

Numerical simulations suffer from a restricted number of bodies N, and consequently of a large random velocity of the binary (that should decrease in $N^{-1/2}$). The binary then wanders in or even out of the loss cone, and the effect of the loss cone depletion does not occur (Makino et al., 1993). Also the 2-body relaxation time is shorter than in the real system, contributing to replenish the cone.

More recent simulations, with increased number of particles, have indeed shown that the hardening of the binary depends of the relaxation time-scale, proportional to the number of particles (Makino & Funato 2004), and therefore another mechanism is required to merge the binary, such as gas accretion.

The ejection out of the core of stars interacting with the binary weakens the stellar cusp, while the binary hardens. In addition, a sinking black hole during a merger, contributes also to form a core (Nakano & Makino 1999, Fig. 5.12). Gas dissipation, and star formation in the central concentration formed, can restore the cusp. This might explain the existence of cores in the center of giant

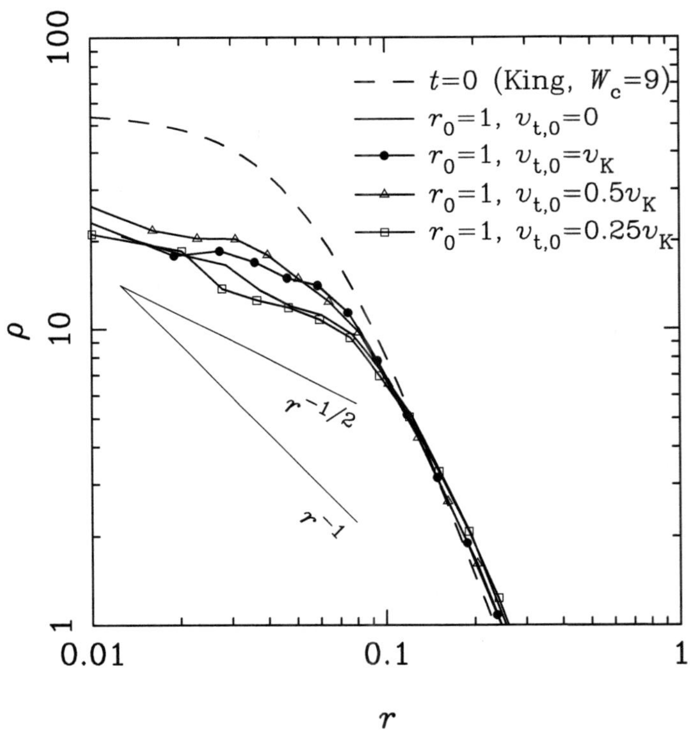

Fig. 5.12. Density profiles obtained through N-body simulations of the spiral-in of a massive BH, from Nakano & Makino (1999). The falling BH mass is 4% of the galaxy mass. The BH has initially only a tangential velocity $v_{t,0}$, function of the Kepler velocity v_K, as indicated on the figure.

ellipticals, having experienced multiple mergers, while small-mass systems have still a cusp. The computation of the deficit of stars in the central profiles, and the formation of cores, appear in agreement with observations (Graham 2004, Volonteri et al., 2003b).

5.5 History of Accretion onto SMBH: X-ray Constraints

It is possible to relate the amount of energy produced in AGN, per comoving volume, to the mass density accreted by black holes, and therefore to their growth history, given an accretion efficiency ϵ (e.g. Soltan 1982). From the M_{bh}-M_{bul} relation, the density of black hole can be estimated to:

$$\rho_{bh} = 1.1\, 10^6 (M_{bh}/M_{bul}/0.002)(\Omega_{bul}/0.002h^{-1}) M_\odot/Mpc^3$$

as a function of the mass density Ω_{bul} in stellar spheroids.

The efficiency of conversion of mass into energy is generally accepted to be around $\epsilon = 0.1$ for quasars, this value being an average between two extremes: for a Schwarzschild black hole, without rotation, this value is low ($\epsilon = 0.054$), and for a Kerr black hole, with maximum rotation, it can reach $\epsilon = 0.37$ (Thorne 1974). It is possible to estimate the present growth of black holes by the optical QSO luminosity function, which yields an accreted mass density of

$$\rho_{accr} = 2 \times 10^5 \frac{0.1}{\epsilon} M_\odot/Mpc^3$$

(Yu & Tremaine 2002). This is however a lower limit, since most AGN light is absorbed at optical wavelengths. The estimation from the far-infrared, assuming a contribution of AGN to the FIR of 30%, is:

$$\rho_{accr} = 7.5 \times 10^5 \frac{0.1}{\epsilon} M_\odot/Mpc^3$$

which confirms that the accretion radiation mainly does not get out in optical light, but is re-radiated by dust.

It has been shown that the optically selected AGN correspond only to one third of the X-ray background (Barger et al., 2003), which is now essentially resolved in individual sources, at least at energy lower than 2–5 kev (Worsley et al., 2005). The accretion density estimated from the X-ray background has been estimated as high as 3–5 $\times 10^5$ $M_\odot/Mpc^3$ (Salucci et al. 1999) and 6–9 $\times 10^5$ $M_\odot/Mpc^3$ (Fabian & Iwasawa 1999). These estimations have now been updated to lower values. Taking into account the hard X-ray selected AGN, and their total corrected bolometric luminosity, Barger et al. (2005) find a strong evolution with redshift of the AGN production rate, in $(1 + z)^\alpha$, with $\alpha = 3.2$ between $z = 0$ and $z = 1$. At higher redshifts the production decreases again (with $\alpha = -1$), but the global integrated production is dominated by the z = 1 objects. The deduced accretion density at z = 0 is

$$\rho_{accr} = 4 \times 10^5 \frac{0.1}{\epsilon} M_\odot / Mpc^3$$

and about 40% of this accretion density is due to the Broad-Line AGN, that are also the most powerful AGN (Steffen et al., 2003). The redshift evolution of the accretion rate is remarkably similar to the star formation history. Both histories reveal a downsizing effect, in the sense that the most active objects assemble mass in the early universe, and are no longer active now, while the remaining activity occurs now in the lower mass objects. Indeed, the most powerful and massive AGN observed at high redshifts have disappeared now, to the benefit of less powerful objects. This is also true for starbursts and ultra-luminous objects. When compared with the density of black hole mass now in galaxies, estimated from the velocity dispersion of early-type galaxies determined from the Sloan Survey and the BH mass to dispersion relation (Yu & Tremaine 2002) or other estimations based on the mass density in the local universe and the $M_{bh} - \sigma$ relation (Aller & Richstone 2002, Marconi et al., 2004), there is a good concordance with the mass expected from accretion luminosity, if the efficiency of accretion is $\epsilon = 0.1$. If the efficiency is higher, then there must exist obscured AGN, not counted in the above balance.

With deep X-ray surveys, it is now possible to draw quite precise conclusions on the AGN redshift evolution, and luminosity functions (Hasinger 2004). There is a clear evolution of luminosity functions versus redshift, which results from both number density evolution and luminosity evolution. The evolution with z depends strongly on luminosity. For low-luminosity AGN, the amplitude of number density evolution is less pronounced, and the maximum occurs at low redshift, while bright quasars reveal a factor 100 increase in density, and the peak occurs at higher z. For $L_X = 10^{42}$–10^{43} erg/s, the peak is at $z \sim 0.5$–0.7, while it is at z $\sim$ 2 for $L_X = 10^{45}$–10^{46} erg/s. This points towards a downsizing effect: rare and bright QSO form very early in the Universe, and then decline by two orders of magnitude, while the more frequent low-luminosity AGN form later, and decline by only a factor 10 in number.

A clear decline at higher redshifts is now detected. The comparison between the X-ray selected and optically-selected AGN is illustrated in Fig. 5.13. The comparison is to be taken with caution, since the precise shape of the number density evolution depends highly on luminosity.

5.6 Conclusion

The BH masses measured in local spheroids and spiral galaxies with bulges, are tightly correlated to the central velocity dispersion, and with more scatter to the bulge luminosity or mass. This has been interpreted as a concomitant formation of stars in the bulge and growth of the black hole at the center. The BH growth occurs by a combination of external gas accretion, and coalescence

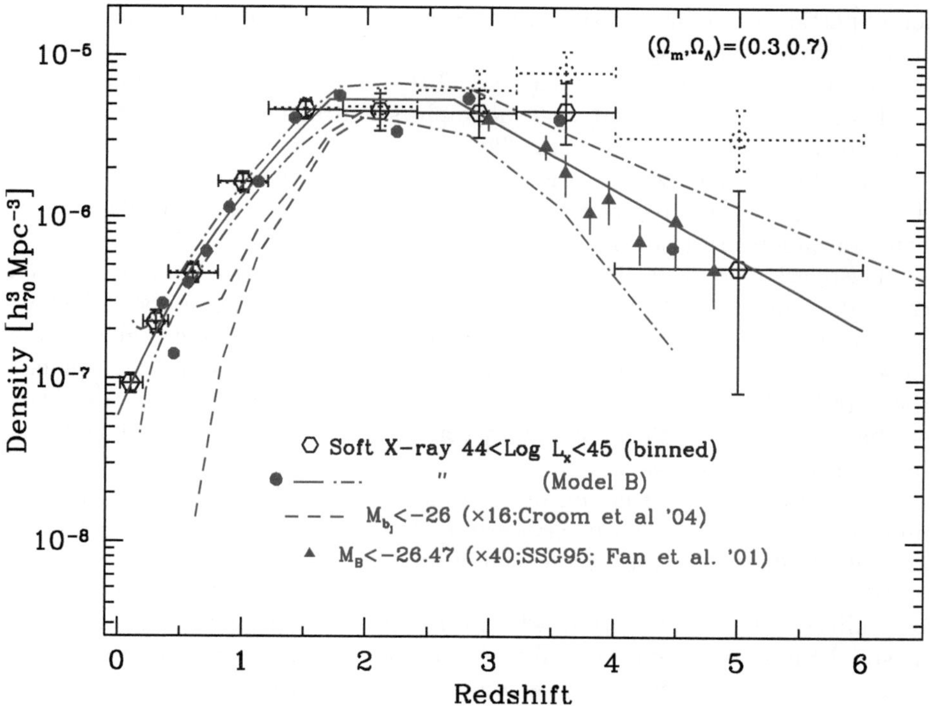

Fig. 5.13. Density of luminous QSO as a function of redshift, from soft X-ray selected sources (empty circles), and optically selected ones (dash lines and triangles), from Hasinger (2004). The $z < 2$ dash curve for optically selected QSO ($M_{bJ} < -26$) is from a combination of 2dF and 6dF surveys (Croom et al., 2004). The triangles at $z > 2.7$ have been renormalised from Schmidt et al. (1995) and Fan et al. (2001).

of binary BHs, during galaxy mergers. From the measured density of BH in the local universe, it is possible to deduce how many AGN have radiated in the past, while accreting and growing these black holes. The comparison of the optical, far-infra-red or X-rays outputs, either in the form of point sources, or unresolved in the background, with the local BH density constrains the radiating efficiency of the AGN; the efficiency should be around $\epsilon \sim 0.1$, in between that expected from Schwarzschild and Kerr black holes.

The discovery of high redshift ($z > 6$) bright QSOs has posed the problem of their formation in a short time-scale. This problem appears related to the anti-hierarchical evolution observed: the brighter AGN form earlier, in a shorter timescale, while the low-luminosity AGN take more time to form, and reveal less evolution amplitude. Their peak number density occurs at lower redshift than for bright QSO. This behaviour might be explained by the much higher gas density at high z, the higher merging rate and the shorter dynamical time-scale.

The most massive black holes are confined in galaxy nuclei. But there must exist an intermediate mass category for BH, between the stellar-mass BHs and the supermassive ones. Those IMBHs with masses around 10^4–10^5 $M_\odot$, are not found particularly in nuclei, since they have not been braked by dynamical friction. There appears now some evidence for the existence of these IMBHs, but it is not yet clear whether they might extend the $M_{bh} - \sigma$ relation, at low masses.

Through the study of AGN demographics and several clues such as their cross-correlation length, the quasar lifetimes are determined to be around a few 10^7 yrs; this is understood as the sum of all activity phases in a single object, if there exist episodic recurrent activity. This lifetime is luminosity dependent, being much longer for low luminosity AGN.

The relation between BH mass and bulge luminosity breaks down for a certain category of AGN, the NLS1 which appear to have formed their stars in the past at a higher rate than growing their black hole. Evidence is found that nuclei in these objects are now accreting mass much above the Eddington rate, although they are barely radiating around the Eddington rate. Gas accretion towards a starburst or a BH are then not exactly concomitant, there could be time delays between the two processes.

Can we trace the formation of supermassive black holes back to the early universe? Small BH can form very early by the collapse of Population III stars, around $z \sim 20$, with masses of a few hundred solar masses. But the low efficiency of dynamical friction on these small masses, make the BH growth through merging unlikely. Most of the BH growth must then be due to gas accretion. There must exist a large range in mass distribution of black holes wandering all across galaxies; when accreting gas, these mini-quasars can contribute to the reionization of the Universe.

Different kinds of feedback processes have been invoked to account for the $M_{bh} - \sigma$ relation, and in particular the energy released in AGN activity, QSO outflows, radiation, that could self-regulate the gas accretion. Such processes are particularly conspicuous in the center of cooling flow clusters, where gas reheating regulate the cooling flow. Episodically, the cooling gas fuels the central AGN, triggering a new activity phase. For low luminosity AGN, dynamical instabilities in galaxy disks, like spirals and bars, are invoked to fuel the central nucleus, and also self-regulate the gas accretion. Bars are destroyed through gas inflow, and this could explain the apparent lack of correlation between nuclear activity and the presence of strong bars.

The existence of binary black holes is a natural consequence of the hierarchical scenario of galaxy formation, if there exists a supermassive black hole dormant in each nucleus. The coalescence of the binaries should occur relatively rapidly, to avoid the loss of SMBH through 3-body interactions. Since dynamical friction on bulge stars is not sufficient, the coalescence must be due to gas accretion by the nucleus. During the hardening of the binary, energy is given to the central stellar population, and any cuspy density distribution can be flattened into a core, by this dynamical heating. A cuspy stellar distribution can later reform around the resulting single black hole.

In spite of significant progress in massive black hole formation in recent years, many questions remain open, such as the evolution of the $M_{bh} - \sigma$ relation with redshift, the local exceptions to the relation (for instance galaxies like M33), the radiative efficiency of the nucleus for a given accretion, etc..

References

Abel T., Bryan G.L., Norman M.L.: The Formation of the First Star in the Universe, 2002, Science **295**, 93

Adelberger K.L., Steidel C.C.: Constraints from Galaxy-AGN Clustering on the Correlation between Galaxy and Black Hole Mass at Redshifts $2 < Z < 3$, 2005, ApJ in press (astro-ph/0505210)

Aller M.C., Richstone D.: The Cosmic Density of Massive Black Holes from Galaxy Velocity Dispersions, 2002, AJ **124**, 3035

Bahcall, J. N., Wolf, R. A.: Star distribution around a massive black hole in a globular cluster, 1976, ApJ **209**, 214

Barger A.J. , Cowie, L. L., Capak, P. et al.: Optical and Infrared Properties of the 2 Ms Chandra Deep Field North X-Ray Sources, 2003, AJ **126**, 632

Barger A.J., Cowie L.L., Mushotzky R.F et al.: The Cosmic Evolution of Hard X-Ray-selected Active Galactic Nuclei, 2005, AJ **129**, 578

Barth, A. J., Greene, J. E.. Ho, L. C.: Dwarf Seyfert 1 Nuclei and the Low-Mass End of the $M_{BH} - \sigma$ Relation, 2005, ApJ **619**, L151

Barth A.J., Ho L.C., Rutledge R.E., Sargent W. L. W.: POX 52: A Dwarf Seyfert 1 Galaxy with an Intermediate-Mass Black Hole, 2004 ApJ **607**, 90

Barnes, J. E., Hernquist, L.: 1992, Dynamics of interacting galaxies, ARAA **30**, 705

Becker, R. H., Fan, X., White, R. L. et al: Evidence for Reionization at $z\sim6$: Detection of a Gunn-Peterson Trough in a z=6.28 Quasar, 2001, AJ **122**, 2850

Begelman M.C., Blandford R.D., Rees M.J.: Massive black hole binaries in active galactic nuclei, 1980, Nature **287**, 307

Binney J., Tabor G.: Evolving Cooling Flows, 1995, MNRAS **276**, 663

Block, D., Bournaud, F., Combes, F.. Puerari, I., Buta, R.: Gravitational torques in spiral galaxies: Gas accretion as a driving mechanism of galactic evolution2002, A&A **394**, L35

Bond, J. R., Arnett, W. D., Carr, B. J.: The evolution and fate of Very Massive Objects, 1984, ApJ **280**, 825

Bournaud F., Combes F.: Gas accretion on spiral galaxies: Bar formation and renewal, 2002, A&A **392**, 83

Bournaud F., Combes F., Semelin B.: 2005, The lifetime of galactic bars: Central Mass Concentrations and gravity torques, MNRAS, sub.

Bromm V., Larson R.B.: The First Stars, 2004, ARAA **42**, 79

Buta R., Combes F.: Galactic Rings, 1996, Fundamental of Cosmic Physics, **17**, 95

Chatterjee, P., Hernquist, L., Loeb, A.: Dynamics of a Massive Black Hole at the Center of a Dense Stellar System, 2002, ApJ **572**, 371

Chatterjee, P., Hernquist, L., Loeb, A.: Effects of Wandering on the Coalescence of Black Hole Binaries in Galactic Centers, 2003, ApJ **592**, 32

Churazov, E., Sunyaev, R., Forman, W., Böhringer, H.: Cooling flows as a calorimeter of active galactic nucleus mechanical power, 2002, MNRAS **332**, 729

Ciotti L., Ostriker J.P.: Cooling Flows and Quasars: Different Aspects of the Same Phenomenon? I. Concepts, 1997, ApJ **487**, L105

Ciotti L., Ostriker J.P.: Cooling Flows and Quasars. II. Detailed Models of Feedback-modulated Accretion Flows, 2001, ApJ **551**, 131

Ciotti, L., Pellegrini, S., Renzini, A., D'Ercole, A.: Winds, outflows, and inflows in X-ray elliptical galaxies, 1991, ApJ **376**, 380

Cipollina, M., Bertin, G.: Analytical study of adiabatic black hole growth in spherical stellar systems, 1994, A&A **288**, 43

Colbert E. J. M., Heckman, T. M. Ptak, A. F. et al. Old and Young X-Ray Point Source Populations in Nearby Galaxies, 2004, ApJ **602**, 231

Collin, S., Kawaguchi, T.: Super-Eddington accretion rates in Narrow Line Seyfert 1 galaxies, 2004, A&A **426**, 797

Combes F.: Fueling of AGN, 2001, in Starburst-AGN Connection, ed. I. Aretxaga et al., World Scientific

Combes F.: AGN Fueling: The Observational Point of View, 2003, in "Active Galactic Nuclei: from Central Engine to Host Galaxy" ASP Conf. Ser., 2003, ed. S. Collin, F. Combes and I. Shlosman, p. 411

Cowie, L. L., Songaila, A., Hu, E. M., Cohen, J. G.: New Insight on Galaxy Formation and Evolution From Keck Spectroscopy of the Hawaii Deep Fields, 1996, AJ **112**, 839

Croom S. M., Smith R. J., Boyle B. J. et al: The 2dF QSO Redshift Survey - XII. The spectroscopic catalogue and luminosity function, 2004, MNRAS **349**, 1397

Dekel A., Woo J: Feedback and the fundamental line of low-luminosity low-surface-brightness/dwarf galaxies, 2003, MNRAS **344**, 1131

de Grijs R., O'Connell R.W., Gallagher J.S.: The Fossil Starburst in M82, 2001, AJ **121**, 768

Di Matteo T., Springel V., Hernquist L.: Energy input from quasars regulates the growth and activity of black holes and their host galaxies, 2005, Nature **433**, 604

Dijkstra, M., Haiman, Z., Loeb, A.: A Limit from the X-Ray Background on the Contribution of Quasars to Reionization, 2004, ApJ **613**, 646

Dubus G., Rutledge R.E.: Chandra observations of the nucleus of M33, 2002, MNRAS **336**, 901

Ebisuzaki, T., Makino, J., Okumura, S. K.: Merging of two galaxies with central black holes, 1991, Nature **354**, 212

Ebisuzaki, T., Makino, J., Tsuru, T.G.: et al: Missing Link Found? The "Runaway" Path to Supermassive Black Holes, 2001, ApJ **562**, L19

Edge A.: The detection of molecular gas in the central galaxies of cooling flow clusters, 2001, MNRAS **328**, 762

Fabian A.C., Iwasawa K.: The mass density in black holes inferred from the X-ray background, 1999, MNRAS **303**, L34

Fabian A.C., Sanders J.S., Allen, S. W. et al.: A deep Chandra observation of the Perseus cluster: shocks and ripples, 2003, MNRAS **344**, L43

Fan X., Strauss M.A., Schneider D.P.: High-Redshift Quasars Found in Sloan Digital Sky Survey Commissioning Data. IV. Luminosity Function from the Fall Equatorial Stripe Sample, 2001, AJ **121**, 54

Filippenko, A. V., Ho, L. C.: A Low-Mass Central Black Hole in the Bulgeless Seyfert 1 Galaxy NGC 4395, 2003, ApJ **588**, L13

Frank J., Rees M.J.: Effects of massive central black holes on dense stellar systems, 1976, MNRAS **176**, 633

Friedli D., Benz W.: Secular evolution of isolated barred galaxies. II. Coupling between stars and interstellar medium via star formation, 1995, A&A **301**, 649

Fu, Y. N., Huang, J. H., Deng, Z. G.: Formation of bulges in very late-type galaxies from super star clusters, 2003, MNRAS **339**, 442

Fukushige, T., Ebisuzaki, T., Makino, J.: Gravitational wave burst produced by merging of central black holes of galaxies, 1992, ApJ **396**, L61

Fuller T.M., Couchman H.M.P.: Simulation of Primordial Object Formation, 2000, ApJ **544**, 6

Garcia-Burillo S., Combes F., Schinnerer E., Boone F., Hunt L.: Molecular Gas in NUclei of GAlaxies (NUGA): IV. Gravitational Torques and AGN Feeding, 2005, A&A, in press (astro-ph/0507070)

Gebhardt K, Bender R., Bower G. et al. A Relationship between Nuclear Black Hole Mass and Galaxy Velocity Dispersion, 2000, ApJ **539**, L13

Gebhardt, K., Lauer, T. R., Kormendy, J. et al.: M33: A Galaxy with No Supermassive Black Hole, 2001, AJ **122**, 2469

Gebhardt, K., Rich, R. M., Ho, L. C.: A 20,000 Msun Black Hole in the Stellar Cluster G1, 2002, ApJ **578**, L41

Ghez, A. M., Duchêne, G., Matthews, K. et al. The First Measurement of Spectral Lines in a Short-Period Star Bound to the Galaxy's Central Black Hole: A Paradox of Youth, 2003, ApJ **586**, L127

Gould, A., Rix, H.-W.: Binary Black Hole Mergers from Planet-like Migrations, 2000, ApJ **532**, L29

Graham A.W.: Core Depletion from Coalescing Supermassive Black Holes, 2004, ApJ **613**, L33

Grupe, D., Mathur, S.: $M_{BH}-\sigma$ Relation for a Complete Sample of Soft X-Ray-selected Active Galactic Nuclei, 2004, ApJ **606**, L41

Haehnelt M.G., Rees M.J.: The formation of nuclei in newly formed galaxies and the evolution of the quasar population, 1993, MNRAS **263**, 168

Haehnelt M.G., Natarajan P., Rees M.J.: High-redshift galaxies, their active nuclei and central black holes, 1998, MNRAS **300**, 817

Haehnelt M.G., Kauffmann G.: The correlation between black hole mass and bulge velocity dispersion in hierarchical galaxy formation models, 2000, MNRAS, **318**, L35

Haiman Z., Abel T., Rees M.J.: The Radiative Feedback of the First Cosmological Objects, 2000, ApJ **534**, 11

Hansen B., Milosavljevic M.: The Need for a Second Black Hole at the Galactic Center, 2003, ApJ **593**, L77

Hasan, H., Pfenniger, D., Norman, C.: Galactic bars with central mass concentrations - Three-dimensional dynamics, 1993, ApJ **409**, 91

Hasinger G.: When Supermassive Black Holes were growing: Clues from Deep X-ray Surveys, 2004, in "Growing Black Holes"; A. Merloni, S. Nayakshin and R. Sunyaev (eds.) (astro-ph/0412576)

Hills J.G.: Possible power source of Seyfert galaxies and QSOs, 1975, Nature **254**, 295

Ho L.C., Filippenko A.V., Sargent W.L.W.: A Search for "Dwarf" Seyfert Nuclei. V. Demographics of Nuclear Activity in Nearby Galaxies, 1997, ApJ **487**, 568

Hopkins P., Hernquist L., Martini P. et al.: A Physical Model for the Origin of Quasar Lifetimes, 2005, ApJ sub (astro-ph/0502241)

Hozumi S., Hernquist L.: Barred Galaxies with Massive Central Black Holes, 1999, in "Galaxy Dynamics", ASP Conf Series, ed. D. R. Merritt, M. Valluri, and J. A. Sellwood, Vol **182**, p.259

Hunt L.K., Malkan M.A.: Morphology of the 12 Micron Seyfert Galaxies. I. Hubble Types, Axial Ratios, Bars, and Rings, 1999 ApJ **516**, 660

Hutchings, J. B., Neff, S. G.: Optical imaging of QSOs with 0.5 arcsec resolution, 1992, AJ **104**, 1

Hutchings, J. B., Morris, S. C.: Imaging of low redshift QSOs with WFPC2, 1995, AJ **109**, 928

Islam R. R., Taylor J. E., Silk J.: Massive black hole remnants of the first stars in galactic haloes, 2003, MNRAS **340**, 647

Islam R. R., Taylor J. E., Silk J.: Massive black hole remnants of the first stars - I. Abundance in present-day galactic haloes, 2004, MNRAS **354**, 427
Jogee, S., Barazza, F. D., Rix, H.-W., et al.: Bar Evolution over the Last 8 Billion Years: A Constant Fraction of Strong Bars in the GEMS Survey, 2004, ApJ **615**, L105
Kauffmann, G., Haehnelt, M.: A unified model for the evolution of galaxies and quasars, 2000, MNRAS **311**, 576
Kawaguchi T., Aoki K., Ohta K., Collin S.: Growth of massive black holes by super-Eddington accretion, 2004, A&A **420**, L23
Kawakatu N., Umemura M.: Radiation drag driven mass accretion in a clumpy interstellar medium: implications for the supermassive black hole-to-bulge relation, 2002, MNRAS **329**, 572
Kawakatu N., Umemura M.: Why Are Massive Black Holes Small in Disk Galaxies?, 2004, ApJ **601**, L21
Kazantzidis, S., Mayer, L., Colpi, M. et al.: The Fate of Supermassive Black Holes and the Evolution of the $M_{BH}-\sigma$ Relation in Merging Galaxies: The Effect of Gaseous Dissipation, 2005, ApJ **623**, L67
King A.R., Davies M.B., Ward M.J. et al.: Ultraluminous X-Ray Sources in External Galaxies, 2001, ApJ **552**, L109
Kormendy J., Gebhardt K.: Supermassive Black Holes in Nuclei of Galaxies, 2001, in 20th Texas Symposium on relativistic astrophysics, AIP Conf. Proceedings (astro-ph/0105230), p.363
Lauer, T. R., Ajhar, E. A., Byun, Y.-I. et al.: The Centers of Early-Type Galaxies with HST.I.An Observational Survey, 1995, AJ **110**, 2622
Laurikainen, E., Salo, H., Buta, R., Vasylyev, S.: Bar-induced perturbation strengths of the galaxies in the Ohio State University Bright Galaxy Survey - I, 2004, MNRAS **355**, 1251
Ligthman A.P., Shapiro S.L.: The distribution and consumption rate of stars around a massive, collapsed object, 1977, ApJ **211**, 244
Liu J-F., Bregman J.: Ultraluminous X-Ray Sources in Nearby Galaxies from ROSAT High Resolution Imager Observations I. Data Analysis, 2005, ApJS **157**, 59
Machacek M. E., Bryan, G. L., Abel, T.: Effects of a soft X-ray background on structure formation at high redshift, 2003, MNRAS **338**, 273
Madau P., Rees, M.: Massive Black Holes as Population III Remnants, 2001 ApJ **551**, L27
Magorrian, J., Tremaine, S., Richstone, D., et al. The Demography of Massive Dark Objects in Galaxy Centers, 1998, AJ **115**, 285
Makino, J., Fukushige, T., Okumura, S. K., Ebisuzaki, T.: The evolution of massive black-hole binaries in merging galaxies. I - Evolution of a binary in a spherical galaxy, 1993, PASJ **45**, 303
Makino, J., Funato, Y.: Evolution of Massive Black Hole Binaries, 2004, ApJ **602**, 93
Makishima Z., Kubota A., Mizuno T. et al: The Nature of Ultraluminous Compact X-Ray Sources in Nearby Spiral Galaxies, 2000, ApJ **535**, 632
Marconi A., Risaliti G., Gilli G. et al.: Local supermassive black holes, relics of active galactic nuclei and the X-ray background, 2004, MNRAS **351**, 169
Martini P.: QSO lifetimes, 2004, in "Carnegie Observatories Astrophysics Series, Vol. 1: Coevolution of Black Holes and Galaxies," ed. L. C. Ho (Cambridge: Cambridge Univ. Press) p170.
Mathur S.: Narrow-line Seyfert 1 galaxies and the evolution of galaxies and active galaxies, 2000, MNRAS, **314**, L17

Mathur S., Grupe D.: Black hole growth by accretion, 2005, A&A **432**, 463
Merloni A.: The anti-hierarchical growth of supermassive black holes, 2004, MNRAS **353**, 1035
Merritt D., Ferrarese L.: The $M_{BH} - \sigma$ Relation for Supermassive Black Holes, 2001, ApJ **547**, 140
Merritt, D., Milosavljevic, M., Favata, M., Hughes, S. A., Holz, D. E.: Consequences of Gravitational Radiation Recoil, 2004, ApJ **607**, L9
Merritt D.: Interaction of Supermassive Black Holes with their Stellar and Dark Matter Environement, 2004, in "Growing Black Holes"; A. Merloni, S. Nayakshin and R. Sunyaev (eds.) (astro-ph/0409290)
Mikkola, S.,Valtonen, M. J.: Evolution of binaries in the field of light particles and the problem of two black holes, 1992, MNRAS 259, 115
Milosavljevic M., Merrit D.: Formation of Galactic Nuclei, 2001, ApJ **563**, 34
Moles, M., Marquez, I., Perez, E.: The relation between dynamical perturbations, morphology, and nuclear activity in spiral galaxies, 1995, ApJ **438**, 604
Nakano T., Makino J.: On the Origin of Density Cusps in Elliptical Galaxies, 1999 ApJ **510**, 155
Peterson B.M., Wandel A.: Evidence for Supermassive Black Holes in Active Galactic Nuclei from Emission-Line Reverberation, 2000, ApJ **540**, L13
Portegies Zwart, S. F., S. F., Dewi J., Maccarone T.: Intermediate mass black holes in accreting binaries: formation, evolution and observational appearance, 2004a, MNRAS **355**, 413
Portegies Zwart, S. F., Baumgardt, H., Hut, P. et al.: Formation of massive black holes through runaway collisions in dense young star clusters, 2004b, Nature **428**, 724
Preto M., Merritt D., Spurzem R.: N-Body Growth of a Bahcall-Wolf Cusp around a Black Hole, 2004, ApJ **613**, L109
Quinlan G.D.: The dynamical evolution of massive black hole binaries I. Hardening in a fixed stellar background, 1996, New A. **1**, 35
Rokaki E., Boisson C., Collin S.: Fitting the broad line spectrum and UV continuum by accretion discs in active galactic nuclei, 1992, A&A **253**, 57
Salomé P., Combes F.: Cold molecular gas in cooling flow clusters of galaxies, 2003, A&A **412**, 657
Salomé P., Combes F.: Mapping the cold molecular gas in a cooling flow cluster: Abell 1795, 2004, A&A **415**, L1
Salucci P., Szuszkiewicz E., Monaco P., Danese L: Mass function of dormant black holes and the evolution of active galactic nuclei, 1999, MNRAS **307**, 637
Saslaw, W. C., Valtonen, M. J., Aarseth, S. J.: The Gravitational Slingshot and the Structure of Extragalactic Radio Sources, 1974, ApJ **190**, 253
Sazonov S. Yu., Ostriker J. P., Sunyaev R. A: Quasars: the characteristic spectrum and the induced radiative heating, 2004, MNRAS **347**, 144
Sazonov S. Yu., Ostriker J. P., Ciotti L., Sunyaev R. A: Radiative feedback from quasars and the growth of massive black holes in stellar spheroids, 2005, MNRAS **358**, 168
Schmidt M., Schneider D. P., Gunn J. E.: Spectrscopic CCD Surveys for Quasars at Large Redshift.IV.Evolution of the Luminosity Function from Quasars Detected by Their Lyman-Alpha Emission, 1995, AJ **110**, 68
Schneider, R., Ferrara, A., Natarajan, P., Omukai, K.: First Stars, Very Massive Black Holes, and Metals, 2002, ApJ **571**, 30
Schödel, R., Ott, T., Genzel, R. et al.: A star in a 15.2-year orbit around the supermassive black hole at the centre of the Milky Way, 2002, Nature, **419**, 694

Shaver, P., Wall, J., Kellermann, K. et al.: Decrease in the space density of quasars at high redshift., 1996, Nature **384**, 439

Shen, J., Sellwood, J. A.: The Destruction of Bars by Central Mass Concentrations, 2004, ApJ **604**, 614

Sheth, R. K., Bernardi, M., Schechter, P. L. et al.: The Velocity Dispersion Function of Early-Type Galaxies, 2003, ApJ **594**, 225

Shields, G., Gebhardt, K., Salviander, S., et al. The Black Hole-Bulge Relationship in Quasars, 2003, ApJ **583**, 124

Silk, J., Rees M.: Quasars and galaxy formation, 1998 A&A **331**, L1

Soltan A.: Masses of quasars, 1982, MNRAS **200**, 115

Steffen A.T., Barger A.J., Cowie L.L., et al.: The Changing Active Galactic Nucleus Population, 2003, ApJ **596**, L23

Tegmark M., Silk J., Rees M.J., Blanchard A., Abel T., Palla F.: How Small Were the First Cosmological Objects?, 1997, ApJ **474**, 1

Terlevich, R., Melnick, J., Moles, M.: The frequency of Low Luminosity AGN, 1987, IAU 121, Kluwer, p.499

Thorne K.S., Disk-Accretion onto a Black Hole. II. Evolution of the Hole, 1974, ApJ **191**, 507

Trejo A., Rodrguez, L. F., Mart, J.: In Search of a Radio Counterpart for the Ultraluminous X-Ray Source at the Nucleus of M33, 2004 RMxAA. **40**, 223

Tremaine, S., Gebhardt, K., Bender, R., et al.: The Slope of the Black Hole Mass versus Velocity Dispersion Correlation, 2002, ApJ **574**, 740

Umemura M.: A Radiation-Hydrodynamical Model for Supermassive Black Hole-to-Bulge Mass Relation and Quasar Formation, 2001 ApJ **560**, L29

van Breugel W., Heckman, T., Miley, G.: Optical line emission associated with the radio galaxy 4C 26.42 in the cluster of galaxies Abell 1795, 1984, ApJ **276**, 79

van der Marel, R.: Intermediate-Mass Black Holes in the Universe: A Review of Formation Theories and Observational Constraints, 2003, in Carnegie Obs. Astrophys. Series, Vol. 1: Coevolution of Black Holes and Galaxies, ed. L. Ho (Cambridge Univ. Press)

Volonteri M., Haardt F., Madau P.: The Assembly and Merging History of Supermassive Black Holes in Hierarchical Models of Galaxy Formation, 2003a, **582**, 559

Volonteri M., Madau P., Haardt F.: The Formation of Galaxy Stellar Cores by the Hierarchical Merging of Supermassive Black Holes, 2003b, **593**, 661

Volonteri M., Perna R.: Dynamical evolution of intermediate-mass black holes and their observable signatures in the nearby Universe, 2005, MNRAS **358**, 913

Wang Y.P., Biermann P.L., Wandel A.: Black hole to bulge mass correlation in Active Galactic Nuclei: a test for the simple unified formation scheme, 2000 A&A **361**, 550

Whalen, D., Abel, T., Norman, M. L.: Radiation Hydrodynamic Evolution of Primordial H II Regions, 2004, ApJ **610**, 14

Worsley M.A., Fabian A.C., Bauer F.E et al.: The unresolved hard X-ray background: the missing source population implied by the Chandra and XMM-Newton deep fields, 2005, MNRAS **357**, 1281

Wyithe, J. S. B., Loeb, A.: A Physical Model for the Luminosity Function of High-Redshift Quasars, 2002, ApJ **581**, 886

Wyithe, J. S. B., Loeb, A.: Self-regulated Growth of Supermassive Black Holes in Galaxies as the Origin of the Optical and X-Ray Luminosity Functions of Quasars, 2003, ApJ **595**, 614

Young P.J.: Numerical models of star clusters with a central black hole. I - Adiabatic models, 1980 ApJ **242**, 1232

Yu Q., Tremaine S.: Observational constraints on growth of massive black holes, 2002, MNRAS **335**, 965

6 Quasars and Their Host Galaxies

M. Lacy

Abstract: This review attempts to describe developments in the fields of quasar and quasar host galaxies in the past five years. In this time period, the Sloan and 2dF quasar surveys have added several tens of thousands of quasars, with Sloan quasars being found to $z > 6$. Obscured, or partially obscured quasars have begun to be found in significant numbers. Black hole mass estimates for quasars, and our confidence in them, have improved significantly, allowing a start on relating quasar properties such as radio jet power to fundamental parameters of the quasar such as black hole mass and accretion rate. Quasar host galaxy studies have allowed us to find and characterize the host galaxies of quasars to $z > 2$. Despite these developments, many questions remain unresolved, in particular the origin of the close relationship between black hole mass and galaxy bulge mass/velocity dispersion seen in local galaxies.

6.1 Introduction

Quasar astronomy has made some significant advances over the past few years. In this review I shall focus on recent developments in the field, and in particular on the links between quasars, their black holes and their host galaxies.

The recent discovery that the mostly dormant black holes in the nuclei of nearby galaxies have masses which correlate with the luminosity and velocity dispersion of their host galaxies has directly linked the quasar phenomenon to galaxy evolution. Black hole mass estimates from host galaxy luminosities can be compared to those which use the results of reverberation mapping and the widths of broad emission lines, with generally consistent results (Laor 1998; Onken et al. 2004). This gives us the possibility of understanding the black hole mass – galaxy mass correlation through a study of the evolution of quasars and their host galaxies.

Reliable black hole mass estimates have also stimulated long-standing debates on how the observational properties of quasars, such as their emission-line spectra, radio-loudness and accretion rates, may (or may not) depend on black hole mass (McLure et al. 1999; Laor 2000; Lacy et al. 2001; Boroson 2002; Woo & Urry 2002). The correlation between black hole mass and the mass of the host galaxy have reinvigorated studies on the links between quasar activity and galaxy formation. In particular, the relationship between galaxy mergers, starbursts and AGN has come under increased scrutiny using both results from Sloan Digital Sky Survey (SDSS) (Kauffmann et al. 2003) and Hubble Space Telescope (HST) studies of quasar host galaxies (McLure et al. 1999; Canalizo & Stockton 2000).

A further recent development, the advent of large, uniform quasar surveys in the optical (the SDSS quasar survey and Two Degree Field (2dF) quasar survey) and the ROSAT X-ray survey, has also helped progress in the subject. One problem that has bedevilled quasar astronomy throughout its history is that the choice of quasar samples has been very prone to selection effects. Optical and (soft) X-ray selection techniques are only sensitive to quasars with little dust or gas in the host galaxy to redden or absorb the quasar light. Radio selection is not sensitive to reddening, but only $\sim 10\%$ of quasars are bright radio emitters, and the selection effects associated with radio quasar samples are only just beginning to be understood. In particular, the SDSS is able to pick objects that would be missed from traditional quasar surveys (Fan 1999; Richards et al. 2001), and also objects showing only narrow lines in the optical (Zakamska et al. 2003). Further red quasar selection using 2MASS, 2MASS/FIRST, hard X-ray and now Spitzer surveys is finally revealing the population of dust-shrouded type-2 quasars, whose existence has long been suspected, but of which, until recently, there were few known examples.

Detection of the faint host galaxies of bright quasars has been a long-standing problem in observational extragalactic astronomy. Although the host galaxy and the bright, unresolved quasar nucleus typically have similar total fluxes, the diffuse nature of the galaxy emission can frequently be confused with the extended wings of a poorly-characterized point spread function (PSF). Cosmological surface brightness dimming makes detecting the host galaxies of high redshift quasars particularly challenging. Breakthroughs in quasar host studies came with the advent of the HST, with its small, stable PSF, and near-infrared array detectors. Early HST studies with the optical instruments on board allowed the detailed study of nearby quasar hosts (Disney et al. 1995; Bahcall et al. 1997). In parallel, ground-based studies in the near-infrared were able to study the quasars at wavelengths where the flux of the quasar was minimized with respect to the flux of the host galaxy (Dunlop et al. 1993). The marriage of HST's small PSF and the near-infrared NICMOS detector allowed routine discoveries of quasar hosts up to $z \sim 2$ (Ridgway et al. 2001; Kukula et al. 2001).

Adaptive optics are a relatively recent addition to the available techniques for quasar host imaging. Although problematic in some respects (principally PSF variability) AO offers the ability to study larger samples than are practical with the limited observing time available with HST, and, through using 10m-class telescopes, better resolution and surface brightness sensitivity.

We end this review with a discussion of some of the remaining open questions in quasar astronomy, and how advances in telescopes, instrumentation and multiwavelength archives may be able to answer at least some of them in the years to come.

When calculating the intrinsic properties of the quasars we assume a cosmology with $H_0 = 70\mathrm{kms^{-1}Mpc^{-1}}$, $\Omega_\Lambda = 0.7$ and $\Omega_\mathrm{M} = 0.3$.

6.2 New Surveys for Quasars

Two major new quasar surveys with many tens of thousands of objects have recently appeared. The 2dF quasar survey (Croom et al. 2004a) uses photographic plate material to select UV-excess objects. It is thus a fairly conventional survey, but, by using fibres from a large-scale survey of part of the southern sky conducted by the 2dF multifibre instrument on the Anglo-Australian Telescope it has been able to find many thousands of objects. The SDSS quasar survey uses a multicolour technique to find point-like objects off the stellar locus in colour space (Fan 1999, Richards et al. 2001). It is thus also able to find samples of lightly dust-reddened quasars and very high redshift ($z > 2$) objects which the UV excess technique cannot (Hall et al. 2002; Fan et al. 2004). The SDSS survey is not as deep as the 2dF, but covers a larger sky area with greater photometric accuracy. The Faint Images of the Radio Sky at Twenty-cm (FIRST) radio survey, a large area radio survey to mJy sensitivity, was combined with digitally-scanned photographic plate data to find quasar candidates, selected with a blue colour cut (but one much less severe than optically-selected surveys), to form the FIRST Bright Quasar Survey (FBQS, White et al. 1999). Although the FBQS is much smaller than the SDSS and 2dF surveys, was interesting as it was the first survey able to find significant numbers of quasars with radio fluxes between those of radio-loud and radio-quiet quasars.

The quasar luminosity function has been measured from both the 2dF and the SDSS surveys. Croom et al. (2004a) have analysed the 2dF quasar survey and find that a double power-law with luminosity evolution provides a good fit to the data. Joint use of the 2dF and SDSS databases and techniques is now being applied to further refine estimates of the luminosity function (Richards et al. 2005).

6.3 Black Hole Masses and their Implications

Perhaps the biggest single advance in the study of quasars in recent years has been our increased confidence in estimates of masses of their black holes. Although a long-term campaign to obtain data for reverberation mapping resulted in estimates of the sizes of broad line regions (BLRs) (and hence, via linewidths and assumptions about cloud dynamics, the masses of the central black holes), it was not until black hole masses could be independently estimated by stellar and gas-dynamical techniques that it became common to use these estimates. Wandel (2002), Peterson et al. (2004) and Onken et al. (2004) showed that black hole masses derived from reverberation mapping and from bulge luminosities of Seyfert-1 galaxies were consistent, and Nelson et al. (2004) showed that black hole masses estimated from the velocity dispersions of Seyfert-1 galaxies were consistent with the reverberation mapping results. Some uncertainties of a factor of a few do remain, e.g. the unknown broad-line region geometry, (Krolik 2001). Unfortunately, reverberation mapping studies are restricted to a few tens of (mostly) low luminosity AGN. Thus the relationship between AGN luminosity and BLR radius is not very well established, and its extension to the high luminosities of high redshift quasars uncertain. Reverberation mapping studies use Hβ, whereas studies of high redshift quasars are usually restricted to UV lines which are redshifted into the optical such as MgII and CIV (McLure & Jarvis 2002; Vestergaard 2002). Despite these problems and uncertainties, the ability to obtain even order-of-magnitude black hole mass estimates has allowed some interesting studies to be made.

A further technique for measuring black hole mass was suggested by Nelson (2000), using the width of narrow [OIII] emission line as a proxy for the bulge velocity dispersion. This seems to work well for radio-quiet quasars, but not for radio-louds, where the narrow-line region dynamics are frequently affected by shocks induced by the radio jets. This technique is easy to apply at moderate redshifts, and near-infrared spectra can be used at higher redshifts (Shields et al. 2003). The discovery of significant blueshifts in the [OIII] line relative to systemic velocities (Boroson 2005) has cast doubt on the accuracy of this technique, however.

By combining black hole masses and measured luminosities, we have another key physical parameter, namely accretion rate relative to the Eddington Limit. Studies seem to indicate that the Eddington Rate is indeed a hard upper limit to the accretion rate (e.g. McLure & Dunlop 2004), with a typical optically-selected quasar having an Eddington Ratio (i.e. the ratio of accretion rate relative to the Eddington Limit) ~ 0.1. One important component to this calculation is the bolometric correction factor to convert quasar luminosities in a given waveband to total accretion luminosities. Most studies still use the Elvis et al. (1994) corrections, but new observations of mostly fainter quasars, in particular with the Spitzer Space Telescope in the mid-infrared and the Galaxy Evolution Explorer (GALEX) in the ultraviolet should enable us to significantly improve these corrections and allow luminosity and redshift effects to be fully taken into account.

Estimates for black hole masses have been used to relate observational correlations to physical properties of the AGN. Boroson (2002) showed that the eigenvector 1 of Boroson & Green (1992) is related to the Eddington Ratio. Objects with high accretion rates tend to be at one extreme, with high FeII emission and low [OIII] emission, while those with low accretion rates are at the other end.

6.4 Radio-loudness of Quasars

Another area in which black hole mass estimates have been used to try to improve our physical understanding of AGN is that of quasar radio emission. Amongst well-defined samples of quasars there is a weak, but significant relation between black hole mass and radio-loudness in the sense that the radio-loud quasars have higher mean black holes masses [Laor (2000), Lacy et al. (2001), Boroson (2002), Jarvis & McLure (2002), Metcalf & Magliochetti (2005)]. Studies of more heterogeneous samples fail to find a correlation of radio-loudness with black hole mass, however (Woo & Urry 2002). When low-luminosity AGN are included amongst the samples being studied, an anti-correlation of radio-loudness with accretion rate relative to the Eddington Limit is seen (Ho 2002), albeit with a lot of scatter. Whether or not the correlation of black hole mass and radio-loudness is real it is surprising that the relationship between radio emission and the physical characteristics of the quasar (accretion rate and black hole mass) is so hard to pin down. Observations of X-ray binary "microquasars" show X-ray emission properties that change significantly during a radio outburst (a hardening of the X-ray spectrum, hypothesized to be due to the temporary formation of an ion-supported torus in the inner part of the accretion disk and a softening during its subsequent collapse, e.g. Fender, Belloni & Gallo 2004), but no obviously analogous behavior is seen in radio-loud quasars. This may be partly a question of timescales of course, but even so one might expect a much bigger difference in the observed properties of radio-loud and radio-quiet quasars were the microquasar analogy to be followed exactly. One promising analogy is between FRI radio galaxies and microquasars in the "low/hard" state with advection-dominated accretion flows, which are able to produce strong outflows (and perhaps radio jets) very efficiently (e.g. Blandford & Begelman 1999). However, directly applying this analogy to high accretion rate radio-loud quasars, which appear to have both highly-efficient relativistic jet production and classic accretion disks, fails.

Black hole spin has long been suspected as being instrumental in radio-loudness. In particular, the Blandford-Znajek mechanism (Blandford & Znajek 1977), in which jets are produced by magnetic fields threading the ergosphere of the black hole requires a rapidly spinning black hole whose rotational kinetic energy is tapped to power the radio jets. X-ray spectroscopy of the relativistic Fe Kα line in Seyfert galaxies can, in principle, be used to measure the radius of the last stable orbit, and hence the spin of the black hole. This is just possible with XMM and Chandra, but to date only one object, MGC-6-30-15, has been studied in sufficient depth to fit the Fe Kα line. Unfortunately, uncertainties in the underlying continuum emission and possible absorption features mean that a definitive statement about whether or not the black hole is spinning is hard to make. Nevertheless, a model with a spinning black hole is preferred by Vaughan & Fabian (2004). Unfortunately for the proponents of the spinning black hole model for the production of radio jets, this Seyfert galaxy is radio-quiet.

Closely related to the question of the mechanism for radio-loudness is the question of whether there exists a dichotomy in the radio-loudness parameter (i.e. the ratio of radio to optical/UV luminosity). Early studies of the PG quasar sample suggested this was the case, but more recent surveys have been more ambiguous. For example, the FBQS shows no dichotomy (e.g. Lacy et al. 2001). The case for dichotomy in the much larger SDSS and 2dF quasar samples is probably not made. Although Ivecic et al. (2002) claim that one exists in the SDSS, the claim is based on indirect evidence. A similar analysis by Cirasuolo et al. (2003) based largely on 2dF quasars showed no evidence for a dichotomy.

It is probably fair to say, however, that pending deep, large-area radio surveys, the jury is still out on both whether there is a correlation of radio-loudness with black hole mass and whether there is a dichotomy in the distribution of radio-loudness.

6.5 Clustering of Quasars

One interesting product of the large-scale quasar surveys is an estimate of quasar clustering and its evolution with redshift. Clustering statistics give information on the masses of the dark matter haloes containing the quasars and the value of the bias parameter (the extent to which objects fail to trace the underlying mass distribution). Croom et al. (2005) use the 2dF survey to analyse quasar clustering and its evolution using a sample of over 20000 quasars. They find that quasars occupy dark matter haloes with masses of a few $\times 10^{12} M_{\odot}$ at all redshifts. An increase in the clustering with redshift is seen, ascribed to an increase in the bias parameter with redshift from close to unity at $z \approx 0$ to about a factor of four at $z \sim 2$. Although quasars at $z \approx 0$ are seen in haloes with the space density of L^* galaxies today, the $\sim 10^{12} M_{\odot}$ haloes hosting quasars where rarer in the past, and the haloes hosting high-z quasars will grow into group and cluster-sized haloes by the present epoch. Croom et al. interpret this as a trend for lower-mass black holes being more active at low redshift. Of course, in any flux limited survey, the characteristic luminosity of the objects will correlate strongly with redshift, and the possibility of an increase of clustering with luminosity, although not seen in the 2dF dataset, remains a viable alternative explanation. Results from the SDSS, with its higher flux limit, should help to determine whether luminosity or evolution effects are responsible for these results.

6.6 High-z Quasars and Reionization

The SDSS has allowed the efficient selection of several $z \sim 6$ quasars. These are important from a cosmological viewpoint for several reasons. The most distant of these show Gunn-Peterson troughs, indicating that at least a small fraction of the intergalactic medium was neutral at $z \approx 6.5$ (Becker et al. 2002; Fan et al. 2003; White et al. 2003). Exactly how much is still the subject of debate, and larger samples of high-z quasars will be needed to establish this (Mesinger, Haiman & Cen 2004). The mere existence of $\sim 10^9 M_{\odot}$ black holes at $z \sim 6$ is itself interesting, particularly as the detection of FeII emission lines in near-infrared spectra shows that the ISM of the host was already significantly enriched with an element which, at least locally, is mostly produced in Type I supernovae, with a long ($\sim$ 0.3Gyr) delay between star formation and the supernova explosions (Freudling et al. 2003; Barth et al. 2003). This implies an early epoch for the first major burst of star formation in the host galaxy ($z \gtrsim 10$). The number density of very high-z quasars can be used to estimate their contribution to reionization. Based on the numbers found so far, it seems that there are not enough of these objects to reionize the Universe (unless the faint end of the quasar luminosity function is very steep), meaning that star-forming galaxies are the most likely culprits.

6.7 Broad Absorption Line Quasars

The past few years have seen significant developments in the study of broad-absorption line quasars (BALs), the quasar population which shows absorption by an ionized wind with velocities up to $0.1c$. The number of iron low-ionization BALs (FeLoBALs) known has increased significantly, due principally to new quasar surveys. The FBQS was the first to find significant numbers of these objects (Becker et al. 2000). Subsequently, several FeLoBALs were found in the SDSS survey using colour-based selection criteria (Hall et al. 2002). These remarkable objects have line blanketing by FeII shortward of ~ 2800Å, frequently almost completely extinguishing the UV light from the quasar. They are also significantly reddened by dust, the combination giving them quite different colours than normal quasars.

The discovery of a number of truly radio-loud BALs (including a couple with the classic FRII morphology [Gregg et al. 2000; Brotherton et al. 2002]) has also disproved a long-standing suggestion that radio-loud quasars are incapable of hosting BAL flows. Further radio studies of the FBQS BALs has also undermined the popular orientation model for BALs, in which the BAL winds arise from the surface of the accretion disk, with the line of sight passing through them. This model predicts that BAL quasars are all seen with their accretion disks nearly edge-on. Becker et al. (2000) showed that the radio spectral indices of BALs range from steep to flat, indicating no special orientation for the BAL population.

Modelling of the gas flows causing broad absorption lines has improved significantly, with careful allowance being made for saturation and partial covering effects (e.g. Arav et al. 2001). Attempts to model the very low ionization outflows in the FeLoBAL quasars have raised interesting questions as to whether the low ionization absorbing gas is very far out, at distances of hundreds of parsec (de Kool et al. 2002), or whether the outflow is comprised of a multi-phase gas with dense, low-ionization clouds embedded in a low density, high ionization wind (Everett, Königl & Arav 2002).

6.8 Finding the Obscured Quasar Population

Recent years have also shown great advances in finding the obscured quasar population. Initially, progress was slow. In the hard X-ray, the small field sizes and depth required to find the obscured quasar population restricted discoveries to a few objects (e.g. Norman et al. 2002, Stern et al. 2002, Padovani et al. 2004). The 2MASS near-infrared survey was able to find some dust-reddened, mostly low-z quasars (Cutri et al. 2002) and by combining 2MASS with the FIRST radio survey, several higher redshift, radio-intermediate reddened quasars were found

(Glikman et al. 2004). However, 2MASS selection was unable to find objects with rest-frame optical reddenings (A_V) greater than a few magnitudes.

Finding the heavily obscured objects, which show no broad lines in the optical and which have rest-frame $A_V \sim 10$–100, has proven difficult, and it is only very recently that significant progress has been made. Zakamska et al. (2003) were able to isolate moderate redshift objects with high-ionization narrow lines characteristic of type-2 AGN in the SDSS survey, and later showed that many of these objects show broad lines in polarized light, confiming their quasar nature. A combination of mid-infrared fluxes from the Infared Space Observatory (ISO) and 2MASS was able to find objects significantly more reddened than those found by 2MASS alone (Leipski et al. 2004, 2005). Using mid-infrared colours from Spitzer, Lacy et al. (2004, 2005) also found examples of this type-2 quasar population. A strategy of searching for Spitzer-detected radio-intermediate objects has been used successfully by Martinez-Sansigre et al. (2005), who have found several $z > 2$ type-2 quasars. Selection effects for all these techniques still need to be fully understood, but it is clear that the type-2 quasar population is at least comparable in number density to the "normal" quasar population, and perhaps 2–3 times larger. Figure 6.1 shows the spectra of a type-1 quasar, a lightly-reddened FeLoBAL quasar and a heavily obscured type-2 quasar normalised to approximately the same emission line fluxes.

An upper bound on the total number of quasars, both obscured and unobscured, can be obtained by comparing the mass in black holes in the centers of galaxies today with the total accretion luminosity of the quasar population, divided by the accretion efficiency, a point orignally made by Soltan (1982). The most up-to-date versions of this calculation (e.g. Yu & Tremaine 2002) leave little room for a very large obscured quasar population, but uncertainties about the exact value of the accretion efficiency (i.e. the ratio of the mass-energy accreted onto the black hole to that liberated as accretion luminosity) still allow a significant obscured (or indeed advection-dominated accreting) population. In the standard accretion disk model, accretion efficiency is determined primarily by the radius of the last stable orbit. For a non-spinning (Schwartzschild) black hole this limits the accretion efficiency to ~ 0.1, but the frame-dragging effect of a spinning (Kerr) black hole allows for a closer-in last stable orbit and accretion efficiencies ~ 0.3. Efficiencies of this order are required to accommodate a significant hidden quasar population.

6.9 Quasar Host Qalaxy Studies

Although ground-based studies of quasar hosts have been proceeding for many years, it required the advent of HST to allow robust determinations of the host galaxy properties. Figure 6.1 shows the regions of the absolute magnitude vs redshift plot covered by the major HST programs. WFPC2 observations of nearby quasars showed that luminous quasars are hosted by massive galaxies, usually early types (Disney et al. 1995; Bahcall et al. 1997; McLure et al. 1999;

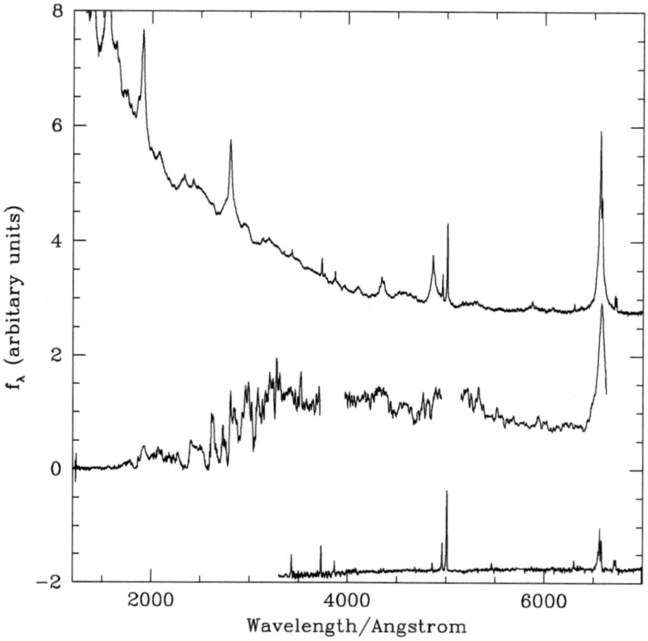

Fig. 6.1. Rest-frame spectra of traditional and non-traditional quasars: top - composite of normal quasars from the SDSS (Vanden Berk et al. (2001); middle – the $z = 2.65$ FeLoBAL quasar FTM J100424.9+122922 (Lacy et al. 2002a); bottom – a composite of type-2 quasars from Spitzer selection from 3200–7000Å (Lacy et al. 2005). The top and bottom spectra have been offset by ± 2 units for clarity. All three spectra have been normalized to approximately the same emission line strength. (The blank regions in the middle spectrum at ≈ 3800 and ≈ 5100Å are zero transparency regions in the near-infrared.)

Dunlop et al. 2003; Floyd et al. 2004). The advent of the NICMOS camera on HST allowed the study of rest-frame optical emission from host galaxies of quasars at high redshifts for the first time. Kukula et al. (2001) and Ridgway et al. (2001) showed that, at $z \sim 2$, the trend for quasars to be hosted in massive galaxies continues, though less-luminous quasars are generally found in less-luminous hosts (Fig. 6.2). The sample of Ridgway et al. contained quasars of similar luminosity to the low-redshift quasars of Dunlop et al., allowing a direct comparison of the stellar masses of the hosts. This indicated that the stellar masses of the high-z quasars were smaller by a factor of a few than their low-z counterparts. Semi-analytic modelling by Kauffmann & Haehnelt (2000) suggested that the hosts of high-z quasars should be lower in stellar mass, though by a larger factor. Whether this difference is significant or not remains to be established, though the results of Kukula et al. on the hosts of slightly more luminous quasars are in more disagreement with the predictions of that model.

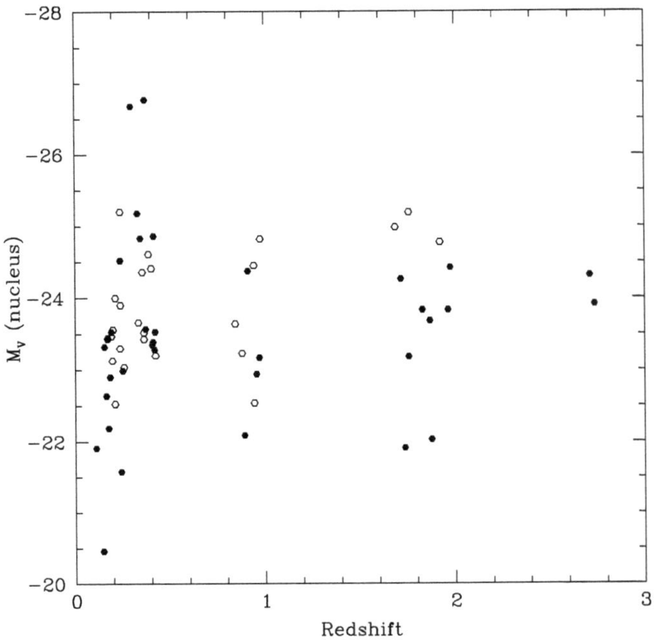

Fig. 6.2. The absolute-magnitude – redshift plane for quasars studied by the largest HST programs at $z > 0.5$ (Kukula et al. 2001; Ridgway et al. 2001), and the two largest programs at low-z (Dunlop et al. 2003; Floyd et al. 2004). All these samples were imaged close to the rest-frame V-band. Although moderate luminosity quasars are well-represented at most redshifts, there is a lack of high-luminosity quasars, partly for practical reasons of the difficulty of PSF subtraction from quasars whose nuclear light is significantly brighter than that from its host galaxy. Radio-loud quasars are shown as open symbols, radio-quiets as filled symbols.

More recent semi-analytic models, which include feedback effects from both supernovae and the AGN, and a different prescription for star formation in massive dark matter haloes at early epochs agree better with observations (e.g. Granato et al. 2003), though a robust, predictive model for galaxy and quasar formation is still lacking. Figure 6.2 also shows that radio-loud quasars tend to be in more massive hosts (at a given quasar luminosity) than radio-quiets at all redshifts (e.g. Kukula et al. 2001)

Spectroscopy of quasar host galaxies has been attempted from the ground. Nolan et al. (2001) found that most quasar host galaxies had evolved stellar populations, $\sim$ 10Gyr old, with only a very small amount of recent star formation. Deeper spectroscopy with Keck by Canalizo & Stockton (2001), however, found evidence of bursts of star formation within the past $\sim$ 100Myr in quasars with far-infrared excesses. Indeed, such populations have also been found in radio galaxies (Tadhunter et al. 2005), and in composite spectra of Seyfert galaxies and low-luminosity quasars from the SDSS (Kauffmann et al. 2003). There is

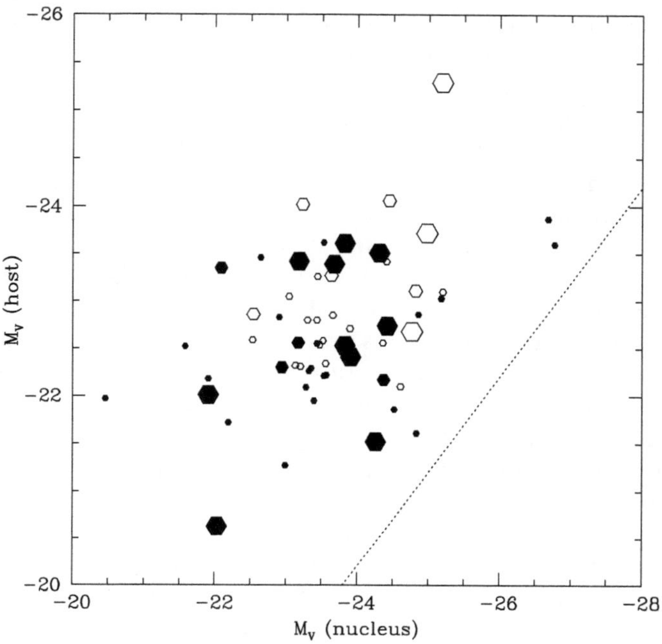

Fig. 6.3. The samples of Fig. 6.2 with absolute magnitude of the host plotted against nuclear absolute magnitude. Small symbols indicate quasars with $z < 0.5$, medium, those with $0.5 < z < 1.5$, and large, those with $z > 1.5$. The dotted line corresponds approximately to the luminosity expected from Eddington-limited accretion [using the van der Marel (1999) relation]. In general, host and quasar luminosity are correlated, even though few quasars closely approach the Eddington Limit. Note also that there is no strong systematic trend for high-z hosts to be fainter than their low-z counterparts [as predicted, for example, by Kauffmann & Haenelt (2000)], but the dispersion in host galaxy magnitudes does seem to be higher at $z \sim 2$, indicating that a wider range of black hole masses may be involved in quasar activity at early epochs. (It should also be borne in mind that the mass:light ratios of young stellar populations are likely to be lower at high-z, so their black holes may be less massive than their position on this plot would indicate.)

much debate about whether the different results are due to better quality spectra taken closer to the nucleus using 8-10m-class telescopes, or whether there are real differences in the host galaxy properties of the different quasar samples studied.

In addition to space-based observations, ground-based observations with adaptive optics (AO) are beginning to be exploited for quasar host galaxy studies (e.g., Stockton, Canalizo & Close 1998, Lacy et al. 2002b, Croom et al. 2004b; Falomo et al. 2005). The high spatial resolution of AO is able to concentrate the light of the nucleus, leaving the host much more visible. Although the PSF is not as stable as that produced by space-based observations, the relative cheapness of ground-based telescope time (particularly on smaller telescopes) allows for large surveys not practical with space-based observatories, and on 8m-class telescopes, the resolution exceeds that from HST by a significant factor. Although it is important to bear in mind that surface brightness sensitivity is fundamental to the detection of quasar hosts, AO does significantly improve the ability to detect hosts through concentrating the quasar light into the center of the galaxy image.With laser guide stars becoming available at several major observatories, coupled with large quasar surveys which allow selection of quasars near bright guide stars, this area of research should have a significant impact in the years to come. Host galaxy studies with 8m-class telescopes without adaptive optics, or with only active optics have also been attempted. Although lacking in spatial resolution, the good surface brightness sensitivity of these telescopes has allowed detections of host galaxies out to $z > 4$ (e.g. Falomo et al. 2001; Hutchings 2003).

An early motivation for quasar host galaxy studies was to search for the trigger mechanism of quasar activity (e.g. Stockton & MacKenty 1987). Mergers, or other galaxy interactions have long been thought of as likely triggers as they are able to disrupt stable gas which might fall onto the black hole. What has never been clear, however, is how disturbances due to mergers or interactions on kpc scales influence the gas motions within the sub-pc inner region of the AGN. Indeed, McLure et al. (1999) conclude that few of their quasar hosts show obvious signs of interaction. However, the presence of moderate age stellar populations in quasar hosts suggests there may be a delay between the large-scale disruption (which would trigger a starburst almost instantaneously) and the onset of quasar activity. One possibility is that dust-reddened and/or type-2 quasars may represent an early stage in quasar evolution. Canalizo & Stockton (2001) find that several of their IR-excess quasars are low-ionization BALs with significant dust reddening, and, although many type-2 quasars have classic high-ionization emission lines in their optical spectra, there are several whose spectra either lack emission lines, or contain only emission lines from starbursts (e.g. the "type-3" quasars of Leipski et al. 2005). These might represent quasars in a very early, completely dust-shrouded phase in their evolution where even the narrow-line region is obscured. On the other hand, an HST-based study of the host galaxies of 2MASS-selected dust-reddened QSOs by Marble et al. (2003) found no significant difference between the hosts of dust-reddened and normal QSOs in terms of the spread in galaxy type and evidence of recent merger activity.

6.10 Open Questions

Over the past few years quasar astronomy has undergone something of a revolution. From surveys of a few hundred quasars we have gone to surveys of several tens of thousands. We have much more confidence in our ability to estimate the important physical parameters of quasars, namely black hole mass and accretion rate. The highest redshift quasar is now at $z > 6$. We have resolved the host galaxies of quasars out to $z > 2$. We are close to being able to place much-improved constraints on the numbers of dust-hidden quasars, which studies suggest at least equal, and perhaps exceed, the numbers of "classical" blue quasars. Although finding the most heavily obscured objects will always be hard, nature seems to be helping us by providing many of these objects with the "obscuring torus" geometry which makes them much more visible in the mid-infrared. Nevertheless, we still have open a number of important questions – indeed, our new information has allowed us to ask new ones. Perhaps the most fundamental is the nature and evolution of the black hole mass – bulge mass relation, as it demonstrates a clear link between quasars and their host galaxies. This link might be established by a mechanism in which the quasar acts as a govenor providing negative feedback to star-forming processes in the host galaxy (e.g. Silk & Rees 1998).

Other questions remain. Why does the evolution of quasars so closely resemble pure luminosity evolution, even though we know high redshift and low redshift quasars must be very different phenomena, one fundamentally linked to galaxy formation and growth, and the other apparently a rare phenomenon which seems to at best reflect the end results of a merger several hundreds of Myr earlier? How are outflow phenomena such as radio-loud quasars and BALs linked to the normal quasar population? Is the obscured quasar population linked to the normal quasar population through evolution, orientation, or both? Exactly how many quasars are hidden by dust?

Future techniques to study quasars will probably involve a shift from the study of pre-selected quasar samples towards survey-based samples. Data mining of public multiwavelength surveys will result in the discovery of objects which occupy unusual parts of parameter space, such as high redshift type-2 quasars (Padovani et al. 2004). New radio telescopes such as the Low Frequency Array (LOFAR) and the Square Kilometer Array, which can image deep enough over a large enough area to measure the radio-luminosity function of radio-quiet quasars will shed new light on the relationship of radio jet power to other quasar properties. With robotic telescopes becoming more common, it becomes feasible to undertake reverberation mapping studies of larger samples of objects for longer periods, including higher luminosity AGN and quasars, which can be used to improve black hole mass estimates. Statistical studies of quasar variability (e.g. Vanden Berk 2004; de Vries et al. 2005) will become easier and more accurate with e.g. the Large Synoptic Survey Telescope, and should open up new ways of investigating the quasar phenomenon.

With the future of HST uncertain, adaptive optics is certain to become a more important tool for the study of quasar host galaxies. Future space missions such as the James Webb Space Telescope and the Terrestrial Planet Finder (Coronagraph) will play an important role, but both missions are some years away. Of the open questions, probably only the nature and numbers of the obscured quasar population is one we will be able to answer soon. The others will require further observations and theoretical modelling before we will be able to make progress.

Acknowledgements

I would like to thank Susan Ridgway for reading through the manuscript, and my other long-term collaborators on studies of quasars: Bob Becker, Gabriela Canalizo, Elinor Gates, Michael Gregg, Steve Rawlings and Margrethe Wold, discussions with whom over the years have helped to shape this review.

References

Arav, N. et al. (1999), Hubble Space Telescope Observations of the Broad Absorption Line Quasar PG 0946+301, ApJ, **516**, 27

Bahcall, J.N., Kirhakos S., Saxe, D.H. & Schnieder,D.P. (1997) Hubble Space Telescope Images of a Sample of 20 Nearby Luminous Quasars, ApJ, **479**, 642

Barth, A.J., Martini, P., Nelson, C.H. & Ho, L.C. (2003) Iron Emission in the $z = 6.4$ Quasar SDSS J114816.64+525150.3, ApJ, **594**, L95

Becker, R.H., White, R.L., Gregg, M.D., Brotherton, M.S., Laurent-Muehleisen, S.A., Arav, N. (2000) Properties of Radio-selected Broad Absorption Line Quasars from the First Bright Quasar Survey, ApJ, **538**, 72

Becker, R.H. et al. (2001) Evidence for Reionization at $z \sim 6$: Detection of a Gunn-Peterson Trough in a $z = 6.28$ Quasar, AJ, **122**, 2850

Blandford, R.D. & Begelman, M.C. (1999) On the fate of gas accreting at a low rate on to a black hole, MNRAS, **303**, L1

Blandford, R.D. & Znajek, R.L. (1977) Electromagnetic extraction of energy from Kerr black holes, MNRAS, **179**, 433

Boroson, T.A. & Green, R.F. (1992) The emission-line properties of low-redshift quasi-stellar objects, ApJS, **80**, 109

Boroson, T.A., (2002) Black Hole Mass and Eddington Ratio as Drivers for the Observable Properties of Radio-loud and Radio-quiet QSOs, ApJ, **565**, 78

Boroson, T. (2005) Blueshifted [O III] Emission: Indications of a Dynamic NLR, AJ, **130**, 381

Brotherton, M.S., Croom, S.M., de Breuck, C., Becker, R.H. & Gregg, M.D. (2002) The Twice-Overlooked, Second Fanaroff-Riley II Broad Absorption Line Quasar LBQS 1138-0126, AJ **124**, 2575

Canalizo, G. & Stockton, A.N. (2001) Quasi-Stellar Objects, Ultraluminous Infrared Galaxies, and Mergers, ApJ, **555**, 719

Cirasuolo, M. Celotti, A., Magliocchetti, M., Danese, L. (2003) Is there a dichotomy in the radio loudness distribution of quasars?, MNRAS, **346**, 447

Croom, S.M. et al. (2004a) The 2dF QSO Redshift Survey – XIV. Structure and evolution from the two-point correlation function, MNRAS, **356**, 415

Croom, S.M., Schade, D., Boyle, B.J., Shanks, T., Miller, L. & Smith, R.J. (2004b) Gemini Imaging of QSO Host Galaxies at $z \sim 2$, ApJ, **606**, 126

Cutri, R.M., Nelson, B.O., Francis, P.J. & Smith, P.S. (2002), The 2MASS Red AGN Survey, in, AGN Surveys, Proceedings of IAU Colloquium 184, ASP Conference Proceedings, Vol. 284. Edited by R.F. Green, E.Ye. Khachikian, and D.B. Sanders. San Francisco, CA: Astronomical Society of the Pacific, p. 127.

de Kool, M. et al., (2002), Keck HIRES Spectroscopy of the Fe II Low-Ionization Broad Absorption Line Quasar FBQS 0840+3633: Evidence for Two Outflows on Different Scales, ApJ, **570**, 514 de Vries, W., Becker, R.H., White, R.L. & Loomis, C. (2005), Structure Function Analysis of Long-Term Quasar Variability, AJ, **129**, 615

Disney, M.J. et al. (1995) Interacting Elliptical Galaxies as Hosts of Intermediate Redshift Quasars, Nat, **376**, 150

Dunlop, J.S., Taylor, G.L., Hughes, D.H. & Robson, E.I. (1993) A near-IR study of the host galaxies of radio-quiet quasars, radio-loud quasars and radio galaxies, MNRAS, **264**, 455

Dunlop, J.S. et al. (2003) Quasars, their host galaxies and their central black holes, MNRAS, **340**, 1095

Elvis, M. et al. (1994) Atlas of quasar energy distributions, ApJS, **95**, 1

Everett, J. Köngl, A. & Arav, N. (2002), Observational Evidence for a Multiphase Outflow in Quasar FIRST J1044+3656, ApJ, **569**, 671

Falomo, R. et al. (2001) Near-Infrared Imaging of the Host Galaxies of Three Radio-loud Quasars at $z \sim 1.5$, ApJ, **547**, 124

Falomo, R. et al. (2005) VLT adaptive optics imaging of QSO host galaxies and their close environment at $z \sim 2.5$: Results from a pilot program, A&A, **434**, 469

Fan, X. (1999) Simulation of Stellar Objects in SDSS Color Space, AJ, **117**, 2528

Fan, X. et al. (2003) Evolution of the Ionizing Background and the Epoch of Reionization from the Spectra of z 6 Quasars, AJ, **123**, 1247

Fan, X. et al. (2004) A Survey of $z > 5.7$ Quasars in the Sloan Digital Sky Survey. III. Discovery of Five Additional Quasars, AJ, **128**, 515

Fender, R.P., Belloni, T.N. & Gallo, E. (2004) Towards a unified model for black hole X-ray binary jets, MNRAS, **355**, 1105

Floyd, D.J.E., et al. (2004) The host galaxies of luminous quasars, MNRAS, **355**, 196

Freudling, W., Corbin, M.R. & Korista, K.T. (2003) Iron Emission in z 6 QSOS, ApJ, **587**, L67

Granato, G.L. et al. (2003) A Physical Model for the Coevolution of QSOs and Their Spheroidal Hosts, ApJ, **600**, 580

Gregg, M.D., Becker, R.H. Brotherton, M.S., Laurent-Muehleisen, S., Lacy, M. & White, R.L. (2000) Discovery of a Classic FR II Broad Absorption Line Quasar from the FIRST Survey, ApJ, **544**, 142

Hall, P.B. et al. (2002) Unusual Broad Absorption Line Quasars from the Sloan Digital Sky Survey, ApJS, **141**, 267

Ho, L. (2002) On the Relationship between Radio Emission and Black Hole Mass in Galactic Nuclei, ApJ, **564**, 120

Hutchings, J.B., (2003), Host Galaxies of $z \sim 4.7$ Quasars, AJ, **125**, 1053

Glikman, E. et al. (2004) FIRST-2Mass Sources below the APM Detection Threshold: A Population of Highly Reddened Quasars, ApJ, **607**, 60

Ivezic, Z. et al. (2002) Optical and Radio Properties of Extragalactic Sources Observed by the FIRST Survey and the Sloan Digital Sky Survey, AJ, **124**, 2364

Kauffmann, G. & Haehnelt, M.G. (2000) A unified model for the evolution of galaxies and quasars, MNRAS, **311**, 576

Kauffmann, G. et al. (2003) The host galaxies of active galactic nuclei, MNRAS, **346**, 1055

Krolik, J.H., (2001) Systematic Errors in the Estimation of Black Hole Masses by Reverberation Mapping, ApJ, **551**, 72

Kukula, M.J., Dunlop, J.S., McLure, R.J., Miller, L., Percival, W.J., Baum, S.A. & O'Dea, C.P. (2001) A NICMOS imaging study of high-z quasar host galaxies, MNRAS, **326**, 1533

Lacy, M., Laurent-Muehleisen, S.A., Ridgway, S.E., Becker, R.H. & White, R.L. (2001) The Radio Luminosity-Black Hole Mass Correlation for Quasars from the FIRST Bright Quasar Survey and a "Unification Scheme" for Radio-loud and Radio-quiet Quasars, ApJ, **551**, L17

Lacy, M. et al. (2002), The Reddest Quasars. II. A Gravitationally Lensed FeLoBAL Quasar, AJ, **123**, 2925

Lacy, M., Gates, E.L., Ridgway, S.E., de Vries, W., Canalizo, G., Lloyd, J.P. & Graham, J.R. (2002) Observations of Quasar Hosts with Adaptive Optics at Lick Observatory, AJ, **124**, 3023

Lacy, M. et al. (2004), Obscured and Unobscured Active Galactic Nuclei in the Spitzer Space Telescope First Look Survey, ApJS, **154**, 166

Lacy, M. et al. (2005), Mid-infrared selection of Quasar-2s in Spitzer's First Look Survey, MmSAI, **76**, 154

Laor, A. (1998) On Quasar Masses and Quasar Host Galaxies, ApJ, **505**, L83

Laor, A. (2000) On Black Hole Masses and Radio Loudness in Active Galactic Nuclei, ApJ, **543**, L111

Leipski, C. et al. (2005), The ISO-2MASS AGN survey: On the type-1 sources, A&A, **440**, L5

Marble, A.R. et al. (2003), A Hubble Space Telescope WFPC2 Snapshot Survey of 2MASS-selected Red QSOs, ApJ, **590**, 707

Martinez-Sansigre, A. et al. (2005), Most supermassive black hole growth is obscured by dust, Nature, in press

McLure, R.J. Kukula, M.J., Dunlop, J.S., Baum, S.A., O'Dea, C.P. & Hughes, D.H. (1999) A comparative HST imaging study of the host galaxies of radio-quiet quasars, radio-loud quasars and radio galaxies – I, MNRAS, **308**, 377

Mesinger, A. Haiman, Z. & Cen, R. (2004) Probing the reionization history using the spectra of high-redshift sources, ApJ, **613**, 23

Metcalf, R.B. & Magliocchetti, M. (2005) The role of black hole mass in quasar radio activity, MNRAS, submitted (astro-ph/0505194)

Nelson, C.H., et al. (2004) The Relationship Between Black Hole Mass and Velocity Dispersion in Seyfert 1 Galaxies, ApJ, **615**, 652

Nelson, C.H. et al. (2000) ApJ, **544**, L91

Nolan, L.A. et al. (2001), The ages of quasar host galaxies, MNRAS, **323**, 308

Norman, C. et al. (2002) A Classic Type 2 QSO, ApJ, **571**, 218

Onken, C.A. et al. (2004), Supermassive Black Holes in Active Galactic Nuclei. II. Calibration of the Black Hole Mass-Velocity Dispersion Relationship for Active Galactic Nuclei, ApJ, **615**, 645

Padovani, P. et al. (2004), Discovery of optically faint obscured quasars with Virtual Observatory tools, A&A, **424**, 545

Peterson, B.M. et al. (2004), Central Masses and Broad-Line Region Sizes of Active Galactic Nuclei. II. A Homogeneous Analysis of a Large Reverberation-Mapping Database, ApJ, **613**, 682

Richards, G.T. et al. (2001) Photometric Redshifts of Quasars, AJ, **121**, 2308

Richards, G.T. et al. (2005) The 2dF-SDSS LRG and QSO (2SLAQ) Survey: the $z < 2.1$ quasar luminosity function from 5645 quasars to $g = 21.85$, MNRAS, **360**, 839

Ridgway, S.E., Heckman, T.M., Calzetti, D. & Lehnert, M. (2001) NICMOS Imaging of the Host Galaxies of $z \sim 2 - 3$ Radio-quiet Quasars, ApJ, **550**, 122

Schneider, D.P. et al. (2003) The Sloan Digital Sky Survey Quasar Catalog. II. First Data Release, AJ, **126**, 2579

Shields, G.A. et al. (2003) The Black HoleBulge Relationship in Quasars, ApJ, **583**, 124

Silk, J. & Rees, M.J. (1998) Quasars and galaxy formation, A&A, **311**, L1

Soltan, A. (1982), Masses of quasars, MNRAS, **200**, 115

Stern, D. et al. (2002), Chandra Detection of a Type II Quasar at $z = 3.288$, ApJ, **568**, 71

Stockton, A. & MacKenty, J.W. (1983) Extended emission-line regions around QSOs, Nat, **305**, 678

Stockton, A., Canalizo, G. & Close, L.M. (1998) PG 1700+518 Revisited: Adaptive-Optics Imaging and a Revised Starburst Age for the Companion, ApJ, **500**, L121

Tadhunter, C.N. et al. 2005, Starbursts and the triggering of the activity in nearby powerful radio galaxies, MNRAS, **356**, 480

Vanden Berk, D.E. et al. (2001), Composite Quasar Spectra from the Sloan Digital Sky Survey, AJ, **122**, 549

Vanden Berk, D.E. et al. (2004), The Ensemble Photometric Variability of ~25,000 Quasars in the Sloan Digital Sky Survey, ApJ, **601**, 692

van der Marel, R.P. (1999), The Black Hole Mass Distribution in Early-Type Galaxies: Cusps in Hubble Space Telescope Photometry Interpreted through Adiabatic Black Hole Growth, AJ, **117**, 744

Vaughan, S. & Fabian, A.C. (2004) A long, hard look at MCG-6-30-15 with XMM-Newton II: detailed EPIC analysis and modelling, MNRAS, **348**, 1415

estergaard, M. (2002) Determining Central Black Hole Masses in Distant Active Galaxies, ApJ, **571**, 733

Wandel, A. (2002) Black Holes of Active and Quiescent Galaxies. I. The Black Hole-Bulge Relation Revisited, ApJ, **565**, 762

White, R.L. et al. (1999) The FIRST Bright Quasar Survey. II. 60 Nights and 1200 Spectra Later, ApJS, **126**, 133

White, R.L., Becker R.H., Fan, X. & Strauss M.A. (2003) Probing the Ionization State of the Universe at $z > 6$, AJ, **126**, 1

Woo, J.-H. & Urry, C.M. (2002) The Independence of Active Galactic Nucleus Black Hole Mass and Radio Loudness, ApJ, **581**, L5

Yu, Q. & Tremaine, S. (2002), Observational constraints on growth of massive black holes, MNRAS, **335**, 965

Zakamska, N., (2003) Candidate Type II Quasars from the Sloan Digital Sky Survey. I. Selection and Optical Properties of a Sample at $0.3 < z < 0.83$, AJ, **126**, 2125

7 Applications of Gravitational Lensing in Cosmology

M. Bartelmann

Abstract: Gravitational lensing originates from the deflection of light by masses, irrespective of their physical state or composition. Since it appears inescapable that most of the matter in the universe is dark, gravitational lensing has developed into one of the primary tools to learn about the amount, composition and distribution of masses in the universe.

The review will summarise the theory of gravitational lensing, starting from Fermat's principle. This will first be applied to isolated lenses like compact objects, galaxies, and galaxy clusters. Cosmologically relevant applications will be described, such as searches for compact dark-matter objects in galactic halos, measurements of the Hubble constant in galaxy lenses, and methods for mapping the dark matter in galaxy clusters. Next, the theory of cosmological lensing will be introduced. The concepts of lensing by large-scale structures and its measurement will be discussed, concluding with an overview of results which have so far been obtained, and an outlook at what can be expected in the near future.

7.1 Introduction

7.1.1 The dark Universe

It seems that we have to accept that most matter in the Universe is dark, and that the majority of the cosmic energy density is not contributed by matter. These conclusions appear as the most straightforward interpretation of numerous cosmological measurements within the class of Friedmann-Lemaître cosmological models.

Those models are derived from Einstein's general theory of relativity using the two symmetry assumptions that the Universe is isotropic around us, and that our position in the Universe is by no means preferred to others. Together, these two assumptions imply that the Universe is isotropic around any of its points, and thus homogeneous.

The assumption that the Universe is isotropic around us seems dubious at first sight, but is very well satisfied if the observable Universe is averaged over sufficiently large scales which are still small compared to its overall size. Assuming our cosmic position not to be preferred compared to any other is a generalisation of the Copernican principle, which said that the Earth is not at the centre of the Solar System.

Given these two symmetry assumptions, Einstein's generally very complicated field equations reduce to two simple, ordinary differential equations for the scale factor of the Universe, the so-called Friedmann equations. They admit families of solutions characterised by a set of parameters like the densities of various energy contributions, the so-called cosmological constant, the current expansion rate of the Universe or Hubble constant, and so forth. Determining the global properties of the Universe is then reduced to measuring these parameters, which used to be the formidable paramount task of modern cosmology.

The past few years have seen a breakthrough. Mainly three types of cosmological measurement have allowed virtually all relevant cosmological parameters to be precisely constrained. Among the most interesting and puzzling of their findings are that the total density of all forms of energy together sums up to a critical value which makes space flat. Radiation contributes negligibly today to this critical density, matter has just about 30 per cent, and the rest is some different, mysterious form of energy. Of the matter, only about 15 per cent is of the "ordinary", baryonic form that we know from Earth, which consists of protons, neutrons and electrons. The remainder of the matter cannot interact with light and is thus called dark matter .

We are thus put into the embarrassing situation of now knowing quite precisely how much energy per volume there is in the Universe, but having no precise idea what the vast majority of this energy may be composed of. Einstein's theory of special relativity has taught us that matter and energy are equivalent. Yet, is does make sense to distinguish dark matter from dark energy: Dark matter can clump and form structures, while dark energy cannot and is thus smoothly distributed.

A possible form of the dark energy could be the cosmological constant which Einstein introduced into his equations when Friedmann found that generally-relativistic cosmological models are unstable and either expand or contract. At that time towards the end of the 1920s, Einstein believed the Universe to be static, which he could achieve only if he included a repulsive term into his equations, which could prevent the Universe from collapsing under its own gravity. This term, which is proportional to the cosmological constant, was soon considered obsolete when Slipher and Hubble found that the Universe is not static but expanding (Hubble, 1929; Hubble & Humason, 1931). However, evidence is mounting that the cosmological constant, or something behaving similarly, is in fact required.

In this situation, which the recent advances of observational cosmology have made quite inescapable, two of the most pressing cosmological questions are, "What is the dark matter composed of, and how is it distributed?", and "What is the dark energy, and how does it evolve in time?" With both those dominant constituents of the Universe not interacting with light, the investigation of these two questions requires methods which reveal structures even if they do not emit or absorb electromagnetic radiation.

7.1.2 The role of gravitational lensing

Another effect following from Einstein's theory of general relativity comes to assist. Masses bend light paths such that they appear curved towards those, very much like light rays passing through convex glass lenses. This effect immediately follows from the equivalence principle , which says that gravitational forces cannot be distinguished from inertial forces experienced in an accelerating frame of reference. This was the heuristic guiding principle for Einstein's construction of general relativity, and it thus became clear very early during the development of the theory that gravitational light deflection would be one of its consequences. It was a triumph for Einstein when the light deflection by the Sun was discovered in 1919 (Eddington, 1919).

For deflecting light, matter (or energy) needs to be inhomogeneously distributed. Perfectly homogeneous matter is thus invisible even for gravitational lensing . However, lensing can also be employed for studying the dark energy, even though it is (or can be) defined as a perfectly smooth background of energy.

If the dark energy is well described as a cosmological constant, it does not evolve in time. However, the cosmological constant is unsatisfactory from the point of view of theoretical physics because its density deviates by 120 orders of magnitude from expectations derived from particle physics. Other models for the dark energy typically predict it to change with time. Thus, it is in fact a question of fundamental physical importance how the dark energy is evolving as the Universe expands. This becomes measurable because the dark energy changes the expansion behaviour of the Universe and thereby the growth of structures in it. Depending on the time evolution of the dark energy, cosmic structures form earlier or later in cosmic history. Thus, using gravitational lensing for investigating matter inhomogeneities in dependence of their cosmic age gives indirect clues as to the nature of the dark energy, even though it does not form structures.

7.2 Basic Principles of Gravitational Lensing

This section briefly summarises the foundation of gravitational lensing theory. Readers not interested in the mathematics can immediately skip to subsection 7.2.4 where the basic physical properties of lensing will be summarised.

7.2.1 Fermat's principle and the deflection angle

The theory of gravitational lensing is most easily built up starting from Fermat's principle, which is well known from geometrical optics. It holds that between a fixed source and a fixed observer, light will choose a path along which its travel time is extremal. Wave optics is typically unimportant for gravitational lensing because the wave length of light is incommensurably smaller than any structures in a gravitational lens.[1]

[1] Wave optics can be relevant, however, for gravitational lensing of gravitational waves.

The light travel time is calculated in geometrical optics using the index of refraction n, which quantifies by how much the light speed is lower in a medium compared to vacuum. In general relativity, the light speed is reduced in presence of a gravitational field.

For by far the most astrophysical applications of gravitational lensing, the gravitational field can be described by a small Newtonian gravitational potential $\Phi \ll c^2$ which moves with velocities small compared to the speed of light. Under these conditions, the index of refraction is

$$n = 1 - \frac{2\Phi}{c^2} , \tag{7.1}$$

from which Fermat's principle yields the deflection angle

$$\hat{\boldsymbol{\alpha}} = \frac{2}{c^2} \int \mathrm{d}l \, \boldsymbol{\nabla}_\perp \Phi , \tag{7.2}$$

where $\boldsymbol{\nabla}_\perp$ is the gradient perpendicular to the light ray along which the integral is to be carried out. However, typical deflection angles being very small, is is permissible to evaluate the integral along the unperturbed, straight light path. This corresponds to Born's approximation familiar from scattering theory.

A point mass M has the gravitational potential

$$\Phi = -\frac{GM}{r} , \tag{7.3}$$

where G is the gravitational constant and r is the distance from the mass. Arranging the coordinate system such that the unperturbed light path propagates into the positive z direction, we find

$$\hat{\boldsymbol{\alpha}} = \frac{4GM}{c^2} \frac{\mathbf{b}}{b^2} , \tag{7.4}$$

where $\mathbf{b}$ is the vector connecting the point mass and the light ray in the x-y plane perpendicular to the light ray containing the mass. This plane is called the lens plane . Lensing by an isolated lens is reduced to a sharp kink in the lens plane of the otherwise straight light path.

The deflection angle is directed towards the lens and proportional to the mass and the inverse separation b between lens and light ray. Since $2GM/c^2$ is the Schwarzschild radius R_s of the point mass, we can write

$$\hat{\alpha} = \frac{2R_\mathrm{s}}{b} . \tag{7.5}$$

The linear dependence of the deflection angle on the lens mass allows it to be easily generalised for extended mass distributions. Projecting the lensing mass distribution into the lens plane gives the surface mass density

$$\Sigma(\mathbf{b}) = \int \mathrm{d}z \, \rho(\mathbf{b}, z) \tag{7.6}$$

from the three-dimensional mass density ρ. The projected lens is then decomposed into infinitesimal mass elements $\Sigma(\mathbf{b}')\mathrm{d}^2b'$ which can individually be considered as point masses. The total deflection angle at a point $\mathbf{b}$ is then the sum of point-mass deflection angles,

$$\hat{\boldsymbol{\alpha}}(\mathbf{b}) = \frac{4G}{c^2}\int \mathrm{d}^2b'\,\frac{\Sigma(\mathbf{b}')(\mathbf{b}-\mathbf{b}')}{|\mathbf{b}-\mathbf{b}'|^2}\,. \tag{7.7}$$

It is convenient to use angular coordinates $\theta_{1,2}$ on the lens plane instead of the physical coordinates $\mathbf{b}$. Introducing the distance D_l of the lens from the observer, $\mathbf{b} = D_\mathrm{l}\boldsymbol{\theta}$ and

$$\hat{\boldsymbol{\alpha}}(\boldsymbol{\theta}) = \frac{4G}{c^2}D_\mathrm{l}\int \mathrm{d}^2\theta'\,\frac{\Sigma(D_\mathrm{l}\boldsymbol{\theta}')(\boldsymbol{\theta}-\boldsymbol{\theta}')}{|\boldsymbol{\theta}-\boldsymbol{\theta}'|^2}\,. \tag{7.8}$$

7.2.2 Lens equation and lensing potential

We now introduce a source plane parallel to the lens plane at distance D_s from the observer, and cover it with angular coordinates $\beta_{1,2}$. If the coordinate origin is suitably chosen, coordinates $\boldsymbol{\beta}$ and $\boldsymbol{\theta}$ are related by

$$D_\mathrm{s}\boldsymbol{\beta} = D_\mathrm{s}\boldsymbol{\theta} - D_\mathrm{ls}\hat{\boldsymbol{\alpha}}(\boldsymbol{\theta})\,, \tag{7.9}$$

where D_ls is the distance between lens and source. Introducing the *reduced* deflection angle

$$\boldsymbol{\alpha}(\boldsymbol{\theta}) := \frac{D_\mathrm{ls}}{D_\mathrm{s}}\hat{\boldsymbol{\alpha}}(\boldsymbol{\theta}) = \frac{4G}{c^2}\frac{D_\mathrm{l}D_\mathrm{ls}}{D_\mathrm{s}}\int \mathrm{d}^2\theta'\,\frac{\Sigma(D_\mathrm{l}\boldsymbol{\theta}')(\boldsymbol{\theta}-\boldsymbol{\theta}')}{|\boldsymbol{\theta}-\boldsymbol{\theta}'|^2}\,, \tag{7.10}$$

the lens equation simply becomes

$$\boldsymbol{\beta} = \boldsymbol{\theta} - \boldsymbol{\alpha}(\boldsymbol{\theta})\,. \tag{7.11}$$

It is the non-linearity of this apparently simple equation due to the dependence of $\boldsymbol{\alpha}$ on $\boldsymbol{\theta}$ which makes it physically interesting. Among other things, it implies that gravitational lenses can produce multiple images of single sources when the inversion of the lens equation, i.e. the determination of image positions $\boldsymbol{\theta}$ from source positions $\boldsymbol{\beta}$, is no longer unique.

Following (7.4), the reduced deflection angle of a point mass is

$$\alpha(\theta) = \frac{4GM}{c^2\theta}\,\frac{D_\mathrm{ls}}{D_\mathrm{l}D_\mathrm{s}}\,. \tag{7.12}$$

A source exactly behind the lens, i.e. at $\boldsymbol{\beta} = 0$, is imaged according to (7.11) as a ring with radius

$$\theta_\mathrm{E} = \left[\frac{4GM}{c^2}\,\frac{D_\mathrm{ls}}{D_\mathrm{l}D_\mathrm{s}}\right]^{1/2} = \left[\frac{2R_\mathrm{s}D_\mathrm{ls}}{D_\mathrm{l}D_\mathrm{s}}\right]^{1/2}\,, \tag{7.13}$$

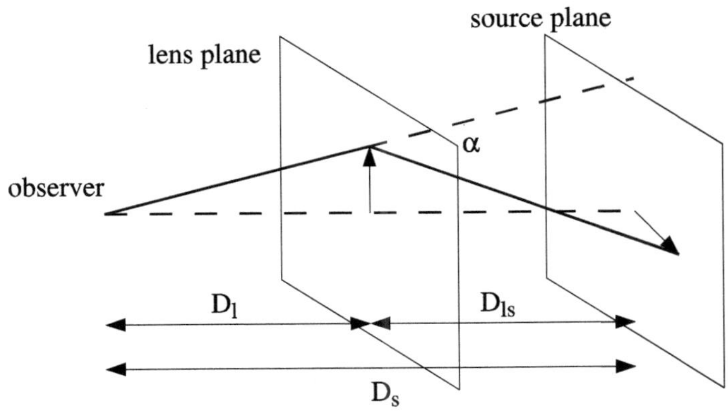

Fig. 7.1. Sketch illustrating the geometry of imaging by an isolated lens

which is called the Einstein radius . It sets a typical scale for the separation of multiple images. Assuming $2D_{\rm ls} \simeq D_{\rm s}$ in typical situations, (7.13) shows that the Einstein radius is typically given by the square root of the ratio between the Schwarzschild radius of the lens and its distance to the observer, $(R_{\rm s}/D_{\rm l})^{1/2}$. The Schwarzschild radius of a star is of order 10^5 cm. If it is within our Galaxy, $D_{\rm l} \simeq 3 \times 10^{22}$ cm, thus $\theta_{\rm E} \simeq 2 \times 10^{-9}$ radians, or $4 \simeq 10^{-4}$ arc seconds. For a galaxy at cosmological distances, $R_{\rm S} \simeq 10^{16}$ cm and $D_{\rm l} \simeq 10^{27}$ cm, thus $\theta_{\rm E} \simeq 3 \times 10^{-6}$ radians or $\simeq 0.6$ arc seconds. Galaxy clusters at cosmological distances have Einstein radii of order 50 larger than that of galaxies, or $\simeq 30$ arc seconds.

The surface-mass density Σ becomes dimension-less when divided by the critical surface mass density

$$\Sigma_{\rm cr} := \frac{c^2}{4\pi G} \frac{D_{\rm s}}{D_{\rm l} D_{\rm ls}} \,. \tag{7.14}$$

Introducing the *convergence* $\kappa(\boldsymbol{\theta})$ as the ratio $\Sigma(D_{\rm l}\boldsymbol{\theta})/\Sigma_{\rm cr}$, the reduced deflection angle reads

$$\boldsymbol{\alpha}(\boldsymbol{\theta}) = \frac{1}{\pi} \int {\rm d}^2\theta' \, \frac{\kappa(\boldsymbol{\theta}')(\boldsymbol{\theta} - \boldsymbol{\theta}')}{|\boldsymbol{\theta} - \boldsymbol{\theta}'|^2} \,, \tag{7.15}$$

Poisson's equation relates the three-dimensional density ρ to the gravitational potential,

$$\boldsymbol{\nabla}^2 \Phi = 4\pi G \rho \,. \tag{7.16}$$

The projection (7.6) implies the Poisson equation for the convergence κ,

$$\frac{\Sigma(\boldsymbol{\theta})}{\Sigma_{\rm cr}} = \kappa(\boldsymbol{\theta}) = \frac{1}{2} \boldsymbol{\nabla}^2 \psi(\boldsymbol{\theta}) \tag{7.17}$$

with the lensing potential

$$\psi(\boldsymbol{\theta}) = \frac{2}{c^2} \frac{D_{\rm ls}}{D_{\rm l} D_{\rm s}} \int {\rm d}z \, \Phi(D_{\rm l}\boldsymbol{\theta}, z) \tag{7.18}$$

Using the Green's function of the two-dimensional Laplacian operator, Eq. (7.17) can be immediately be solved for $\psi(\boldsymbol{\theta})$,

$$\psi(\boldsymbol{\theta}) = \frac{1}{\pi} \int \mathrm{d}^2\theta' \, \kappa(\boldsymbol{\theta}') \ln |\boldsymbol{\theta} - \boldsymbol{\theta}'| \, . \tag{7.19}$$

A comparison with (7.15) shows that the (reduced) deflection angle is the gradient of the (reduced) lensing potential,

$$\boldsymbol{\alpha}(\boldsymbol{\theta}) = \boldsymbol{\nabla}\psi(\boldsymbol{\theta}) \, . \tag{7.20}$$

As in geometrical optics, light deflection is described as a consequence of reduced light speed, either in a medium like glass or in a gravitational field. The index of refraction introduced above quantifies the actual light speed c' in terms of the light speed c in vacuum, $c' = c/n$. The reduced speed leads to an enhanced travel time (Shapiro delay) in a gravitational field,

$$t' = \int \frac{\mathrm{d}z}{c'} = \frac{1}{c} \int \mathrm{d}z \, n = \frac{1}{c} \int \mathrm{d}z \left(1 - \frac{2\Phi}{c^2}\right) = t - \frac{2}{c^3} \int \mathrm{d}z \, \Phi \, , \tag{7.21}$$

where t is the travel time in vacuum.

Due to the light deflection, the light path in presence of a gravitational field is also geometrically longer. Thus, gravitational lenses cause a two-fold time delay which is composed of a gravitational and a geometrical contribution. Travel times from the source to the observer along different paths belonging to multiple images are thus typically different. This time delay becomes observable with variable sources. Features in their light curves such as sudden rises or drops appear first in the image with the shortest travel time, and then repeat with a certain delay in the other images.

These relations were obtained assuming an isolated lens, i.e. a lensing mass distribution which is small compared to the cosmological distances which typically separate sources, lenses and the observer. When lensing by mass distributions on cosmological scales needs to be taken into account, the mathematical

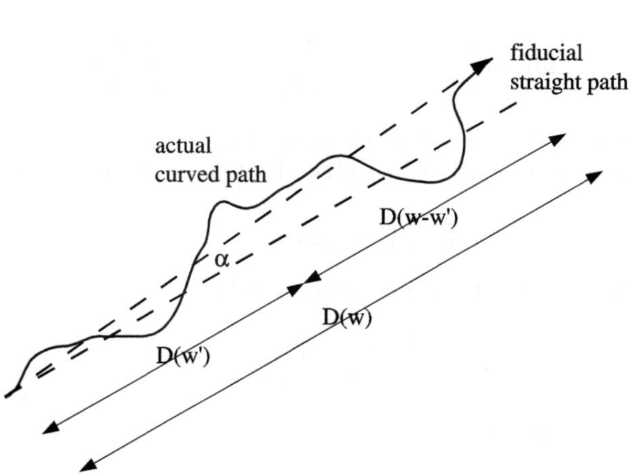

Fig. 7.2. Illustration of lensing by large-scale structures

description complicates considerably. However, the end result is that an effective lensing potential

$$\psi(\boldsymbol{\theta}) = \frac{2}{c^2} \int_0^w \mathrm{d}w' \, \frac{D(w - w')}{D(w) D(w')} \, \Phi[D(w')\boldsymbol{\theta}, w'] \tag{7.22}$$

can be introduced such that the total deflection angle experienced by a light ray is again $\boldsymbol{\nabla}\psi$, and an effective convergence is given by Poisson's equation (7.16). In (7.22), w and $D(w)$ are the comoving coordinate and angular diameter distances in the possibly curved background Universe. If space is flat, $w = D(w)$. The sources are assumed at distance w, and the integration includes all lenses at lower distances w'. The similarity to (7.18) is evident.

These are not merely formal developments. The fact that the lensing deflection angle is the gradient of a scalar potential has profound implications for the astrophysical applications of gravitational lensing, as we shall see shortly.

7.2.3 Local imaging properties

Typical sources are small compared to the lenses. Imaging by gravitational lensing can then be described by the linearised lens equation. Let the centre of the source be at $\boldsymbol{\beta}_0$ and its image at $\boldsymbol{\theta}_0$, where the deflection angle be $\boldsymbol{\alpha}_0 := \boldsymbol{\alpha}(\boldsymbol{\theta}_0)$. Then, the deflection angle at a neighbouring image point $\boldsymbol{\theta} = \boldsymbol{\theta}_0 + \delta\boldsymbol{\theta}$ is, to first approximation,

$$\boldsymbol{\alpha} = \boldsymbol{\alpha}_0 + \left.\frac{\partial \boldsymbol{\alpha}}{\partial \boldsymbol{\theta}}\right|_{\boldsymbol{\theta}_0} \delta\boldsymbol{\theta} \;, \tag{7.23}$$

and the lens equation simplifies to

$$\delta\boldsymbol{\beta} = \boldsymbol{\beta} - \boldsymbol{\beta}_0 = \delta\boldsymbol{\theta} - \left.\frac{\partial \boldsymbol{\alpha}}{\partial \boldsymbol{\theta}}\right|_{\boldsymbol{\theta}_0} \delta\boldsymbol{\theta} = \mathcal{A}\,\delta\boldsymbol{\theta} \;, \tag{7.24}$$

where $\mathcal{A}$ is the Jacobian matrix of the lens equation with components

$$\mathcal{A}_{ij} = \frac{\partial \beta_i}{\partial \theta_j} = \delta_{ij} - \frac{\partial \alpha_i}{\partial \theta_j} = \delta_{ij} - \frac{\partial^2 \psi}{\partial \theta_i \partial \theta_j} \;, \tag{7.25}$$

where we have used that $\boldsymbol{\alpha}$ is the gradient of ψ.

The matrix $\mathcal{A}$ is evidently symmetric. Its trace is related to the convergence κ by

$$\mathrm{tr}\mathcal{A} = \mathcal{A}_{11} + \mathcal{A}_{22} = 2(1 - \kappa) \;. \tag{7.26}$$

Accordingly, the matrix $\mathcal{A}$ can be decomposed as

$$\mathcal{A} = \begin{pmatrix} 1 - \kappa - \gamma_1 & -\gamma_2 \\ -\gamma_2 & 1 - \kappa + \gamma_1 \end{pmatrix} \;, \tag{7.27}$$

with the *shear components*

$$\gamma_1 := \frac{1}{2}\left(\frac{\partial^2 \psi}{\partial \theta_1^2} - \frac{\partial^2 \psi}{\partial \theta_2^2}\right) \;, \quad \gamma_2 := \frac{\partial^2 \psi}{\partial \theta_1 \partial \theta_2} \;, \tag{7.28}$$

which form the trace-free, symmetric shear matrix.

If there is no shear, $\gamma_{1,2} = 0$, the matrix $\mathcal{A}$ becomes proportional to the unit matrix. Then, lensed sources appear isotropically stretched or shrunk, but undistorted. Conversely, the shear (γ_1, γ_2) is responsible for image distortions.

Typically, the images will cover a different solid angle than the source. This is quantified by the determinant of $\mathcal{A}$, whose inverse is the magnification factor,

$$\mu = \frac{1}{\det \mathcal{A}} = \frac{1}{(1-\kappa)^2 - \gamma_1^2 - \gamma_2^2} \; . \tag{7.29}$$

The magnification will become very large where $\det \mathcal{A} \to 0$. Points $\boldsymbol{\theta}_{\rm c}$ in the lens plane where $\det \mathcal{A} = 0$ are called critical points. They form closed, so-called *critical curves*. Their images in the source plane,

$$\boldsymbol{\beta}_{\rm c} = \boldsymbol{\theta}_{\rm c} - \boldsymbol{\alpha}(\boldsymbol{\theta}_{\rm c}) \; , \tag{7.30}$$

are called *caustic curves*, or *caustics*. A source close to a caustic has highly magnified images next to a critical curve.

Apart from identifying regions of highest magnification in the source or lens planes, critical curves and caustics also separate areas of different image multiplicity. Far away from a lens, a single source must have a single image. When the source crosses a caustic on its way towards the (projected) lens centre, its image number increases by two. These additional images appear next to the critical curve belonging to the caustic. As a corollary, this implies that odd numbers of images are expected from gravitational lenses. The fact that even image numbers are typically observed allows interesting conclusions on the lensing mass distributions.

7.2.4 Basic properties of gravitational lensing

Summarising the preceding subsections, we see that gravitational lensing leads to a variety of phenomena which can qualitatively be understood in an intuitive way. First, the reduced light speed in a gravitational field causes light to be deflected much as in ordinary geometrical optics. This gravitational light deflection is differential, which means that neighbouring light rays experience slightly different deflections. Consequently, images typically appear enlarged or shrunk and distorted compared to the sources. These imaging properties are described by the convergence κ, which is proportional to the surface mass density of the lens and responsible for isotropic stretching or shrinking of images, and the two-component shear $\gamma_{1,2}$, which is responsible for image distortions. Together, convergence and shear cause the image magnification, i.e. the change of the total solid angle covered by an image compared to source.

Due to their deflection, light rays from a single source can reach the observer along multiple paths, leading to multiple images. Gravitational time delay and the geometrically longer deflected light paths cause a net time delay which is generally different for each of the images, leading to time delays between multiple images of a single source.

All properties of gravitational lensing can be summarised by a scalar potential ψ whose gradient is the deflection angle. This statement holds true for individual, localised lenses which are small compared to the distance from the source to the observer such as stars, galaxies or galaxy clusters, but also for extended lenses of cosmological scale such as the large-scale structure of the Universe.

7.2.5 Simple lens models

Important mass distributions often used for understanding basic lensing properties of astrophysical objects, or for modelling image configurations of observed lenses, are the point mass, the isothermal sphere, and the (generalised) Navarro-Frenk-White (NFW) profile. They are intrinsically axially symmetric, but can easily be distorted to mimic asymmetric mass distributions.

The point mass was already introduced. Its lensing potential is

$$\psi(\theta) = \frac{4GM}{c^2} \frac{D_{\rm ds}}{D_{\rm d} D_{\rm s}} \ln |\theta| \; . \tag{7.31}$$

Point masses are an essential tool especially for microlensing studies.

The isothermal sphere has a three-dimensional density profile falling as r^{-2}. It may be singular or have a core with finite density within a core radius $r_{\rm c}$. Its lensing potential is

$$\psi(\theta) = \frac{4\pi\sigma^2}{c^2} \frac{D_{\rm ds}}{D_{\rm s}} \sqrt{\theta_1^2 + \theta_2^2 + \theta_{\rm c}^2} \; , \tag{7.32}$$

where the angular core radius $\theta_{\rm c} = r_{\rm c}/D_{\rm l}$ may vanish. Since circular velocities in isothermal spheres do not depend on radius, they naturally reproduce flat rotation curves, which is why they are often used for modelling galaxy lenses.

Numerical simulations showed that dark-matter halos consistently have density singular profiles which are shallow inside and steep outside a scale radius $r_{\rm s}$. Generally,

$$\rho(r) = \frac{\rho_{\rm s}}{x^\alpha (1+x)^{3-\alpha}} \tag{7.33}$$

with $x := r/r_{\rm s}$. In its original form (suggested by Bavarri et al. 1996, 1997), $\alpha = 1$. Other simulations found steeper central slopes, $\alpha \simeq 1.5$. More recent simulations show that α does not converge to a fixed value down to the smallest resolved scales, but rather gently approaches unity towards the halo centres. The lensing potential of the NFW profile ($\alpha = 1$) reads

$$\psi(\theta) = 4\kappa_{\rm s} \left[\frac{1}{2} \ln^2 \frac{x}{2} - 2 \mathrm{arctanh}^2 \sqrt{\frac{1-x}{1+x}} \right] \; , \tag{7.34}$$

with $\kappa_{\rm s} := \rho_{\rm s} r_{\rm s} \, \Sigma_{\rm cr}^{-1}$ (Meneghetti et al., 2003b).

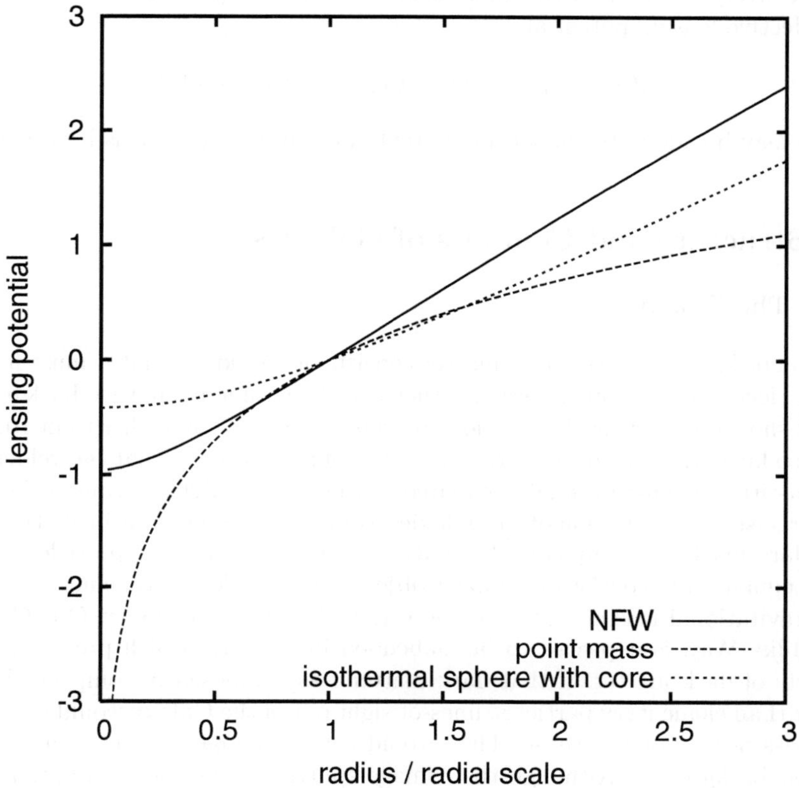

Fig. 7.3. Radial dependence of three different lensing potentials: the NFW profile (solid line), the point mass (dashed curve), and the non-singular isothermal sphere (dotted curve). All curves are arbitrarily normalised such as to pass through zero at a dimension-less radius of unity.

Deviations from axial symmetry are generally necessary to explain observed image configurations. The simplest approach is to replace the radius θ by

$$\theta' := \left[(1-\epsilon)\theta_1^2 + \frac{\theta_2^2}{1-\epsilon}\right]^{1/2} \tag{7.35}$$

in the lensing potential. This elliptically deformed lensing potential has the disadvantage that its Laplacian is dumbbell-shaped and not non-negative for moderate and large ellipticities ϵ, which implies an unphysical surface-mass density. Similar elliptical deformations of the convergence κ are thus preferred, albeit much more difficult to study.

It is often necessary to embed lenses into an environment which supplies constant shear, and possibly also constant convergence. Examples are stars in galaxies, or galaxies in galaxy clusters. In both cases, the lenses are small compared to the scale on which the properties of their surroundings change. Constant convergence κ_0 and constant shear components $\gamma_{01,2}$ can be described by the effective lensing potential

$$\psi_0(\theta) = (\kappa_0 + \gamma_{01})\theta_1^2 + (\kappa_0 - \gamma_{01})\theta_2^2 + \gamma_{02}\theta_1\theta_2 , \tag{7.36}$$

which may be added to the lensing potential of an individual lens if needed.

7.3 Structure and Contents of Galaxies

7.3.1 The Galaxy

We do not know what the dark matter consists of. We know that it must not interact electromagnetically, because otherwise the cosmic microwave background would show temperature fluctuations on the level of 10^{-3} K rather than 10^{-5} K. We also know that the dark matter must be cold in the sense that the velocity of its constituents must be small compared to the speed of light, because otherwise the large-scale distribution of the galaxies would be different. It is likely that this cold dark matter is composed of weakly-interacting elementary particles, but it could equally well consist of compact objects like, e.g. low-mass black holes.

Gravitational lensing provides one way to test this possibility. Our Galaxy, the Milky Way, is expected to be embedded into a halo which predominantly consists of dark matter. If that dark matter was composed of compact objects rather than elementary particles, lines-of-sight out of the Galaxy would occasionally pass nearby one of those. They would act as point-mass lenses on sources in their background. Although their image splitting would be substantially below the detection threshold, they would cause a well-measurable magnification (Paczyński, 1986).

Quite independent on the mass spectrum of these hypothetical compact objects (called MACHOs[2]), the probability of any one of them causing a microlensing event at any instant in time is of order $(v/c)^2$, where $v \simeq 220\,\mathrm{km\,s^{-1}\,Mpc^{-1}}$ is a typical velocity for the stars in the Galaxy. Consequently, this microlensing optical depth is of order 10^{-6}. Finding its magnification signature thus requires of order 10^6 light curves to be monitored.

Originally perceived more like science fiction, projects were carried out which observed sufficient numbers of stars in the Large and Small Magellanic Clouds (LMC and SMC, respectively) with sufficient accuracy and time sampling for detecting some microlensing events among the overwhelming signal from variable stars.

Analysing data taken from 11.9 million stars over 5.7 years, the MACHO project (Alcock et al., 2000) found 13–17 events, while between 2 and 4 were expected from known stellar populations in the Milky Way and the LMC. The microlensing optical depth deduced from lensing events lasting between 2 and 400 days is $\tau = 1.2^{+0.4}_{-0.3} \times 10^{-7}$. This implies that between 8% and 50% of the Milky Way's halo can be composed of MACHOs (at 95% confidence), whose

[2]acronym for massive compact halo objects

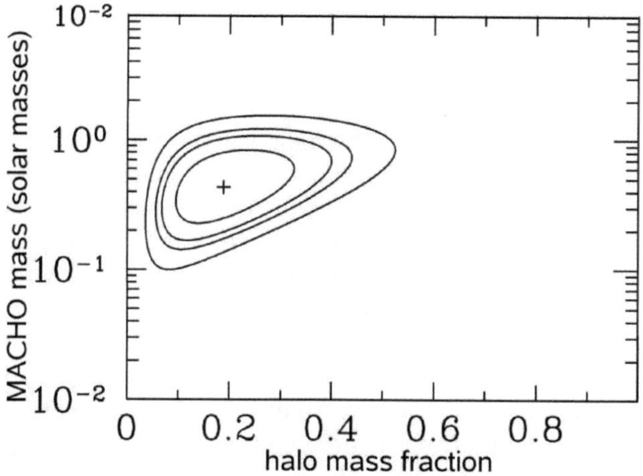

Fig. 7.4. Example of likelihood contours obtained from the MACHO experiment for one specific model for the Milky Way halo. The abscissa is the fraction of the halo mass contained in MACHOs, the ordinate is the MACHO mass. The contours show the 60%, 90%, 95%, and 99% confidence levels. (adapted from Alcock et al., 2000)

most likely mass ranges between 0.15 and $0.9\,M_\odot$ (cf. Fig. 7.4). Consistently, the EROS project (Lasserre et al., 2000) found, based on observations of LMC and SMC, that MACHOs cannot dominate the Galactic halo if their masses are $\lesssim 1\,M_\odot$. They find that the halo mass fraction in MACHOs is $< 20\%$ for MACHO masses between $10^{-7}\,M_\odot$ and $0.1\,M_\odot$ (at 95% confidence).

Thus, although MACHOs have been detected between us and the Magellanic Clouds, they are insufficient for explaining all of the Milky Way's dark mass. These MACHOs can in principle be anywhere between the source stars and the observer, i.e. in the dark halos of the Milky Way or of the Magellanic Clouds (Gould, 1993; Sahu, 1994; Wu, 1994b; Gould, 1995; Evans & Kerins, 2000). It had been speculated that self-lensing within the LMC might suffice for explaining the observed optical depth (Aubourg et al., 1999), but later studies showed that certainly not all of the LMC lensing events can be explained as being due to stars in the LMC. Rather, the LMC needs to be embedded into an extended halo (Gyuk et al., 2000; Jetzer et al., 2002; Mancini et al., 2004). Microlensing experiments have thus confirmed that the Galaxy and the Magellanic Clouds must have extended dark halos, only a fraction of which can be composed of compact objects of stellar and sub-stellar mass.

Microlensing monitoring programs are now targeting the Andromeda galaxy M 31. Few early detections of microlensing candidates (Paulin-Henriksson et al., 2003; Riffeser et al., 2003) proved the feasibility of such surveys, and data are now accumulating beginning to allow constraining the halo fraction of MACHOs in M 31 (de Jong et al., 2004; Uglesich et al., 2004), supporting results consistent with those found in our Galaxy.

7.3.2 Galaxies

Density profiles

Well over 70 cases of strong lensing by galaxies are now known. Most of them have two or four images, but a few have higher image numbers. Image splittings, typically of order an arc second, allow the projected lens mass to be constrained which is enclosed by the images. However, it turns out to be surprisingly difficult to constrain the mass profile. Essentially, multiple images constrain the average surface-mass density in an annulus bounded by the images.

Using many lens systems, and assuming their mass profiles to be self-similar, it becomes possible to trace the average surface mass density at different radii, and thus to map out the density profile. Analysing 22 galaxy lenses, and adapting a mass model composed of a concentrated component representing the light and a power-law component representing the dark matter, Rusin et al. (2003) found that the slope of the density profile is very nearly isothermal, with a double-logarithmic slope of $n = 2.07 \pm 0.13$ (isothermal has $n = 2$; cf. Sect. 7.2.5); cf. Fig. 7.5. Models in which the mass traces the light and is therefore more centrally concentrated fail at the 99% confidence level. If the dark matter follows the NFW density profile, (22±10)% of the matter inside two effective radii has to

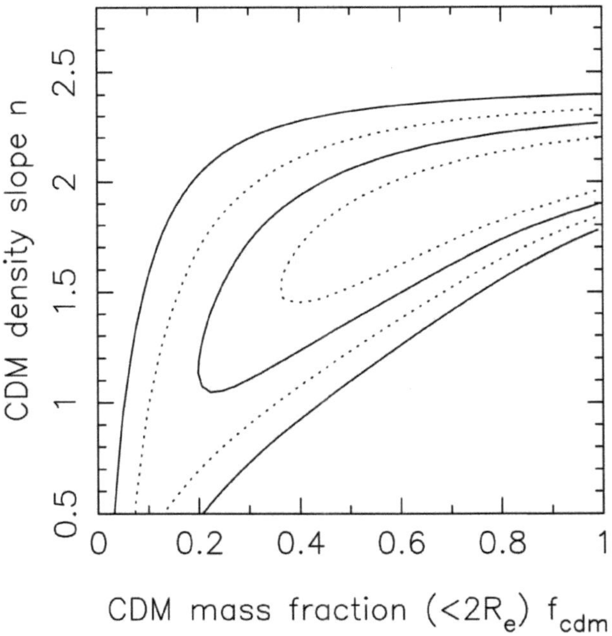

Fig. 7.5. Likelihood contours in the plane spanned by the dark mass fraction in lensing galaxies (abscissa) and the slope of their density profile (ordinate). The solid and dotted contours show the 68% and 95% confidence levels for two and one free parameter, respectively. (adapted from Rusin et al. 2003)

be dark. They also find a weak trend of the mass-to-light ratio, $M/L \propto L^{0.14^{+0.16}_{-0.12}}$, consistent with the fundamental plane of elliptical galaxies. Treu & Koopmans (2004) agree that the density profiles are nearly isothermal, but find a somewhat larger scatter. They confirm that lensing galaxies in which light traces mass are ruled out, and find a dark-matter fraction of between 15% and 65% within the effective radius.

We have seen in the introduction that gravitational lenses are expected to produce an odd number of images. In contrast, all but very few observed galaxy-lens systems have an even image number, most of them either two or four. The missing images are expected to be faint if the central density profile of the lensing galaxies is steep enough, thus their absence can be used for constraining the central concentration of the lensing mass distributions. Based on this argument, Rusin & Ma (2001) find that inner mass distributions of lensing galaxies cannot be much shallower than isothermal. Conversely, Winn et al. (2004) use a lens system in which a faint, central image has been found to constrain the mass of the central black hole in the lensing galaxy to be $< 2 \times 10^8 \, h^{-1} \, M_\odot$. From the general absence of faint, central images, Keeton (2001) concludes that the central mass profiles of lensing galaxies must be more concentrated than CDM alone predicts. Central black holes may reconcile CDM density profiles with even image numbers only if they are about an order of magnitude more massive than expected from the relation between black-hole and bulge masses.

Time delays and the Hubble constant

Time-delay measurements in multiple-image systems promise constraints on the Hubble constant, provided a sufficiently accurate mass model for the lens is known. Conversely, considering the Hubble constant as known, time-delay measurements can be used as further constraints on the lensing density profile. Values for the Hubble constant derived this way tended to be lower than those, e.g. obtained from the HST Key Project (e.g. Fassnacht et al. 2002), but lens models constructed upon a large number of constraints yield values which are very well in agreement with other determinations. For instance, Koopmans et al. (2003) find $H_0 = 75^{+7}_{-6} \, \mathrm{km\,s^{-1}\,Mpc^{-1}}$ from time delays measured in the quadruply lensed quasar B 1608+656.

An interesting problem with interpreting time delays in galaxy-lens systems is pointed out by Kochanek (2003). It turns out that the time delays between multiple images is essentially determined by the mean surface-mass density in an annulus around the lens centre bounded by the images. An estimate for that mass density can also be obtained converting the visible light to mass, assuming typical values for the fraction f_b of matter that condenses into stars. Adopting $f_\mathrm{b} \simeq 0.02$ in accordance with local observations works well with near-isothermal mass models, but yields substantially too low values for the Hubble constant, $H_0 = (48 \pm 5) \, \mathrm{km\,s^{-1}\,Mpc^{-1}}$. Conversely, values for the Hubble constant agreeing with the HST Key Project result, $H_0 = (72 \pm 8) \, \mathrm{km\,s^{-1}\,Mpc^{-1}}$, are compatible with the measured time delays only if lens models with constant mass-to-light ratios are adopted, which are otherwise ruled out. There seems to be an as yet unexplained discrepancy between measurements of H_0 and the measured time delays within the CDM framework.

Substructure in lensing galaxies

Interestingly, the number of observed lenses with four images (quadruples) are about as abundant as such with two images (doubles), while they should only contribute 25% to 30% of the galaxy lenses. The fraction of quadruples can be enhanced by satellite galaxies orbiting the main lens galaxies (Cohn & Kochanek, 2004) or by matter in their larger-scale environments (Keeton & Zabludoff, 2004), although the latter explanation is potentially problematic because it also tends to lower inferred values of the Hubble constant.

Axially-symmetric lens models are insufficient for modelling observed multiple-image systems. At least elliptical lens models are necessary. Embedding the lenses into additional external shear fields helps fitting observed image configurations, but typically more shear is required (10%–15%) than the average large-scale structure can provide (1%–3%, Keeton et al. 1997). This hints at the presence of angular structure in the lensing galaxies.

It is an interesting problem which caused much recent discussion that lens models are typically very successful in reproducing image positions, but fail in a large fraction of lens systems to explain the flux ratios between different images. A particularly obstinate and well-known case is B 1422+231 (Kormann et al., 1994; Hogg & Blandford, 1994). This is most striking in situations where the source falls just inside a cusp, in which case the sum of the signed magnifications of the three related images should vanish exactly. This expectation is frequently violated in real lens systems.

Mao & Schneider (1998) first suggested that substructure in the lensing galaxy could account for these anomalous flux ratios. While microlensing by the stars in the lens would less affect radio than optical fluxes because of their larger size, lensing by larger-scale substructures would equally change radio and optical flux ratios. Alternatively, CDM galaxy halos should contain sub-halos which may also account for anomalous flux ratios. Bradač et al. (2002) found similar modelling problems for images produced by a simulated lensing galaxy as for B 1422+231. Chiba (2002) discussed that the sub-halo population of CDM halos produces perturbations of the magnitude required for explaining anomalous flux ratios. Metcalf & Zhao (2002) estimated that $\gtrsim 5\%$ of the lensing halo mass must be contained in substructures and argued that elliptically deformed power-law models embedded into external shear are insufficient for most lenses. Similarly, Dalal & Kochanek (2002) concluded that substructure comprising 0.6% to 7% of the lens mass, with a median at 2%, was necessary for reproducing the observed anomalous flux ratios, in excellent agreement with CDM halo simulations. They also estimated that the sub-halos should have masses in the range $(10^6 \ldots 10^9)\,h^{-1}\,M_\odot$. Metcalf (2002) found that halo substructures with masses within $(10^5 \ldots 10^7)\,h^{-1}\,M_\odot$ may explain the curved radio jet in B 1152+199.

In contrast to these arguments, Evans & Witt (2003) explicitly constructed smoothly deformed lens models which could well reproduce image configurations

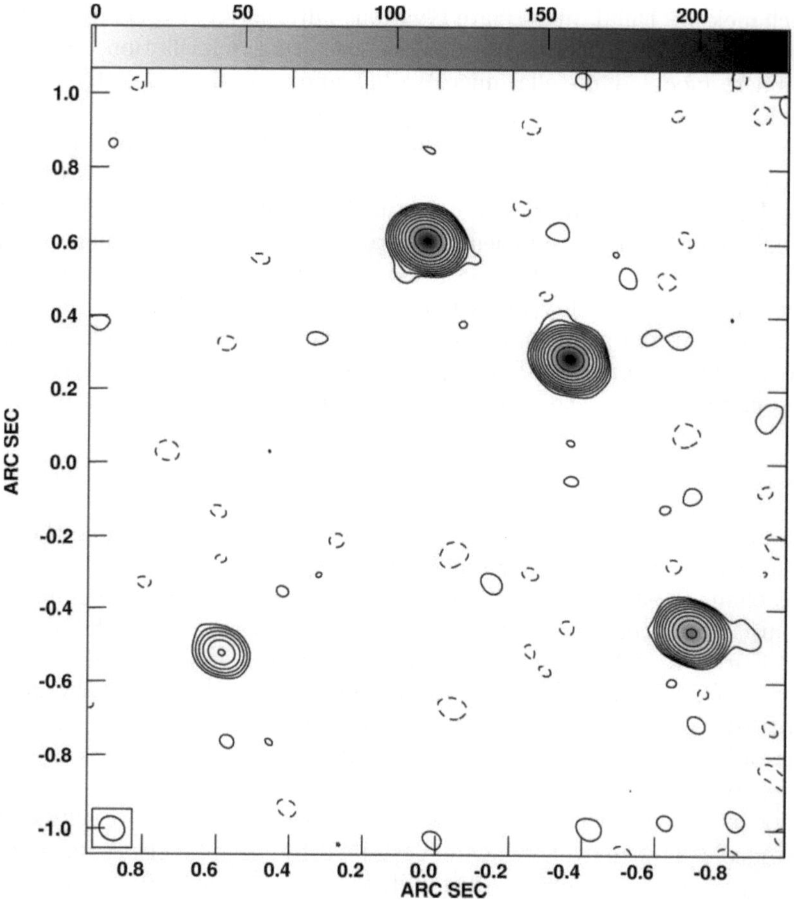

Fig. 7.6. MERLIN radio image of the multiply-imaged quasar B 1422+231 (courtesy of the JVAS/CLASS team). If image by a single, unperturbed lens, the two outer of the three bright images should together be as bright as the one between them.

and flux ratios for most lens systems and argued that substructure in the lensing galaxies and smoothly deformed lenses are both viable explanations for the anomalous flux ratios. Along the same line, Möller et al. (2003) and Quadri et al. (2003) showed that disks in lens galaxies can alter image magnification ratios considerably, while Chen et al. (2003) pointed out that halos projected onto the main lens galaxy may also cause the observed magnification perturbations.

From a somewhat different perspective, Schechter & Wambsganss (2002) discussed that decomposing lensing galaxies into microlenses has the most prominent effect when part of the lensing mass remains smooth. Specifically, they showed that so-called saddle-point images can be substantially demagnified in presence of microlenses. Schechter et al. (2004) added that the magnification distribution of the macro-images depends on the mass spectrum of the microlenses, in contrast to earlier expectations.

Kochanek & Dalal investigated various alternative explanations for the anomalous flux ratios, such as absorption, scattering, scintillation, uncertainties in the macro-model, and stellar microlensing, and arrived again at the conclusion that halo substructures remain as the most likely reason. Bradač et al. (2004) verified that numerically simulated galaxies can produce anomalous flux ratios as observed and emphasised the importance of the demagnification of saddle-point images. However, the statistics remains unclear. Although it appears doubtless that CDM halos contain sufficient substructure for sufficiently perturbing image flux ratios, such sub-halos must also appear projected onto at least one of the images. The probability for that is low. Mao et al. (2004) find in numerical simulations that the probability of finding suitably massive sub-halos in front of macro-images is only $\lesssim 0.5\%$, which is probably too low for explaining the anomalous flux ratios. The situation thus remains interestingly confused.

Lens statistics

The abundance of galaxy lenses has often been used for constraining the cosmological constant Λ. While early studies typically found *upper* limits of $\Lambda \lesssim 0.7$ (e.g. Kochanek 1996; Falco et al. 1998), more recent investigations find values which are better compatible with other determinations, (e.g. Chiba & Yoshii 1999; Chae et al. 2002), finding spatially-flat model universes with low matter density ($\Omega_0 \simeq 0.3$) preferred. The reason for this change is that gradually more realistic galaxy luminosity functions were used for estimating the expected number of lenses, rather than error-prone extrapolations of local galaxy number densities towards high redshift (Keeton, 2002).

Halos are expected to have a continuous mass spectrum in universes dominated by cold dark matter, which is described by mass functions such as those derived by Press & Schechter (1974), Sheth & Thormen (2002) and Jenkins et al. (2001). Thus, one would expect a continuous distribution of splitting angles between fractions of an arc second to several ten arc seconds. Narayan & White (1988) investigated whether the observed image-splitting distribution was consistent with expectations from CDM. They found observation and theory by and large agreed if selection effects were taken into account. Kochanek (1995) found that the splitting-angle distribution in CDM grossly incompatible with microwave-background constraints in a model universe with high matter density and vanishing cosmological constant, but that both could be comfortably reconciled in a spatially-flat, low-density CDM model.

Occasionally, therefore, lens systems should be detected with splitting angles of ten or more arc seconds. Phillips et al. (2001) interpreted the absence of wide-separation lenses in the CLASS survey as a being due to low central mass concentrations in group- and cluster-sized halos. It was perceived as a further confirmation of the CDM paradigm when a quadruply imaged quasar was detected in the Sloan Digital Sky Survey with a splitting angle of 14.62 arc seconds (Inada et al., 2003), for which Williams & Saha (2004) derived a lens mass of $(5 \pm 1) \times 10^{13}\, h^{-1}\, M_\odot$ within a radius of $100\, h^{-1}$ kpc based on a non-parametric lens model. Oguri & Keeton (2004) noted that the triaxiality of CDM halos must be taken into account in probability and mass estimates for the formation of wide-separation lens systems, a theme which is familiar from studies of strong lensing in galaxy clusters.

Galaxy-galaxy lensing

Less distant galaxies can act as weak gravitational lenses on more distant galaxies. Their shear imprints a feeble tangential distortion pattern on the images of background galaxies which appear projected close to them. This weak signal is superposed on the intrinsic ellipticities and irregularities of the background-galaxy images and thus requires statistical techniques for its extraction. Brainerd et al. (1996) first discussed the principal features of this effect and searched for it in a sample of galaxies, in which they separated background from foreground galaxies according to their apparent brightness. They could already infer that the shear profile of brighter galaxies was compatible with an isothermal mass profile with a circular velocity of $v_c = (220 \pm 80)\,\mathrm{km\,s^{-1}}$. They also placed a lower limit $r_* \gtrsim 100\,h^{-1}$ kpc on the halo size of the lensing galaxies. dell'Antonio & Tyson (1996) searched for galaxy-galaxy lensing in the Hubble Deep Field (North) and found a mean velocity dispersion for the lensing halos of $\sigma_v = 185^{+30}_{-35}\,\mathrm{km\,s^{-1}}$, and a weak lower limit on the halo radius.

Schneider & Rix (1997) devised a maximum-likelihood technique for efficient analysis of galaxy-galaxy lensing data which specifically took the redshift distributions of foreground and background galaxies into account. They applied this technique to numerically simulated data and calibrated its performance. Natarajan & Kneib (1997) and Geiger & Schneider (1998) developed methods for detecting the weak-lensing signal of galaxies embedded in galaxy clusters. Applying their technique to the cluster Cl 0939+4713, Geiger & Schneider (1999) detected the shear signal of individual massive cluster galaxies. More recently, Natarajan et al. (2002a) compared the weak-lensing signal of early-type, L_* galaxies in clusters and in the field and found evidence for the cluster galaxies to be truncated, with a truncation radius shrinking with the density of the environment. Hoekstra et al. (2004) combined weak-lensing data on galaxy halos to show that they are flattened.

Recent wide-field surveys also triggered an exciting development of galaxy-galaxy lensing. Fischer et al. (2000) used the Commissioning Data of the SDSS to infer that the tangential shear profile is compatible with a power law with exponent between 0.7 and 1.1, i.e. close to isothermal. They found a best-fitting circular velocity of $v_c = (150\ldots190)\,\mathrm{km\,s^{-1}}$ and a lower limit to the physical halo radius of $260\,h^{-1}$ kpc. From the Las Campanas Redshift Survey, Smith et al. (2001) deduced an isothermal tangential shear profile within $200\,h^{-1}$ kpc and a circular velocity of $v_c = (164 \pm 20)\,\mathrm{km\,s^{-1}}$ for L_* field galaxies. They found a virial mass for the dark halo of a typical L_* galaxy of $(2.7 \pm 0.6) \times 10^{11}\,h^{-1}\,M_\odot$. Wilson et al. (2001) used data taken with the UH8K camera at the Canada-France-Hawaii telescope to measure galaxy-galaxy lensing. They also found tangential shear profiles compatible with an isothermal slope and a rotation velocity of

$v_c = 238^{+27}_{-30}\,\mathrm{km\,s^{-1}}$ for L_* galaxies. They concluded that the mass-to-light ratio of L_* galaxies in the B band is $M/L \simeq (121 \pm 28)\,h\,L_\odot/M_\odot$.

Combining the galaxy-galaxy weak-lensing signal obtained from SDSS data with the Tully-Fisher and fundamental-plane relations for late- and early-type galaxies, respectively, Seljak (2002) found that the galaxy velocity profile must drop substantially towards the virial radius, which indicates a steep dark-matter profile. Guzik & Seljak (2002) compared theoretically motivated CDM halo models with SDSS data and constrained the halo properties of galaxies with luminosities $\gtrsim L_*$. They constrained the virial mass of an early-type L_* halo to $M_{200} = (5 \ldots 10) \times 10^{11}\,h^{-1}\,M_\odot$, and somewhat less for late-type galaxies, depending on the colour. They found a gentle increase of the mass-to-light ratio with luminosity, with $M/L \simeq 17\,h\,M_\odot/L_\odot$ for late-type and $M/L \simeq 45\,h\,M_\odot/L_\odot$ for early-type L_* galaxies.

Sheldon et al. (2004) studied the cross-correlation between galaxies and mass from the galaxy-galaxy lensing signal detected in SDSS data. The wide area covered by the survey allowed constraining the correlation function out $10\,h^{-1}\,\mathrm{Mpc}$. They find a power law with a correlation length of $r_0 \simeq (5.4 \pm 0.7)\,h^{-1}\,\mathrm{Mpc}$ and an exponent of 1.79 ± 0.05. The bias parameter turns out to be approximately scale-independent (see also Hoekstra et al. 2001), while Hoekstra et al. (2002b) find the bias parameters to be gently increasing from Mpc to larger scales. Comparisons with theoretical expectations for the galaxy distribution relative to the dark matter find overall good agreement (Weinberg et al., 2004), except that the simulated mass-to-light ratio is somewhat too high (Yang et al., 2003). Satellite galaxies orbiting the lensing galaxies could be physically aligned with their hosts and thus mimic a weak galaxy-galaxy lensing signal. Hirata et al. (2004) estimated this possible contamination and constrained it to less than 15% at the relevant scales.

7.4 Galaxy Clusters

7.4.1 Strong lensing

Qualitative conclusions

Strong lensing in galaxy clusters was first detected by Soucail et al. (1987) and Lynds & Petrosian (1989). They found extended, arc-like images in the galaxy clusters A 370 and Cl 2244. Several explanations were proposed for these objects, among them gravitational lensing of background galaxies (Paczyński, 1987), which was confirmed when the redshift of the arc in A 370 was measured and found to be substantially higher than the cluster's (Soucail et al., 1988).

It was quickly recognised that gravitational arcs provided important information on the structure of galaxy clusters. It was unclear at the time how the dark matter was distributed and whether the X-ray surface-brightness profiles, which typically show a flat core of $\simeq 200\,h^{-1}\,\mathrm{kpc}$ radius, were representative for the dark-matter profiles. Arcs were soon found to reveal the following about

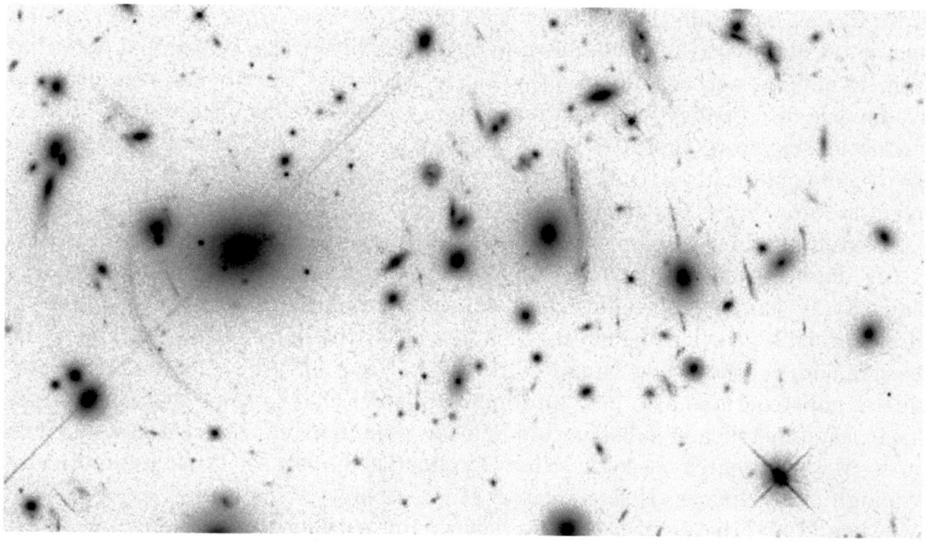

Fig. 7.7. HST image of the cluster Abell 2390 (courtesy Jean-Paul Kneib). The "straight arc" near the centre directly demonstrates the presence of substantial amounts of dark matter in this cluster.

clusters: (1) Cluster mass distributions cannot typically be axially symmetric, because large counter-arcs would otherwise be expected (Grossman & Narayan, 1988; Kovner, 1989). (2) The substantial amounts of dark matter in galaxy clusters cannot be attached to the galaxies because arcs would then have much smaller curvature radii (Hammer et al., 1989; Bergmann et al., 1990). Particularly striking were the detections of "straight arcs" in two clusters (Pelló et al., 1991; Mathez et al., 1992; Pierre et al., 1996) because they visually demonstrated the need for substantial concentrations of dark matter with very high mass-to-light ratio (Kassiola et al., 1992). (3) Clusters need to have steep density profiles, because arcs would be substantially thicker otherwise (Hammer & Rigaut, 1989). For clusters to be strong lenses, their central convergence κ has to be close to unity, but for arcs to be thin, the convergence at their locations has to be around 0.5. From cluster centres to the arc radii of typically $10'' \ldots 30''$, the κ profile must thus fall by approximately a factor of two. Cluster core radii, if they exist, must thus be substantially smaller than the X-ray core radii, which was also confirmed by the detection of "radial arcs" (Fort et al., 1992; Miralda-Escudé, 1993; Mellier et al., 1993).

Cluster masses

Arcs allow cluster masses to be easily estimated. It was soon discovered that the masses obtained this way are very close to mass estimates derived from the X-ray temperature and surface-brightness profile. This is not obvious because gravitational lensing is sensitive to the mass in whatever form and physical state it may be, while the interpretation of X-ray data requires assumptions on symmetry and hydrostatic equilibrium of the gas with the gravitational potential well, if not on isothermality of the intracluster gas. This being reassuring, a systematic discrepancy was soon revealed in the sense that masses derived from strong lensing were typically higher by factors of $\simeq 2\ldots3$ than X-ray masses (Wu, 1994a; Miralda-Escudé & Babul, 1995; Wu & Fang, 1996); more recent examples are, e.g. Chen et al. (2003) and Ota et al. (2004).

Bartehnmann & Steinmetz (1996) used numerical simulations to show that X-ray mass estimates can be systematically lower in merging clusters because their X-ray gas is still cooler than expected from their total mass, which is already seen by the lensing effect. This seems to explain the mass discrepancy at least in some clusters (e.g. Smail et al. 1995; Ota et al. 1998). Asymmetries and cluster substructures also play an important role. Due to their relatively larger shear, asymmetric and substructured clusters are more efficient lenses at a given mass. Mass estimates based on axially symmetric models are thus systematically too high (Bartelmann, 1995a; Hattori et al., 1998).

Allen (1998) distinguished clusters with and without cooling flows and found an appreciable mass discrepancy in clusters without, but good agreement of X-ray and lensing mass estimates in clusters with cooling-flow. This supports the concept that well-relaxed clusters which had sufficient unperturbed time to develop a cooling flow are well-described by simple, axially-symmetric models for lensing and the X-ray emission, while dynamically more active clusters tend to give discrepant mass estimates; this was confirmed by Wu (2000). Makino & Asano (1999) noted that the mass discrepancy is reduced if cluster density profiles are steeper than inferred from the X-ray emission.

Cluster mass profiles

Assuming mass profiles with cores, tangential arcs require small core radii as described above, but radial arcs require the cores to be finite (Le Fevre et al., 1994; Luppino et al., 1999). Numerical simulations of CDM halos, however, show that density profiles flatten towards the core, but do not develop flat cores (Navarro et al., 1996, 1997). Bartelmann (1996) showed that radial arcs can also be formed by halos with such "cuspy" density profiles, provided the central cusp is not too steep.

In principle, the relative abundances and positions of radial compared to tangential arcs in clusters provide important constraints on the central density profile in clusters (Miralda-Escudé, 1995; Molikawa & Hattori, 2001; Oguri et al., 2001). Radial arcs are still too rare for successfully exploiting this method. Being much closer to the cluster cores than tangential arcs, they are also more likely to be confused with or hidden by the light of the cluster galaxies. Following Miralda-Escudé (1995), Sand et al. (2004) compiled a sample of clusters containing radial and tangential arcs and added constraints on the central mass profile from velocity-dispersion measurements in the central cluster galaxies. They demonstrated that, assuming axially-symmetric mass distributions, central density profiles have to be substantially flatter than those found in CDM simulations.

However, Bartelmann & Meneghetti (2004) showed that even small deviations from axial symmetry can invalidate this conclusion and establish agreement between these observations and CDM density profiles.

Attempts at modelling arcs with isothermal mass distributions are typically remarkably successful (see Kneib et al. (1996) for an impressive example). This is all the more surprising as numerical simulations consistently find density profiles which are flatter than isothermal within the scale radius and steeper outside. In a very detailed analysis, Gavazzi et al. (2003) find that an isothermal core profile for the cluster MS 2137 is preferred compared to the flatter NFW profile. Smith et al. (2001) constrain the core density profile in A 383 using X-ray, weak-, and strong-lensing data and find it more peaked than the NFW profile, but argue that this may be due to the density profile of the cD galaxy. Conversely, Kneib et al. (2003) find in a combined weak- and strong-lensing analysis of Cl 0024+1654 that an isothermal mass profile can be rejected, while the NFW profile fits the data well. The most likely explanation is that the innermost cluster density profiles can be significantly influenced and steepened by baryonic physics.

Arc abundance and statistics

The mean density profile of galaxy clusters can also be constrained statistically because the probability for a cluster to become a strong lens depends sensitively on the mass concentration in its core (Wu & Hammer, 1993). Miralda-Escudé (1993) suggested that the core densities of strong-lensing clusters could be enhanced by projection of elongated clusters along the line-of-sight. Bartelmann & Weiss (1994) used a numerically simulated galaxy cluster to show that asymmetric, substructured cluster models are significantly more efficient strong lenses than axially-symmetric mass distributions because of their substantially larger shear field. Averaging over a sample of simulated clusters, Bartelmann et al. (1995) quantified that the cross sections for arc formation could be up to two orders higher for asymmetric than for axially symmetric cluster models of the same mass.

Hattori et al. (1997) confirmed that structured lenses help understanding the observational results of Le Fevre et al. (1994), who detected six arcs in a sample of 16 clusters selected for their high X-ray luminosity as measured by the EMSS satellite, but argued that even more concentrated mass profiles than those used by Bartelmann et al. (1995) are necessary for explaining them quantitatively. Bartelmann et al. (1998) used samples of numerically simulated clusters to estimate the total arc-formation probability in different cosmological models. Comparing their results with the data from Le Fevre et al. (1994), they concluded that only their cluster sample taken from a simulation with low matter density ($\Omega_0 = 0.3$) and no cosmological constant could well reproduce the measured high arc abundance, but the other three models failed badly. In particular, a flat cosmological model with $\Omega_0 = 0.3$ and $\Omega_\Lambda = 0.7$ produced an order of magnitude less arcs than observed.

This so-called arc-statistics problem was disputed based on calculations using analytic models for cluster lenses (Cooray, 1999; Kaufmann & Straumann, 2000), which failed to reproduce the strong dependence on the cosmological constant claimed by Bartelmann et al. (1998). The possible influence of cluster galaxies on the arc-formation efficiency of cluster lenses was investigated by Flores et al. (2000) and Meneghetti et al. (2000), but found to be negligible. Molikawa et al. (1999) confirmed that axially-symmetric mass models adapted to the X-ray emission do not produce a sufficient number of arcs. They found that using NFW profiles for the dark-matter profile helped, but the profiles required too high masses, and proposed that substructured mass distributions could be the solution. Meneghetti et al. (2003b) adapted elliptically distorted lenses with NFW mass profile (see also Golse & Kneib 2002) to numerically simulated clusters and found the analytic models inadequate for quantitative arc statistics despite the asymmetry, demonstrating the importance of substructures.

Oguri et al. (2003) studied the strong-lensing properties of triaxial (rather than ellipsoidal) halos and found that they may well explain the high arc abundance (Le Fevre et al., 1994; Luppino et al., 1999), provided their central density slopes are steep enough, with a double-logarithmic slope near -1.5. Wambsganss et al. (2004) simulated the magnification probability for light rays propagating through a section of the Universe and found that the abundance of high-magnification events depends strongly on the source redshift. They attributed this to the exponential mass function of massive halos, which leads to a steep increase with source redshift in the number of halos suitable for strong lensing. Identifying the probability for highly magnified light bundles on random patches of the sky with the probability for finding arcs in massive galaxy clusters, they suggested this result as the resolution for the arc statistics problem. Dalal et al. (2004) used numerical cluster simulations to estimate arc cross sections and found reasonable agreement with the earlier results of Bartelmann et al. (1998), but arrived at a higher expected arc abundance because they inserted a higher normalisation for the number density of both X-ray clusters and background sources.

Williams et al. (1999) noted that the arc radii in clusters depend only weakly on clusters mass and suggested that massive cD galaxies may be the reason. However, Meneghetti et al. (2003a) studied the effect of cD galaxies on the overall arc abundance and found it insufficient to remove the arc statistics problem. If the cosmological constant is replaced by some form of dynamical dark energy, structures tend to form earlier during cosmic history. Since cluster core densities reflect the mean cosmic density at their formation time, clusters thus tend to be more concentrated in dark-energy compared to cosmological-constant models. Bartelmann et al. (2003) estimated the effect of higher cluster concentrations on arc statistics by analytic means. They found that dark energy may in fact increase arc abundances noticeably, but again not sufficiently for solving the arc statistics problem.

The problem has experienced an interesting recent twist. Galaxy clusters at high redshifts are found to be remarkably efficient lenses (Gladders et al., 2003; Zaritsky & Gonzalez, 2003) even though they should be by far not massive enough for producing large arcs. A particularly impressive example is the cluster RX J105343+5735 at $z = 1.263$ which contains a large arc from a source at

$z = 2.577$ (Thompson et al., 2001). In this respect, it is interesting that the strong-lensing efficiency of clusters can be increased substantially and on a short timescale during a major merger (Torri et al., 2004). As a subcluster approaches a cluster, the tidal field is increased, leading to a first maximum of the cross section approximately when the two virial regions touch. The cross section then slightly decreases and approaches a second maximum when the separation of cluster and subcluster is minimal. A third peak corresponding to the first is formed when the subcluster leaves the virial region again after the merger. During that process, the arc cross section can change by an order of magnitude or more on a time scale of $\simeq 0.1\,\mathrm{Gyr}$. It thus appears that strong lensing can be a transient phenomenon at least in some clusters which would otherwise be not massive or concentrated enough. The dependence of the main merger epoch on cosmic history would then establish an interesting link between high-redshift, strong cluster lenses and the cosmological framework model.

7.4.2 Other applications of strong cluster lensing

If a cluster produces arcs from multiple sources at different redshifts, the lensing mass distribution remains the same, but the geometrical lensing efficiency is different for the arcs. Since this depends on cosmological parameters, these can thus be purely geometrically constrained from multiple-arc systems (Link & Pierce, 1998; Gautret et al., 2000). Soucail et al. (2004) applied this technique to multiple arcs in the cluster A 2218 and found that a universe with critical matter density and no cosmological constant is excluded at $> 4\,\sigma$ confidence from this single cluster.

For alleviating potential problems e.g. with the abundance of satellite galaxies, it was proposed that the dark-matter particles might interact with each other in another way than through gravity. Such a self-interaction would act as a source of isotropic pressure and thus symmetrise and smooth mass distributions (Miralda-Escudé, 2002). Strong gravitational lensing, being very sensitive to cluster asymmetries, places a tight limit on the interaction cross section. Using numerical simulations, Meneghetti et al. (2001) showed that the strong-lensing efficiency of galaxy clusters would abruptly disappear if the specific self-interaction cross section was $\gtrsim 0.1\,\mathrm{cm}^2\,\mathrm{g}^{-1}$.

Finally, lensing clusters are frequently being used as cosmic telescopes, magnifying distant sources above the limits for photometry or spectroscopy. To give a few examples, Casoli et al. (1996) used the magnification by A 370 to detect CO lines in arc sources; Frye & Broadhurst (1998) and Pelló et al. (1999) identified sources at $z = 4.04$ lensed by A 2390, Kneib et al. (2004) found an object with $z \simeq 7$ lensed by A 2218, and Pelló et al. (2004) claim to have detected an object with $z = 10.0$ magnified by A 1835.

7.4.3 Weak cluster lensing

Cluster inversion

Apart from the occasional spectacular strong-lensing effects, clusters imprint a coherent weak distortion pattern onto the many faint and distant galaxies in their background. Those galaxies reach number densities of $\simeq 40$ per square arc minute in typical images taken with large ground-based telescopes. The virial region of a typical galaxy cluster thus covers of order 10^3 galaxies. Due to their intrinsically irregular shapes, lensing-induced distortions cannot be inferred from individual galaxies. Averaging over a few galaxies, however, the intrinsic ellipticities should average to zero, leaving the ellipticity caused by the gravitational shear as the average signal.

As shown in the introduction, shear and convergence are both related through the scalar lensing potential. Knowing the shear thus allows the scaled surface-mass density to be reconstructed. Kaiser & Squires (1993) were the first to show that cluster convergence maps could be obtained by convolving the measured shear signal with a simple kernel, opening the way to systematic, parameter-free, two-dimensional cluster studies. Their technique was immediately applied to the cluster MS 1224, for which Fahlman et al. (1994) found a surprisingly high mass-to-light ratio of $\simeq 800\, h\, M_\odot/L_\odot$ in solar units, about a factor of four times the typical cluster value.

Weaknesses in the convolution algorithm by Kaiser & Squires (1993), such as its limitation to formally infinite data fields and weak shear, were discussed and removed by, e.g. Seitz & Schneider (1995, 1996). Another technique for recovering cluster mass maps based on a maximum-likelihood approach was proposed by Bartelmann et al. (1996), and later augmented with maximum-entropy regularisation (Seitz et al., 1998) and further developed by Marshall et al. (2002). An algorithm for measuring the weak shear signal from data fields was described and implemented by Kaiser et al. (1995).

Distributions of mass and light

These inversion techniques for cluster lenses have by now been applied to numerous clusters. For most of them, the mass-to-light ratios turned out to be quite normal, i.e. $M/L \simeq 250\ldots300$ in blue and $M/L \simeq 150\ldots200$ in red colour bands, respectively. Some examples are Clowe et al. (1998); Hoekstra et al. (2002a) and Gavazzi et al. (2004). The high mass-to-light ratio found by Fahlman et al. (1994), however, was qualitatively confirmed by Fischer (1999). Thus, mass and light generally appear well correlated in weak-lensing clusters, although some interesting deviations have been observed. For instance, while mass follows light well in two of the three clusters in the A 901/902 super-cluster field, the third cluster shows a significant offset between the mass and the light (Gray et al. 2002; see Fig. 7.8).

Similar phenomena appear in comparisons between the X-ray surface-brightness and the weak-lensing mass contours. While the X-ray emission follows the

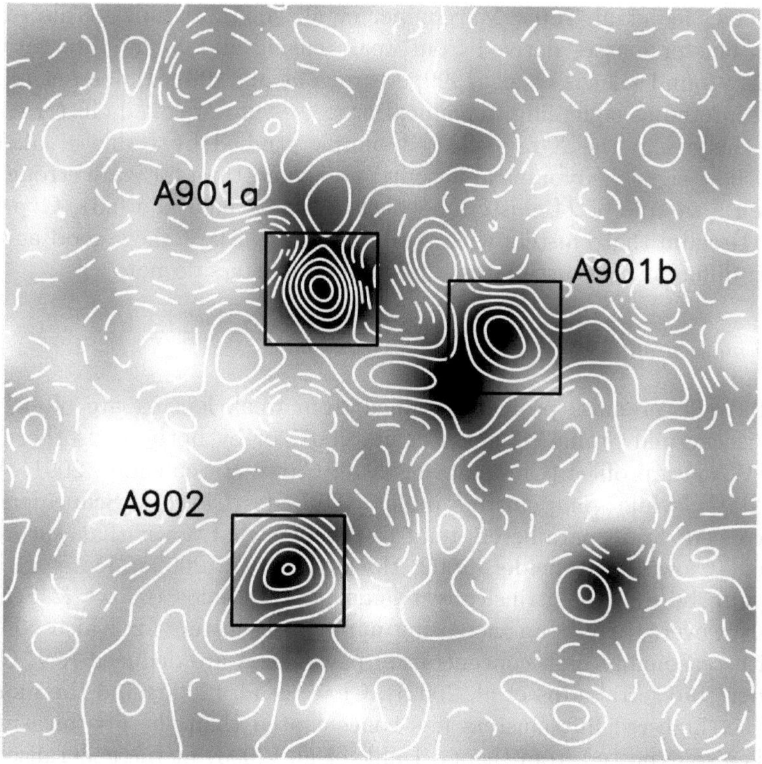

Fig. 7.8. The supercluster field around A 901/902. The contours indicate the reconstructed convergence κ, the grey scale shows the smoothed distribution of the luminosity of early-type galaxies. (adapted from Gray et al. 2002)

mass in many clusters (see Gioia et al. (1999); Clowe et al. (2000); Clowe & Schneider (2002); Hoekstra et al. (2000) for examples), interesting deviations have recently been discovered. Machacek et al. (2002) find good agreement between surface-density and X-ray contours in the outer parts of A 2218, but deviations near the cluster centre, which they interpret as a sign of dynamical activity in the cluster. Several recent studies find the X-ray gas lagging behind the dark matter in merging clusters (Clowe et al., 2004; Markevitch et al., 2004; Jee et al., 2005), as expected for hot gas embedded into collision-less dark-matter halos.

If the dark-matter particles interacted with each other, such a separation between gas and dark matter would be suppressed. Thus, from gas lagging behind the dark matter in merging clusters, and from small dark-matter core radii, limits could be obtained for the self-interaction cross section of the dark-matter particles, typically finding values $\lesssim (0.1 \ldots 1)\,\mathrm{cm}^2\,\mathrm{g}^{-1}$ (Arabadjis et al., 2002; Natarajan et al., 2002b; Markevitch et al., 2004), comparable to what Meneghetti et al. (2001) concluded from strong cluster lensing. Although different mass estimates agree well in some clusters (e.g. Clowe & Schneider 2002; Irgens et al. 2002; Jia et al. 2004; Jee et al. 2005; Margoniner et al. 2005), significant discrepancies between cluster masses derived from weak lensing and X-ray observations are frequently found (Lewis et al., 1999; Athreya et al., 2002; Holden et al., 2002) and interpreted as signalling dynamical processes in unrelaxed cluster cores. Of the 38 clusters in the X-ray selected sample studied by Dahle et al. (2002), $\simeq 30\%$ show signs of dynamical activity, and more than 50% are strong lenses. Based on a sample of 24 clusters between redshifts 0.05 and 0.31, Cypriano et al. (2004) claim that clusters with temperatures $\lesssim 8\,\mathrm{keV}$ show good agreement between different mass estimates, while hotter clusters do not.

Cluster mass profiles

Projected cluster mass profiles obtained from weak lensing are often well fit by the isothermal profile (Sheldon et al., 2001) or by both the isothermal and the NFW mass profiles (Clowe et al., 2000; Clowe & Schneider, 2001; Athreya et al., 2002), while some clusters prefer NFW fits (Clowe & Schneider, 2001). NFW concentration parameters tend to be somewhat lower than theoretically expected (Clowe & Schneider, 2002; Hoekstra et al., 2002a; Jee et al., 2005), which may be due to intrinsically triaxial cluster halos (Clowe & Schneider, 2002). However, there is an increasing number of clusters for which NFW profiles with reasonably high concentration parameters are deduced (e.g. Clowe & Schneider 2001; Arabadjis et al. 2002)). Clowe et al. (2000) find the more massive of six high-redshift clusters less concentrated than the less massive ones, which is also expected from theory. Dahle et al. (2003) fit the generalised NFW profile to six massive clusters at $z \simeq 0.3$, finding a central double-logarithmic slope $\alpha = -0.9 \ldots -1.6$ at 68% confidence. Assuming $\alpha = -1$, the concentration parameters are well in the expected range, i.e. $5 \ldots 10$ depending on cluster mass.

Large-scale structure in front of and behind galaxy clusters is projected onto them and can affect weak-lensing mass determinations. Using large-scale structure simulations, Metzler et al. (1999) estimate that weak-lensing mass estimates exceed real cluster masses by several tens of per-cents due to the added large-scale structure. Hoekstra (2003) estimated that projected large-scale structure approximately doubles the error budget for weak-lensing cluster mass estimates. However, cluster mass profiles are affected by cluster substructures and asymmetries only at the per-cent level (King et al., 2001; Clowe et al., 2004).

We have seen in the discussion of strong cluster lensing that clusters at moderate and high redshifts, $z \gtrsim 0.8$, are already remarkably efficient strong lenses. The first weak-lensing mass map of a cluster at such high redshift (MS 1054−03 at $z = 0.83$) was produced by Luppino & Kaiser (1997). The weak-lensing signal of many similarly distant clusters was measured since, typically confirming the presence of well-developed, massive and compact clusters at that epoch (Clowe et al., 1998; Gioia et al., 1999; Margoniner et al., 2005), but also frequently indicating violent dynamical activity in cluster cores (Hoekstra et al., 2000; Holden et al., 2002; Jee et al., 2005).

Dark clusters?

Occasional detections of clusters with very high mass-to-light ratios (e.g. Fahlman et al. (1994); Fischer (1999); Gray et al. (2002) as discussed above) raise the question whether cluster-sized dark-matter halos may exist which are so inefficient in producing stellar or X-ray emission that they are invisible for anything but gravitational lensing. Erben et al. (2000) detected a peak in the weak-lensing signal 7 arc minutes south of the cluster A 1942 where no optical or infrared emission could be found (Gray et al., 2001). A similarly dark weak-lensing signal peak was discovered next to the high-redshift cluster Cl 1604+4304 by Umetsu & Futamase (2000). Another tangential shear alignment potentially revealing a dark halo (Miralles et al., 2002) was meanwhile found to be spurious (Erben et al., 2003).

Schneider (1996) introduced the aperture-mass statistic specifically for detecting dark-matter halos through their weak-lensing signal. The aperture mass is a weighted integral within a circular aperture over the shear component tangentially oriented with respect to the aperture centre. When applied to numerical simulations, the aperture-mass statistic turned out to be highly efficient in finding group- and cluster-sized halos, although the completeness of the resulting halo catalogues has to be balanced against the frequency of spurious detections by carefully choosing the signal-to-noise threshold (Reblinsky & Bartelmann, 1999; White et al., 2002). Wittman et al. (2001) report the first detection of a galaxy cluster through weak lensing, which was confirmed later through its optical signal. Conversely, Schirmer et al. (2004) use the aperture-mass technique for confirming the weak-lensing signal of clusters found optically in the ESO Imaging Survey.

Bartelmann et al. (2001) showed that the detection efficiency of the aperture-mass technique varies strongly with the density profile of the dark-matter halos, allowing a statistical discrimination between isothermal and NFW profiles. Miyazaki et al. (2002) found 4.9 ± 2.3 dark-matter halos in a field of 2.1 square degrees taken with the Subaru telescope, which is consistent with expectations based on CDM models and NFW density profiles (Kruse & Schneider, 1999).

As mentioned before, cosmological models with dynamical dark energy cause dark-matter halos to be more concentrated compared to models with cosmological constant. While this should in principle lead to a higher number of weak-lensing halo detections in dark-energy cosmologies and thereby provide a way for discriminating cosmological-constant from dark-energy models, the expected sensitivity is very weak due to competing effects (Bartelmann et al., 2002; Weinberg & Kamionkowski, 2003). Weinberg & Kamionkowski (2002) argue that clusters in formation, which are not virialised yet and thus under-luminous, may be detected through weak lensing. They suggest this as an explanation for the potential dark clusters found by Erben et al. (2000) and Umetsu & Futamase (2000) and argue that cosmological constraints could be placed by comparing the numbers of visible and dark clusters.

7.5 Large-scale Structures

Weak gravitational lensing by large-scale structures is covered by several dedicated reviews, highlighting different aspects of this rich and quickly developing subject (Mellier, 1999; Bartelmann & Schneider, 2001; Hoekstra et al., 2002c; Mellier & van Waerbeke, 2002; Refregier, 2003). We can only summarise the most important aspects here and refer the interested reader to those reviews for further detail.

7.5.1 Expectations and measurements

Being inhomogeneously distributed in the Universe, matter on scales even larger than galaxy clusters must also gravitationally lens background sources. Early studies (Blandford et al., 1991; Miralda-Escudé, 1991; Kaiser, 1992) calculated the ellipticities and ellipticity correlations expected to be imprinted on the images of background galaxies, and found them to be of order a few per cent on arc-minute angular scales. In a first attempt at measuring this weak cosmological lensing signal, Mould et al. (1994) could place an upper limit in agreement with theoretical expectations.

Since weak cosmological lensing is highly sensitive to the non-linear evolution of the large-scale structures (Jain & Seljak, 1997), numerical simulations had to be carried out for precisely estimating the amplitude of the signal and the shape of the ellipticity correlation function (e.g. Bartelmann & Schneider 1992; Jain et al. 2000; Hamana & Mellier 2001; Vale & White 2003). The cosmological potential of large weak-lensing surveys was quickly pointed out (Bernardeau et al., 1997; Kaiser, 1998; van Waerbeke et al., 1999), emphasising the possibility of measuring in particular the matter density parameter Ω_0 and the amplitude σ_8 of the dark-matter power spectrum.

Schneider et al. (1998) announced the detection of a coherent shear signal in the field of the radio galaxy PKS 1508−05 which they interpreted as being caused by large-scale structure lensing. The breakthrough came soon thereafter, when several different groups almost simultaneously reported the measurement of the cosmic-shear correlation function (Bacon et al., 2000; Van Waerbeke et al., 2000; Wittman et al., 2000; Maoli et al., 2001). Given the difficulty of the measurement and the different telescopes, cameras, and analysis techniques used, the agreement between these results and their compatibility with theoretical expectations was exciting and encouraging.

Cosmological parameters were soon derived from these first cosmic-shear measurements (Van Waerbeke et al., 2001b), finding $\sigma_8 \gtrsim 0.7$ and $\Omega_0 \lesssim 0.4$ for spatially-flat cosmological models. Two-point statistics of the cosmic shear are approximately proportional to the product $\sigma_8 \Omega_0^2$, i.e. they are degenerate in these two parameters. This degeneracy can be lifted using third-order statistics such as the skewness (Van Waerbeke et al., 2001a), which arises because the non-linear evolution of cosmic structures leads to non-Gaussianity in the weak-lensing signal. Non-Gaussianities were first detected by Bernardeau et al. (2002) in the Virmos-Descart weak-lensing survey.

Much effort was devoted to calibrating weak-lensing measurements, to designing optimal cosmic-shear estimators and studying their noise properties. Erben et al. (2001) used numerical simulations to show that relative accuracies of 10%...15% can be reached by cosmic-shear measurements. A method for estimating the weak-lensing power spectrum inspired by the CMB data analysis was proposed by Hu & White (2001). Cooray & Hu (2001) investigated how non-Gaussianity can affect parameter estimates from the cosmic-shear power spectrum. Different estimators for the two-point statistics of cosmic shear and their correlation matrices were dicussed by Schneider et al. (2002a).

Numerous weak-lensing surveys have meanwhile been conducted. A non-exhaustive selection of the results obtained on σ_8 for fixed $\Omega_0 = 0.3$ is given in Tab. 7.1. Although most values of σ_8 agree within the error bars, the scatter is still substantial. This is at least partially due to remaining systematics in the data analysis, as will now be discussed.

Table 7.1. Non-exhaustive selection of results for σ_8 extracted from weak-lensing surveys. $\Omega_0 = 0.3$ is adopted throughout.

σ_8	reference
1.04 ± 0.05	Maoli et al. (2001)
$0.81^{+0.14}_{-0.19}$	Hoekstra et al. (2002d)
0.97 ± 0.13	Bacon et al. (2003)
0.72 ± 0.09	Brown et al. (2003)
0.97 ± 0.35	Hamana et al. (2003)
$0.71^{+0.12}_{-0.16}$	Jarvis et al. (2003)
1.02 ± 0.16	Rhodes et al. (2004)
0.83 ± 0.07	Van Waerbeke et al. (2005)

7.5.2 Potential systematics

As the effects of weak lensing can be summarised by a scalar potential, only such distortion patterns can be caused by weak lensing which can be described by derivatives of a scalar potential. Possible patterns in the data resembling the derivatives of a vector potential thus signal the presence of distortions of other than lensing origin. Alluding to electrodynamics, where the electric field **E** is the gradient of a scalar potential and the magnetic field **B** is the curl of a vector potential, measured distortion patterns are decomposed into so-called E and B modes, of which only the E modes can be caused by gravitational lensing. Significant B modes in the data are interpreted as remainders of undetected or incompletely removed systematics.

More or less significant B modes have been found in almost all weak-lensing surveys. Until recently, their origin was unclear. Schneider et al. (2002b) showed that source clustering could cause a B-mode contribution to the distortion, but

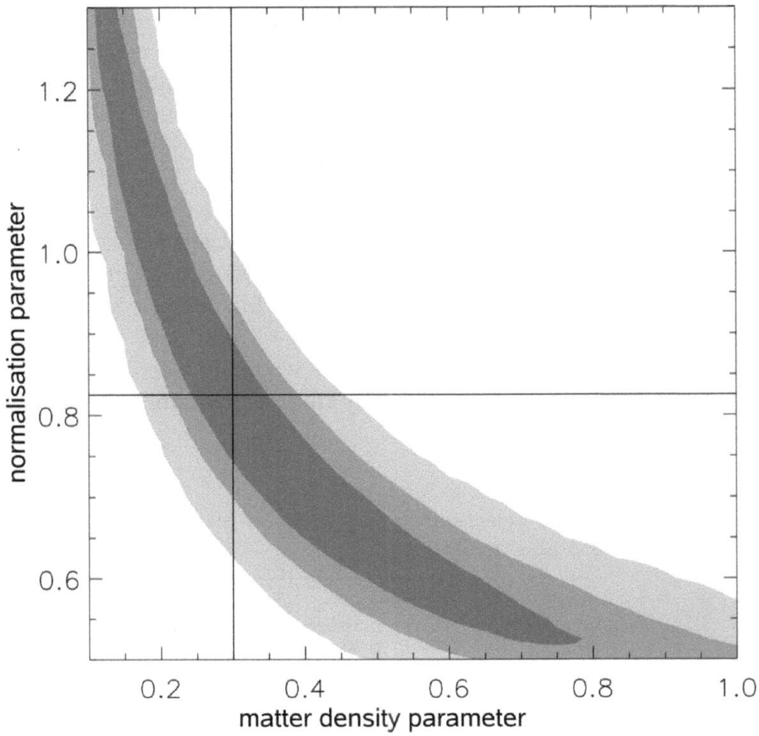

Fig. 7.9. Likelihood contours (at 68%, 95%, and 99.9% confidence) obtained from analyses of the cosmological weak shear in the plane spanned by the matter density parameter $\Omega_{\rm m}$ and the normalisation parameter σ_8. The solid lines indicate the value of σ_8 corresponding to $\Omega_{\rm m} = 0.3$. (adapted from Van Waerbeke et al. 2005)

not of sufficient strength on arc-minute scales to fully account for the observations. Hoekstra (2004) showed that incomplete correction for the anisotropies in the point-spread function of the imaging system could give rise to a substantial B mode. Application of an improved model for the point-spread function to the Virmos-Descart weak-lensing survey caused the B mode to disappear (Van Waerbeke et al. 2005; see also Fig. 7.9). It thus appears that the B-mode problem, which was discussed at length in the literature, was due to insufficient correction for the distortions imprinted by the imaging system.

Analyses of cosmic-shear measurements assume that intrinsic galaxy ellipticities are uncorrelated such that they average to zero when several images are combined. However, galaxies being physically close to each other are also expected to have their shapes aligned, e.g. by the tidal field of the large-scale matter distribution into which they are embedded. The potential effect of intrinsic rather than lensing-induced galaxy alignments depends obviously on the depth of the survey. Deep surveys project galaxy images along light paths which are substantially longer than any large-scale structure correlation scale and thus suppress any spurious signal due to intrinsinc alignments of physically neighbouring galaxies. In shallow surveys, however, intrinsic source alignments may substantially contaminate any weak-shear signal (Croft & Metzler, 2000; Heavens et al., 2000; Catelan et al., 2001; Crittenden et al., 2001). Measurements of intrinsic alignments were found to agree well with these predictions (Lee & Pen, 2002; Brown et al., 2002). Recent numerical (Jing, 2002) and analytic studies (Lee, 2004) claim strong intrinsic alignments of galaxy halos.

Possibilities for removing the signal contamination due to intrinsic alignments were discussed extensively. They advocate using photometric redshifts to remove physically close pairs of source galaxies from the analysis (King & Schneider, 2002; Heymans & Heavens, 2003). An application of this technique to the multi-colour Combo-17 survey (Heymans et al., 2004) showed intrinsic alignments near the lower end of the theoretical predictions and slightly lowered the value of σ_8 previously derived from these data (Brown et al., 2003) to 0.67 ± 0.1 for $\Omega_0 = 0.3$. Hirata & Seljak (2004) point out that foreground galaxies are aligned with the large-scale structures lensing background galaxies, thus giving rise to a higher-order alignment between galaxies at different redshifts.

7.5.3 Perspectives

For several years now, cosmological parameters are obtained with high accuracy mainly from combined analyses of CMB data, large-scale galaxy surveys and the dimming of type-Ia supernovae with redshift. What is the role of weak gravitational lensing in this context? It should be emphasised that parameter constraints from the CMB alone suffer from degeneracies which can only be broken using additional information. Measuring directly the dark-matter density and the normalisation of its fluctuations, gravitational lensing adds constraints which substantially narrow the parameter ranges allowed by the CMB, as Hu & Tegmark (1999) pointed out. Contaldi et al. (2003) combined weak-lensing data from the Red-sequence Cluster Survey with CMB data from the WMAP satellite to find $\sigma_8 = 0.89 \pm 0.05$, $\Omega_0 = 0.30 \pm 0.03$ and a Hubble constant of $H_0 = (70 \pm 3)\,\mathrm{km\,s^{-1}\,Mpc^{-1}}$. Tereno et al. (2005) predicted the accuracy of joint cosmological parameter estimates using CMB data in combination with the weak-lensing constraints from the CFHT Legacy Survey.

The exploitation of higher than second-order statistics will become increasingly important for breaking degeneracies in the weak-lensing parameter estimates and the investigation of non-Gaussianity developing due to non-linear structure growth (Zhang et al., 2003; Takada & Jain, 2003; Dolney et al., 2004). Much valuable information must also be contained in the morphology of the two-dimensional weak-lensing pattern.

Among the most exciting perspectives of weak lensing is its potential to study the three-dimensional distribution of dark structures. Although lensing observables measure the two-dimensional, projected density distribution, selecting sources at different distances allows structures along the line-of-sight to be resolved. Sufficiently accurate information on the distance to the source galaxies is provided by photometric redshifts. Simon et al. (2004) show that the accuracy of cosmic-shear parameter estimates can be improved by a factor of $5 \ldots 10$

by splitting the source galaxies into only four redshift bins. Heavens (2003) developed an elegant formalism for extracting three-dimensional information from weak-lensing data (see also Pen 2004a), and Taylor et al. (2004) recovered the three-dimensional matter distribution in the field of the Combo-17 survey, using photometric redshifts as distance indicator for the source galaxies.

The possibility to extract three-dimensional information from weak-lensing data opens a way for studying the growth of cosmic structures along the line-of-sight from the distant and past Universe. This, in turn, is sensitive to the change of the dark-energy density with time. Thus, sufficiently accurate and wide-field weak-lensing surveys will allow the dark energy to be constrained (Huterer, 2002). The sensitivity of such experiments is very promising (Munshi & Wang, 2003; Benabed & van Waerbeke, 2004), in particular if higher-order statistics are considered. This is one of the strongest motivations for a weak-lensing survey from space (Massey et al., 2004; Refregier et al., 2004).

Planned wide-area surveys of the sky in radio wavebands are expected to yield information on both cosmology and the distribution of neutral hydrogen during reionisation (Pen, 2004b). Weak lensing in the data of the FIRST radio survey undertaken at 20 cm wavelength was recently detected by (Chang et al., 2004).

7.5.4 Cosmic magnification

Magnification bias

Besides distorting the images of distant galaxies, large-scale structures also magnify background sources. To linear order, the power spectrum of cosmic magnification is simply four times the power spectrum of the cosmic shear, i.e. both contain the same amount of cosmological information. Cosmic shear, however, is much more easily measurable than cosmic magnification because it can reasonably be assumed that the ellipticities of distant galaxies average to zero. The intrinsic flux of any given source being unknown, cosmic magnification is much harder to identify.

Currently the most promising (and perhaps the only) method for detecting cosmic magnification is the magnification bias. If a population of distant sources, e.g. quasars, is observed within a solid angle $\delta\Omega$ on the sky where the magnification is μ, fainter sources become visible there, but their number density is reduced because the solid angle is stretched by the magnification. The net effect depends on how many more sources the magnification lifts above the flux threshold of the observation. If the number-count function of the sources is sufficiently steep, the dilution is outweighed and the magnification causes more sources to be visible, while sources with flat number counts are diluted and appear less numerous than without the magnification.

The spatial galaxy distribution follows the dark-matter structures which act as lenses on background sources. Bright quasars, for example, have a steep number-count function and thus appear more numerous behind magnifying dark-matter overdensities. The correlation of foreground galaxies with the lensing matter then leads to a positive cross-correlation on angular scales of arc minutes and larger between distant QSOs and foreground galaxies which are physically separated by cosmological distances.

The cross-correlation function between background sources and foreground galaxies was derived in linear approximation by Bartelmann (1995b) and Dolag & Bartelmann (1997); see also Guimarães et al. (2001). However, magnification is non-linear in shear and convergence. Computing the theoretical expectation of the cosmic magnification accurately is thus considerably more complicated than for the cosmic shear. Ménard et al. (2003b) derived second-order corrections to the linear cosmic-magnification statistics and found that the linear approximation was systematically low by 20% ... 30%, which was confirmed by Takada & Hamana (2003) in a fully non-linear treatment based on the halo model of the dark-matter distribution. The accuracy of analytic calculations can be calibrated using numerical simulations of light propagation through large-scale structures (e.g. Barber & Taylor 2003; Ménard et al. 2003b). Sufficiently accurate analytic and numerical calculations of cosmic magnification yield typical magnifications of $\lesssim 10\%$ for point sources at redhifts $z_\mathrm{s} \simeq 1$. While most emphasis was put on cross-correlations between QSOs and galaxies, cosmic magnification may also induce detectable cross-correlations between background and foreground galaxies (Moessner & Jain, 1998; Moessner et al., 1998). Jain (2002) discussed survey strategies for detecting cosmic magnification.

Observations

Bartelmann & Schneider (1993a) confirmed that the QSO-galaxy correlations detected by Seldner & Peebles (1979) and Fugmann (1990) can indeed be explained in terms of gravitational lensing by large-scale structures. Subsequent more detailed correlation studies confirmed that the correlations showed the significance variations expected from the lensing hypothesis (Bartelmann & Schneider, 1993b), and revealed cross-correlations of distant QSOs with infrared (Bartelmann & Schneider, 1994; Bartsch et al., 1997) and X-ray emission (Bartelmann et al., 1994). While the *existence* of QSO-galaxy cross-correlations was thus firmly established, their *amplitude* and *angular scale* was typically found to be substantially too high (Williams & Irwin, 1998; Norman & Williams, 2000; Benítez et al., 2001; Gaztañaga, 2003), although some analyses obtained the theoretically expected results (Rodrigues-Williams & Hogan, 1994; Norman & Impey, 2001). The observational situation was thus utterly confused. Fort et al. (1996) found evidence for cosmic shear in the vicinity of five distant QSOs. Dust absorption in foreground galaxies may be responsible for similar anti-correlations as expected from weak lensing of faint background QSOs (e.g. Croom 2001), and selection effects may cause correlations as well as anti-correlations (e.g. Ferreras et al. 1997). The detailed relation between dark-matter halos and their galaxy content adds further uncertainty (Jain et al., 2003).

Due to the weakness of the signal and the possibility of confusing it with other effects, large multi-colour surveys such as the SDSS were expected to be necessary for an unambiguous detection of cosmic magnification and its exploitation in terms of cosmological parameters (Benítez & Sanz, 1999; Ménard & Bartelmann, 2002). The degeneracy between the density and bias parameters Ω_0 and b can be broken determining the three-point cross-correlation between QSOs and galaxy pairs (Ménard et al., 2003a). Clear and unambiguous evidence for cosmic magnification was finally detected by Scranton et al. (2005) who measured the cross-correlation between a colour-selected sample of $\simeq 200,000$ distant QSOs and foreground galaxies in $\simeq 3,800$ square degrees of the SDSS data and found a signal significant at the 8-σ level and in complete agreement with theoretical expectations.

A potentially very promising manifestation of cosmic magnification was recently discussed by Ménard & Péroux (2003). Distant QSOs magnified by intervening matter are more likely to show absorption features in their spectra due to the gas associated with the lensing structures. Using the QSO sample from the 2dF survey, Ménard & Péroux (2003) demonstrated evidence for lensing by MgII and FeII absorbers along the lines-of-sight to the QSOs.

7.6 Summary

Many are the applications of gravitational lensing to cosmology, and the results are numerous, as the preceding discussion has shown. A review like this must be based on a subjective selection which is necessarily biased to some degree. Within these limitations, I summarise the results as follows:

- Microlensing experiments in the Galaxy have shown that, although massive compact objects exist in its halo, they insufficient to make up all the dark matter in the Galactic halo. These studies are now being extended towards the Andromeda galaxy, M 31 (Sect. 7.3.1).
- Central density profiles of lensing galaxies are approximately isothermal and thus more concentrated than CDM predicts. This indicates that galaxy density profiles have been steepened by baryonic physics (Sect. 7.3.2).
- The isothermal density profiles and the characteristic velocity dispersions of galaxies are confirmed by weak galaxy-galaxy lensing, which finds very large halo sizes with radii of $\gtrsim 200\,h^{-1}\,\mathrm{kpc}$. Halos of cluster galaxies seem to be smaller, as expected (Sect. 7.3.2).
- The biasing of galaxies relative to the dark-matter distribution is found by galaxy-galaxy lensing to be almost scale-independent, or gently increasing with scale (Sect. 7.3.2).
- Galaxies have to be structured in order to explain multiple-image geometries and the high fraction of quadruple compared to double images. Anomalous flux ratios of quadruple images seem to be best explained by lensing, but it is yet unclear whether they are caused by the galactic subhalos expected in CDM (Sect. 7.3.2).
- Measured time delays between multiple images lead to an interesting and as yet unresolved conflict between the lensing mass distribution and the Hubble constant: Isothermal profiles yield Hubble constants which are substantially

too low, and lens models giving compatible Hubble constants have too steep mass profiles (Sect. 7.3.2).

- The statistics of distant sources multiply imaged by galaxies is sensitive to the cosmological parameters. Recent applications of this method showed agreement with a low-density universe with cosmological constant (Sect. 7.3.2).
- Galaxy clusters have to be asymmetric, and they must be dominated by dark matter which is more broadly distributed than the cluster light. Cores in the dark-matter distribution must be small or absent (Sect. 7.4.1, 7.4.1).
- Frequent and substantial discrepancies between lensing and X-ray mass determinations are most likely signalling violent dynamical activity in clusters (Sect. 7.4.1, 7.4.3).
- It remains unclear whether galaxy clusters in the "concordance", low-density spatially-flat cosmological models can explain the observed abundance of gravitational arcs. Clusters need to be highly substructured and asymmetric, and their dynamics may temporarily boost their strong-lensing efficiency (Sect. 7.4.1).
- Although cluster density profiles inferred from strong and weak lensing do typically not contradict expectations from CDM, isothermal density profiles are not ruled out. Claims of flat central profiles are not supported by reasonably asymmetric models (Sect. 7.4.1, 7.4.3).
- Typical mass-to-light ratios derived from weak cluster lensing range around $\simeq 200$ in solar units, but very high values have been found. While this may indicate a separation of gas from dark matter in cluster mergers, the possible existence of dark clusters is intriguing (Sect. 7.4.3, 7.4.3).
- Surprisingly massive and compact clusters which are significant weak and powerful strong lenses exist at redshifts $z \simeq 0.8$ and above (Sect. 7.4.1, 7.4.3).
- Cosmic shear, i.e. the distortion of background-galaxy images due to weak gravitational lensing by large-scale structures, has been detected and found to be in remarkable agreement with theoretical expectations. It has allowed constraints to be placed on the matter-density parameter and the amplitude of the dark-matter fluctuations (Sect. 7.5.1).
- Systematic effects such as possible intrinsic alignments of source galaxies and the so-called B mode which is evidently not of lensing origin now seem to be well under control (Sect. 7.5.2).
- Joint analyses of weak lensing and CMB data allow parameter degeneracies in both types of experiment to be lifted. When combined with photometric redshifts of source galaxies, three-dimensional reconstructions of the large-scale matter distribution become possible. This will also allow the dark energy to be constrained (Sect. 7.5.3).
- Cosmic magnification, which is more complicated to measure than cosmic shear, can be quantified by the magnification bias. It has been detected, and most recent measurements are also in excellent agreement with theoretical expectations (Sect. 7.5.4).

References

Alcock, C., Allsman, R. A., Alves, D. R., et al. 2000, ApJ, 542, 281
Allen, S. W. 1998, MNRAS, 296, 392
Arabadjis, J. S., Bautz, M. W., & Garmire, G. P. 2002, ApJ, 572, 66
Athreya, R. M., Mellier, Y., van Waerbeke, L., et al. 2002, A&A, 384, 743
Aubourg, É., Palanque-Delabrouille, N., Salati, P., Spiro, M., & Taillet, R. 1999, A&A, 347, 850
Bacon, D. J., Massey, R. J., Refregier, A. R., & Ellis, R. S. 2003, MNRAS, 344, 673
Bacon, D. J., Refregier, A. R., & Ellis, R. S. 2000, MNRAS, 318, 625
Barber, A. J. & Taylor, A. N. 2003, MNRAS, 344, 789
Bartelmann, M. 1995a, A&A, 299, 11
Bartelmann, M. 1995b, A&A, 298, 661
Bartelmann, M. 1996, A&A, 313, 697
Bartelmann, M., Huss, A., Colberg, J. M., Jenkins, A., & Pearce, F. R. 1998, A&A, 330, 1
Bartelmann, M., King, L. J., & Schneider, P. 2001, A&A, 378, 361
Bartelmann, M. & Meneghetti, M. 2004, A&A, 418, 413
Bartelmann, M., Meneghetti, M., Perrotta, F., Baccigalupi, C., & Moscardini, L. 2003, A&A, 409, 449
Bartelmann, M., Narayan, R., Seitz, S., & Schneider, P. 1996, ApJL, 464, L115+
Bartelmann, M., Perrotta, F., & Baccigalupi, C. 2002, A&A, 396, 21
Bartelmann, M. & Schneider, P. 1992, A&A, 259, 413
Bartelmann, M. & Schneider, P. 1993a, A&A, 268, 1
Bartelmann, M. & Schneider, P. 1993b, A&A, 271, 421
Bartelmann, M. & Schneider, P. 1994, A&A, 284, 1
Bartelmann, M. & Schneider, P. 2001, Phys. Rep., 340, 291
Bartelmann, M., Schneider, P., & Hasinger, G. 1994, A&A, 290, 399
Bartelmann, M. & Steinmetz, M. 1996, MNRAS, 283, 431
Bartelmann, M., Steinmetz, M., & Weiss, A. 1995, A&A, 297, 1
Bartelmann, M. & Weiss, A. 1994, A&A, 287, 1
Bartsch, A., Schneider, P., & Bartelmann, M. 1997, A&A, 319, 375
Benabed, K. & van Waerbeke, L. 2004, Phys. Rev. D, 70, 123515
Benítez, N. & Sanz, J. L. 1999, ApJL, 525, L1
Benítez, N., Sanz, J. L., & Martínez-González, E. 2001, MNRAS, 320, 241
Bergmann, A. G., Petrosian, V., & Lynds, R. 1990, ApJ, 350, 23
Bernardeau, F., Mellier, Y., & van Waerbeke, L. 2002, A&A, 389, L28
Bernardeau, F., van Waerbeke, L., & Mellier, Y. 1997, A&A, 322, 1
Blandford, R. D., Saust, A. B., Brainerd, T. G., & Villumsen, J. V. 1991, MNRAS, 251, 600
Bradač, M., Schneider, P., Lombardi, M., et al. 2004, A&A, 423, 797
Bradač, M., Schneider, P., Steinmetz, M., et al. 2002, A&A, 388, 373
Brainerd, T. G., Blandford, R. D., & Smail, I. 1996, ApJ, 466, 623
Brown, M. L., Taylor, A. N., Bacon, D. J., et al. 2003, MNRAS, 341, 100
Brown, M. L., Taylor, A. N., Hambly, N. C., & Dye, S. 2002, MNRAS, 333, 501
Casoli, F., Encrenaz, P., Fort, B., Boisse, P., & Mellier, Y. 1996, A&A, 306, L41+
Catelan, P., Kamionkowski, M., & Blandford, R. D. 2001, MNRAS, 320, L7
Chae, K.-H., Biggs, A. D., Blandford, R. D., et al. 2002, Physical Review Letters, 89, 151301
Chang, T., Refregier, A., & Helfand, D. J. 2004, ApJ, 617, 794

Chen, J., Kravtsov, A. V., & Keeton, C. R. 2003, ApJ, 592, 24
Chiba, M. 2002, ApJ, 565, 17
Chiba, M. & Yoshii, Y. 1999, ApJ, 510, 42
Clowe, D., De Lucia, G., & King, L. 2004, MNRAS, 350, 1038
Clowe, D., Luppino, G. A., Kaiser, N., & Gioia, I. M. 2000, ApJ, 539, 540
Clowe, D., Luppino, G. A., Kaiser, N., Henry, J. P., & Gioia, I. M. 1998, ApJL, 497, L61+
Clowe, D. & Schneider, P. 2001, A&A, 379, 384
Clowe, D. & Schneider, P. 2002, A&A, 395, 385
Cohn, J. D. & Kochanek, C. S. 2004, ApJ, 608, 25
Contaldi, C. R., Hoekstra, H., & Lewis, A. 2003, Physical Review Letters, 90, 221303
Cooray, A. & Hu, W. 2001, ApJ, 554, 56
Cooray, A. R. 1999, A&A, 341, 653
Crittenden, R. G., Natarajan, P., Pen, U., & Theuns, T. 2001, ApJ, 559, 552
Croft, R. A. C. & Metzler, C. A. 2000, ApJ, 545, 561
Croom, S. M. 2001, Publications of the Astronomical Society of Australia, 18, 169
Cypriano, E. S., Sodré, L. J., Kneib, J., & Campusano, L. E. 2004, ApJ, 613, 95
Dahle, H., Kaiser, N., Irgens, R. J., Lilje, P. B., & Maddox, S. J. 2002, ApJS, 139, 313
Dahle, H., Pedersen, K., Lilje, P. B., Maddox, S. J., & Kaiser, N. 2003, ApJ, 591, 662
Dalal, N., Holder, G., & Hennawi, J. F. 2004, ApJ, 609, 50
Dalal, N. & Kochanek, C. S. 2002, ApJ, 572, 25
de Jong, J. T. A., Kuijken, K., Crotts, A. P. S., et al. 2004, A&A, 417, 461
dell'Antonio, I. P. & Tyson, J. A. 1996, ApJL, 473, L17+
Dolag, K. & Bartelmann, M. 1997, MNRAS, 291, 446
Dolney, D., Jain, B., & Takada, M. 2004, MNRAS, 352, 1019
Eddington, A. S. 1919, The Observatory, 42, 119
Erben, T., Miralles, J. M., Clowe, D., et al. 2003, A&A, 410, 45
Erben, T., Van Waerbeke, L., Bertin, E., Mellier, Y., & Schneider, P. 2001, A&A, 366, 717
Erben, T., van Waerbeke, L., Mellier, Y., et al. 2000, A&A, 355, 23
Evans, N. W. & Kerins, E. 2000, ApJ, 529, 917
Evans, N. W. & Witt, H. J. 2003, MNRAS, 345, 1351
Fahlman, G., Kaiser, N., Squires, G., & Woods, D. 1994, ApJ, 437, 56
Falco, E. E., Kochanek, C. S., & Munoz, J. A. 1998, ApJ, 494, 47
Fassnacht, C. D., Xanthopoulos, E., Koopmans, L. V. E., & Rusin, D. 2002, ApJ, 581, 823
Ferreras, I., Benitez, N., & Martinez-Gonzalez, E. 1997, AJ, 114, 1728
Fischer, P. 1999, AJ, 117, 2024
Fischer, P., McKay, T. A., Sheldon, E., et al. 2000, AJ, 120, 1198
Flores, R. A., Maller, A. H., & Primack, J. R. 2000, ApJ, 535, 555
Fort, B., Le Fevre, O., Hammer, F., & Cailloux, M. 1992, ApJL, 399, L125
Fort, B., Mellier, Y., Dantel-Fort, M., Bonnet, H., & Kneib, J.-P. 1996, A&A, 310, 705
Frye, B. & Broadhurst, T. 1998, ApJL, 499, L115+
Fugmann, W. 1990, A&A, 240, 11
Gautret, L., Fort, B., & Mellier, Y. 2000, A&A, 353, 10
Gavazzi, R., Fort, B., Mellier, Y., Pelló, R., & Dantel-Fort, M. 2003, A&A, 403, 11
Gavazzi, R., Mellier, Y., Fort, B., Cuillandre, J.-C., & Dantel-Fort, M. 2004, A&A, 422, 407
Gaztañaga, E. 2003, ApJ, 589, 82
Geiger, B. & Schneider, P. 1998, MNRAS, 295, 497

Geiger, B. & Schneider, P. 1999, MNRAS, 302, 118
Gioia, I. M., Henry, J. P., Mullis, C. R., Ebeling, H., & Wolter, A. 1999, AJ, 117, 2608
Gladders, M. D., Hoekstra, H., Yee, H. K. C., Hall, P. B., & Barrientos, L. F. 2003, ApJ, 593, 48
Golse, G. & Kneib, J.-P. 2002, A&A, 390, 821
Gould, A. 1993, ApJ, 404, 451
Gould, A. 1995, ApJ, 441, 77
Gray, M. E., Ellis, R. S., Lewis, J. R., McMahon, R. G., & Firth, A. E. 2001, MNRAS, 325, 111
Gray, M. E., Taylor, A. N., Meisenheimer, K., et al. 2002, ApJ, 568, 141
Grossman, S. A. & Narayan, R. 1988, ApJL, 324, L37
Guimarães, A. C. C., van De Bruck, C., & Brandenberger, R. H. 2001, MNRAS, 325, 278
Guzik, J. & Seljak, U. 2002, MNRAS, 335, 311
Gyuk, G., Dalal, N., & Griest, K. 2000, ApJ, 535, 90
Hamana, T. & Mellier, Y. 2001, MNRAS, 327, 169
Hamana, T., Miyazaki, S., Shimasaku, K., et al. 2003, ApJ, 597, 98
Hammer, F. & Rigaut, F. 1989, A&A, 226, 45
Hammer, F., Rigaut, F., Le Fevre, O., Jones, J., & Soucail, G. 1989, A&A, 208, L7
Hattori, M., Matuzawa, H., Morikawa, K., et al. 1998, ApJ, 503, 593
Hattori, M., Watanabe, K., & Yamashita, K. 1997, A&A, 319, 764
Heavens, A. 2003, MNRAS, 343, 1327
Heavens, A., Refregier, A., & Heymans, C. 2000, MNRAS, 319, 649
Heymans, C., Brown, M., Heavens, A., et al. 2004, MNRAS, 347, 895
Heymans, C. & Heavens, A. 2003, MNRAS, 339, 711
Hirata, C. M., Mandelbaum, R., Seljak, U., et al. 2004, MNRAS, 353, 529
Hirata, C. M. & Seljak, U. 2004, Phys. Rev. D, 70, 063526
Hoekstra, H. 2003, MNRAS, 339, 1155
Hoekstra, H. 2004, MNRAS, 347, 1337
Hoekstra, H., Franx, M., & Kuijken, K. 2000, ApJ, 532, 88
Hoekstra, H., Franx, M., Kuijken, K., & van Dokkum, P. G. 2002a, MNRAS, 333, 911
Hoekstra, H., van Waerbeke, L., Gladders, M. D., Mellier, Y., & Yee, H. K. C. 2002b, ApJ, 577, 604
Hoekstra, H., Yee, H. K. C., & Gladders, M. D. 2001, ApJL, 558, L11
Hoekstra, H., Yee, H. K. C., & Gladders, M. D. 2002c, New Astronomy Review, 46, 767
Hoekstra, H., Yee, H. K. C., & Gladders, M. D. 2004, ApJ, 606, 67
Hoekstra, H., Yee, H. K. C., Gladders, M. D., et al. 2002d, ApJ, 572, 55
Hogg, D. W. & Blandford, R. D. 1994, MNRAS, 268, 889
Holden, B. P., Stanford, S. A., Squires, G. K., et al. 2002, AJ, 124, 33
Hu, W. & Tegmark, M. 1999, ApJL, 514, L65
Hu, W. & White, M. 2001, ApJ, 554, 67
Hubble, E. 1929, Proceedings of the National Academy of Science, 15, 168
Hubble, E. & Humason, M. L. 1931, ApJ, 74, 43
Huterer, D. 2002, Phys. Rev. D, 65, 063001
Inada, N., Oguri, M., Pindor, B., et al. 2003, Nat, 426, 810
Irgens, R. J., Lilje, P. B., Dahle, H., & Maddox, S. J. 2002, ApJ, 579, 227
Jain, B. 2002, ApJL, 580, L3
Jain, B., Scranton, R., & Sheth, R. K. 2003, MNRAS, 345, 62
Jain, B. & Seljak, U. 1997, ApJ, 484, 560

Jain, B., Seljak, U., & White, S. 2000, ApJ, 530, 547
Jarvis, M., Bernstein, G. M., Fischer, P., et al. 2003, AJ, 125, 1014
Jee, M. J., White, R. L., Benítez, N., et al. 2005, ApJ, 618, 46
Jenkins, A., Frenk, C. S., White, S. D. M., et al. 2001, MNRAS, 321, 372
Jetzer, P., Mancini, L., & Scarpetta, G. 2002, A&A, 393, 129
Jia, S. M., Chen, Y., Lu, F. J., Chen, L., & Xiang, F. 2004, A&A, 423, 65
Jing, Y. P. 2002, MNRAS, 335, L89
Kaiser, N. 1992, ApJ, 388, 272
Kaiser, N. 1998, ApJ, 498, 26
Kaiser, N. & Squires, G. 1993, ApJ, 404, 441
Kaiser, N., Squires, G., & Broadhurst, T. 1995, ApJ, 449, 460
Kassiola, A., Kovner, I., & Blandford, R. D. 1992, ApJ, 396, 10
Kaufmann, R. & Straumann, N. 2000, Annalen der Physik, 9, 384
Keeton, C. R. 2001, ApJ, 561, 46
Keeton, C. R. 2002, ApJL, 575, L1
Keeton, C. R., Kochanek, C. S., & Seljak, U. 1997, ApJ, 482, 604
Keeton, C. R. & Zabludoff, A. I. 2004, ApJ, 612, 660
King, L. & Schneider, P. 2002, A&A, 396, 411
King, L. J., Schneider, P., & Springel, V. 2001, A&A, 378, 748
Kneib, J., Ellis, R. S., Santos, M. R., & Richard, J. 2004, ApJ, 607, 697
Kneib, J., Hudelot, P., Ellis, R. S., et al. 2003, ApJ, 598, 804
Kneib, J.-P., Ellis, R. S., Smail, I., Couch, W. J., & Sharples, R. M. 1996, ApJ, 471, 643
Kochanek, C. S. 1995, ApJ, 453, 545
Kochanek, C. S. 1996, ApJ, 466, 638
Kochanek, C. S. 2003, ApJ, 583, 49
Kochanek, C. S. & Dalal, N. 2004, ApJ, 610, 69
Koopmans, L. V. E., Treu, T., Fassnacht, C. D., Blandford, R. D., & Surpi, G. 2003, ApJ, 599, 70
Kormann, R., Schneider, P., & Bartelmann, M. 1994, A&A, 286, 357
Kovner, I. 1989, ApJ, 337, 621
Kruse, G. & Schneider, P. 1999, MNRAS, 302, 821
Lasserre, T., Afonso, C., Albert, J. N., et al. 2000, A&A, 355, L39
Le Fevre, O., Hammer, F., Angonin, M. C., Gioia, I. M., & Luppino, G. A. 1994, ApJL, 422, L5
Lee, J. 2004, ApJL, 614, L1
Lee, J. & Pen, U. 2002, ApJL, 567, L111
Lewis, A. D., Ellingson, E., Morris, S. L., & Carlberg, R. G. 1999, ApJ, 517, 587
Link, R. & Pierce, M. J. 1998, ApJ, 502, 63
Luppino, G. A., Gioia, I. M., Hammer, F., Le Fèvre, O., & Annis, J. A. 1999, A&AS, 136, 117
Luppino, G. A. & Kaiser, N. 1997, ApJ, 475, 20
Lynds, R. & Petrosian, V. 1989, ApJ, 336, 1
Ménard, B. & Bartelmann, M. 2002, A&A, 386, 784
Ménard, B., Bartelmann, M., & Mellier, Y. 2003a, A&A, 409, 411
Ménard, B., Hamana, T., Bartelmann, M., & Yoshida, N. 2003b, A&A, 403, 817
Ménard, B. & Péroux, C. 2003, A&A, 410, 33
Möller, O., Hewett, P., & Blain, A. W. 2003, MNRAS, 345, 1
Machacek, M. E., Bautz, M. W., Canizares, C., & Garmire, G. P. 2002, ApJ, 567, 188
Makino, N. & Asano, K. 1999, ApJ, 512, 9

Mancini, L., Calchi Novati, S., Jetzer, P., & Scarpetta, G. 2004, A&A, 427, 61
Mao, S., Jing, Y., Ostriker, J. P., & Weller, J. 2004, ApJL, 604, L5
Mao, S. & Schneider, P. 1998, MNRAS, 295, 587
Maoli, R., Van Waerbeke, L., Mellier, Y., et al. 2001, A&A, 368, 766
Margoniner, V. E., Lubin, L. M., Wittman, D. M., & Squires, G. K. 2005, AJ, 129, 20
Markevitch, M., Gonzalez, A. H., Clowe, D., et al. 2004, ApJ, 606, 819
Marshall, P. J., Hobson, M. P., Gull, S. F., & Bridle, S. L. 2002, MNRAS, 335, 1037
Massey, R., Rhodes, J., Refregier, A., et al. 2004, AJ, 127, 3089
Mathez, G., Fort, B., Mellier, Y., Picat, J.-P., & Soucail, G. 1992, A&A, 256, 343
Mellier, Y. 1999, ARA&A, 37, 127
Mellier, Y., Fort, B., & Kneib, J. 1993, ApJ, 407, 33
Mellier, Y. & van Waerbeke, L. 2002, Classical and Quantum Gravity, 19, 3505
Meneghetti, M., Bartelmann, M., & Moscardini, L. 2003a, MNRAS, 346, 67
Meneghetti, M., Bartelmann, M., & Moscardini, L. 2003b, MNRAS, 340, 105
Meneghetti, M., Bolzonella, M., Bartelmann, M., Moscardini, L., & Tormen, G. 2000, MNRAS, 314, 338
Meneghetti, M., Yoshida, N., Bartelmann, M., et al. 2001, MNRAS, 325, 435
Metcalf, R. B. 2002, ApJ, 580, 696
Metcalf, R. B. & Zhao, H. 2002, ApJL, 567, L5
Metzler, C. A., White, M., Norman, M., & Loken, C. 1999, ApJL, 520, L9
Miralda-Escudé, J. 1991, ApJ, 380, 1
Miralda-Escudé, J. 1993, ApJ, 403, 497
Miralda-Escudé, J. 1995, ApJ, 438, 514
Miralda-Escudé, J. 2002, ApJ, 564, 60
Miralda-Escudé, J. & Babul, A. 1995, ApJ, 449, 18
Miralles, J.-M., Erben, T., Hämmerle, H., et al. 2002, A&A, 388, 68
Miyazaki, S., Hamana, T., Shimasaku, K., et al. 2002, ApJL, 580, L97
Moessner, R. & Jain, B. 1998, MNRAS, 294, L18
Moessner, R., Jain, B., & Villumsen, J. V. 1998, MNRAS, 294, 291
Molikawa, K. & Hattori, M. 2001, ApJ, 559, 544
Molikawa, K., Hattori, M., Kneib, J., & Yamashita, K. 1999, A&A, 351, 413
Mould, J., Blandford, R., Villumsen, J., et al. 1994, MNRAS, 271, 31
Munshi, D. & Wang, Y. 2003, ApJ, 583, 566
Narayan, R. & White, S. D. M. 1988, MNRAS, 231, 97P
Natarajan, P. & Kneib, J. 1997, MNRAS, 287, 833
Natarajan, P., Kneib, J., & Smail, I. 2002a, ApJL, 580, L11
Natarajan, P., Loeb, A., Kneib, J., & Smail, I. 2002b, ApJL, 580, L17
Navarro, J. F., Frenk, C. S., & White, S. D. M. 1996, ApJ, 462, 563
Navarro, J. F., Frenk, C. S., & White, S. D. M. 1997, ApJ, 490, 493
Norman, D. J. & Impey, C. D. 2001, AJ, 121, 2392
Norman, D. J. & Williams, L. L. R. 2000, AJ, 119, 2060
Oguri, M. & Keeton, C. R. 2004, ApJ, 610, 663
Oguri, M., Lee, J., & Suto, Y. 2003, ApJ, 599, 7
Oguri, M., Taruya, A., & Suto, Y. 2001, ApJ, 559, 572
Ota, N., Mitsuda, K., & Fukazawa, Y. 1998, ApJ, 495, 170
Ota, N., Pointecouteau, E., Hattori, M., & Mitsuda, K. 2004, ApJ, 601, 120
Paczyński, B. 1986, ApJ, 304, 1
Paczyński, B. 1987, Nat, 325, 572
Paulin-Henriksson, S., Baillon, P., Bouquet, A., et al. 2003, A&A, 405, 15
Pelló, R., Kneib, J. P., Le Borgne, J. F., et al. 1999, A&A, 346, 359

Pelló, R., Sanahuja, B., Le Borgne, J., Soucail, G., & Mellier, Y. 1991, ApJ, 366, 405
Pelló, R., Schaerer, D., Richard, J., Le Borgne, J.-F., & Kneib, J.-P. 2004, A&A, 416, L35
Pen, U. 2004a, MNRAS, 350, 1445
Pen, U. 2004b, New Astronomy, 9, 417
Phillips, P. M., Browne, I. W. A., Jackson, N. J., et al. 2001, MNRAS, 328, 1001
Pierre, M., Le Borgne, J. F., Soucail, G., & Kneib, J. P. 1996, A&A, 311, 413
Press, W. H. & Schechter, P. 1974, ApJ, 187, 425
Quadri, R., Möller, O., & Natarajan, P. 2003, ApJ, 597, 659
Reblinsky, K. & Bartelmann, M. 1999, A&A, 345, 1
Refregier, A. 2003, ARA&A, 41, 645
Refregier, A., Massey, R., Rhodes, J., et al. 2004, AJ, 127, 3102
Rhodes, J., Refregier, A., Collins, N. R., et al. 2004, ApJ, 605, 29
Riffeser, A., Fliri, J., Bender, R., Seitz, S., & Gössl, C. A. 2003, ApJL, 599, L17
Rodrigues-Williams, L. L. & Hogan, C. J. 1994, AJ, 107, 451
Rusin, D., Kochanek, C. S., & Keeton, C. R. 2003, ApJ, 595, 29
Rusin, D. & Ma, C. 2001, ApJL, 549, L33
Sahu, K. C. 1994, PASP, 106, 942
Sand, D. J., Treu, T., Smith, G. P., & Ellis, R. S. 2004, ApJ, 604, 88
Schechter, P. L. & Wambsganss, J. 2002, ApJ, 580, 685
Schechter, P. L., Wambsganss, J., & Lewis, G. F. 2004, ApJ, 613, 77
Schirmer, M., Erben, T., Schneider, P., Wolf, C., & Meisenheimer, K. 2004, A&A, 420, 75
Schneider, P. 1996, MNRAS, 283, 837
Schneider, P. & Rix, H. 1997, ApJ, 474, 25
Schneider, P., van Waerbeke, L., Kilbinger, M., & Mellier, Y. 2002a, A&A, 396, 1
Schneider, P., van Waerbeke, L., & Mellier, Y. 2002b, A&A, 389, 729
Schneider, P., van Waerbeke, L., Mellier, Y., et al. 1998, A&A, 333, 767
Scranton, R., Ménard, B., Richards, G., et al. 2005, preprint
Seitz, C. & Schneider, P. 1995, A&A, 297, 287
Seitz, S. & Schneider, P. 1996, A&A, 305, 383
Seitz, S., Schneider, P., & Bartelmann, M. 1998, A&A, 337, 325
Seldner, M. & Peebles, P. J. E. 1979, ApJ, 227, 30
Seljak, U. 2002, MNRAS, 334, 797
Sheldon, E. S., Annis, J., Böhringer, H., et al. 2001, ApJ, 554, 881
Sheldon, E. S., Johnston, D. E., Frieman, J. A., et al. 2004, AJ, 127, 2544
Sheth, R. K. & Tormen, G. 2002, MNRAS, 329, 61
Simon, P., King, L. J., & Schneider, P. 2004, A&A, 417, 873
Smail, I., Hogg, D. W., Blandford, R., et al. 1995, MNRAS, 277, 1
Smith, D. R., Bernstein, G. M., Fischer, P., & Jarvis, M. 2001, ApJ, 551, 643
Soucail, G., Fort, B., Mellier, Y., & Picat, J. P. 1987, A&A, 172, L14
Soucail, G., Kneib, J.-P., & Golse, G. 2004, A&A, 417, L33
Soucail, G., Mellier, Y., Fort, B., Mathez, G., & Cailloux, M. 1988, A&A, 191, L19
Takada, M. & Hamana, T. 2003, MNRAS, 346, 949
Takada, M. & Jain, B. 2003, MNRAS, 344, 857
Taylor, A. N., Bacon, D. J., Gray, M. E., et al. 2004, MNRAS, 353, 1176
Tereno, I., Doré, O., van Waerbeke, L., & Mellier, Y. 2005, A&A, 429, 383
Thompson, D., Pozzetti, L., Hasinger, G., et al. 2001, A&A, 377, 778
Torri, E., Meneghetti, M., Bartelmann, M., et al. 2004, MNRAS, 349, 476
Treu, T. & Koopmans, L. V. E. 2004, ApJ, 611, 739

Uglesich, R. R., Crotts, A. P. S., Baltz, E. A., et al. 2004, ApJ, 612, 877
Umetsu, K. & Futamase, T. 2000, ApJL, 539, L5
Vale, C. & White, M. 2003, ApJ, 592, 699
van Waerbeke, L., Bernardeau, F., & Mellier, Y. 1999, A&A, 342, 15
Van Waerbeke, L., Hamana, T., Scoccimarro, R., Colombi, S., & Bernardeau, F. 2001a, MNRAS, 322, 918
Van Waerbeke, L., Mellier, Y., Erben, T., et al. 2000, A&A, 358, 30
Van Waerbeke, L., Mellier, Y., & Hoekstra, H. 2005, A&A, 429, 75
Van Waerbeke, L., Mellier, Y., Radovich, M., et al. 2001b, A&A, 374, 757
Wambsganss, J., Bode, P., & Ostriker, J. P. 2004, ApJL, 606, L93
Weinberg, D. H., Davé, R., Katz, N., & Hernquist, L. 2004, ApJ, 601, 1
Weinberg, N. N. & Kamionkowski, M. 2002, MNRAS, 337, 1269
Weinberg, N. N. & Kamionkowski, M. 2003, MNRAS, 341, 251
White, M., van Waerbeke, L., & Mackey, J. 2002, ApJ, 575, 640
Williams, L. L. R. & Irwin, M. 1998, MNRAS, 298, 378
Williams, L. L. R., Navarro, J. F., & Bartelmann, M. 1999, ApJ, 527, 535
Williams, L. L. R. & Saha, P. 2004, AJ, 128, 2631
Wilson, G., Kaiser, N., Luppino, G. A., & Cowie, L. L. 2001, ApJ, 555, 572
Winn, J. N., Rusin, D., & Kochanek, C. S. 2004, Nat, 427, 613
Wittman, D., Tyson, J. A., Margoniner, V. E., Cohen, J. G., & Dell'Antonio, I. P. 2001, ApJL, 557, L89
Wittman, D. M., Tyson, J. A., Kirkman, D., Dell'Antonio, I., & Bernstein, G. 2000, Nat, 405, 143
Wu, X. 1994a, ApJL, 436, L115
Wu, X. 1994b, ApJ, 435, 66
Wu, X. 2000, MNRAS, 316, 299
Wu, X. & Fang, L. 1996, ApJL, 467, L45+
Wu, X. & Hammer, F. 1993, MNRAS, 262, 187
Yang, X. H., Mo, H. J., Kauffmann, G., & Chu, Y. Q. 2003, MNRAS, 339, 387
Zaritsky, D. & Gonzalez, A. H. 2003, ApJ, 584, 691
Zhang, T., Pen, U., Zhang, P., & Dubinski, J. 2003, ApJ, 598, 818

8 The Oldest Stars

T.C. Beers

Abstract: I review the techniques used to identify the oldest "living" stars in the Universe, concentrating on two large modern surveys: the HK survey of Beers and colleagues and the Hamburg/ESO survey of Christlieb and collaborators. I then consider the knowledge that has been gained recently concerning the the distribution of measured stellar metallicities from these samples, the so-called Metallicity Distribution Function of the halo of the Milky Way. A summary of some of the most exciting results from recent high-resolution spectroscopy of very metal-poor stars identified in these samples is provided. Special attention is given to the nature of r-process-enhanced metal-poor stars, and what they reveal about the operation and possible astrophysical site(s) of the r-process in the early Universe. Finally, I discuss plans for next-generation surveys for extremely metal-poor stars.

8.1 Introduction

The oldest "living" stars in the Universe are inferred to be low-mass objects, still shining from nuclear fusion in their interiors, that inherited their observed low abundances of elements from the earliest generations of (presumably quite massive) metal-producing objects. These massive progenitors, often referred to as Population III stars, have been suggested by a number of studies (see Bromm & Larson 2004 for a recent review) to be the first stable objects to have formed in the early Universe, shortly after the Big Bang. The main-sequence lifetimes of the low-mass ($\leq 0.8 M_{\odot}$) stars that may have formed at or near the time of the Population III stars are sufficiently long ($>$ 14 Gyrs) that they can still be identified today. Such stars can be thought of as the "scribes" of the nucleosynthesis that took place in the early Galaxy and Universe, as their atmospheres have recorded the results of the processes involved in the production of the first elements; they make it feasible to reconstruct a increasingly coherent picture of the story of cosmic chemical evolution. Figure 8.1 depicts one model for the formation of low-mass stars in the early Universe.

The great interest in the origin of the elements, and the nature of the objects that produced them, has been fueled by the results of several modern surveys for the lowest metallicity stars, in particular the HK objective prism survey of Beers

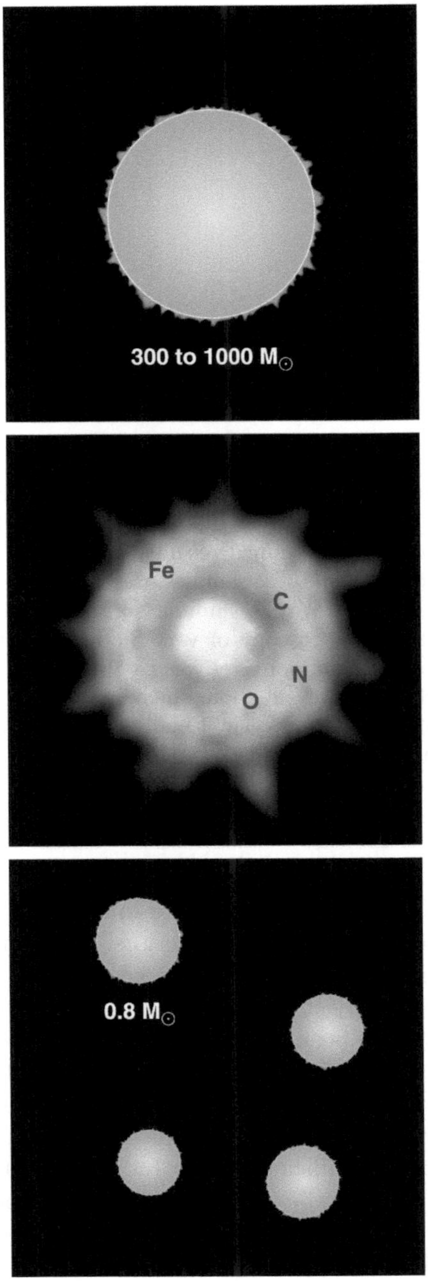

Fig. 8.1. The first generations of stars (Population III) probably included objects that were extremely massive and short-lived. When they exploded at the end of their lives, the elements created by Population III stars were incorporated into lower-mass second-generation (Population II) stars that may still be detectable in the Universe today. $M_{\odot}$ indicates the solar mass. This figure is courtesy of Preston Huey/Science Magazine.

and collaborators (Beers, Preston, & Shectman 1985, 1992; Beers 1999) and the Hamburg/ESO objective-prism survey of Christlieb and colleagues (HES; Christlieb 2003), as well as by the detailed high-resolution spectroscopic studies that have been made possible with the current generation of 8m-10m optical telescopes and their sensitive instruments. Important clues have also come from spectroscopic observations in the near ultraviolet with the Hubble Space Telescope. In this brief contribution we summarize the present status of the surveys for low metallicity stars, the nature of the elemental abundance patterns that have been revealed from high-resolution spectroscopic follow-up efforts, and describe the constraints that these observations place on early element production. We close with a glimpse of where the field is headed in the decade to come, describing new surveys, and new survey techniques, that are just now getting underway. These new surveys will enlarge the numbers of recognized low-metallicity stars by several orders of magnitude.

Emphasis is placed on the most recent results. Readers interested in obtaining more detailed information on a number of the topics considered herein may wish to read the review article by Beers & Christlieb (2005).

8.2 Low-Metallicity Stars

Low-metallicity stars exhibit atmospheric abundances of their heavy elements that are significantly below that of the Sun. For historical reasons, and due to the large number of iron absorption lines that are found in high-resolution stellar spectra of stars with intermediate temperatures, the nomenclature used to quantify the metallicity of stars is usually given as a comparison of the ratio of the number of iron atoms relative to the number of hydrogen atoms in a given star to the same ratios measured in the Sun, i.e., $[\mathrm{Fe/H}] \equiv \log_{10}(N_{\mathrm{Fe}}/N_{\mathrm{H}})_* - \log_{10}(N_{\mathrm{Fe}}/N_{\mathrm{H}})_{\odot}$, where N_{Fe} and N_H refer to the numbers of atoms of the elements Fe, and H, respectively. This nomenclature is readily generalized to refer to the abundance ratios of any two species, as compared to their observed ratios in the Sun.

In general, the abundances of other metallic species track that of iron; as the observed amount of iron in a stellar atmosphere declines, so do the observed amounts of other elements, such as Ca, Mg, Co, etc., although not necessarily in a one-to-one fashion. Indeed, the detailed behavior of elemental abundance ratios in metal-poor stars, for instance, [Ca/Fe], [Mg/Fe], [Co/Fe], etc., provide crucial clues as to the nucleosynthetic pathways involved in their formation. Examples are provided below.

Stars are presently known that cover the range of observed metallicity from somewhat above the solar iron abundance, $[\mathrm{Fe/H}] \simeq +0.5$ (sometimes referred to as super metal-rich stars), to the lowest iron abundance record set by the recently discovered star HE 1327−2326 (Frebel et al. 2005), with $[\mathrm{Fe/H}] = -5.6$, a factor

Table 8.1. Nomenclature for Stars of Different Metallicity

[Fe/H]	Term	Acronym
$> +0.5$	Super Metal-Rich	SMR
~ 0.0	Solar	–
< -1.0	Metal-Poor	MP
< -2.0	Very Metal-Poor	VMP
< -3.0	Extremely Metal-Poor	EMP
< -4.0	Ultra Metal-Poor	UMP
< -5.0	Hyper Metal-Poor	HMP
< -6.0	Mega Metal-Poor	MMP

of almost 400,000 times lower than the Sun.[1] For clarity of the terminology, Beers & Christlieb (2005) have suggested that one adopt the modifiers listed in Table 8.1 for the description of varying degrees of metal deficiency.

Although there are no MMP stars yet known, this last category is included in anticipation of their eventual discovery.

8.3 Ongoing Surveys for Metal-Poor Stars, and the Nature of the Halo Metallicity Distribution Function

8.3.1 The HK (and HK-II) prism surveys

The most efficient technique that has been developed to date for the identification of low-metallicity stars is a (somewhat) modernized version of the methods that were employed by astronomers over a century ago for the first spectral classifications of stars. The objective-prism technique makes use of wide-field Schmidt telescopes (which can typically cover on the order of 25 to 36 square degrees of sky at once) to photographically image low-resolution spectra of hundreds to thousands of stars at once. The spectra are formed by a prism that is placed over the objective end of the telescope.

In the HK survey of Beers and collaborators, the spectra are "widened" at the telescope in order to enable later visual inspection. The insertion of an interference filter at the focal plane passes only a limited amount of stellar spectrum, centered on the absorption features of singly ionized calcium, the CaII K (3933 Å) and the CaII H (3969 Å) lines, which are the strongest metallic species in the optical spectra of many stars. The cartoon shown in Figure 8.2 illustrates the appearance of the HK survey spectra, as seen by a visual classifier through a 10-X microscope. Over a number of years the classifiers (originally G. Preston,

[1]The abundances that are quoted here, unless otherwise noted, are those obtained using one-dimensional, static, stellar atmosphere models, where absorption lines are assumed to form under conditions of Local Thermodynamic Equilibrium (LTE). The development of fully three-dimensional, convective atmosphere models, and attempts to take into account the effects of non-LTE, is ongoing at present, and will someday (perhaps soon) be the method of choice for abundance determinations (Asplund 2005).

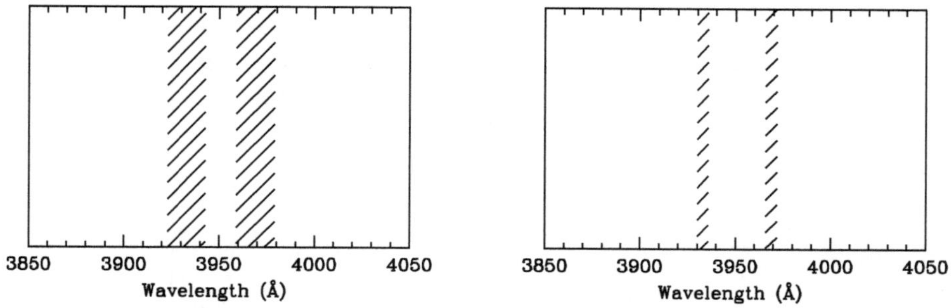

Fig. 8.2. Illustration of the appearance of the CaII K and H lines in the prism spectra of the HK survey, as they appear to the classifier. In the left panel is shown an example of a solar-metallicity star of intermediate temperature. The right panel shows a star of similar temperature, but with substantially lower metallicity. At even lower metallicity, the CaII lines disappear completely.

later T. Beers) were able to identify on the order of 10,000 metal-poor candidates from the HK prism plates. Although the selection of candidates in this manner is less than ideal, in particular because of the absence of color information for the stars, the HK survey has been remarkably effective at the identification of large numbers of the most metal-deficient stars.

One consequence of the lack of color information for the selection of the HK survey metal-poor candidates is the introduction of a quite severe temperature bias into the target list. The CaII K lines for warmer stars (and hence bluer colors) of a given metallicity are weaker than stars of the same metallicity but with lower temperatures (and hence redder colors). As a result, the list of candidates is biased toward the selection of objects near the main-sequence turnoff of the halo population, and under-represents the cooler giants. Below *some* metallicity, presently estimated to be around [Fe/H] $= -2.0$, these biases become less important, as the CaII K lines of even the cooler stars become sufficiently weak to be readily identified during the visual inspection.

Many of the limitations of the visual selection approach have now been overcome. As part of his thesis work, Rhee (2000) and collaborators carried out machine scans of the HK survey plates (using the PDS machine in Cambridge, UK), from which one could construct quantitative estimates of the strengths of the CaII K (and H) lines. Figure 8.3 shows examples of these scanned spectra.

Broadband near-infrared *JHK* photometry is now available for essentially all of the stars on the HK survey plates from the 2MASS Point Source Catalog (Cutri et al. 2003). This information is being used to obtain a new set of candidate metal-poor stars from the HK survey, which is now referred to as the HK-II survey. Rhee et al. (in preparation) have identified numerous cool metal-poor stars that were overlooked in the original HK survey, as confirmed by medium-resolution spectroscopic follow-up of some 1000 HK-II candidates conducted to date. It is expected that medium-resolution spectroscopy of several thousand additional HK-II candidates will be completed over the course of the next few years.

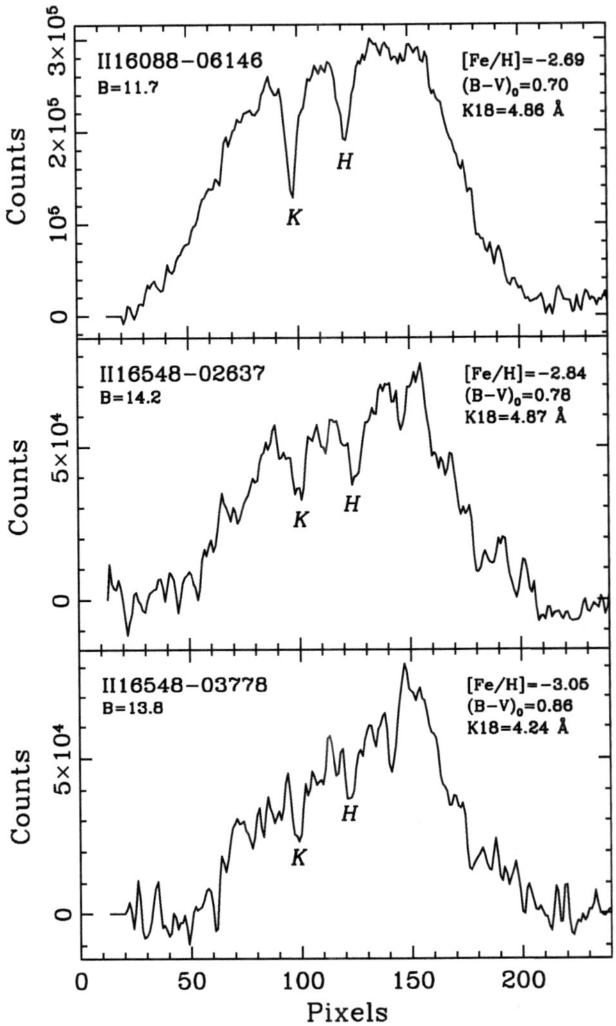

Fig. 8.3. Examples of HK-II survey scanned prism spectra. The estimated metallicity, [Fe/H], estimated color, $(B - V)_o$, and strength of the CaII K line in an 18 Å wide band, K18, are listed for each star.

8.3.2 The Hamburg/ESO prism survey

The HES is now making it possible to enlarge the number of known EMP ([Fe/H] < -3.0) stars by at least an order of magnitude over the numbers found in the HK survey, and is finding many more VMP ([Fe/H] < -2.0) stars as well. In part, this is because the HES reaches about two magnitudes deeper than the HK survey ($B \sim 17.5$ vs. $B \sim 15.5$), and also covers regions of the sky that were not included in the HK survey (see Fig. 8.4).

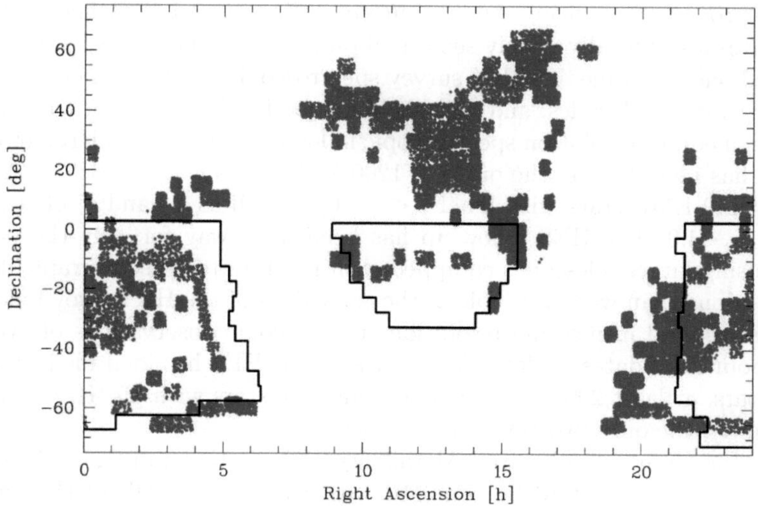

Fig. 8.4. Comparison of the HES area (framed) with the HK survey area. The shaded areas denote the locations of HK-survey candidates. The HK survey covers about 2800 $\deg^2$ in the northern sky and about 4100 $\deg^2$ in the south, and includes a number of plates at lower Galactic latitude than the HES. In total, the HES covers about 8225 $\deg^2$ of southern sky at Galactic latitudes $|b| > 30^o$.

The HES plates were taken with the ESO 1m-Schmidt telescope on La Silla, and were machine-scanned and reduced at the Hamburger Sternwarte (Germany). The spectra are not widened in the HES, as the prism plates were designed to be digitally scanned. The HES covers a much wider spectral range than the HK survey, from about 3300 Å to 5400 Å .

The digitized HES spectra were then analyzed with an automated spectral classification approach (Christlieb, Wisotzki, & Grasshoff 2002). It was also feasible to determine an approximate $B - V$ color directly from the prism spectra, with an accuracy on the order of ~ 0.1 mag. The resulting list of some 10,000 VMP candidates is thus not expected to suffer from the temperature-related biases of the original HK survey; the HES should be a rich source of low-metallicity giants as well as warmer turnoff stars.

8.3.3 The halo metallicity distribution function

Once suitable candidate low-metallicity stars are identified, medium-resolution (1–2 Å) follow-up spectroscopy is obtained for as many candidates as possible, using telescopes of 1.5m–4m apertures. Estimates of metallicity based on these spectra are obtained by a variety of methods, the most important of which is the calibration of Beers et al. (1999), which provides an measure of the metallicity of a given star based on the strength of the CaII K line and (where available) a broadband color, such as $B - V$ or $J - K$. Such estimates of metallicity are typically accurate to on the order of δ [Fe/H] ~ 0.2 dex.

The process of acquiring the medium-resolution spectra is rather slow, as the candidates are typically widely separated on the sky, hence the spectra must be obtained one at a time. The HK survey spectroscopic follow-up has been underway for over two decades, and will likely not be finished for another few years. To date, medium-resolution spectroscopic follow-up of over 6000 HK survey candidates has identified on the order of 1200 VMP stars with [Fe/H] < -2.0, and roughly 140 EMP stars with [Fe/H] < -3.0, as well as a handful of stars near [Fe/H] $= -4.0$. The HES follow-up has been underway for only the past five years (using larger telescopes equipped with much faster spectrograph/detector combinations than were available in the early days of the HK survey follow-up), and has resulted in medium-resolution spectroscopic observations of over 3500 metal-poor candidates to date. Among these, the HES has identified some 1500 VMP stars, at least 200 EMP stars, a number of stars with [Fe/H] ~ -4.0, and signficantly, the only two HMP stars known.

The "as-observed" observed Metallicity Distribution Functions (MDFs) in these two surveys (not to be confused with the *actual* MDF of the halo) are shown in Fig. 8.5. The distinctly bi-modal appearance of these two samples is the result of the imperfect selection of candidate low-metallicity targets; the large number of stars with [Fe/H > -1.5 are the "mistakes". The stars with metallicities [Fe/H] ≤ -2.0 are the sought-after objects. Note that even though the total number of targets in the HES follow-up to date is only slightly more than half of the HK survey stars with available medium-resolution spectroscopy, the HES follow-up has turned up more VMP stars than the HK survey. The far greater efficiency of the HES in the selection stage has also greatly reduced the number of mistakes with [Fe/H] > -1.5.

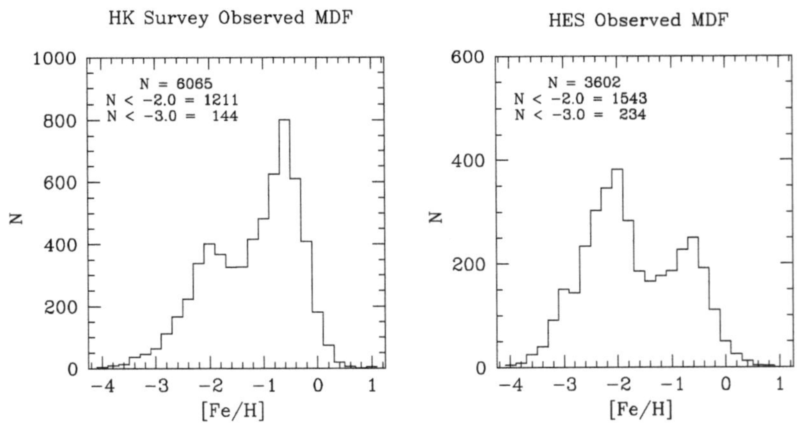

Fig. 8.5. The as-observed Metallicity Distribution Functions for the HK survey (left panel) and the HES (right panel) candidate metal-poor stars. The numbers of stars with [Fe/H] < -2.0 and [Fe/H] < -3.0 in these surveys are listed individually. The selection efficiency of VMP stars in the HES is clearly higher than that of the HK survey.

Details of the halo MDF are clearly important to quantify. To accomplish this, in particular for the low-metallicity tail of the MDF, it is necessary to validate that the selection of input stars is unbiased below a given metallicity. A set of tests to meet this objective, by comparison of the HK-survey and HES MDFs with kinematically-selected samples of stars (e.g., Ryan & Norris 1991), as well as through a set of complete observations of stars in the HES classified as likely metal-poor candidates, are presently being carried out. Since the metal-poor candidates from the HES were selected by well-understood criteria, it is possible to investigate any selection biases quantitatively. Once these exercises have been completed, a "true" halo MDF (at the low-metallicity end of the function) will be extracted. Hernandez & Ferrara (2001) provide an example of how the nature of the low-metallicity tail of the halo MDF can be used to obtain information about the nature of the first metal-producing stars in the Galaxy.

8.4 Recent Results from High-Resolution Spectroscopic Follow-Up

One of the primary reasons for carrying out large surveys for metal-poor stars is to provide targets for detailed high-resolution spectroscopic studies of the most interesting objects using 8m-10m class telescopes. Such studies are required in order to obtain accurate estimates of the individual elemental abundances for significant numbers of VMP stars. Over the past few years we have begun to reap this harvest of information on the nature of the chemically most primitive, and by inference, oldest stars in the Galaxy. In the published literature, the interested reader is directed to the series of papers on the "First Stars" Large Program conducted with the VLT (e.g., Cayrel et al. 2004), as well as the "0Z" program being conducted with the Keck telescope (e.g., Cohen et al. 2004), and the studies being carried out with the Subaru telescope (e.g., Aoki et al. 2002a; Honda et al. 2004). Typically, such programs obtain exquisite high signal-to-noise spectra for on the order of 20-40 stars at a time. Most recently, a new technique, referred to by the authors as "snapshot" spectroscopy, has been exploited to obtain moderately high-resolution ($R = 20,000$), moderate signal-to-noise ($S/N \sim 30/1 - 50/1$) spectra of several *hundred* VMP giants (Christlieb et al. 2004; Barklem et al. 2005). This effort is described in more detail below.

8.4.1 Trends and scatter of elemental abundances

Two of the important results that are emerging from these high-resolution spectroscopic analyses are the identification of well-defined trends of elemental abundances with [Fe/H] (or [Mg/H]), as well as quantitative determination of the cosmic scatter of these abundances about the general trends. Although the character of the trends (at least for elements in common) has not changed much compared to what was known a decade ago (see, e.g., McWilliam et al. 1995; Norris, Ryan, & Beers 1996), there has recently been a rather dramatic change in the reported scatter about these trends.

In particular, the new studies have shown that much of the previously claimed scatter for alpha- and iron-peak elements at low metallicities was spurious, and due to limitations in the quality of the available data, or the addition of "observer scatter" arising from the practice of combining small individual samples from multiple sources (and hence importing differences in the analysis techniques used by the various studies). The new (much higher-quality) data have revealed that the observed scatter for many elements in VMP stars is at or near the level of the observational error, i.e., consistent with zero. For example, the reported scatter in [Cr/Fe] by Cayrel et al. (2004), based on a large sample of VMP giants, is only 0.07 dex. See Fig. 8.6.

Arnone et al. (2005) report, based on high-resolution spectroscopy of a sample of VMP main-sequence turnoff stars, a scatter in [Mg/Fe] of only 0.06 dex. This measure of scatter is a factor of two smaller than the 0.13 dex result reported

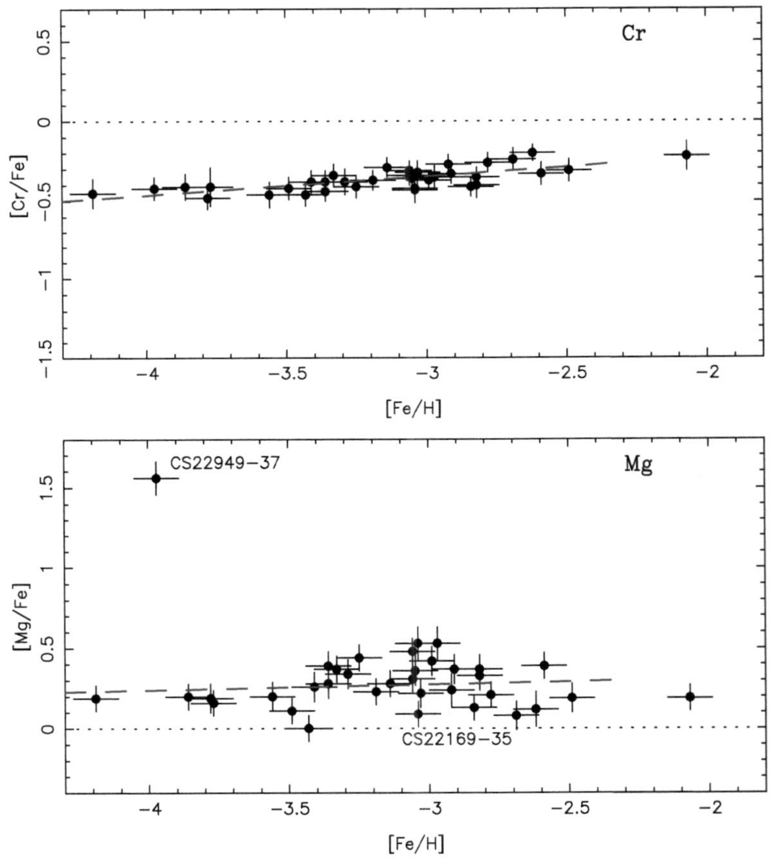

Fig. 8.6. (Upper Panel) [Cr/Fe] as a function of [Fe/H] for 35 VMP giants from the HK survey observed with VLT/UVES (Cayel et al. 2004). The error bars are one-sigma estimates. Note the extremely small scatter about the trend line. (Lower Panel) [Mg/Fe] vs. [Fe/H] for the same sample of stars.

by Cayrel et al. (2004), and is likely due to the fact that the Arnone et al. sample of stars were chosen (originally for the study of primordial lithium) to be quite similar to one another in temperature and surface gravity. As a result, the accuracy of the derived [Mg/Fe] ratios from star to star will not vary as much as the Cayrel et al. giants, which cover a much wider range of atmospheric properties.

The implications of these results for models of supernovae yields, early star formation, and Galactic chemical evolution, is expected to be profound (see, e.g., the discussion by Prantzos 2004).

8.4.2 Carbon-enhanced metal-poor stars

Several recent papers (e.g., Norris, Ryan, & Beers 1997a; Rossi, Beers, & Sneden 1999) have pointed out that the frequency of carbon-enhanced stars in the Galaxy appears to increase at the lowest metal abundances. The ongoing medium-resolution spectroscopic follow-up of candidate low-metallicity stars from the HK and HES surveys has indicated that the actual fraction of VMP stars with carbon enhancements in excess of [C/Fe] $\simeq +1.0$ may in fact be even higher than previously suspected. It should be cautioned, however, that a quantitative test of the fraction of carbon-enhanced metal-poor (hereafter, CEMP) stars as a function of [Fe/H] has not yet been carried out. Rossi et al. (2005) describe a new set of methods for the estimation of [Fe/H] and [C/Fe], based on a calibration of the KP and GP indices of Beers et al. (1999) with $J-K$ colors from the 2MASS Point Source Catalog (Cutri et al. 2003) . Application of these methods to the large databases of spectra from the HK and HES surveys should prove illuminating. Medium-resolution stellar spectra obtained as part of the Sloan Digital Sky Survey (SDSS; York et al. 2000) will be another valuable sample for examination of this question. In addition, recent large samples of high-resolution observations of HES stars obtained by the HERES program (described below) can be used to provide useful constraints as well.

One current challenge is to understand the nature of the phenomena responsible for the apparently high frequency of CEMP stars at low metallicity, and to assess its impact on the early chemical evolution of the Galaxy. Lucatello et al. (2005b), for instance, have argued that the large fraction of CEMP stars at low metallicities provides evidence for an alteration in the Initial Mass Function (IMF) during these epochs to include a greater number of intermediate mass stars than are formed from the present-day IMF. The connection, if any, to the significant amounts of ionized carbon in the intergalactic medium, detected in observations toward distant quasars (e.g., Ellison et al 2000; Pettini et al. 2003), may also hold important clues to the production of carbon in the early Universe.

High-resolution abundance analyses for a number of CEMP stars (Barbuy et al. 1997; Norris, Ryan, & Beers 1997b; Bonifacio et al. 1998; Aoki et al. 2000, 2001, 2002a,b,c; Norris et al. 2002; Depagne et al. 2002; Sivarani et al. 2004; Barbuy et al. 2005) indicates that a variety of carbon-production mechanisms may need to be invoked to account for the observed range of elemental abundance patterns in these stars (e.g., mass-transfer from former asymptotic giant-branch

(AGB) companions, self-pollution via mixing of CNO-processed material in individual stars, the possible existence of zero-abundance "hypernovae" which may produce large amounts of carbon, etc.). Meynet, Maeder, & Ekstrom (2005) have produced models of low-metallicity massive (60 $M_{\odot}$) stars which, by inclusion of a significant initial rotation, are shown to very efficiently produce and mix the light elements C, N, and O to the stellar surface. The addition of large amounts of these elements dramatically increases the mass loss via stellar winds, as compared to non-rotating models, and hence would provide a viable mechanism for rapildy increasing the amount of carbon available for incorporation by subsequent generations of stars.

Many of the CEMP stars have been shown to be members of binary systems (Preston & Sneden 2001; Tsangarides, Ryan, & Beers 2004; Lucatello et al. 2005a). The majority, but not all, of the CEMP stars are associated with neutron-capture-element enhancement (in particular via the s-process; see Aoki et al. 2003). At least one member of the growing class of highly r-process-enhanced, metal-poor stars, CS 22892-052 (Sneden et al. 2003) also exhibits large C (and N) overabundances relative to the solar ratios, although the origin of the carbon enhancement and the r-process enhancement may well be decoupled from one another.

Beers & Christlieb (2005) provide a table of the twelve known EMP stars with [Fe/H] < -3.5 and with published abundance analyses based on high-resolution spectroscopy. Five of these stars are CEMP stars (42%). It is surely not a coincidence that the two most iron-deficient stars yet identified, HE 0107-5240 (Christlieb et al. 2002; Christlieb et al. 2004), and HE 1327-2326 (Frebel et al. 2005) with [Fe/H] $= -5.3$ and -5.6, respectively, also exhibit carbon overabundances that are the highest yet reported amongst extremely metal-poor stars, on the order of [C/Fe] $\simeq +4.0$. Thus, the fraction of HMP stars that are CEMP stars is 100%. Clearly, much remains to be learned about the nature, origin, and evolution of the many classes of CEMP stars in the Galaxy.

8.4.3 Neutron-capture-element-rich, metal-poor stars

Over the past decade, the identification and analysis of metal-poor stars that exhibit large over-abundances of elements beyond the iron peak, species that are almost exclusively formed as the result of neutron captures, have taken a leading role in many investigations. Some of the most exciting new results, and the questions raised by them, are described below.

The s-process-enhanced stars

The slow neutron-capture process (s-process) occurs in astrophysical environments where the flux of neutrons is sufficiently low to allow time for beta decays to occur between neutron captures. One such site is thought to be associated with the so-call "main component" of the $s-$process, the thermal pulsing episodes in intermediate mass AGB stars (see Gallino et al. 1998; Busso et al. 2001; Herwig 2004a; Herwig 2004b).

Recent large-scale abundance studies of stars have revealed the presence of s-process patterns of elements in stars with metallicities as low as [Fe/H] $= -2.6$ (Simmerer et al. 2004). A handful of stars have been found to exhibit such signatures at even lower metallicities, for example, CS 29497-030, a CEMP star with [Fe/H] $= -2.8$ (Sivarani et al. 2004), and the current record-holder, CS 22183-015, a CEMP star with [Fe/H] $= -3.1$ (Johnson & Bolte 2002). The existence of such stars indicates that the s-process can apparently occur quite early in the nucleosynthetic history of the Galaxy. Although previous modeling (for intermediate and solar-metallicity AGB stars) indicated that one might require the presence of iron-peak "seed" elements in order for significant s-process elements to be produced, such low-metallicity stars with detectable s-process patterns suggests that this requirement may not hold at all metallicities.

Models for the operation of the s-process in metal-poor stars (e.g., Gallino et al. 1998; Goriely & Mowlavi 2000; Busso et al. 2001) have indicated that, as a result of the high neutron-to-seed ratios in such stars, one might expect efficient production of the heaviest elements such as Pb. Recently, Aoki et al. (2000) and Van Eck et al. (2001) have identified VMP stars in which large amounts of Pb are indeed present. These so-called "lead stars" now comprise more than twenty examples, with the most extreme case, CS 29497-030, exhibiting [Pb/Fe]$= +3.5$, a factor of over 3000 times greater than the solar ratio of these elements (Sivirani et al. 2004). Very recently, observations of this same star by Ivans et al. (2005) enabled the first detection of the heavy element Bi (only one step away from Pb in the periodic table) in a metal-poor star .

The r-process-enhanced stars

In the rapid neutron-capture process, the flux of neutrons is so large that the nuclei involved do not have time to undergo beta decays before additional neutrons are added; as a result the r-process isotopes accept neutrons until they are quite far from the valley of stability before decaying back. See Cowan & Thielemann (2004) for a recent review of our present knowledge of the operation of the astrophysical r-process.

Astronomers (and their nuclear astrophysics colleagues) are fortunate that nature has chosen to supply nearly ideal probes of the operation of the r-process in astrophysical environments. The so-called r-process-enhanced, metal-poor stars exhibit, in spite of their overall low values of [Fe/H], r-process elements (such as Eu, Os, and Pt) that exhibit abundance ratios relative to Fe which are much larger than observed in the Sun. If such stars did not exist, studies of the operation of the r-process in the early Galaxy would be made much more difficult, since the absorption lines of r-process elements would become undetectably weak for most species.

One such star, CS 22892-052, an HK-survey giant with [Fe/H] $= -3.1$, exhibits a ratio of its r-process elements relative to iron $[r/\mathrm{Fe}] \approx +1.7$, a factor of fifty times the ratio of these elements seen in the Sun (Sneden et al. 1994, 1996). This star has been studied intensively with essentially all 4m and 8m-class telescopes (equipped with high-resolution spectrographs) in the world, as well as over the course of 60 orbits with the Hubble Space Telescope (Sneden et al. 2003). As a result, 57 elements (to date) have been measured for CS 22892-052, making this object the spectroscopically best-studied star other than the Sun.

Shortly after the discovery of CS 22892-052, another HK-survey star was found that has had equally profound impact, CS 31082-001. This star, a giant with [Fe/H] = −2.9, exhibits a similar excess of r-process elements as CS 22892-052, [r/Fe] ≈ +1.6 (Hill et al. 2002). Because this star exhibits a much lower abundance of carbon (which can mask important neutron-capture species) than CS 22892-052, this was the first VMP star with a measurable abundance of the radioactive element uranium (Cayrel et al. 2001). Not only did CS 31082-001 establish the class of highly r-process-enhanced MP stars (referred to collectively as r-II stars, having [r/Fe] > +1.0), it also enabled the first application of a valuable new cosmo-chronometer, U/Th (Hill et al. 2002; Schatz et al. 2002).

However, the "uranium star" CS 31082-001 also raised a number of unresolved questions concerning the operation of the r-process in the early Galaxy. Of greatest significance, the enhancement of the actinides Th and U is on the order of +0.4 dex higher, compared to the level of enhancement of other stable r-process elements (such as Eu), than observed in CS 22892-052. Although application of the U/Th chronometer yields age estimates of this star that are consistent with expectation, ≈ 15.5 ± 3.2 Gyrs (Schatz et al. 2002), application of the Th/Eu chronometer for CS 31082-001 yields ages that beg credulity (no more than a few Gyrs). Recently, additional stars have been noted that display the "actinide boost" phenomenon (Honda et al. 2004).

At least one additional r-II star with a clear signature of uranium has been reported, CS 29497-004, a giant with [Fe/H] = −2.8 and [Eu/Fe] = +1.6 (Christlieb et al. 2004). This star is neither carbon-enhanced, nor does it appear to suffer from the actinide boost phenomenon. A detailed analysis of the derived U abundance, and application of the U/Th and Th/Eu chronometers for this star is being carried out at present (Hill et al., in preparation).

The interest in the nature of the astrophysical r-process has also led to the recent analysis of a larger number of metal-poor stars with moderately enhanced r-process elements (which are referred to collectively as r-I stars, having +0.3 < [r/Fe] < +1.0). A few particularly important examples include HD 115444 (Westin et al. 2000) and BD+17:3248 (Cowan et al. 2002), the first metal-poor star in which gold was detected. There are on the order of ten r-I stars discussed in the present literature. Many of them have measurable abundances of Th, and estimates of ages based on the Th/Eu chronometer have been made that fall in the range 9–18 Gyrs (Johnson & Bolte 2001). Cowan *et al.* (2002) claimed the detection of uranium in BD+17:3248, at the ragged edge of believability, from which an lower limit on the age of 13.4 Gyr (with a large error bar) was derived from the U/Th chronometer.

The small number of r-I and r-II stars studied to date all exhibit a pattern of "heavy" r-process elements that closely follow the scaled solar pattern of r-process elements in the range of atomic number 56 < Z < 76 (Cowan & Thielemann 2004). This immediately suggests the possibility that, at least for these elements, the astrophysical conditions leading to the r-process operated in a consistent manner throughout the history of the Galaxy. This may indicate that the so-called "strong" r-process is associated with a unique site (e.g., SN-II explosions of similar-mass progenitors). This apparent regularity does not apply to the observed patterns of the lighter r-process elements, in the range $40 < Z < 50$. These elements tend to fall below the scaled solar pattern, and may (although the sample is still too small to be certain) exhibit star-to-star scatter even for objects of identical overall metallicity. These observations support the idea (originally indicated from meteoritic isotopic analyses; Wasserburg, Busso, & Gallino 1996) that one may have to invoke the additional operation of a "weak" r-process.

Finally, a handful of r-process-rich metal-poor stars have been identified that exhibit elemental abundance patterns with apparent contributions from *both* the s-process and the r-process (Hill et al. 2000; Cohen et al. 2003; Barbuy et al. 2005). Full understanding of the nucleosynthetic events that led to these stars (referred to collectively as r/s stars) is still lacking, but interesting possibilities have been discussed in the recent literature (Qian & Wasserburg 2003; Zijlstra 2004; Barbuy et al. 2005; Ivans et al. 2005)

Understanding the nature of the *r*-process

The numbers of stars in each recognized class of r-process-enhanced objects are so few that astronomers speculate about the astrophysical origins for all of these classes, and their possible inter-relationships. The recently completed Hamburg/ESO R-process Enhanced Star (HERES) survey (Barklem et al. 2005), described in detail below, has considerably enlarged the numbers of known r-II (8 newly-discovered objects) and r-I stars (35 newly-discovered objects), and several critical questions concerning the astrophysical processes involved in the production of the r-process elements are now beginning to be addressed. These questions include:

1. What is the frequency of r-process-enhanced stars as a function of [Fe/H] ?

   The first two r-II stars reported in the literature both have $[Fe/H] \approx -3.0$. Is the lack of r-II stars with [Fe/H] significantly above -3.0 real, or just a result of the small samples obtained to date? If real, it may reflect the possibility that these large enhancements are only associated with the earliest r-process-generating events. Similar questions must be answered about the r-I and r/s stars, which apparently include objects of higher [Fe/H].

   The HERES survey results place some interesting constraints on this question. The frequency of r-II stars identified from HERES is 4%-5% of metal-poor giants with $[Fe/H] < -2.5$, slightly higher than had been estimated from the smaller previous samples. The r-I stars comprise at least 14% of HERES giants with $[Fe/H] < -1.5$. Note that the low signal-to-noise of the HERES spectra precludes the detection of Eu features in some stars, hence the frequency of the r-I stars is a lower limit.

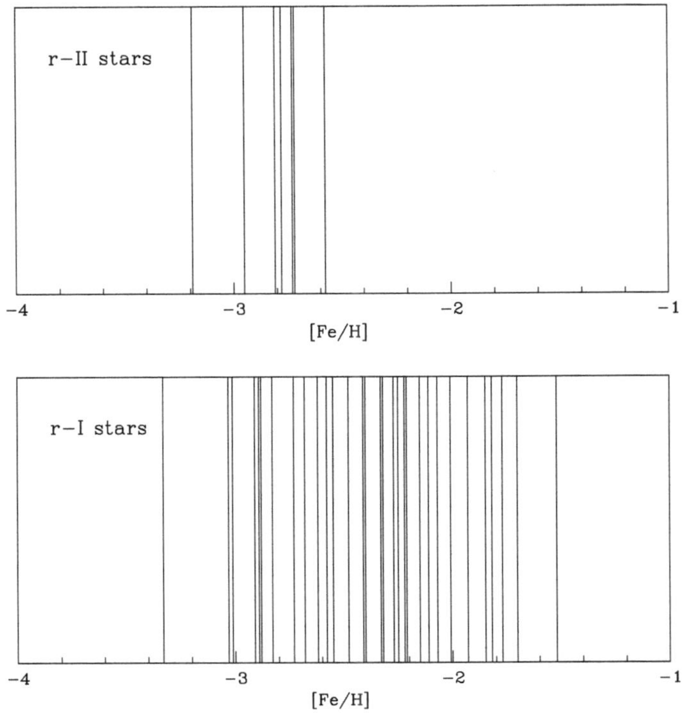

Fig. 8.7. Stripe density plots of the distribution of [Fe/H] for HERES stars classified as r-II (upper panel) and r-I (lower panel), respectively. Note the clear difference in their distributions.

Figure 8.7 shows the distribution of [Fe/H] for the r-II and r-I stars that were detected by the HERES survey. As is clear from this Figure, the r-II stars (including the two previously known such stars) cluster quite tightly around a mean metallicity of [Fe/H] $= -2.81$, with a very small scatter of 0.16 dex, while the r-I stars are found across a very wide range of metal abundance, $-3.2 \leq$ [Fe/H] ≤ -1.5. This may be the first evidence that the r-II phenomenon might be associated with progenitors of a very similar nature, quite possibly with a narrow range of stellar masses.

2. What is the distribution of r-process enrichment for the r-process-enhanced stars ?

Although for simplicity we have separated the classes of r-I and r-II stars, it is possible that this distinction has no physical basis, and that there exists a continuous distribution of the level of r-process-enhancement amongst metal-poor stars in the Galaxy. Or, could the distribution be bi-modal, as the present data suggests? This question (combined with the one above) is essentially the same as that put forward by Wasserburg & Qian (2000) and Fields et al. (2002), who sought to test phenomenological models for the

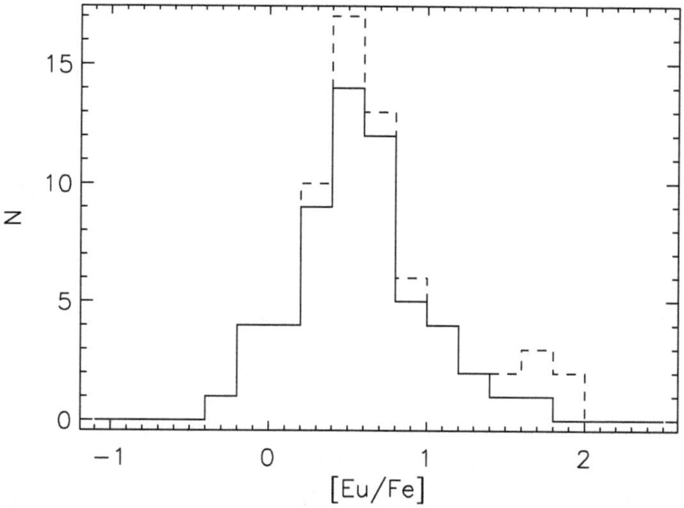

Fig. 8.8. Distribution of [Eu/Fe] for stars in the HERES sample. The dashed line is the distribution for all stars in which Eu was detected, including the HERES pilot sample, which incorporated several previously known r-process-enhanced stars. The solid line includes the pure r-process stars (i.e., [Ba/Eu] < 0). Note that this distribution is biased against stars with very weak [Eu/Fe] since Eu was not detectable in all HERES stars.

origin of the r-process. Its resolution clearly requires much larger samples, so that consideration of the range of r-process enhancements may be studied in detail.

Again, HERES has produced some interesting hints. Figure 8.8 shows the distribution of [Eu/Fe] for the HERES sample stars with detectable Eu. It appears that this distribution includes a "heavy tail" due to the presence of the r-II stars. Although larger samples are still required, the evidence from HERES suggests that their may indeed be a physically meaningful separation in the level of r-process enhancements between the r-I and r-II stars.

3. How stable is the pattern of r-process-element abundances in the range of atomic numbers $56 < Z < 76$?

The r-I and r-II stars exhibit abundance patterns that are well-matched to the scaled solar r-process pattern in this range. Although this piece of information is clearly fundamental, it would be of interest to better quantify it by consideration of the scatter about the solar pattern on an element-by-element basis. This would provide a testable target for models of the astrophysical production of r-process elements. A first attempt this was made by Johnson & Bolte (2001), albeit for a small sample of stars.

The new large samples of r-I and r-II stars from HERES will permit, once higher signal-to-noise high-resolution spectra are obtained for them, a much improved estimate of the scatter (or lack thereof) in the r-process elemental abundance patterns.

4. What is the range of abundances exhibited by r-process elements in the range of atomic numbers $40 < Z < 50$?

   It has been suggested that a number of r-I and r-II stars exhibit different star-to-star abundances of their light r-process elements, such as Sr, Y, Zr, Mo, Pd, and Ag, even at similar metallicity. To better constrain the origin of these elements, and to establish whether or not their production is consistent with a hypothesized "weak" r-process, or other alternatives (Travaglio et al. 2004) a sufficiently large sample from which the level of variation might be confidently established is required.

   The HERES results only provide measurements for the lighter r-process elements Sr, Y, and Zr. Amongst these, both Sr and Y exhibit a possibly real small amount of scatter amongst the r-II stars. The scatter in Zr amongst the r-II stars appears larger, although better-quality data should be obtained to be certain on all of these points. Clearly, the addition of other light element abundances would be useful as well.

5. What is the range of r-process-enhancement for the third r-process-peak elements, $Z > 76$, and for the actinides Th and U ?

   This question goes to the very heart of the application of cosmo-chronometers involving the actinides, as well as their decay products. None of the eight newly discovered r-II stars from HERES have significant carbon enhancement, hence with suitably high-quality spectra, a determination of the fraction of these objects that exhibit the actinide boost phenomenon should be feasible in the near future.

8.4.4 The lowest metallicity stars

As noted above, Beers & Christlieb (2005) have compiled a list of the twelve stars in the published literature that have $[Fe/H] < -3.5$, based on high-resolution abundance analyses. These stars exhibit a number of interesting similarities. For example, forty percent of this sample (5 of 12) are strongly enhanced in carbon ($[C/Fe] > +1.0$), including the two known HMP stars, HE 0107-5240 and HE 1327-2326. One of the lowest [Fe/H] stars, the dwarf carbon star G 77-61, which was once claimed to have $[Fe/H] = -5.6$ (Gass, Wehrse, & Liebert 1988), has been recently shown by Plez & Cohen (2005) to have $[Fe/H] = -4.0$.

Two of the most Fe-poor stars, CS 22949-037 and CS 28948-043, exhibit strong overabundances, not only of C, N, and O, but also of the elements Mg and Si (e.g., Aoki et al. 2004; Israelian et al. 2004). The astrophysical scenarios that produced these patterns are still the subject of debate, but possibilities include the explosion of massive Population III stars (Fryer, Woosly, & Heger 2001), the partial ejection of supernova mantles (Woosley & Weaver 1995; Norris, Ryan, &

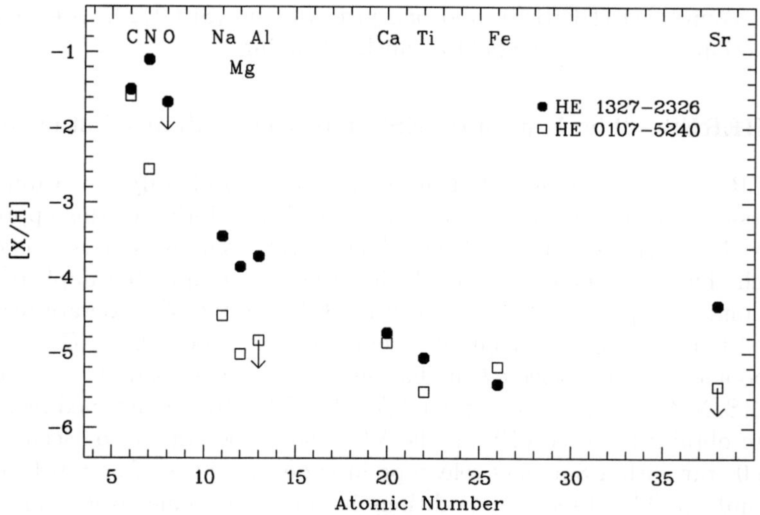

Fig. 8.9. Observed distributions of elemental abundances in the two known HMP stars (taken from Frebel et al. 2005). These distributions are unlike those found in other halo stars of higher [Fe/H].

Beers 2001), and models of low explosion energy Type II SN, with mixing and fallback onto a compact object (Umeda & Nomoto 2003, 2005)

The two HMP stars exhibit rather similar abundance patterns of their lighter elements, as well as a number of remarkable differences (see Fig. 8.9). Of particular interest is the detection of the neutron-capture element Sr in HE 1327-2326 (but not in HE 0107-5240), and the non-detection of Li in this same star, which would be expected to yield a measurable line if the lithium content were comparable to the primordial level. Details of the numerous models put forth to account for the origin of the observed abundance patterns are discussed in detail in Beers & Christlieb (2005).

8.5 Next Generation Surveys for Low-Metallicity Stars

Detailed investigations of the nature of low-metallicity stars, although pursued for many years, have only recently achieved the stage where sufficiently large samples of objects have begun to be assembled. The advent of 8m–10m class telescopes, and their sensitive spectrographs, along with the large catalogs of candidate low-metallicity stars obtained from the HK and HES survey efforts, are the prime reason for the progress obtained over the past few years. However, some of the most interesting objects, such as the carbon-enhanced stars, the neutron-capture-enhanced stars, and the lowest metallicity stars are sufficiently rare that efforts to obtain larger numbers of stars with well-measured elemental abundances is of increasing importance. Below we describe two recent projects, one just completed (HERES), and one just starting (SEGUE), as examples of the directions we believe the field is quickly heading.

8.5.1 HERES – The Hamburg/ESO r-process enhanced star survey

The HERES survey was initiated in order to *quickly* identify new examples of r-process-enhanced metal-poor stars. A detailed description of this approach is provided by Christlieb et al. (2004). The HERES survey adopts a two-stage approach. The first step consists of the identification of a large sample of metal-poor giants with [Fe/H] < -2.0 in the HES, by means of moderate-resolution (~ 2 Å) follow-up spectroscopy of several thousand cool ($0.5 < B-V < 1.2$) metal-poor candidates selected in that survey. In the second step, "snapshot" spectra (S/N $> 30/1$ per pixel at 4100 Å ; R $\approx 20,000$) of confirmed metal-poor stars are obtained with UVES on the VLT. Such spectra can be secured for a $B = 15.0$ star with a 8m-class telescope in exposure times of only 900 seconds, during unfavorable observing conditions in terms of seeing, bright moonlight, cloud coverage, and airmass.

During the HERES program a total of 373 candidate low-metallicity giants were observed. Given the large number of spectra to be processed, it was mandatory that automated techniques for abundance analysis be employed, as described in detail by Barklem et al. (2005). The snapshot spectra allow one to readily identify stars with enhancements of r-process elements, using the Eu II 4129.73 Å line as an indicator.

As a by-product of the search for r-process enhanced stars, HERES has obtained abundance measurements of light elements, such as C, the alpha-elements, such as Mg, Ca, and Ti, and of iron-peak elements such as Cr, Mn, Fe, Co, Ni, and Zn, as well as other species, for the entire set of low-metallicity giants. Figure 8.10 shows some examples of the trends of elemental abundance ratios with [Fe/H] from HERES.

The importance of the snapshot approach is that very large samples can be assembled within a relatively short time (HERES consumed a total of 350 hours of VLT time, roughly 40 nights). It goes further than this. Another great advantage is that such samples can be analysed in a homogeneous fashion, which avoids the potential problems that ensue when observers attempt to construct large samples from the assembly of results from different (and often differently analysed) samples from multiple sources. That is to say, the information content of the resulting abundance determinations is much higher. Furthermore, since the majority of the HERES sample assembly was blind to any other criterion other than low [Fe/H], the relative frequencies of the abundance patterns from star to star are preserved in an unbiased fashion. This is extremely important for placing constraints on models Galactic chemical evolution. Not only must one seek models that are able to reproduce the general trends and scatter, but also the frequencies of the "unusual" objects that are identified along the way, such as the CEMP stars, the neutron-capture-enhanced stars, and the lowest metallicity stars.

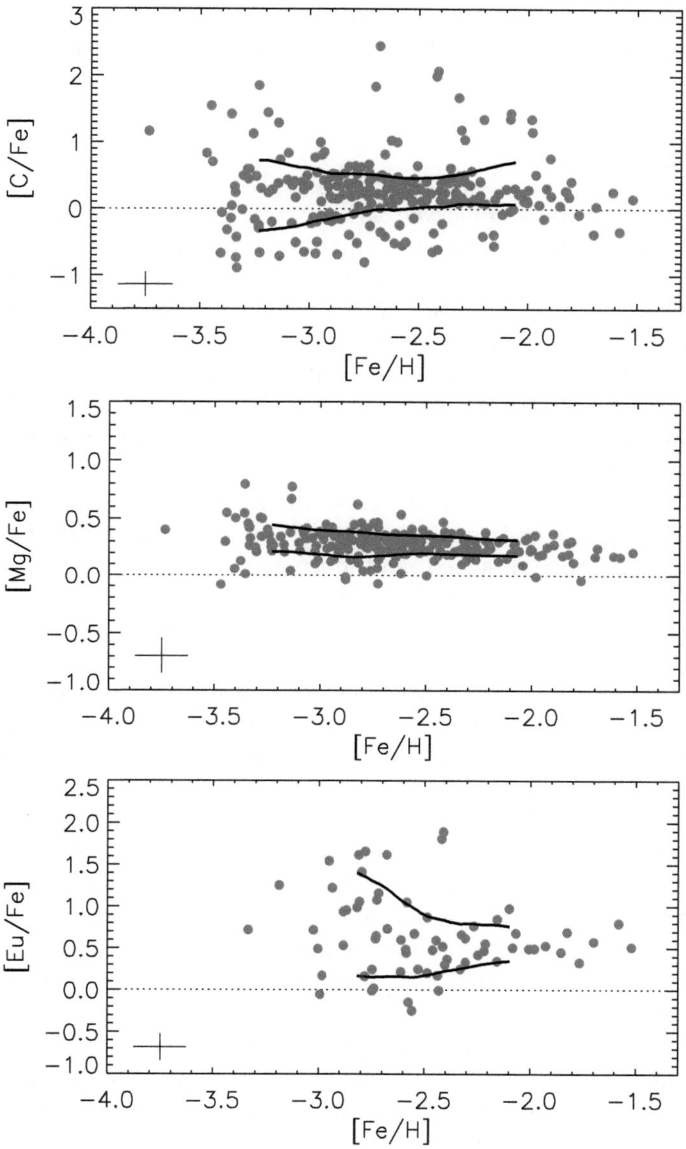

Fig. 8.10. (Top Panel) [C/Fe] as a function of [Fe/H] for 253 metal-poor stars from the HERES survey (Barklem et al. 2005). (Middle Panel) [Mg/Fe] vs. [Fe/H] for the same sample. (Bottom Panel) [Eu/Fe] vs. [Fe/H] for the same sample. Note that Eu is not detected for many of the HERES stars due to the low S/N of the snapshot spectra. The solid lines indicate the estimated one-sigma scatter in the observed distributions. The average relative error bars are shown in the lower left corner of each panel.

For all of the above reasons it is expected that snapshot surveys like HERES will become part of the arsenal of techniques used over the course of the next decades.

8.5.2 SDSS-II: SEGUE

The SDSS was originally conceived as an extragalactic survey, and it has certainly proven successful in meeting many of its original goals. Nevertheless, several of the most important insights from SDSS have come from studies of the "foregound" objects, i.e., the stars of the Milky Way galaxy. Among these are the identification of large numbers of horizontal-branch and other A-type stars (such as blue stragglers) associated with the Sagittarius stream (Newberg et al. 2002), the discovery of the Monoceros stream (Yanny et al. 2003), which is quite likely the debris from an ongoing interaction with the Canis Major dwarf (e.g., Martin et al. 2004).

However, the SDSS ended in July 2005. Members of the extended SDSS community (including original consortium members as well as newly participating external partners) have designed, and are now funded to execute, a three-year extension of SDSS, referred to as SDSS-II, to be carried out through July 2008. A major part of this extension is SEGUE, the Sloan Extension for Galactic Understanding and Exploration. SEGUE will obtain additional low-Galactic-latitude *ugriz* imaging for some 3500 square degrees of sky at lower Galactic latitudes than the original SDSS, and in directions of particular interest, such as along the Sagittarius debris trail. In addition, SEGUE will carry out dedicated spectroscopic follow-up of stellar targets over a well-sampled region of the sky accessible from Apache Point Observatory, with the goal of unraveling the structure, formation history, kinematical and dynamical evolution, chemical evolution, and dark matter distribution of the Milky Way. These results underpin our knowledge of the formation and evolution of the Milky Way galaxy, and will provide the cornerstone of our understanding of galaxy formation processes in general.

SEGUE will obtain some 250,000 medium-resolution ($R = 2000$) spectra distributed over 200 individual directions, spaced roughly 20 degrees apart on the northern sky (including the original SDSS footprint). The SEGUE target selection algorithm has been designed to obtain spectra for stars in the magnitude range $14 \leq g \leq 21$ and with different luminosities (F turnoff stars, G dwarfs, G and K giants, and horizontal-branch stars) so as to populate samples of the stellar populations of the Galaxy from a few kpc out to on the order of 100 kpc. Additional observations will be obtained in locations of star formation, globular and open clusters, Galactic satellites, and other regions of high interest. Many thousands of high proper-motion stars chosen from the LSPM catalogs (e.g., Lepine & Shara 2004) will also be targeted. Accurate proper motions for all SEGUE stars within $\sim 5 - 10$ kpc will be available from astrometric comparisons of the new imaging with previous photographic surveys (e.g., Munn et al. 2004).

As part of the preparation for SEGUE, a substantial amount of test data have already obtained. These observations are being used to verify the SEGUE target-selection algorithm, and to optimize a color-based selection of low-metallicity stars (see, e.g., Lenz et al. 1998; Helmi et al. 2003) that enables efficient identification of stars with [Fe/H] < -1.0. Depending on the shape of the metallicity distribution function of the halo, it is expected that SEGUE will obtain on the order of 20,000 stars VMP stars with [Fe/H] < -2.0, 2000 EMP stars with [Fe/H] < -3.0, 200 UMP stars with [Fe/H] < -4.0, etc. By the time it is completed, the SEGUE sample of low-metallicity stars will be *ten times* the size of the *cumulative number* of such stars identified during the previous half century of dedicated efforts (primarily from the HK survey and the HES).

The brighter VMP stars (g < 17.5) discovered during SEGUE will form the input sample for a "HERES-like" snapshot high-resolution spectroscopic survey to be carried out with the Hobby Eberly Telescope. This effort, tests of which have already started, should produce data of similar quality to HERES for on the order of 3000 VMP giants. Extrapolating from the known frequencies of the *r*-process-enhanced obtained by HERES, the Hobby Eberly Telescope R-process Enhanced Star (HETRES) survey should be able to identify a minimum of 100-150 *r*-II stars, and 300-500 *r*-I stars, as well as numerous other *r*/*s*- and *s*-process-enhanced stars. The elemental abundance determinations (for N $\sim 15 - 20$ elements, as was achieved with HERES) of the complete sample will provide unprecedented knowledge of trends and scatter of these patterns, as a function of metallicity, in the early Galaxy.

References

Aoki, W., et al. 2003, Subaru/HDS Studies of Carbon-Rich Very Metal-Poor Stars, in Elemental Abundances in Old Stars and Damped Lyman-α Systems, 25th meeting of the IAU, Joint Discussion 15, Sydney, Australia, **15**, 19

Aoki, W., et al. 2000, Detection of Lead in the Carbon-rich, Very Metal-poor Star LP 625-44: A Strong Constraint on S-Process Nucleosynthesis at Low Metallicity, The Astrophysical Journal Letters, **536**, 97

Aoki, W., et al. 2002a, The Chemical Composition of Carbon-rich, Very Metal Poor Stars: A New Class of Mildly Carbon Rich Objects without Excess of Neutron-Capture Elements, The Astrophysical Journal, **567**, 1166

Aoki, W., et al. 2002b, A Subaru/High Dispersion Spectrograph Study of Lead (Pb) Abundances in Eight s-Process Element-rich, Metal-poor Stars, The Astrophysical Journal, **580**, 1149

Aoki, W., et al. 2002c, Chemical Composition of the Carbon-rich, Extremely Metal Poor Star CS 29498-043: A New Class of Extremely Metal Poor Stars with Excesses of Magnesium and Silicon, The Astrophysical Journal Letters, **576**, 141

Aoki, W., et al. 2001, Neutron Capture Elements in s-Process-Rich, Very Metal-Poor Stars, The Astrophysical Journal, **561**, 346

Aoki, W., et al. 2003, Measurement of the Europium Isotope Ratio for the Extremely Metal poor, r-Process-enhanced Star CS 31082-001, The Astrophysical Journal, **586**, 506

Aoki, W., et al. 2003, Europium Isotope Ratios in s-Process Element-enhanced Metal-poor Stars: A New Probe of the ^{151}Sm Branching, The Astrophysical Journal Letters, **592**, 67

Aoki, W., et al. 2004, Oxygen Overabundance in the Extremely Iron-poor Star CS 29498-043, The Astrophysical Journal, **608**, 971

Arnone, E., et al. 2005, Mg abundances in Metal-Poor Halo Stars as a Tracer of Early Galactic Mixing, Astronomy and Astrophysics, **430**, 507

Asplund, M. 2005, Annual Reviews of Astronomy and Astrophysics, New Light on Stellar Abundance Analyses: Departures from LTE and Homogeneity, **43**, 481

Barbuy, B., et al. 1997, Analysis of Two CH/CN-Strong Very Metal-Poor Stars, Astronomy and Astrophysics, **317**, 63

Barbuy, B., et al. 2005, New Analysis of the Two Carbon-Rich Stars CS 22948-27 and CS 29497-34: Binarity and Neutron Capture Elements, Astronomy and Astrophysics, **429**, 1031

Barklem, P.S., et al. 2005, The Hamburg/ESO R-process Enhanced Star Survey (HERES) II. Spectroscopic Analysis of the Survey Sample, Astronomy and Astrophysics, **439**, 129

Beers, T.C. 1999, Low-Metallicity and Horizontal-Branch Stars in the Halo of the Galaxy, in Third Stromlo Symposium: The Galactic Halo, eds. B. Gibson, T. Axelrod, & M. Putman (ASP: San Francisco), **165**, p. 206

Beers, T.C., & Christlieb, N. 2005, The Discovery and Analysis of Very Metal-Poor Stars in the Galaxy, Annual Reviews of Astronomy and Astrophysics, **43**, 531

Beers, T.C., Preston, G.W., & Shectman, S.A. 1985, A Search for Stars of Very Low Metal Abundance. I, The Astronomical Journal, **90**, 2085

Beers, T.C., Preston, G.W., & Shectman, S.A. 1992, A Search for Stars of Very Low Metal Abundance. II, The Astronomical Journal, **103**, 1987

Beers, T.C., et al. 1999, Estimation of Stellar Metal Abundance. II. A Recalibration of the Ca II K Technique, and the Autocorrelation Function Method, The Astronomical Journal, **117**, 981

Bonifacio, P., et al. 1998, CS 22957-027: A Carbon-Rich Extremely-Metal-Poor Star, Astronomy and Astrophysics, **332**, 672

Bromm, V., & Larson, R.B. 2004, The First Stars, Annual Reviews of Astronomy and Astrophysics, **42**, 79

Busso, M., et al. 2001, Nucleosynthesis and Mixing on the Asymptotic Giant Branch. III. Predicted and Observed s-Process Abundances, The Astrophysical Journal, **557**, 802

Cayrel, R., et al. 2001, Measurement of Stellar Age from Uranium Decay, Nature, **409**, 691

Cayrel, R. et al. 2004, First stars. V. Abundance Patterns from C to Zn and Supernova Yields in the Early Galaxy, Astronomy and Astrophysics, **416**, 1117

Christlieb, N. 2003, Finding the Most Metal-poor Stars of the Galactic Halo with the Hamburg/ESO Objective-prism Survey, Reviews in Modern Astronomy, **16**, 191

Christlieb, N. Wisotzki, L., & Grasshoff, G. 2002, Statistical Methods of Automatic Spectral Classification and Their Application to the Hamburg/ESO Survey, Astronomy and Astrophysics, **391**, 397

Christlieb, N. et al. 2002, A Stellar Relic from the Early Milky Way, Nature, **419**, 904

Christlieb, N. et al. 2004, HE 0107-5240, a Chemically Ancient Star. I. A Detailed Abundance Analysis, The Astrophysical Journal, **603**, 708

Christlieb, N., et al. 2004, The Hamburg/ESO R-process Enhanced Star Survey (HERES). I. Project Description, and Discovery of Two Stars with Strong Enhancements of Neutron-Capture Elements, Astronomy and Astrophysics, **428**, 1027

Cohen, J.G., et al. 2004, Abundances In Very Metal-Poor Dwarf Stars, The Astrophysical Journal, **612**, 1107

Cohen, J.G. et al. 2003, Abundance Analysis of HE 2148-1247, A Star with Extremely Enhanced Neutron Capture Elements, The Astrophysical Journal, **588**, 1082

Cowan, J.J., & Thielemann, F.-K. 2004, R-process Nucleosynthesis in Supernovae, Physics Today, **57**, 47

Cowan, J.J., et al. 2002, The Chemical Composition and Age of the Metal-poor Halo Star BD+17:3248, The Astrophysical Journal, **572**, 861

Cutri, R. M., et al. 2003, 2MASS All Sky Catalog of Point Sources, VizieR Online Data Catalog, 2246

Depagne, E., et al. 2002, First Stars. II. Elemental Abundances in the Extremely Metal-Poor Star CS 22949-037. A Diagnostic of Early Massive Supernovae, Astronomy and Astrophysics, **390**, 187

Ellison, S.L., et al. 2000, The Enrichment History of the Intergalactic Medium – Measuring the C IV/HI Ratio in the Ly-alpha Forest, The Astronomical Journal, **120**, 1175

Fields, B.D., Truran, J.W., & Cowan, J.J. 2002, A Simple Model for r-Process Scatter and Halo Evolution, The Astrophysical Journal, **575**, 845

Frebel, A., et al. 2005, Nucleosynthetic Signatures of the First Stars, Nature, **434**, 871

Fryer, C.L., Woosley, S.E., & Heger, A. 2001, Pair-Instability Supernovae, Gravity Waves, and Gamma-Ray Transients, The Astrophysical Journal, **550**, 372

Gallino, R., et al. 1998, Evolution and Nucleosynthesis in Low-Mass Asymptotic Giant Branch Stars. II. Neutron Capture and the s-Process, The Astrophysical Journal, **497**, 388

Gass, H., Wehrse, R., & Liebert, J. 1988, Spectrum Analysis of the Extremely Metal-Poor Carbon Dwarf Star G 77-61, Astronomy and Astrophysics, **189**, 194

Goriely, S., & Mowlavi, N. 2000, Neutron-Capture Nucleosynthesis in AGB Stars, Astronomy and Astrophysics, **362**, 599

Helmi, A., et al. 2003, Selection of Metal-poor Giant Stars Using the Sloan Digital Sky Survey Photometric System,The Astrophysical Journal, **586**, 195

Hernandez, X., & Ferrara, A. 2001, Cosmological Origin of the Lowest Metallicity Halo Stars, Monthly Notices of the Royal Astronomical Society, **324**, 484

Herwig, F. 2004a, Dredge-up and Envelope Burning in Intermediate-Mass Giants of Very Low Metallicity, The Astrophysical Journal, **605**, 425

Herwig, F. 2004b, Evolution and Yields of Extremely Metal-Poor Intermediate-Mass Stars, The Astrophysical Journal Supplements, **155**, 651

Hill, V., et al. 2002, First Stars. I. The Extreme r-Element Rich, Iron-Poor Halo Giant CS 31082-001. Implications for the r-Process Site(s) and Radioactive Cosmochronology, Astronomy and Astrophysics, **387**, 560

Honda, S., et al. 2004, Spectroscopic Studies of Extremely Metal-Poor Stars with the Subaru High Dispersion Spectrograph. II. The r-Process Elements, Including Thorium, The Astrophysical Journal, **607**, 474

Israelian, G., et al. 2004, Oxygen and Magnesium Abundance in the Ultra-Metal-Poor Giants CS 22949-037 and CS 29498-043: Challenges in Models of Atmospheres, Astronomy and Astrophysics, **419**, 1095

Ivans, I. et al. 2005, Near-UV Observations of CS 29497-030: New Constraints on Neutron-Capture Nucleosynthesis Processes, The Astrophysical Journal Letters, **627**, 145

Johnson, J.A., & Bolte, M. 2002, Abundances in the Very Metal Poor s-Process-Rich Star CS 22183-015, The Astrophysical Journal Letters, **579**, 87

Johnson, J.A., & Bolte, M. 2001, Thorium Ages for Metal-poor Stars, The Astrophysical Journal, **554**, 888

Lepine, S. & Shara, M.M. 2004, A Catalog of Northern Stars with Annual Proper Motions Larger than 0.15" (LSPM-NORTH Catalog), The Astronomical Journal, **129**, 1483

Lenz, D.D. et al. 1998, Photometric Separation of Stellar Properties Using SDSS Filters, The Astrophysical Journal Supplements, **119**, 121

Lucatello, S., et al. 2005a, The Binary Frequency Among Carbon-enhanced, s-Process-rich, Metal-poor Stars, The Astrophysical Journal, **625**, 833

Lucatello, S., et al. 2005b, Observational Evidence for a Different Initial Mass Function in the Early Galaxy, The Astrophysical Journal, **625**, 833

Martin, N.F., et al. 2004, A Dwarf Galaxy Remnant in Canis Major: the Fossil of an In-Plane Accretion on to the Milky Way, Monthly Notices of the Royal Astronomical Society, **348**, 12

McWilliam, A., et al. 1995, Spectroscopic Analysis of 33 of the Most Metal Poor Stars. II., The Astronomical Journal, **109**, 2757

Meynet, G., Maeder, A., & Ekstrom, S. 2005, Evolution at Very Low and Zero Z, in The Fate of the Most Massive Stars, eds. R. Humphreys & K. Stanek (ASP: San Francisco), **332**, p. 232

Munn, J.A., et al. 2004, An Improved Proper-Motion Catalog Combining USNO-B and the Sloan Digital Sky Survey, The Astronomical Journal, **127**, 3034

Newberg, H.J., et al. 2002, The Ghost of Sagittarius and Lumps in the Halo of the Milky Way, The Astrophysical Journal, **569**, 245

Norris, J.E., Ryan, S.G., & Beers, T.C. 1996, Extremely Metal-poor Stars. II. Elemental Abundances and the Early Chemical Enrichment of the Galaxy,The Astrophysical Journal, **471**, 254

Norris, J.E., Ryan. S.G., & Beers, T.C. 1997a, Extremely Metal-poor Stars. IV. The Carbon-rich Objects, The Astrophysical Journal, **488**, 350

Norris, J.E., Ryan, S.G., & Beers, T.C. 1997b, Extremely Metal-poor Stars. The Carbon-rich, Neutron Capture Element-Poor Object CS 22957-027, The Astrophysical Journal Letters, **489**, 169

Norris, J.E., Ryan, S.G., & Beers, T.C. 2001, Extremely Metal-Poor Stars. VIII. High-Resolution, High Signal-to-Noise Ratio Analysis of Five Stars with [Fe/H] < -3.5,The Astrophysical Journal, **561**, 1034

Norris, J.E., et al. 2002, Extremely Metal-poor Stars. IX. CS 22949-037 and the Role of Hypernovae, The Astrophysical Journal Letters, **569**, 107

Pettini, M., et al. 2003, The C IV Mass Density of the Universe at Redshift 5, The Astrophysical Journal, **594**, 695

Plez, B., & Cohen, J.G. 2005, Analysis of the Carbon-Rich Very Metal-Poor Dwarf G 77-61, Astronomy and Astrophysics, **434**, 1117

Prantzos, N. 2004, The Early Phases of Milky Way's Chemical Evolution, Nuclear Physics A, **758**, p. 249

Preston, G.W., & Sneden, C. 2001, The Incidence of Binaries among Very Metal-poor Carbon Stars, The Astronomical Journal, **122**, 1545

Qian, Y.-Z., & Wasserburg, G.W. 2003, Stellar Sources for Heavy r-Process Nuclei, The Astrophysical Journal, **588**, 1099

Rhee, J. 2000, Ph.D. Thesis, Michigan State University

Rossi, S., Beers. T.C., & Sneden, C. 1999, Carbon Abundances for Metal-Poor Stars Based on Medium-Resolution Spectra, in Third Stromlo Symposium: The Galactic Halo, eds. B. Gibson, T. Axelrod, & M. Putman (ASP: San Francisco), **165**, p. 264

Rossi, S., et al. 2005, Estimation of Carbon Abundances in Metal-Poor Star. I. Application to the "Strong G-Band" Stars of Beers, Preston, & Shectman, The Astronomical Journal, in press (astro-ph/0508202)

Ryan, S.G., & Norris, J.E. 1991, Subdwarf Studies. III. The Halo Metallicity Distribution, The Astronomical Journal, **101**, 1865

Schatz, H., et al. 2002, Thorium and Uranium Chronometers Applied to CS 31082-001, The Astrophysical Journal, **579**, 626

Simmerer, J. et al. 2004, The Rise of the s-Process in the Galaxy, The Astrophysical Journal, **617**, 1091

Sivarani, T., et al. 2004, First Stars IV. CS 29497-030: Evidence for Operation of the s-Process at Very Low Metallicity, Astronomy and Astrophysics, **413**, 1073

Sneden, C. et al. 1994, Ultra Metal-Poor Halo Stars: The Remarkable Spectrum of CS 22892-052, The Astrophysical Journal Letters, **431**, 27

Sneden, C., et al. 1996, The Ultra-Metal-Poor, Neutron-Capture-Rich Giant Star CS 22892-052, The Astrophysical Journal, **467**, 819

Sneden, C., et al. 2003, The Extremely Metal-Poor, Neutron Capture-Rich Star CS 22892-052: A Comprehensive Abundance Analysis, The Astrophysical Journal, **591**, 936

Sneden, C., et al. 2002, Europium Isotopic Abundances in Very Metal Poor Stars, The Astrophysical Journal Letters, **566**, 25

Travaglio, C., et al. 2004, Galactic Evolution of Sr, Y, and Zr: A Multiplicity of Nucleosynthetic Processes, The Astrophysical Journal, **601**, 864

Tsangarides, S., Ryan, S.G., & Beers 2004, T.C., On the Binarity of Carbon-Enhanced, Metal-Poor Stars, Mem. Soc. Ast. It., **72**, 775

Umeda, H. & Nomoto, K. 2003, First-Generation Black-Hole-Forming Supernovae and the Metal Abundance Pattern of a Very Iron-Poor Star, Nature, **422**, 871

Umeda, H. & Nomoto, K. 2005, Variations in the Abundance Pattern of Extremely Metal-Poor Stars and Nucleosynthesis in Population III Supernovae, The Astrophysical Journal, **619**, 427

Van Eck, S. et al. 2001, Discovery of Three Lead-Rich Stars, Nature, **412**, 793

Van Eck, S., et al. 2003, More Lead Stars, Astronomy and Astrophysics, **404**, 291

Wasserburg, G.J. & Qian, Y.-Z. 2000, Prompt Iron Enrichment, Two r-Process Components, and Abundances in Very Metal-Poor Stars, The Astrophysical Journal Letters, **529**, 21

Wasserburg, G.J., Busso, M., & Gallino, R. 1996, Abundances of Actinides and Short-lived Nonactinides in the Interstellar Medium: Diverse Supernova Sources for the r-Processes, The Astrophysical Journal Letters, **466**, 109

Westin, J., et al. 2000, The r-Process-Enriched Low-Metallicity Giant HD 115444, The Astrophysical Journal, **530**, 783

Woosley, S.E. & Weaver, T.A. 1995, The Evolution and Explosion of Massive Stars. II. Explosive Hydrodynamics and Nucleosynthesis, The Astrophysical Journal Supplements, **101**, 181

Yanny, B., et al. 2003, A Low-Latitude Halo Stream Around the Milky Way, The Astrophysical Journal, **588**, 824

York, D.G. et al. 2000, The Sloan Digital Sky Survey: Technical Summary, The Astronomical Journal, **120**, 1579

Zijlstra, A. A. 2004, Low-Mass Supernovae in the Early Galactic Halo: Source of the Double r/s-Process Enriched Halo Stars?, Monthly Notices of the Royal Astronomical Society, **348**, 23

9 Ultraluminous Infrared Galaxies

C.J. Lonsdale, D. Farrah & H.E. Smith

Abstract: Ever since their discovery in the 1970's, UltraLuminous InfraRed Galaxies (ULIRGs; classically $L_{ir} > 10^{12} L_{\odot}$) have fascinated astronomers with their immense luminosities, and frustrated them due to their singularly opaque nature, almost in equal measure. Over the last decade, however, comprehensive observations from the X-ray through to the radio have produced a consensus picture of local ULIRGs, showing that they are mergers between gas rich galaxies, where the interaction triggers some combination of dust-enshrouded starburst and AGN activity, with the starburst usually dominating. Very recent results have thrown ULIRGs even further to the fore. Originally they were thought of as little more than a local oddity, but the latest IR surveys have shown that ULIRGs are vastly more numerous at high redshift, and tantalizing suggestions of physical differences between high and low redshift ULIRGs hint at differences in their formation modes and local environment. In this review we look at recent progress on understanding the physics and evolution of local ULIRGs, the contribution of high redshift ULIRGs to the cosmic infrared background and the global history of star formation, and the role of ULIRGs as diagnostics of the formation of massive galaxies and large-scale structures.

9.1 Introduction

9.1.1 The biggest and the brightest

In human endeavour there's a fascination with the "biggest and the best", or the "best and the brightest". It's a matter for the social and psychological sciences to speculate on the reasons we feel driven to give Oscars to the best movies and to climb the highest mountains, but being human (well, most of them anyway), astrophysicists are not immune to the desire to search for the Universe's own brand of the biggest and the brightest. We tend to give them extreme names such as *ultra*-this or *hyper*-that. In their rarity however, unusual objects and environments teach us about the most extreme physical processes in the Universe.

Which brings us to ULIRGs (or ULIGs to some) - Ultra Luminous InfraRed Galaxies. There are also HLIRGs (or HiLIRGs or HyLIRGs, or indeed HyLIGs): Hyper Luminous InfraRed Galaxies. These Oscar contenders have historically been defined simply in terms of luminosity: $L_{8-1000\mu m}$ between 10^{12} and $10^{13} L_\odot$ for the ULIRGs and $> 10^{13}\ L_\odot$ for HLIRGs [1]. What are these dramatic objects, and why are they among the brightest objects in the Universe? Does this also imply they are the biggest in term of mass or size, or are they just superficial fireworks that leave little lasting impression? What can understanding these enigmatic objects teach us about the evolution of galaxies and how the the Universe came to look as it now does?

The last few years have seen a dramatic shift in our perceptions of ULIRGs. Once believed by many to be a rare curiosity – certainly interesting, yes, but perhaps no more than a local oddity – they're beginning to see center stage much more frequently. Ironically, this increased interest in these rare objects has arisen because we now realize that ULIRGs were once not nearly so rare as we find them to be in the local Universe: pioneering submillimeter and millimeter surveys have demonstrated that ULIRGs are many hundreds of times more numerous at z>1 than they are locally. This in turn suggests that they played a much more important role in galaxy formation and evolution than we imagined, and so understanding them becomes of prime importance. Fortunately, this rise in the fortunes of the ULIRG has occurred in an era when many new observing capabilities are coming on line. Foremost among these is the Spitzer Space Telescope, with its suite of deep imaging infrared cameras and its sensitive infrared spectrograph. Spitzer's extensive first results on ULIRGs are just now beginning to be published. Submillimeter and millimeter cameras are also improving dramatically; upcoming facilities include AzTEC, SCUBA-2, Herschel, and ALMA. We can also look forward to major new insights from mid- and near-IR facilities such as ASTRO-F, WISE, and JWST, as well as radio facilities such as the Square Kilometer Array, and new X-ray facilities with high hard X-ray sensitivities such as Con-X and XEUS. The timeliness of this subject is exemplified by recent reviews of The Cosmic InfraRed Background (Lagache, Puget & Dole, 2005), Interacting Galaxies (van Gorkum & Hibbard, 2005), Megamasers (Lo, 2005), Galactic Winds (Veilleux, Cecil & Bland-Hawthorn, 2005), and High Redshift Molecular Gas (Vandem Bout & Solomon, 2005), all of relevance to ULIRGs.

9.1.2 Overview

ULIRGs were first discovered in large numbers by the Infrared Astronomical Satellite in 1983, and were found to be comparatively rare locally, with a space density several orders of magnitude lower than that of normal galaxies, and possibly a factor of a few higher than QSOs. Followup observations show that most, if not all ULIRGs are found in major disk mergers, and that the central few hundred pc of their nuclear regions harbour very large masses of gas and dust. The power source behind the IR emission is some combination of a large popu-

[1]The supporting cast consists of LIRGs (or LIGs), the much more common lower luminosity understudies of the prima donnas, with $L_{8-1000\mu m}$ between 10^{11} and $10^{12} L_\odot$

lation of hot young stars (a 'starburst'[2]) or a very massive black hole accreting matter at a rapid rate (which for the remainder of this review we refer to as an 'AGN'). Though distinguishing between the two initially (and even now) proved to be very difficult, it is now thought that, at least locally, ULIRGs are mainly powered by a starburst, but frequently with a significant AGN contribution. Local ULIRGs reside in relatively low-density environments (not unexpectedly, since relative velocities are thought to be too high for mergers to occur in rich, virialized environments), and are expected to evolve into spheroidal systems as the galaxy mergers that appear to trigger ULIRG activity progresses.

Even IRAS was sensitive enough to determine that there has been very strong evolution in the ULIRG (and LIRG) population with redshift out to at least $z \sim 0.5$, with approximate form $(1+z)^4$. IRAS also found ULIRGs out to extremely high redshifts, including the famous, lensed, F10214+4724 at z=2.286. This strong evolution was confirmed with results from the Infrared Space Observatory which, although covering much smaller areas than IRAS, could probe this evolution out to $z \sim 1$ due to its greater sensitivity (Fig. 9.1). This evolution was later recast as the now ubiquitous 'star formation history of the Universe' figures, which show that LIRGs rather than ULIRGs are responsible for the bulk of the evolution seen since z$\sim$1 in IR galaxies. ULIRGs, however, did not slink into the shadows; on the contrary they returned triumphant with the advent of submillimeter imaging surveys, which came shortly after ISO and can in principle probe IR-luminous galaxies up to $z \sim 7$. These sub-mm surveys showed that ULIRGs are orders of magnitude more numerous at $z > 1$ than locally, outnumbering optically bright QSOs at those redshifts by a large margin. Followup observations showed that these distant ULIRGs bear many similarities to their local cousins, but also exhibit some key differences, and that they may signpost the obscured phases of the very dramatic events suspected of building the most massive galaxies seen in the local Universe.

When considered within the framework of modern theories for the formation of galaxies and large-scale structure, it seems initially surprising that there are many more ULIRGs at high redshift than locally, because in early implementations of the 'hierarchical buildup' paradigm, large galaxies build up slowly from the mergers of smaller systems. This is in contrast to the early 'monolithic collapse' models (Eggen, Lynden-Bell & Sandage, 1962), where ellipticals formed early in a dramatic burst of star formation, which had been largely supplanted in favour of hierarchical models. The discovery of so many ULIRGs at high redshifts caused hierarchical models of the time major difficulties in making enough distant systems with such high star formation rates. The basic dark matter halo growth theory, described by an extended/modified Press-Schechter formalism, does however allow for rapid baryon accumulation in very massive dark matter halos, and recent galaxy formation models are having greater success in producing the observed number of ULIRGs in sub-mm surveys, albeit with some stringent assumptions.

---

[2]For our purposes defined as a star forming event with a gas exhaustion timescale very short compared to the Hubble time

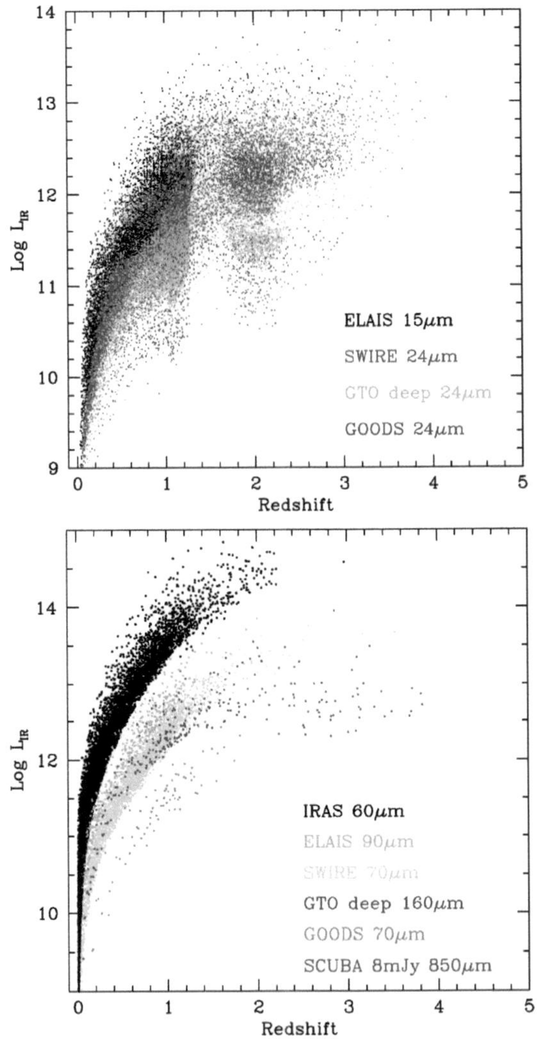

Fig. 9.1. Simulation of the ability of recent infrared surveys to discover ULIRGs, with shorter wavelength ($\lambda < 25\mu$m) surveys plotted in the upper panel and longer wavelength surveys below. The band in the upper figure at $z \sim 1.5$ is produced by the the 10μm silicate absorption feature which falls in the Spitzer 24μm band at that redshift. Wide shallow surveys have the largest volume for discovering the most luminous ULIRGs, while narrow deep surveys can of course find the most distant ones, though in much smaller number. Tiered ('wedding cake') surveys are thus required to construct complete luminosity functions within a given redshift interval (vertical slices). Based on the simulations of Xu et al. (2003), these simulations can fit IRAS, ISO, Spitzer and submillimeter counts and redshift distributions. We plot 1/5 of the objects expected within the area and depth of each survey: ISO ELAIS 15μm (Vaccari et al., 2005); SWIRE 24 & 70μm (Surace et al., 2005); Spitzer Guaranteed Time Observer deep 24μm (Pérez-González et al., 2005); GOODS 24μm (Chary et al., 2004); IRAS 60μm (Lonsdale et al., 1990); SCUBA 850μm (Scott et al. 2002); GOODS 70μm (D. Frayer, priv. comm.); Spitzer GTO deep 160μm (Dole et al., 2004).

In this review, we will therefore focus on selected key topics: (1) our understanding of the astrophysics of local ULIRGs, and in particular the relative importance of star formation versus AGN in powering ULIRGs, (2) similarities and differences between local and high-redshift ULIRGs, and (3) the relationship between ULIRGs and the formation of large-scale structure and of galaxies as a function of redshift. Since the study of ULIRGs clearly connects to many major disciplines of observational and theoretical extragalactic astronomy we cannot hope to cover all topics of relevance to them in this review. Nor can we completely review all recent published studies of ULIRGs; excellent ULIRG papers simply abound. We therefore highlight the most recent advances and our perspective on the most important questions concerning their study within the framework of galaxy and structure formation.

In Sect. 9.2 we provide brief historical context to the discovery of ULIRGs and their evolutionary role. In Sect. 9.3 we review current understanding of the astrophysics of local ULIRGs by wavelength, and in Sect. 9.4 we summarize local studies into a picture of ULIRG nature and evolution in the local Universe. In Sect. 9.5 we review observations of ULIRGs at higher redshifts, based primarily on data from ISO, SCUBA, HST and Spitzer. Section 9.6 places these studies into the context of structure formation and reviews their role within galaxy formation scenarios. Finally, Sect. 9.7 highlights the key open questions and our perspectives of where the answers are likely to come from.

For the remainder of this review we usually refer to ULIRGs and HLIRGs combined as ULIRGs, because many of the earlier works used the term "ULIRG" to refer to all objects above 10^{12} in $L_\odot$ and the term HLIRG has been in only recent and inconsistent use. We assume $H_0 = 70$ km s^{-1} Mpc^{-1}, $\Omega = 1$, and $\Omega_\Lambda = 0.7$. Luminosities are quoted in units of bolometric solar luminosities, where $L_\odot = 3.826 \times 10^{26}$ Watts. Unless otherwise stated, the term 'IR' or 'infrared' luminosity refers to the integrated rest-frame luminosity over 1–1000 or 8–1000μm (which differ very little for most SEDs).

9.2 The Discovery of ULIRGs

It is often stated that UltraLuminous InfraRed Galaxies were discovered by the InfraRed Astronomy Satellite (IRAS), which was launched in 1983. Strictly speaking this may be true since the first objects to meet the now generally acknowledged criteria which define a ULIRG – $L_{8-1000\mu m} > 10^{12}$ L$_\odot$ and, optionally, exceptionally large ratios of $L_{infrared}/L_{optical}$ – can indeed be considered to have been discovered with the publication by Houck et al. (1985) of a sample of 9 IRAS sources invisible or exceptionally faint on the Palomar Sky Survey plates, and exhibiting L_{ir}/L_{opt} ratios over 50. We must go back further, however, to appreciate the origins of the ULIRG phenomenon. Galaxies with an unusually high amount of mid- or far-infrared emission compared to their optical output had been discovered by the pioneers of infrared astronomy in the 10 to 15 years preceding IRAS, and had been recognised to represent luminous, probably short-lived, active events of some sort within a galaxy.

Several excellent reviews have described the IRAS and pre-IRAS legacy, including Rieke and Lebofsky (1979), Soifer et al. (1987), Sanders & Mirabel (1996), and Mirabel (2002) so we present here only the brief highlights relevant to ULIRG discovery and ULIRG samples.

The first infrared observations of galaxies were undertaken in the late sixties (Low & Kleinmann, 1968; Kleinmann & Low, 1970), and a population of galaxies with infrared-dominant emission from the nuclear regions was discussed by Rieke & Low (1972). The well known radio-infared relation (de Jong et al., 1985; Helou et al., 1985) was originally suggested by van der Kruit (1971). An early debate ensued as to the thermal vs non-thermal origin of the infrared emission from galaxies (Rees et al., 1969; Burbidge & Stein, 1970) with the conclusion that most systems are powered by re-radiation of starlight by dust (Rieke & Lebofsky 1979 and references therein). The other key contribution of the pre-IRAS days was the role of interactions in triggering nuclear and starburst activity (Toomre & Toomre, 1972; Larson & Tinsley, 1978), which was confirmed for the strongly interacting system Arp 299 by Gehrz et al. (1983), and several other interacting and merging systems by Lonsdale, Persson & Matthews (1984); Joseph & Wright (1985).

IRAS scanned almost the entire sky in the thermal infrared[3], observing in four bands centered at 12, 25, 60 and 100μm and opening up an unprecedented volume of space to study in this wavelength regime. This allowed IRAS to find exceedingly rare objects and to demonstrate the importance of this new class of exceptionally infrared-luminous objects (Fig. 9.1). One of the most exciting early discoveries was that IRAS would not be limited to exploring only the very local Universe due to its relatively bright flux limits, but it could reach out to significant distances thanks to the existence of large numbers of exceptionally infrared-luminous sources. Strong evolution was demonstrated for the IRAS deep north ecliptic hole field by Hacking, Houck & Condon (1987) and confirmed over the sky by Lonsdale et al. (1990) and Saunders et al. (1990). Extensive redshift surveys (Oliver et al., 1996) led to the discovery of the first z>2 IRAS HLIRG, IRAS FSC 10214+4724 at z=2.86 (Rowan-Robinson et al., 1991) which is a lensed system. The extremely high evolution rates, modeled, for example, as $L_{IR}{\sim}(1+z)^{\sim 4}$, were actually anticipated, based on the previously known strong evolution of starburst-related sub-mJy radio sources (Hacking, Houck & Condon, 1987).

ULIRGs were found to have comparable space densities to those of PG QSOs of similar luminosity by Soifer et al. (1987); however, significant incompleteness in the PG QSO sample has since been demonstrated (Wisotzki et al., 2000) so this result needs re-visiting. The best known samples of IRAS luminous and ultraluminous galaxies, all selected at 60μm, are the Bright Galaxy Sample of Soifer et al. (1987), recently significantly updated into the Revised Bright Galaxy

[3]Defined as the wavelength region over which dust grains can thermally re-radiate emission, ranging from the dust sublimation temperature on the short wavelength side ($\sim$ 1μm) to $\sim$ 1000μm on the long wavelength side where non-thermal processes often begin to dominate

Sample (RBGS; Sanders et al. 2005), and the complete flux-limited IRAS 1 Jy sample (Kim & Sanders, 1998). Also notable are the 2Jy sample of Strauss et al. (1990) and the FIRST/IRAS sample of Standford et al. (2000). The RBGS contains 629 IRAS 60μm galaxies brighter than 5.24 Jy and Galactic latitude >5 degrees and contains 20 ULIRGs; the most luminous being Mrk 231 with $L_{ir} = 3.2 \times 10^{12} L_{\odot}$, the highest redshift being IRAS 07251-0248 at z=0.0876, and the closest being Arp 220 at z=0.018. The 1 Jy sample consists of 118 ULIRGs drawn from the IRAS Faint Source Catalog, with declination >-40 degrees and Galactic latitude $|b|$ >30 degrees. This sample also has a warm 60/100μm>0.3 colour selection, which introduces a bias against cooler objects.

Whilst dramatic in nature, LIRGs and ULIRGs in the local Universe are rare, and they contribute only ~6% of the total infrared luminous energy density (Soifer & Neugebauer, 1991), and about ~3% of the total energy density. It was found that both the IRAS 60/100μm colour and the L_{IR}/L_{opt} ratio increased with luminosity (Soifer et al., 1987), reaching ~ 100 in the most extreme systems. This indicates that the more luminous systems must have an increasing contribution from an additional warm source compared to the relatively cool emission from modest levels of star formation seen in spiral disks (de Jong et al., 1985). IRAS ULIRGs have been separated into warm and cool subsamples based on the 25/60 and 60/100μm IRAS colours (Sanders et al., 1988b; Surace et al., 1998), with the warmer objects more likely to host an AGN (de Grijp et al., 1985). The spectral energy distributions of IRAS ULIRGS were reviewed by Sanders and Mirabel (1996), illustrating these trends with colour and luminosity. These reviewers also compared the SEDs of QSOs and Blazars with those of ULIRGs.

It should be remembered that although the IRAS datasets are now nearly a quarter of a century old, they are very under-explored. As of publication date there are >37,000 infrared-bright galaxies in the IRAS Faint Source Catalog (FSC) that have never been observed with any other instrument and reported in any journal article. Only 43% of the 64,606 IRAS extragalactic FSC sources[4] have been included in any sort of publication (J. Mazzarella, priv. comm.).

9.3 The Physics of Local ULIRGs

The complexity of the ULIRG phenomenon requires a multi-wavelength approach and we have learned an enormous amount from detailed studies at many wavelengths. For this review we focus on those observational areas which we believe are key to understanding the ULIRG phenomenon, in part because they represent the spectral regions of lowest optical depth, and in which key advances have been made in the years since the Sanders & Mirabel review. We stress, however, that even the relatively low-optical depth radio, mid-infrared, and X-ray regimes may still show substantial obscuration due to starburst-related free-free absorption (Condon et al., 1991), compact nuclear molecular clouds or molecular

[4]As reported by NED, the NASA/IPAC Extragalactic Database

tori, which may be optically thick at wavelengths as long as 30μm, or Compton-thick absorbing columns with $N_H \gtrsim 10^{24}$ cm^{-2}.

The compactness of the IR-emitting regions in luminous IR galaxies (*c.f.* Condon et al. 1991) suggests two possible origins for the high ULIRG luminosities: compact nuclear starbursts and/or highly obscured AGN activity. Much early effort was given to determining which phenomenon powers ULIRGs; with the understanding that starburst and AGN emission are frequently found together in luminous galaxies, the emphasis has shifted to determining which mechanism is dominant and to understanding the relationship between co-existing AGN & starburst emission.

9.3.1 Optical to mid-IR imaging

Early imaging surveys of IR-luminous sources spanned a wide range in luminosity (from $< 10^9 L_\odot$ up to $\sim 10^{12} L_\odot$, and revealed an interesting picture (Soifer et al., 1984; Rieke & Lebofsky, 1986). Sources with $L_{ir} < 10^9 L_\odot$ are almost exclusively confined to undisturbed E and S0 systems, with few spirals. The fraction of spirals increases sharply with increasing IR luminosity however, with most systems in the range $10^{10} L_\odot < L_{ir} < 10^{11} L_\odot$ being Sb or Sc type systems. At luminosities above about $10^{11} L_\odot$, the majority of systems are still spirals, but an increasing fraction (up to $\sim 25\%$) appear to be involved in interactions, or to show signs of morphological disturbance.

This apparent increase in the number of interacting IRAS systems was thrown into sharp relief by early ULIRG imaging studies, showing that interactions and mergers are much more common amongst ULIRGs than in lower luminosity systems, though the exact fraction in ongoing interactions remained contentious for some years. The first imaging surveys (Armus, Heckman & Miley, 1987) showed that at least 70% of systems with ULIRG or near-ULIRG luminosities are interacting, with morphologies expected from the collision of two disk galaxies. Later optical and near-IR studies of ULIRGs (Melnick & Mirabel, 1990; Hutchings & Neff, 1991; Clements et al., 1996) found a higher fraction of ULIRGs involved in interactions, at least 90%, and that there are a wide range of merger stages present in the ULIRG population, from widely separated systems to advanced mergers (Murphy et al., 1996). Other studies, however, found a much lower fraction of ULIRGs involved in interactions, fewer than 70% (Lawrence et al., 1989; Zou et al., 1991; Leech et al., 1994).

These results demonstrate that interactions and mergers play an important role in the ULIRG population. This provided a plausible trigger for the immense IR luminosities seen in ULIRGs. In order to both fuel and enshroud the power sources in ULIRGs (irrespective of whether they are starbursts, AGN, or both), a large quantity of gas and dust needs to be channeled into a small volume, most plausibly sited in the nucleus of the host galaxy. Results from N-body modeling of galaxy collisions (see Barnes & Hernquist 1992 for a review) suggest that this is readily achieved during the course of a merger (Barnes, 1989; Barnes & Hernquist, 1996; Mihos & Hernquist, 1996; Dubinski, Mihos & Hernquist, 1999), though details depend on many variables (e.g. angle of approach, relative velocity, disc inclinations, bulge size and gas and dark matter masses). The majority of mergers involve an initial close approach, followed by a maximum separation of up to $\sim$ 50kpc, reached after $\sim 2.5 \times 10^8$ years, then a second close approach a few tens of millions of years later, which is rapidly followed by coalescence, and relaxation towards an elliptical profile. Total times scales to coalescence range from $\sim 7 \times 10^8$ years to $\sim 2 \times 10^9$ years, depending on the parameters of the encounter. Gas and dust can be channeled into the nuclear regions of the progenitors in one of two ways; either before coalescence, when tidal forces during the first close approach form bars in one or both progenitors (particularly if the progenitors have small bulge components) which are very efficient at channeling material into the central regions of a galaxy, or during coalescence, when shocks drive very large quantities of gas and dust into the nuclear regions. More recent N-body simulations have in general supported these conclusions, but added an intriguing possibility; under certain conditions (particularly if the progenitors are gas dominated), a merger between two disk galaxies can result in a disklike remnant, rather than an elliptical (Naab & Burkert, 2003; Springel & Hernquist, 2005; Robertson et al., 2005). It seemed therefore that galaxy mergers could provide all the necessary physical conditions for triggering a ULIRG, and that this implicated ULIRGs in the formation of elliptical galaxies. Furthermore, these simulations could explain a result that had been puzzling; that ULIRG activity could apparently be triggered in mergers when the progenitors are still physically separate.

Optical imaging from the Hubble Space Telescope (HST) offered enhanced resolution and sensitivity over ground based facilities. An early HST study of ULIRGs using WFPC2 (Surace et al., 1998) focused on a small sample with 'warm' infrared colours[5], which biases towards systems likely to contain an obscured AGN. All of the sample were found to be interacting, with complex structures in their nuclear regions. Several systems showed a large number of compact bright 'knots' a few hundred pc in diameter, whose ages suggest they result from the merger. A complementary ground-based survey of ULIRGs with 'cool' IR colours (Surace et al., 2000) found a similar picture; all the galaxies show signs of interactions, from early to late stage, and many systems harbour 'knots' similar to those seen in the warm sample. The optical magnitudes are in most cases relatively modest (at least compared to the enormous IR luminosities); most systems have $\sim L^*$ luminosities, with very few being substantially brighter. A later HST survey of a larger, unbiased ULIRG sample (Farrah et al., 2001) found similar results; nearly all of the sample are interacting, with a wide range of merger stages. A small number host optical QSOs whose host galaxies are either interacting, or elliptical-like. Other authors have, on the basis of HST data, suggested that some ULIRGs show evidence for mergers between more than two galaxies, suggesting that ULIRGs may be the remnants of compact galaxy groups (Borne et al., 2000) HST imaging of ULIRGs in the near-infrared

[5]i.e. systems with $f_{25}/f_{60} > 0.2$, where f_{25} and f_{60} are the 25μm and 60μm IRAS fluxes respectively

with NICMOS (Colina et al., 2001; Bushouse et al., 2002) has produced very similar results to those from optical imaging, with at least 90% being interactions between two or (possibly) more progenitors over a wide range of merger stages, and that very few are much brighter than L^*.

To date however, the largest optical/NIR imaging survey of ULIRGs has been done from the ground; Veilleux, Kim & Sanders (2002) present R and K images for 118 ULIRGs from the IRAS 1Jy spectroscopic survey, and find that virtually all show signs of interactions, though very few showed definite signs of interactions between more than two progenitors. Most of the systems appear to be late-stage mergers, especially for the more luminous systems and/or those with spectroscopic signatures of an AGN, and typically have luminosities (in the K band) of L^* or greater. The most advanced mergers show evidence for emerging elliptical profiles.

Near diffraction-limited Keck mid-infrared imaging of ULIRGs has been obtained by Soifer et al. (1999; 2000) who find extremely compact structures, with spatial scales smaller than $0''.3$ in six of the seven ULIRGs observed. These compact sources emit between 30% – 100% of the mid-infrared energy from these galaxies. In Mrk 231, IRAS 05189-2524 and IRAS 08572+3915 there is strong evidence that the source size increases with increasing wavelength, suggesting heating by a central rather than extended source, consistent with the optical classification as an AGN. Spitzer mid-IR imaging programs of over 200 nearby ULIRGs (ongoing programs of J. Surace (Fig. 9.2) and J. Mazzarella) will provide great sensitivity in this important wavelength region, if limited spatial resolution.

9.3.2 Optical & near-IR spectroscopy

It might be suspected that spectroscopy at wavelengths shortward of a few microns would be useful mainly for redshift surveys, given the dust-enshrouded nature of ULIRGs. This, however, is a misconception; optical spectroscopic signatures of starburst and AGN activity are still apparent even when such activity is moderately obscured (up to and surpassing $A_V \sim 10$ depending on the observations), and starbursts and AGN can be detected in polarized light even when the direct line of sight is completely hidden. Furthermore, spectroscopic diagnostics from the UV through to the near-IR are, on the whole, more mature than those at longer wavelengths (Baldwin, Phillips & Terlevich, 1981; Veilleux & Osterbrock, 1987; Osterbrock, Tran & Veilleux, 1992; Dopita et al., 2000), and line strengths, ratios and profile shapes can provide powerful constraints on the nature of the source of excitation. Recent advanced spectral synthesis codes (e.g. Leitherer et al., 1999; Kewley et al., 2001) allow for insightful diagnostics of starburst events, particularly in the UV where direct emission from hot young star photospheres, rather than reprocessed light from dust, is being sampled.

Early optical spectroscopic surveys of ULIRGs generally showed that they were mostly starburst-like in the optical (Elston, Cornell & Lebofsky, 1985), while samples with 'warmer' infrared colours appeared more biased towards Seyferts or LINERS (de Grijp et al., 1985; Osterbrock & de Robertis, 1985),

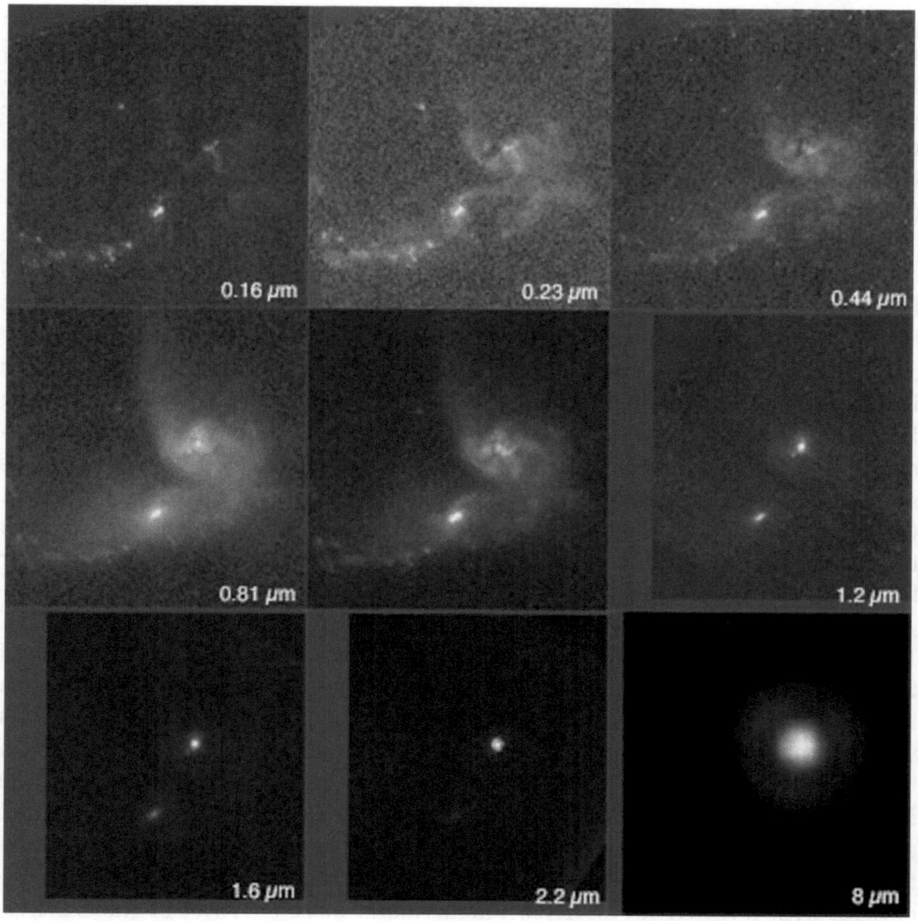

Fig. 9.2. IRAS 08572+3915, as seen at many wavelengths. The colour composite at the center encodes wavelengths from the far-UV to the near-IR as blue-to-red, and illustrates the complex composite nature of ULIRG systems. The ultraviolet emission is dominated by one of the two merger nuclei and by young super-star clusters along the leading edge of one of the tidal tails. Almost no ultraviolet or blue emission is seen from the NW nucleus. At optical wavelengths, increasingly dominated by old stars, we see the merger galaxy body, which appears to be two spirals, accompanied by tidal tails. The NW nucleus has a complex structure of dust lanes. As we proceed into the thermal infrared, all emission sources other than a single compact emission source in the NW nucleus fade away. The images from 0.16-2.2μm are from STIS, WFPC2, and NICMOS (Goldader et al., 2002; Surace et al., 1998, 2000; Scoville et al., 2000). The 8μm image is from Spitzer (Surace et al, in prep), and has a beam size $\sim$ 5 times greater than the HST data, but is sensitive enough to exclude much emission from the SE nucleus. Keck observations (Soifer et al., 2000) at similar wavelengths constrain the size of the nucleus to < 0.3 arcseconds, or 300 pc. (J. Surace, priv. comm.).

or even optical QSOs (Beichman et al., 1986). Conversely, studies of ULIRGs with 'cool' IR colours generally indicated the presence of starbursts (Heckman, Armus & Miley, 1987; Armus, Heckman & Miley, 1988), sometimes accompanied by spectral signatures of Wolf-Rayet stars, indicating that the starburst was likely only a few Myr old. Also noted at this time was a tendency for more IR-luminous objects to exhibit Seyfert spectra, with narrow lines in direct light (Cutri et al., 1994) and broad lines in polarized light (e.g. Hines et al., 1995).

To gain a complete picture of the spectroscopic properties of ULIRGs, however, requires large-scale surveys, which were soon forthcoming as part of IRAS followup. One of the largest to date is the IRAS 1Jy survey (Kim, Veilleux & Sanders, 1998; Veilleux, Kim & Sanders, 1999a), which showed that the majority of ULIRGs have optical spectra reminiscent of starbursts, but with a systematic increase in the fraction of ULIRGs with Seyfert (1 or 2) spectra with increasing IR luminosity. Most of the ULIRGs with Seyfert spectra however also show evidence for ongoing or recent star formation. Approximately 30% of the Seyferts are Sy1's, with a systematic increase compared to Sy2's with IR luminosity. Other notable findings included that optical reddening generally decreases with increasing distance from the nuclear regions and that the optically derived star formation rates are in most cases many times lower than those derived from mid-/far-IR data. Followup spectroscopy in the near-IR (Veilleux, Sanders & Kim, 1997, 1999b) refined and extended this picture, showing that, overall, around 25% of ULIRGs show evidence for an AGN and that this fraction increases with increasing IR luminosity (reaching $\sim 50\%$ at $L_{ir} > 10^{12.3} L_{\odot}$). Those ULIRGs with 'warm' IR colours are more likely to show broad lines in the near-IR than 'cool' ULIRGs, and there is no observed correlation between extinction in the NLR and the presence of broad lines in the near-IR, suggesting that the NLR and BLR do not lie along the same line of sight. An intriguing further result is that, of all the objects that show broad lines in the near-IR, all are Sy2's in the optical, with no LINERs or HII's. Furthermore, most ($\sim 70\%$) of the ULIRGs that show a Sy2 spectrum in the optical show broad lines in the near-IR.

Very recent spectroscopy of ULIRGs from the UV through to the near-IR has revealed some further important details. High spatial resolution UV and optical spectroscopy using the Space Telescope Imaging Spectrograph onboard HST of four 'warm' ULIRGs (Farrah et al., 2005) has shown that the 'knots' seen in optical imaging in many cases harbour very luminous starbursts and AGN, implying that these optically bright knots may also be the sites of the heavily obscured power sources behind the IR emission. The spectral properties of some of these knots also suggest further links between ULIRGs and (low ionization) broad absorption line (BAL) QSOs, and the forming cores of elliptical galaxies. The starbursts in these knots were all observed to be young, with ages of 4Myr-20Myr, supersolar metallicities, and an IMF (Salpeter) slope of less than about 3.3. Near-IR spectroscopy with the Very Large Telescope (VLT) in Chile, and with Keck (Genzel et al., 2001; Tacconi et al., 2002) has shown that the host galaxy kinematics of ULIRGs resemble those of ellipticals with luminosities of $\sim L^*$ (but with a large scatter), and that the host galaxy properties of those ULIRGs that contain an AGN and those without are very similar.

Finally, there is strengthening evidence that nuclear and galactic scale outflows may be common in ULIRGs (Heckman, Armus & Miley, 1987; Wilman, Crawford & Abraham, 1999), and high resolution spectroscopy has discovered galactic-scale outflows with high ejection efficiencies in many ULIRGs that are dominated by star formation (Rupke, Veilleux & Sanders, 2002). Lipari et al. (2003) noted that relatively low velocity outflows are present in starburst dominated ULIRGs, but that higher velocity outflows are present in systems that show evidence for both a starburst and an AGN, though even in composite systems, the starburst is still probably the dominant mechanism behind galactic-scale outflows (Rupke, Veilleux & Sanders, 2005).

9.3.3 Mid-infrared spectroscopy

An area where the Infrared Space Observatory, ISO (Kessler et al., 1996), excelled in the study of ULIRGs was in spectroscopy of nearby systems using SWS (de Graauw et al., 1996), LWS (Clegg et al., 1996), ISOCAM (Cesarsky et al., 1996) and ISOPHOT-S (Lemke et al., 1996), because for the first time high-sensitivity and high-resolution spectroscopy could be obtained in a wavelength region with minimal effects of dust extinction, at least compared to the optical. Moreover the mid-infrared region holds several diagnostic lines which are very useful for characterization of the major power source for ULIRGs. The mid-infrared continuum shape can be a powerful diagnostic itself, betraying the existence of warm dust in the close vicinity of an AGN. We refer the reader to several previous reviews on ISO's legacy on star forming galaxies, AGN and ULIRGs (Genzel & Cesarsky, 2000; Elbaz, 2005; Verma et al., 2005; Oliver & Pozzi, 2005) for more details.

Genzel et al. (1998) demonstrated the power of SWS spectroscopy to separate AGN from starbursts by comparing the strength of the 7.7μm PAH equivalent width to high/low excitation line ratios, such as [OIV]25.9μm/[NeII]12.8μm; PAH molecules will be destroyed by high-intensity AGN radiation fields so low PAH/continuum ratios should correlate with strong [OIV]/[NeII] ratios. They concluded that at leat half of their sample of 15 local ULIRGs have simultaneous starburst and AGN activity, but 70–80% of the sample is predominantly powered by star formation (a result mirrored by mid/far-IR SED fitting, Farrah et al. (2003)). The method requires high sensitivity to determine the fine structure line ratios, and the PAH emission can also be contaminated by strong silicate absorption at 9.7μm. Moreover dense nuclear environments can hide AGN activity even at mid-IR wavelengths, as discovered by Clavel et al. (2000) in Seyfert 2 nuclei in which the the mid-infrared continuum from the AGN is sufficiently absorbed to allow extranuclear PAH emission to dominate the spectrum. Laurent et al. (2000) developed a diagnostic diagram based on the 6.2μm PAH feature relative to continuum strength versus mid-IR continuum colour, which could be used for systems lacking fine-structure line spectroscopy, and Peeters et al. (2004) extended this approach to include far-infrared continuum colours. Lutz et al. (1998); Rigopoulou et al. (1999) and Tran et al. (2001) applied these techniques to large ULIRG samples studied with ISOPHOT-S, concluding, in agreement with Genzel et al. (1998), that starbursts predominantly power these systems, though the presence of heavily obscured AGN cannot be ruled out.

Soifer et al. (2002) obtained low spectral resolution, but high angular resolution, Keck mid-IR spectra of five LIRGs–ULIRGs demonstrating that PAH emission, when present, generally is circumnuclear in origin, extended over scales of 100-500pc. The silicate optical depths in these sources can be as high as 15, suggesting that even mid-IR spectroscopy may not be probing the true nuclei in the most compact sources.

Mid-IR spectral classifications, however, generally agree with optical line classifications of AGN vs star formation power, although some ULIRGs with LINER-like optical spectra were interpreted as starburst dominated in the mid-IR, attributed to starburst wind-driven ionising shocks (Lutz, Veilleux & Genzel, 1999; Sugai & Malkan, 2000) instead of to low-level AGN. However, the LINER situation is complex as demonstrated by subsequent comparison of ISO-SWS fine structure line spectroscopy with Chandra X-ray imaging of a sample of LINERs. (Satyapal et al., 2004) has shown that LINERs are intermediate between starbursts and AGN in mid-infrared line excitation, and that most LINERs contain a compact hard X-ray source characteristic of an AGN. They also found some anti-correlations between mid-IR fine structure line diagnostics and hard X-ray AGN diagnostics: their highest excitation mid-IR spectrum source NGC 404 shows only weak soft X-ray emission while the low mid-IR excitation LINER NGC 6240 shows an extremely luminous binary X-ray AGN (Komossa et al., 2003). The most likely explanation for objects of this kind are extremely high optical depths even at mid-infrared and X-ray wavelengths, and potentially different lines-of-sight to the AGN core at different wavelengths (Risaliti et al., 2000), and/or unusual ratios of gas-to-dust optical depths.

Sturm et al. (2002) have developed diagnostic diagrams using [OIV]25.9μm, [SiII]34μm, [NeVI]7.65μm and [NeII]12.8μm and other fine structure lines, coupled with Brackett β at 2.63μm, to separate high from low excitation systems and estimate the fractional contribution of star formation and AGN to the integrated light using mixture lines, in analogy to the well-known optical line diagnostic diagrams (Veilleux & Osterbrock, 1987). Most ULIRGs were not detectable by ISO in these lines, though Arp 220 was shown to exhibit low excitation in these diagrams.

Spinoglio, Andreani & Malkan (2002) proposed an analogous far-infrared approach using [CII]158μm, [OI]63μm and [OII]88μm, though again very few ULIRGs were detectable in these lines by ISO. Arp 220 has been observed extensively by ISO, as the nearest and brightest ULIRG. Arp 220's FIR spectrum is dominated by molecular absorption from species common to Galactic photo-dissociation regions (PDR): OH, H_2O, CH, NH, NH_3 (Fischer et al., 1999) and is very weak in the fine structure emission lines, with [OI] 63μm in absorption. The [CII]158μm fine structure transition decreases with increasing IR luminosity (and more strongly with FIR colour – $S_{60\mu m}/S_{100\mu m}$) in infrared galaxies (Malhotra et al., 1997), and in ULIRGs is only 10% of that measured in less luminous starbursts (Fischer et al. 1999; Luhman et al., 2003). Since [CII] is the primary coolant in the ISM of normal galaxies it was expected to be strong in high-star formation systems, and its apparent lack is interpreted to indicate a stronger radiation field in luminous starbursts producing charged grains which result in a lower heating rate in PDRs (Malhotra et al., 2001). González et al. (2004) have modeled the FIR spectrum of Arp 220 over the spectral range 40–200μm. Their model requires three components: the compact nuclei, modeled as a unit, with $T = 10^6 K$ which are optically-thick throughout the infrared, an extended region which is dominated by PDR emission at $T \sim 40$–$90K$, and a halo which produces absorption from the low-lying levels of OH and H_2O plus the CH. Given that the two nuclei show distinct mid-IR spectra (Soifer et al., 1999) it is unclear what effect modeling the nuclei as a single source may have. In this model the [CII] emission is produced in the extended (PDR) region with little [CII] emission from the nucleus, which has distinctly non-PDR conditions.

For the majority of ULIRGs for which spectroscopy is not available broad-band colours have been used to characterize the broad spectral type and major infrared energy source, following the early IRAS schemes (Klaas et al., 2001; Spinoglio, Andreani & Malkan, 2002; Farrah et al., 2003). ULIRGs are found to fall into two general classes by these methods: "cool" systems dominated by star formation, which show a relatively flat mid-infrared spectral shape and a steep rise towards longer wavelengths, and "warm" systems with a red power-law-like SED through the mid-infrared, usually assumed to be AGN-dominated, although compact HII regions can also show a warm mid-IR SED (Dopita et al., 2005). Warm ULIRGs tend to have Seyfert 1 like optical spectral features while cool ones have starburst or Seyfert 2 optical spectra. The two clases were found to be indistinguishable in the far-infrared (Klaas et al., 2001), suggesting that at these wavelengths the emission may not be dominated by an AGN in either class, however Peeters et al. (2004) find ULIRGs to have more prominent far-infrared emission than AGN or lower luminosity starbursts such as M82.

Several lines of evidence indicate that AGN become more bolometrically significant with increasing IR luminosity in ULIRGs. The incidence of both AGN optical line indicators and mid-infrared continuum and line AGN diagnostics were found to increase with increasing infrared luminosity (Shier, Rieke & Rieke, 1996; Genzel et al., 1998; Lutz et al., 1998; Rigopoulou et al., 1999; Tran et al., 2001). However, these ISO HLIRG samples are biased towards previously known AGN so Spitzer confirmation of these result is important, using unbiased samples of infrared-selected HLIRGs.

Observations with the ISO satellite greatly expanded our understanding of the mid-infrared spectra of ULIRGs, however, many ULIRGs were beyond the reach of many of the diagnostic methods until the advent of Spitzer. Several extensive Spitzer programs are underway to adequately sample the local ULIRG population, the most comprehensive being a study of the mid-infrared spectra of a large number (> 100) of ULIRGs having $0.02 < z < 0.93$ with the Infrared Spectrograph (IRS), as part of the IRS guaranteed time program. These sources are chosen primarily from the IRAS 1-Jy (Kim & Sanders 1998), 2-Jy (Strauss et al., 1992), and the FIRST/IRAS radio-far-IR sample of Stanford et al. (2000). In Fig. 9.3 we show IRS spectra for three nearby ULIRGs (Mrk 1014, UGC 5101, and NGC 6240) whose spectra serve to highlight the range in properties

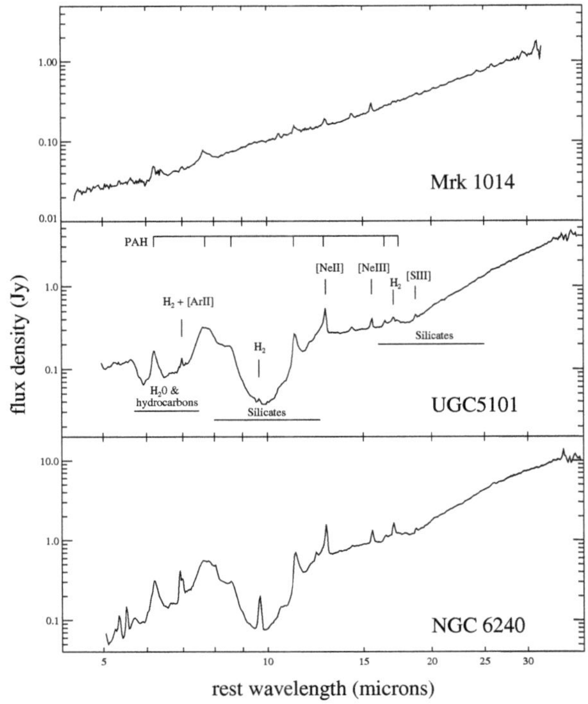

Fig. 9.3. IRS low-resolution spectra of Mrk 1014, UGC 5101 and NGC 6240. The positions of prominent emission features and absorption bands (the latter indicated by horizontal bars) are marked on the UGC 5101 spectrum. Figure courtesy of L. Armus and J. Houck.

seen throughout much of the sample, and the power of the IRS in wavelength coverage and sensitivity for studies of the nuclei and interstellar media in dusty galaxies (Armus et al., 2004; 2005).

Mrk 1014 ($z = 0.1631$) is a radio-quiet, infrared luminous QSO with broad optical emission lines and twin tidal tails indicative of a recent interaction (MacKenty & Stockton 1984). UGC 5101 ($z = 0.039$) has a single, very red nucleus within a disturbed morphology suggestive of a recent interaction. Optically, UGC 5101 is classified as a LINER (Veilleux et al., 1995). It has a high brightness temperature ($T > 10^7$K) radio nucleus at 1.6 GHz which is resolved with the VLBA (Lonsdale et al., 1995). ISO SWS and PHT-S spectroscopy (Genzel el al., 1998) indicate a powerful, circumnuclear starburst. Based upon its IRAS colours, UGC 5101 is classified as a cold, starburst-dominated, far-infrared source. However, XMM data indicate an obscured, but luminous, hard X-ray source with L_x(2–10 keV)$\sim 5 \times 10^{42}$ erg s^{-1} and L_x(2–10 keV)$/L_{IR} \sim 0.002$ suggestive of a buried AGN (Imanishi et al., 2003). NGC 6240 ($z = 0.0245$), is a double-nucleus, merging galaxy (Fosbury & Wall 1979), with an 8–1000μm luminosity of $\sim 7 \times 10^{11} L_\odot$. The optical nuclear spectrum of NGC 6240 is classified as a

LINER (Armus, Heckman & Miley 1987), and the extended optical nebula reveals the presence of a starburst-driven superwind (Heckman, Armus & Miley 1987). X-ray observations with ASCA (Turner et al., 1998), Beppo-Sax (Vignati et al., 1999), Chandra (Komossa et al., 2003; Ptak et al., 2003), XMM-Newton (Netzer et al., 2005) provide clear evidence for the presence of one (or two) AGN behind significant columns of absorbing material ($N_H = 1\text{–}2 \times 10^{24}$ cm^{-2}). Relatively strong [OIV] 25.89μm line emission in the ISO SWS spectrum of NGC 6240 led Lutz et al. (2003) to suggest that up to 50% of the infrared energy emitted by NGC 6240 could be powered by a buried AGN.

The spectra of Mrk 1014, UGC 5101, and NGC 6240 are strikingly different. Mrk 1014 has a steeply rising mid-infrared spectrum with weak emission features and little or no silicate absorption. The spectra of UGC 5101 and NGC 6240, on the other hand, are dominated by strong silicate absorption at 9.7μm and 18μm, and PAH emission at 6.2, 7.7, 11.3, and 12.7μm (Armus et al. 2004, 2005). The extinctions toward the nuclei in UGC 5101 and NGC 6240, as estimated from the depths of the silicate absorption, are at least A_V = 15–35 and A_V = 60 mag, respectively. UGC 5101 also shows strong absorption between 5–7.5μm from water ice and hydrocarbons. As suggested by Spoon et al. (2002), the water ice features may indicate the presence of shielded molecular clouds along the line of sight to the nucleus. NGC 6240 has little or no water ice absorption, but very strong emission lines from warm ($T \sim$ 300–400K) H_2. The mass of this warm gas is estimated to be approximately $1.4 \times 10^8 M_\odot$ – about 1% of the cold molecular gas mass derived from single-dish millimeter CO line measurements (Solomon et al. 1997), but up to 3–7% of the cold molecular gas within the central 1 kpc measured by Tacconi et al. (1999).

The IRS high-resolution (R=650) spectra (not shown) provide important diagnostic measures of the dominant ionizing sources in the ULIRGs because it is possible to accurately measure unresolved atomic, fine-structure lines of Ne, O, Si, and S, covering a large range in ionization potential. The [NeV] 14.3 / [NeII] 12.8 and [OIV] 25.9 / [NeII] 12.8 line flux ratios in Mrk 1014 (0.9 and 1.7, respectively) suggests that nearly all the ionizing flux comes from the central AGN, although the obvious presence of PAH emission suggests some extranuclear star formation. In UGC 5101 and NGC 6240 weak [NeV] emission has been detected, with 14.3μm line fluxes of 5–6 $\times 10^{-21}$W cm^{-2}, indicating buried AGN. While the [NeV] / [NeII] and the [OIV] / [NeII] line flux ratios imply an AGN contribution of $< 10\%$ to the total luminosity in both sources (Armus et al. 2004, 2005), the large optical depth to the nuclei, as evidenced by the deep silicate absorption and X-ray columns, leaves open the possibility that the true contribution of the AGN to the bolometric power output in UGC 5101 and NGC 6240 may be much larger than revealed by the mid-IR emission lines. In NGC 6240, the extinction-corrected hard X-ray data are consistent with the buried AGN producing 50–100% of the luminosity. An inclined, dusty torus, patchy extinction, and/or a low covering factor for the [NeV]-emitting clouds, could reconcile these apparently discrepant estimates – an explanation often evoked to explain observations of other type-2 AGN.

Spoon et al. (2004) observed the more distant ULIRG IRAS F00183-7111 at z=0.327, detecting strong absorption from CO_2 & CO gas, water ice, hydrocarbons and silicates, indicating high obscuration, a complex line of sight, and the presence of very warm, dense gas (720K). Direct signs of an obscured AGN are not found but Spoon et al. conclude that an obscured AGN probably accounts for most of the luminosity based on similarities to other systems with highly obscured AGN.

9.3.4 Radio continuum studies: AGN *vs* luminous radio supernovae

The existence of a tight correlation between integrated far-infrared flux-density and radio continuum emission – the "Radio-FIR Relation" (Helou et al., 1985; Yun, Reddy & Condon, 2001) – over four orders of magnitude in IR luminosity, allows the use of high-resolution radio interferometric techniques to study the compact nuclei of luminous IR galaxies in one of the few spectral regions with relatively low optical depth. The Radio-FIR relation is not well understood, but is believed to be produced by nonthermal radiation from relativistic electrons in, or leaking out of, starburst-related supernova remnants. The seminal study of a complete sample of luminous IR galaxies by Condon et al. (1991) with the VLA at 1.49 and 8.44GHz demonstrated the extreme compactness of luminous IR nuclei and concluded that most of their 40 galaxies – 6 of which are ULIRGs – are consistent with a starburst-related origin for the radio emission.

Lonsdale et al. (1993) used a global VLBI array to detect 18cm high brightness temperature, $T_b \geq 10^6 K$, emission cores from 17 of 31 luminous (log $L_{FIR} \gtrsim 11.25(L_\odot)$) infrared galaxies, consistent with such cores existing in all such LIRGs at ~10% of the total 18cm flux density. This result indicates either obscured radio-quiet AGN in most LIRGs and ULIRGs, possibly energetically dominant (Lonsdale et al., 1995), or clumps of starburst-related luminous radio supernovae (LRSN) and remnants (Smith et al., 1998). Similar studies in the South (Norris et al., 1990; Kewley et al., 2000) detected compact nuclear radio cores with lower frequency, but their results are consistent, allowing for differences in sample selection and sensitivity.

Subsequent VLBI imaging dramatically revealed a cluster of luminous RSN in the nearest ULIRG, Arp 220 (Smith et al., 1998) – clear evidence for a high star formation rate, estimated at that time as $\sim 100 M_\odot$, yr^{-1} by Smith, Lonsdale & Lonsdale (1998), which would be enough to power the infrared luminosity by star formation without resorting to an obscured AGN. Momjian et al. (2003); Neff, Ulvestad & Teng (2004); Bondi et al. (2005) have reported possible "supernova factories" in IRAS 17208-0014, Mrk 299 and Mrk 273. respectively. The original dozen LRSN in Arp 220 have been monitored with VLBI (Rovilos et al., 2005) and more sensitive VLBI imaging studies (LOnsdale et al., 2006a) have revealed a number of new and fainter LRSN. These Luminous RSN were originally modeled after RSN1986J in NGC 891 (Weiler et al., 1990), which exhibited a maximum radio power, log $P_{1.5GHz}^{1986J} = 21.15$. The LRSN in Arp 220, however, exhibit much slower decay in their radio light curves (Rovilos et al., 2005) than RSN 1986J; longer RSN lifetimes reduce the inferred star formation rate, therefore it is not clear from the VLBI imaging whether a starburst is the chief power source for Arp 220.

On the other hand (Lonsdale et al., 2003a) demonstrated that the radio emission in UGC 5101, Mrk 231 and NGC 7469 is AGN dominated. Mrk 231, in some senses the classical "infrared quasar", has been studied with the VLBI by Ulvestad et al. (1999). Their images (see also Lonsdale et al. 2003a) show a triple structure, with a core and two lobes which classify it as a Compact Symmetric Object (CSO). It has been suggested that CSOs are young, $\tau \ll 10^6$yr, with the hot spots representing the working surface of a relativistic jet upon the ambient medium (Readhead et al., 1996). If the southern (primary) lobe/hot-spot in Mrk 231 is confined by ram pressure, (Lonsdale et al., 2003a) estimate a lobe advance speed, $v_a \sim 10^{-4}c$ and an age for the jet/compact source, $\tau < 10^6$ yr. Despite the clear evidence for AGN domination of the radio structure in these systems, the radio power is small compared to the infrared/bolometric emission, which may still therefore be starburst dominated. In Mrk 231, for example, several studies (Carilli et al., 1998; Lonsdale et al., 2003a; Farrah et al., 2003) suggest that more than half of the total luminosity comes from a circumnuclear starburst in a molecular ring rather than from the AGN.

9.3.5 Molecular gas: CO & HCN observations

The interpretation of Luminous Infrared Galaxies as starburst systems was strengthened soon after their identification with IRAS by studies of neutral hydrogen (Mirabel & Sanders, 1988) molecular gas, principally CO (Sanders et al., 1986, 1991) but more recently HCN (Gao & Solomon, 2004) and OH maser emission (Baan et al., 1982; Baan, 1989). These early studies demonstrated that ULIRGs as a class exhibit compact nuclear reservoirs of high-density gas, and, with mass estimates of order 10^9–$10^{10} M_\odot$ in HI and in H_2, are consistent with the interpretation that star formation accounts for a substantial fraction of the FIR luminosity in Luminous IR Galaxies.

Millimeter-wave interferometer measurements of CO emission (e.g. Scoville et al. 1991; Solomon et al. 1997; Downes & Solomon 1998; Bryant & Scoville 1999) have demonstrated that nearly half of the CO mass in Luminous IR Galaxies is contained within the central regions, $r < 0.5 - 1$ kpc, with as much as $10^{10}\, M_\odot$ of molecular gas distributed in nuclear disks of radius a few hundred pc, but thicknesses of one-tenth the radius and densities over 10^4 cm^{-3} (Bryant & Scoville, 1999). The mean molecular surface densities in these structures may exceed $10^4 M_\odot$ pc^{-2} with the molecular gas providing a large fraction of the dynamical mass. The picture of molecular gas in ULIRGs is very different from that in the Galaxy, with much higher surface densities and inferred optical depths ($A_v \gtrsim 10^2 - 10^3$). Early CO studies assumed a Galactic conversion ratio between CO Luminosity and H_2 mass, $M_{H_2}/L'_CO = 4.6\, M_\odot$ (K km s^{-1} pc^{-2}), however, recent analyses suggest that this may overestimate the molecular mass, either owing to high-brightness temperatures in the CO emission (*e.g.* Mrk 231; Bryant & Scoville 1999) or because the line width/velocity dispersion reflects the total dynamical mass in the central nuclear region, rather than that within virialized molecular clouds (Solomon et al., 1997).

Again, Arp 220 provides a convenient laboratory to study molecular gas, if not always in a typical ULIRG environment. Sakamoto et al. (1999) have studied the Arp 220 nuclei in the CO(2-1) transition and accompanying 1mm continuum. Steep velocity gradients are found in the CO associated with each of the two nuclei, which are misaligned with each other and with the outer CO disk. Sakamoto et al. interpret these as molecular disks associated with the merging nuclei, counter-rotating with respect to each other and with respect to the outer disk. The dynamical masses inferred from CO kinematics are of the order of $2 \times 10^9 M_\odot$. The central molecular gas is inferred to have high filling factor, more like a uniform disk than individual clouds, which will have an inward accretion rate of approximately $100 M_\odot$ yr^{-1} (Scoville, Yun & Bryant, 1997) which is similar to the star formation rate of Arp 220 inferred from the FIR Luminosity (Smith, Lonsdale & Lonsdale, 1998; Farrah et al., 2003).

Early studies placed considerable emphasis on the ratio of FIR Luminosity to CO Luminosity – often called the Star Formation Efficiency – demonstrating that the ratio of FIR/CO increases at higher L_{FIR}. This is interpreted as either an increase in Star-Formation Rate per unit molecular gas mass, *i.e.* more efficient star formation, or as possible evidence for an alternative, AGN, contribution to the FIR luminosity (Sanders & Mirabel 1996). More recently, studies of HCN, which, owing to the higher dipole-moment of the HCN molecule is a tracer of warmer ($T_{kin} \sim 60 - 90$K), higher-density ($n_{H_2} \sim 10^5 - 10^7$ cm^{-3}) gas, shows a much higher ratio of HCN/CO luminosity in Luminous IR Galaxies than in quiescent spirals like the Milky Way (Solomon, Downes & Radford, 1992). Furthermore the FIR/HCN luminosity (HCN Star Formation Efficiency, as above) shows an approximately linear relationship in Luminous IR Galaxies with L_{fir} from $10^{10} - 10^{13} L_\odot$ indicating the presence of an abundant warm, high-density molecular environment within ULIRGs (Gao & Solomon, 2004). ULIRGs thus have a much greater dense, warm medium like that of the star-forming cores of Galactic molecular clouds – perhaps the central regions of the molecular disks, although little spatial information is available about the HCN distribution in ULIRGs. Since there is little doubt that the lower luminosity systems amongst this sample are dominated by star formation, the continuity of this relation up to luminosities characteristic of HLIRGs is one of the most compelling pieces of evidence, albeit statistical, that most ULIRGs are dominated by starburst power rather than AGN power.

9.3.6 Maser emission

Luminous IR Galaxies have been known to be strong emitters of OH maser emission since the discovery of OH 1667MHz emission in Arp 220 by Baan et al. (1982), who dubbed it a "megamaser", having an OH luminosity roughly 10^6 times that of masers in the Galaxy. H_2O maser emission, which has been used to map torus/disklike structures around compact supermassive AGN nuclei such as

NGC 4258 (Miyoshi et al., 1996) is not generally detected in classical Luminous IR Galaxies (see Lo 2005 for a review).

OH megamasers are found preferentially in the most luminous IR galaxies (Baan, 1989), and a rough dependence of the OH luminosity on the square of the FIR luminosity was originally found, suggesting that the FIR emission provided the pumping mechanism to provide the necessary population inversion (*e.g.* Henkel & Wilson 1990) within a foreground molecular screen amplifying the diffuse radio continuum, which is in turn correlated with the FIR emission. The OH gas may trace very high density regions ($n_{H_2} = 10^{5-7}\mathrm{cm}^{-3}$), although an apparently separate component of the OH megamasers is associated with a much lower-density high-velocity outflow (Baan, Haschick & Henkel, 1989).

The initial interpretation of the maser phenomenon of a foreground, low-gain molecular screen amplifying the diffuse radio continuum in Luminous IR Galaxies, however, has been questioned by the VLBI observations of Arp 220 (Diamond et al., 1989; Lonsdale et al., 1994, 1998; Rovilos et al., 2003) demonstrating that over two-thirds of the masing gas in Arp 220 is compact, produced in structures a few pc^3 in volume with amplification ratios of order 10^3 or higher. These compact masers have complex spatial and velocity structure arising in clouds generally within the Arp 220 nuclei, but not coincident with nuclear radio continuum (or any other detectable radio continuum). The southern component within the western nucleus (the one with the majority of LRSN) shows a velocity gradient in excess of 18,000 $\mathrm{km\,s^{-1}}$ (Rovilos et al., 2003), which, interpreted as rotation, implies a mass of order $2 \times 10^7 M_\odot$.

Similar compact maser emission has been detected with Global VLBI in IIIZw35 and IRAS17208-0014 (Diamond et al., 1999), and in 12032+1707 (Pihlstrom et al., 2005). On the other hand Mrk 231 (Lonsdale et al., 2003a; Klöckner et al., 2003), Mrk 273 (Klöckner & Baan, 2004) and 14070+0525 (Pihlstrom et al., 2005) show more extended emission in a circumnuclear disk with conditions more like the classical maser model. IIIZw35 also shows diffuse ring of maser emission with radius $\sim$ 20pc and central mass $\sim 7 \times 10^6 M_\odot$ (Pihlstrom et al., 2001). In this case the compact emission occurs at the tangent points of the ring, and the maser structure has been modeled by multiple high density clouds within the diffuse ring. Such a structure does not, however, appear to be able to explain the complex, compact maser emission in Arp 220.

In the case of Mrk 231, the OH emission appears to be the central portion of a circumnuclear HI disk detected in absorption by Carilli et al. (1998), though it is misaligned with the central CO distribution (Klöckner & Baan, 2004). Lonsdale et al. (2003a) speculate that the ignition of Mrk 231's Seyfert nucleus may have disrupted compact maser emission in the nucleus itself.

An extensive survey for OH maser emission in over 300 IRAS galaxies from the PSCz Survey (Saunders et al., 1990) with $z > 0.1$ has been carried out with the upgraded Arecibo Telescope by Darling & Giovanelli (2002 and references therein). This survey has expanded the sample of known megamasers in luminous infrared galaxies to over 100 with detections in nearly 20% of the galaxies surveyed. Their analysis suggests few correlations between OH and FIR properties. A re-analysis of the $L_{OH} - L_{FIR}$ relationship in their data suggests a much flatter relation $L_{OH} \propto L_{FIR}^{1.2}$ consistent with other recent analyses (Kandalian, 1996) which finds a slope of approximately 1.4, suggesting a mixture of extended, unsaturated emission plus compact, saturated clouds. One source in the Darling & Giovanelli (2002) sample, IRAS 21272+2514, shows apparent variability which the authors interpret as due to interstellar scintillation. This interpretation requires $\sim$ 30–60% of the OH maser emission to originate in saturated clouds of dimension smaller than $\sim$ 2pc.

A satisfactory model for the complexities of OH masers in ULIRGs has yet to be proposed. Almost certainly the diffuse emission follows the classical model with largely unsaturated OH maser clouds, pumped by the strong FIR radiation field, amplifying the nuclear 1.6GHz radio continuum. The compact, pc-scale structures in Arp 220 and other galaxies, coupled with the lack of detectable associated 1.6GHz continuum emission, requires a substantial saturated maser component. These structures subtend solid-angles too small to intercept sufficient FIR photons to excite the OH molecules, suggesting that collisional excitation must be important (Lonsdale et al., 1998). A tentative correlation between OH line-width and X-ray luminosity (Kandalyan, 2003) may indicate that X-ray heating of the molecular gas plays a role in collisional excitation.The possible roles of shocks and/or AGN activity remain to be explored.

9.3.7 X-ray emission

The importance of X-rays has been recognized not only because of the diagnostic ability of the X-ray to discriminate between AGN and Starburst emission (Rieke, 1988), but also because models of the X-ray background (XRB) require substantial populations of highly-obscured AGN at redshifts, $z \sim 0.5$–1.5, to reproduce the observed XRB spectrum (Worsley et al. 2005). Rieke's early analysis from HEAO A-1 demonstrated that ULIRGs are underluminous in the 2-10keV band compared to classical Seyfert galaxies or QSOs. It has only been recently with the Chandra and XMM-Newton X-ray observatories that the implications of this early discovery could be investigated in significant ULIRG samples with sufficient resolution and signal-to-noise to provide reliable diagnostics of the nature of the nuclear X-ray sources.

Further evidence of the weakness of X-ray emission in ULIRGs comes from a ROSAT survey (Boller et al., 1998) of 323 ULIRGs which detected fewer than 10%. Probably the most significant starburst-related discovery of the ROSAT soft X-ray satellite is the detection of extended thermal outflows, first in nearer, lower-luminosity starbursts, and then in Arp 220, dubbed "superwinds" (Heckman et al., 1996). Similar thermal components appear to be common in ULIRGs and they are described as "ubiquitous" in starbursts with star formation rates $> 10^{-1} M_{\odot}$ yr^{-1} kpc^{-2} in Heckman's (2001) review; see also the recent review by Veilleux, Cecil & Bland-Hawthorn (2005). The superwinds are believed to be driven by supernova supplied kinetic energy with outflow rates comparable to the star formation rates in these galaxies. Heckman speculates that these superwinds may be the principal "polluters" of metals and dust into the IGM.

Further Chandra studies (McDowell et al., 2003) of the Arp 220 superwind reveal extended, faint, edge-brightened, soft X-ray lobes outside the optical galaxy out to a distance of 10–15 kpc. Bright plumes inside the optical isophotes coincide with the optical line emission (Colina et al., 2004) and extend 11 kpc from end to end across the nucleus. The data for the plumes cannot be fitted by a single-temperature plasma and display a range of temperatures from 0.2 to 1 keV. There is a close morphological correspondence between the Hα and soft X-ray emission on all spatial scales.

There have been three recent ULIRG X-Ray surveys with XMM-Newton (Franceschini et al., 2003a) which observed 10 ULIRGs, and with Chandra (Ptak et al., 2003; Teng et al., 2005) which observed 8 and 14 respectively. All of the ULIRGs surveyed were detected, with X-ray luminosities typically $L_{2-10keV} < 10^{42}$–10^{43} erg s^{-1}. These luminosities represent $< 1\%$ of the infrared luminosities in these systems, confirming that ULIRGs are much less luminous in the X-ray than classical AGN. Furthermore, the soft X-ray emission from all systems is dominated by extended, thermal emission with $kT \sim 0.7$keV, and is uncorrelated with IR luminosity. In at least two XMM-Newton systems the emission is extended on scales of 10s of kpc, suggesting a superwind origin. In at least 5 galaxies a hard X-ray (2–10keV) component and/or the presence of 6.4keV Fe K-line emission suggests the presence of an AGN which is not energetically dominant (Franceschini et al., 2003a). Similarly, in the nearer Chandra sample of Ptak et al. (2003) all galaxies exhibit hard components which are interpreted as AGN sources. 5 galaxies in the Chandra samples which are classed as AGN (including IRAS 05189-2524, Mrk 231, Mrk 273) show order of magnitude greater X-ray luminosities than the Starburst ULIRGs (UGC 5101, IRAS 17208-0014, IRAS 20551-4250, IRAS 23128-5919) with NGC 6240 being an intermediate case[6]. As already noted UGC 5101 and Mrk231 are believed to harbor obscured AGN based on their Spitzer spectra. Mrk 273, NGC 6240 and UGC 5101 exhibit Fe K-line emission, but sufficiently weak that the authors argue against X-rays reflected from a Compton-thick ($N_H > 10^{24}$ cm^{-2}) absorber as the origin of the observed X-ray emission. Teng et al. (2005) use hardness ratios to estimate the X-Ray spectral properties of their fainter galaxies. The photon indices for the combined Chandra samples peak in the range $1 < \Gamma < 1.5$, with a tendency for Seyfert ULIRGs to have steeper spectra, $\Gamma > 2$. Although the X-ray properties of this fainter sample are consistent with a Starburst origin, the presence of Compton-thick AGN cannot be ruled out and may be expected in many cases given the results discussed above.

Although the situation is complicated, certainly some ULIRGs must harbor luminous X-ray AGN, though perhaps behind large absorbing columns and thus only visible in hard X-rays. In Mrk 231 BeppoSAX revealed a highly absorbed ($N_H \sim 2 \times 10^{24}$ cm^{-2}) power-law component (Braito et al., 2004) and analysis of XMM-Newton data indicates that below 10 keV only scattered or reflected X-rays escape. In a 40ks observation of of Mrk 231, Gallagher et al. (2002)

[6]Mrk 231, IRAS 17208-0014 and IRAS 23128-5919 are included in both the Ptak et al. (2003) and Franceschini et al. (2003a) samples

find the majority of the X-ray luminosity is emitted from an unresolved nuclear point source with a very hard spectrum, the majority of the flux emitted above 2 keV. The source is also variable on a timescale of a few hours. They argue against a Compton-thick reflection model of Mrk 231 (Maloney & Reynolds, 2000), proposing a Compton-thick absorber which allows scattered light from multiple lines of sight to be detected. Highly absorbed AGN power-law hard X-ray sources are also reported in Mrk 273 (Xia et al., 2002; Balestra et al., 2005), NGC 6240 (Iwasawa & Comastri, 1998; Vignati et al., 1999; Netzer et al., 2005), and Mrk 1014 (Boller et al., 2002) although the starbursts in these galaxies dominate in softer X-rays, and possibly also bolometrically.

Again the nearby system, Arp 220, provides more details, as well as unanswered questions. BeppoSAX observations (Iwasawa et al., 2001) placed severe constraints on an energetically significant AGN in Arp 220, requiring an absorbing column, $N_H > 10^{25}$ cm^{-2} to hide an X-ray AGN. They suggested X-ray binaries as the source of Arp 220's hard X-rays. Clements et al. (2002) detected several sources near Arp 220's nucleus, including a mildly absorbed point source with a hard spectrum coincident with the western radio nucleus, plus a fainter source which may coincide with the eastern nucleus. A classical X-ray AGN cannot be ruled out, but again, columns greater that 5×10^{24} cm^{-2} would be required to hide it. Iwasawa et al. (2005) have reported the detection of Fe K emission in the XMM-Newton spectrum of Arp 220. A supernova shocked bubble as suggested by the VLBI observations of radio supernovae could provide an explanation. However, the apparent lack of emission from X-ray binaries is incompatible with the high supernova rate ($\sim$2 SNe yr^{-1}) required.

9.4 The Emerging Picture of ULIRGs: Local Universe

Taken as a whole, the observations of local Luminous Infrared Galaxies suggest that, at lower luminosities at least, they are dominated by starburst emission. The continuity of the $L_{HCN}\, vs\, L_{FIR}$ relation (Gao & Solomon, 2004) provides circumstantial evidence that most, if not all, Luminous IR Galaxies are starburst powered. Their sample, however, as well as many others quoted above, suffers from the low space density of ULIRGs, with only 6 classical ULIRGs ($L_{IR} > 10^{12}\, L_\odot$) and only two with $L_{IR} > 10^{12.2}\, L_\odot$). The LRSN discovered in compact ULIRGs would seem to indicate starburst dominance in these systems, however we do not understand enough about these LRSN to establish a star-formation rate or a starburst bolometric luminosity, leaving open the possibility of significant if not dominant AGN power even in the best studied source Arp 220. ULIRGs are generally under-luminous in the X-rays compared to classical AGN, requiring sensitive, high energy, observations to detect them. Evidence for very high X-ray obscuring columns is mounting, which could explain the low observed flux levels, and estimates of the importance of the X-ray-detected, obscuration-corrected AGN power to ULIRG energetics range from minor to dominant. Spectroscopy from the optical through mid-infrared suggests that the incidence of AGN and their strength relative to the ubiquitous starbursts increases with FIR Luminosity and that AGN may dominate for $L_{IR} > 10^{12.5} L_{\odot}$, but there are very few such ULIRGs in most local samples. Mid-infrared high excitation diagnostic lines are a very promising line of investigation, however very deep silicate and ice absorptions are also detected in some ULIRGs, so even these mid-IR diagnostics may severely under-estimate AGN energetic contributions due to extinction.

And so the debate continues, and the very fact that it continues two decades after IRAS highlights that most ULIRGs almost certainly contain both a starburst and a monster, and that the key question really concerns the connection between the two, be it evolutionary (one evolves into the other), causal (one triggers the other somehow) or coincidental (another influence triggers both). We cannot address the wealth of literature on this topic here, involving as it does evolutionary schemes of all kinds and their implications, and extensive comparative AGN and starburst population studies and demographics, and so we outline here only a few key points. One popular scenario is an evolutionary sequence in which major gas rich galaxy mergers first result in a massive cool starburst-dominated ULIRG, followed by a warm ULIRG as a QSO turns on inside the dust cocoon and heats the surrounding dust, and then finally the QSO emerges in an optically bright phase when it blows away the surrounding dust cocoon, and the resulting stellar system resembles a spheroid (Sanders et al. 1988a; Kormendy and Sanders 1992; Joseph 1999; Fabian 1999; Lipari et al. 2003). Another important scenario is unification-by-orientation of AGN, in which a broad-line (type I) active nucleus is highly obscured by a dusty molecular torus (or other non-symmetrical geometry) when viewed off-axis, so that an optically bright QSO viewed off-axis would appear as a ULIRG. Both of these scenarios predict close relationships between starbursts and AGN.

The evolutionary scenarios have recently received a boost from a sequence of papers reporting high resolution hydrodynamic simulations of gas rich major mergers (Di Matteo et al., 2005; Springel et al., 2005; Hopkins et al., 2005a,b,c,d), motivated by linking the growth of spheroid masses and supermassive black hole (SMBH) masses in order to explain the the observed correlations between SMBH mass and bulge mass (Magorrian et al. 1998) or velocity dispersion (Ferrarese and Merritt 2000, Gebhardt et al. 2000) of local spheroids. Accretion rates are predicted to be highest at late merger stages, when the SMBH grows exponentially, followed by the most luminous optically-visible QSO phase when the active QSO essentially explosively drives out all remaining material in the system (Hopkins et al. 2005a). The period of high obscuration during the high accretion rate phase would correspond to an obscured QSO, ie an AGN-powered ULIRG. Starburst events occur earlier in the lifetime of the merger when gas is still plentiful, and a starburst-ULIRG phase could occur when the gas is centrally concentrated into a dense compact region at relatively late stages.

Hopkins et al. (2005e) have made approximate predictions of ULIRG space densities based on these simulations, estimating at z=0.15, 3×10^{-7} and 9×10^{-8} Mpc^{-3} at infrared luminosities of 1.6 and $2.5 \times 10^{12} L_{\odot}$, respectively, which is in good agreement with the 1Jy survey of Kim and Sanders (1998). Hopkins et al. also predict a decrease by a factor of 1.5 to the lower redshift, z=0.04, which also agrees well with the Kim and Sanders work, and an increase to z$\sim$1–3 of $\Phi(L > 10^{11} L_{\odot})$ of 1–3 $\times$ 10^{-5} Mpc^{-3}, which they compare to the sub-mm number counts of Barger et al. (2005). Chapman et al. (2005) have derived the first estimate of ULIRG luminosity functions in this redshift range, though sensitive only to systems above about $10^{12.3} L_{\odot}$. They find space densities >6 $\times 10^{-6}$ Mpc^{-3} in their luminosity range, which could be significantly higher than the Hopkins et al. predictions, depending on the unmeasured break of the IR ULIRG luminosity function at these redshifts. Deep Spitzer surveys should help extend the ULIRG luminosity function to lower luminosities in this redshift range.

These ULIRG-QSO evolutionary sequences are appealing and seem likely to be correct for at least a fraction of the ULIRG population. Many predictions remain to be verified, however, and alternative scenarios are also popular. A key test is the host morphology of ULIRGs and optically-bright QSOs. As we have described in Sect. 9.3, not all ULIRGs are found at late stages of a merger, although most are indeed found to be associated with the merger process. Farrah et al. (2001) have proposed an alternative merger-evolution scheme in which the stage at which a ULIRG occurs depends chiefly on the morphological type of the two interacting systems. Deep HST imaging of QSO hosts is now revealing that significant numbers also appear to inhabit disturbed systems (Lim & Ho, 1999; Percival et al., 2001; Sánchez et al., 2004), supporting an evolutionary connection between starbursts and QSOs, although some QSOs appear to occur in spiral hosts with very little evidence of recent disturbance. This latter result has been taken to indicate that they could not have been triggered by a recent major merger, however recent simulations have demonstrated that major gas-rich mergers can result in disk systems under certain circumstances (Springel & Hernquist, 2005; Robertson et al., 2005).

Genzel et al. (2001) and Tacconi et al. (2002) observed NIR structural properties and stellar dynamics of 18 z<0.18 ULIRGs, showing that ULIRG hosts lie on the fundamental plane of elliptical galaxies and thus are very likely to evolve into ellipticals, but that they have significantly lower mass hosts than a luminosity-matched sample of radio-loud and radio-quiet QSOs from Dunlop et al. (2003). A difficulty with this approach is that both starbursts and QSOs are expected to have rapidly varying luminosity during the merger sequence, so it is not clear that samples should be luminosity matched to compare their host masses. These authors also reviewed the cluster environment of local ULIRGs, QSOs (from the study of McLure & Dunlop 2002) and ellipticals, finding that while the ellipticals and quasars are found in all environments, none of the 117 ULIRGs that were investigated are located in an environment richer than a small group, which provides some statistical evidence that at least those quasars found locally in rich environments may not have evolved via a ULIRG phase.

Haas et al. (2003) searched for a red transition population between obscured and unobscured QSO populations amongst Palomar-Green (PG) QSOs. Optically-selected QSOs as a class have a very high incidence rate of luminous FIR and submm emission, consistent with most of them qualifying as ULIRGs by IR luminosity level (Haas et al., 2003; Polletta et al., 2000; Hatziminaoglou et al., 2005). Haas et al. suggested a sequence of near-, mid- and far-infrared SEDs for their PG QSO sample which could plausibly represent the transition from a young dust-obscured QSO through the stages of re-distribution of the dust and settling of the dust into a torus, and which is less well explained by simple orientation effects alone. Late evolution-stage systems may be expected to have low fuel supplies, perhaps as seen some moderate redshift HLIRGs with low CO masses and L_{IR}/M_{gas} ratios or which lack cool dust components (*eg.* Yun & Scoville 1998; Verma et al., 2002). Another suggested young QSO transitory class are the broad absorption line (BAL) QSOs, especially those with strong FeII and weak [OIII] emission, which are preferentially found to be infrared luminous and which could therefore be young dusty QSOs (Voit et al., 1993; Lipari, 1994; Egami et al., 1996; Canalizo & Stockton 2001, Lipari et al., 2003). FeLoBAL QSOs could alternatively be older QSOs viewed preferentially along the radial surface of the torus (Elvis 2000).

Another class of potentially young broad-line QSOs are the red 2MASS QSOs, selected to have J-K>2 and expected to be significantly dust-obscured (Cutri et al. 2002). These objects tend to be type-1 AGN with moderate luminosities and z<0.8, the relatively low luminosities and redshifts being due to a selection bias due to the near-infrared k-correction. Leipski et al. (2005) extended this red 2MASS QSO search using ISO 6.7μm data, and found a surface density of 1.5 times that of SDSS QSOs to the same optical magnitude depths, and also that the 2MASS-ISO QSOs are significantly redder than SDSS QSOs, and that SDSS QSO colour selection criteria would have missed about 1/3 of these red QSOs.

Most detailed imaging and morphological studies so far have focused on relatively low redshift systems. At higher redshifts we might expect different triggering mechanisms under different environmental circumstances, which we discuss further in Sect. 9.5.

9.5 ULIRGs at High Redshift

9.5.1 The cosmic infrared background

The existence of a significant population of IR-luminous galaxies at high redshift, though hinted at in followup of IRAS, was thrown to the fore of astronomical debate by results from the Cosmic Background Explorer (COBE) satellite. Launched on November 18, 1989, COBE was responsible for two of the most significant astronomical results in the 20th century. The first was measuring the spectral shape and level of anisotropy in the Cosmic Microwave Background (CMB, peaking at $\sim 2000\mu$m), providing overwhelming evidence for the Hot Big Bang model and giving a glimpse at the early structures that would eventually evolve into todays galaxies and clusters. The second result, more relevant for this review, was the discovery of a Cosmic Infrared Background (CIB) with FIRAS at 240μm (Puget et al., 1996), and at 140μm and 240μm with DIRBE (Schlegel,

Finkbeiner & Davis, 1998; Hauser et al., 1998; Fixsen et al., 1998) (later detections with DIRBE of the CIB at 2.4μm and 3.5μm have been published, see Hauser & Dwek, 2001; and Kashlinsky, 2005 for reviews). Such an infrared background had been predicted many years previously (Partridge & Peebles, 1967), but had proven fiendishly difficult to detect, remaining invisible both to rocket-borne IR observatories (Kawada et al., 1994), and to IRAS (Rowan-Robinson et al., 1990; Oliver, Rowan-Robinson & Saunders, 1992). COBE was the first observatory with both the right instrumentation and sufficient sensitivity to detect the CIB.

The cosmological implications from the discovery of the CIB were profound (Dwek et al., 1998). The total background detected by COBE between 140 and 5000 μm is $\sim$ 16 nW m^{-2} sr^{-1}, or 20%–50% of the total background light expected from energy release by nucleosynthesis over the entire history of the Universe, implying that 5%–15% of all baryons are or have been parts of stars. And while the CIB itself amounts to less than about 2% of the CMB, the intensity of the CIB is still surprisingly high, comparable to, or exceeding, the integrated optical light from the galaxies in the Hubble Deep Field (Hauser et al., 1998).

When compared to the cosmic history of star formation derived from optical and UV surveys (Madau et al., 1996) a serious discrepancy became apparent; the CIB detected by COBE requires at least a factor of two more star formation than was apparent in optical and UV surveys, meaning that the integrated star formation rate at $z \sim 1.5$ must be higher than that implied from UV/optical observations by a comparable factor, and that this star formation must be largely surrounded by dust. COBE however could provide little further constraint on the form of the star formation history implied by the CIB; in principle there could be large numbers of galaxies at high redshift that are faint in the IR, or a few extremely IR-luminous sources[7]. Further progress required that the CIB be resolved into its constituent sources.

9.5.2 Resolving the CIB: LIRGs and ULIRGs at $0 < z < 1.5$

The first major steps in resolving the CIB came from extragalactic surveys carried out by ISO at 7μm and 15μm with ISOCAM, and at 90μm and 170μm with ISOPHOT, most notably observations of the HDF (Oliver et al., 1997; Rowan-Robinson et al., 1997), the European Large Area ISO Survey (ELAIS; Oliver et al., 2000; Rowan-Robinson et al., 2004), and the FIRBACK survey (Puget et al., 1999; Dole et al., 2001) (see also reviews in Elbaz, 2005; Verma et al., 2005; Oliver & Pozzi, 2005). The 15μm surveys (Elbaz et al., 2002) were particularly successful; the sources seen in these surveys could, with considerable extrapolation of the SED shape at longer wavelengths, account for around 80% of the CIB. Followup observations showed these sources have $\langle z \rangle \sim 0.8$, and that the comoving density of infrared light due to these 15μm sources is at least 40

[7]Though even at this stage source count models that could explain the CIB all invoked a population of high-redshift IR-luminous starbursts (Guiderdoni et al., 1998; Blain et al., 1999a; Rowan-Robinson, 2001; Xu et al., 2001)

times greater at $z \sim 1$ than in the local Universe (compare this to the B band luminosity density, which is only about three times the local value at $z \sim 1$). Not to be outdone, the FIRBACK survey found that the 170μm source counts show very strong evolution with redshift (reaching z$\sim$1), directly resolve 5% of the CIB (Dole et al., 2001), and are responsible for a dramatic rise in the integrated star formation rate, with a value at least ten times that seen locally at $z \sim 1$ (Rowan-Robinson et al., 1997; Flores et al., 1999; Pozzi et al., 2004). Overall, therefore, the ISO deep surveys put the IRAS discovery of strong IR galaxy evolution onto a very firm footing. Out to z$\sim$1 the ISO sources are chiefly LIRGs, not ULIRGs, and they are generally similar in many ways to lower redshift LIRGs, although possibly with lower average metallicity (Franceschini et al., 2003b; Elbaz, 2005; Liang et al., 2004).

The 170μm surveys also resulted in another interesting discovery; a large number of the 170μm-selected objects have cooler dust temperatures and larger dust masses than those seen in starburst galaxies selected at mid-IR wavelengths. This is not unexpected for a long wavelength-selected survey, since a large luminosity from a cool dust component requires a larger dust mass than a warmer source. ISO however was the first observatory to show conclusively that large masses of 'cool' dust existed in many galaxies, from local spirals to distant ULIRGs. In local spirals this cold "cirrus" component is expected to be diffuse dust heated by the interstellar radiation field from later type stars. In ULIRGs however this cold component could alternatively be a compact dusty starburst with colder than average dust. This discovery has interesting implications for studies of ULIRGs at higher redshifts, as we will discuss in Sect. 9.5.3.

Focusing on ULIRGs; the largest ISO survey, and therefore the one most likely to discover ULIRGs in any number, was ELAIS (see Fig. 9.1). The most sensitive ELAIS band for LIRGs and ULIRGs was 15μm, resulting in the detection of just under 100 ULIRGs, comprising >10% of the 15μm sample. Around 10 of these are HLIRGs; the first of these to be identified being a QSO at z=1.01 (Morel et al., 2001). These ULIRGs range in redshift up to $z \gtrsim 3$ and many are fit well in colour by an Arp 220-like SED. The longer wavelength ELAIS surveys result in many fewer sources due to decreased sensitivities; (Taylor et al., 2005) have identified 4 likely ULIRGs in the FIRBACK-ELAIS N2 (FN2) 170μm population, while (Dennefeld et al., 2005) identify 1 or 2 in the FIRBACK-ELAIS N1 (FN1) field. (Sajina et al., 2003) identified a population of z$\sim$0.5–1 ULIRGs, representing $\sim 1/6$ of the total 170μm FN1 sample, and Chapman et al. (2002) identified two of these at z=0.5 and 0.9 to be unusually cold systems with merger morphologies. Several sources in each field remained optically unidentified and are likely to be ULIRGs at moderate redshifts (0.5–1). Several ULIRGs and HLIRGs have been discovered in other ISO surveys, including a z=1.7 FeLoBAL QSO behind the z=0.56 cluster J1888.16CL (Duc et al., 2002), and a radio detected ERO at z=1.5 (Pierre et al., 2001), probably a QSO with very hot dust.

Follow-up of $z \gtrsim 0.5$ ULIRGs revealed that many are involved in ongoing interactions, but that interestingly a significant number are QSOs with very massive ($> 2L^*$) elliptical host galaxies (Farrah et al., 2002a). Detailed SED modeling (Rowan-Robinson & Crawford, 1989; Rowan-Robinson, 2000; Verma et al., 2002; Farrah et al., 2002b) showed evidence for both starburst and AGN activity in most sources, with colossal implied star formation rates of up to $\sim 1000 M_{\odot}$ yr^{-1}. X-ray observations however found unexpectedly weak X-ray emission, implying either no obscured AGN or extremely high obscuration levels (Wilman et al., 1998); the latter interpretation is favored by deeper observations which found evidence for Compton thick AGN in some (but not all) of these systems (Wilman et al., 2003). Very recently, Iwasawa et al. (2005) have used XMM to detect faint X-ray emission with a hint of the 6.4 keV Fe Kα line from F15307+3252, which they attribute to a Compton thick AGN with L(2–10kev)$> 10^{45}$ erg/s, which can account for a significant fraction of the infrared luminosity.

The picture that emerged from ISO, therefore, was one where very IR-luminous galaxies become substantially more numerous with increasing redshift, making $z \sim 1$ LIRGs and ULIRGs a cosmologically significant population. The $0.5 < z < 1.0$ LIRG population seems quite similar to LIRGs at lower redshift, but also revealed a population of moderate redshift LIRGs with relatively cool dust emission. The higher redshift ULIRG population exhibits a more prevalent level of QSO activity than found in their local cousins, a result that could be either luminosity or redshift related since the most luminous ULIRGs are found at the higher redshifts.

9.5.3 Resolving the CIB: ULIRGs at $z > 1.5$

The CIB had another startling surprise in store. Though ISO had proven itself remarkably adept at resolving much of the CIB, there remained a significant shortfall between the CIB as measured by COBE, and the sources detected by ISO. The background levels at $\lambda > 200\mu$m determined by FIRAS onboard COBE (Fixsen et al., 1998) implied a population of "colder" sources, probably $z \gtrsim 1$ systems with IR emission redshifted to longer wavelengths than the sources detectable in shallower, $\lambda < 200\mu$m, ISO surveys.

Fortunately, the perfect instrument to search for just this kind of source had just been commissioned at the James Clerk Maxwell Telescope, namely the Sub-millimetre Common User Bolometer Array (SCUBA, Holland et al. 1999). SCUBA heralded the advent of bolometer arrays for sub-mm observations, and was one of the first sub-mm instruments that could map large areas of sky relatively quickly in the sub-mm; operating at substantially longer wavelengths than ISO, namely 850μm and 450μm, SCUBA had the sensitivity to detect these distant 'cold' ULIRGs. Furthermore, the strongly negative k-correction at sub-mm wavelengths (Blain & Longaie, 1993) meant that a ULIRG at $z = 5$ is (depending on the choice of cosmology) almost as easy to detect as a ULIRG of comparable luminosity at $z = 1$. Other important sub-mm bolometer arrays include the 350μm optimized SHARC-II camera and the 1.1mm Bolocam instrument, both at the Caltech Sub-mm Observatory, and the MAx-Planck Millimetre Bolometer (MAMBO) at the Institut de Radioastronomie Millimetrique (IRAM) 30m telescope, which operates at 1.2mm.

The advent of sub-mm array instruments therefore prompted a plethora of surveys try to find these distant "cold" sources, ranging from ultra-deep surveys of lensing cluster fields (Smail, Ivison & Blain, 1997; Smail et al., 2002; Ivison et al., 2000) through deep blank-field surveys of small areas (Hughes et al., 1998; Barger et al., 1998), to wider area blank-field surveys to shallower depths (Eales et al., 2000; Scott et al., 2002; Borys et al., 2003; Greve et al., 2004; Laurent et al., 2005). These surveys found a huge population of sub-mm bright, optically faint sources, with source counts of 320^{+80}_{-100} and 180 ± 60 per square degree at $S_{850} > 10$mJy (Scott et al., 2002). Taken together, the sources found in the various sub-mm & mm surveys can directly account for around 50% of the CIB detected by FIRAS, and with reasonable extrapolation, can account for all of it (Barger et al., 1999b, Blain et al., 1999b).

The first step was to determine whether these sub-mm sources really are distant ULIRGs, or some other class of object. If, as is very plausible, the sub-mm emission from distant galaxies can be modeled with a simple modified blackbody, where the dust temperature distribution is characterised by T, then the flux at frequency ν, $F(\nu)$, is given by $F(\nu) \propto \nu^{\beta} B(\nu, T)$, where β is the spectral index (also referred to as the emissivity), and $B(\nu, T)$ is the Planck function. As a virtue of the shape of this function, sources with a redshift of $z \gtrsim 1$, a dust temperature of $\gtrsim 25$K, an 850μm flux of $S_{850} > 1$mJy, and a reasonable choice of emissivity will, under most circumstances, be ULIRGs. Simple photometric redshift estimates (Bertoldi et al., 2000; Fox et al., 2002; Borys et al., 2004; Webb et al., 2003) place virtually all the sources from sub-mm surveys at $z > 1$, and most at $z > 2$, meaning that the sub-mm sources are likely to be extremely luminous ULIRGs, with star formation rates substantially exceeding 1000$M_{\odot}$ yr^{-1}. This implies that there are of order a few hundred ULIRGs per square degree at $z > 1$, compared to about one ULIRG every four square degrees locally; though contributing less than 1% of the total extragalactic background light locally, ULIRGs were therefore shown to be a major, perhaps even dominant contributor at $z > 1$.

Before proceeding further, there is an important cautionary note. A single sub-mm flux cannot be used to infer dust mass or source luminosity without assuming a dust grain size and temperature distribution (Hildebrand, 1983), thus, for example, large cold disks can be confused with compact warmer starbursts of very different bolometric luminosity (Kaviani, Haehnelt & Kauffmann, 2003; Efstathiou & Rowan-Robinson, 2003; Farrah et al., 2004a; Almaini et al., 2005). If the redshift is also unknown, a much lower luminosity foreground cold disk or even a Galactic dust cloud could easily mimic a high-redshift ULIRG (Lawrence, 2001). As we describe below, however, there is now good evidence that at least $\sim$70% of sub-mm sources are indeed $z > 1$ ULIRGs, but within the remaining 30% there remains the possibility of significant sample contamination. Moreover, a strong temperature selection bias does exist in these very long-wavelength surveys; for example, a ULIRG at $2 < z < 3$ with $L_{ir} \sim 10^{13} L_{\odot}$ and a dust temperature of 60K would not be detected in the wide, shallow surveys with SCUBA (Blain et al., 2004b). Such "hot" sources do appear to exist (Chapman et al., 2004b; Lutz et al., 2005), and could be responsible for up to 1/3 of the

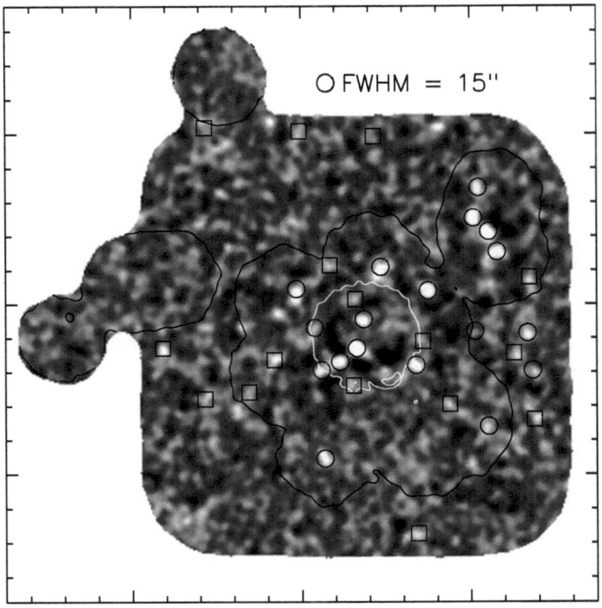

Fig. 9.4. An example of a recent submm survey; in this case the HDF-N SCUBA "Supermap" (Borys et al., 2003). circles = $> 4\sigma$, squares = $3.5 - 4D\sigma$. The white line encompasses the Hughes et al. (1998) SCUBA-HDF observations.

DIRBE detected CIB. Therefore, sub-mm surveys alone do not necessarily provide a hyaline view of dust-shrouded star formation, or even of ULIRG activity, at high redshift. Spitzer is proving to be extremely important for discovering the warmest high-redshift ULIRGs (see Fig. 9.1), as we shall describe further below.

Due to the coarse angular resolution of SCUBA (and indeed most sub-mm & mm observatories up to now) and the optical faintness of most of these sources, determining reliable ID's for follow-up proved to be extremely difficult (see Dunlop et al. 2004, for a perfect example). Nevertheless, since 1997, progress has slowly but surely been made in following up Sub-mm Galaxies (SMGs), and a clear picture of their nature has now started to emerge. A prime question is: how well do these distant systems resemble local ULIRGs? We would not necessarily expect them to be closely similar in nature because conditions are very different at high redshift, notably an earlier stage in the clustering evolution of the underlying dark matter, and larger gas fractions.

Optical and near-IR imaging (Smail et al., 1999; Webb et al., 2003; Smail et al., 2004; Pope et al., 2005) has shown that SMGs have a diverse range in optical properties, ranging from optically bright sources, to sources undetected in even the deepest optical imaging. Radio observations, particularly with the VLA, provided a major step forward in followup of SMGs (Ivison et al., 2002), detecting around 70% of the SMGs found in wide-field surveys with sufficiently high spatial resolution to allow reliable identification of optical/near-IR counterparts. Later, very high resolution radio studies have shown, intriguingly, that the radio emission in a surprisingly high fraction of SMGs, around 70%, is spatially extended on scales of ~ 10 kpc (Chapman et al., 2004a), significantly larger than the dust emission seen in most local ULIRGs, which are generally less than a kpc across (Downes & Solomon, 1998). The beam shape, especially when beam smearing is accounted for, is probably not well enough known to provide much evidence on the intrinsic shape of these radio sources, so possible explanations include large scale star formation over large disks, multiple compact starburst sites in a large disk or a coalescing group, or to AGN jet-induced star formation as seen in some distant radio galaxies (van Breugel et al., 2004). Another class of z>2 large scale systems inferred to have ULIRG-level luminosities are the so-called Lyα blobs (Geach et al., 2005).

Early efforts to obtain spectroscopic redshifts for SMGs had been mostly stymied (but not entirely, see Barger et al., 1999a), not only by uncertain IDs and extreme optical faintness, but also because many SMGs lie within the so-called redshift 'desert', $1 < z < 2$, within which the bandpasses of most optical spectrographs could only sample weak absorption features, from which redshifts are hard to determine. But with high resolution radio ID's in hand, and significant tenacity, spectroscopic redshifts for SMGs were soon forthcoming (Chapman et al., 2005; Simpson et al., 2004). These surveys showed that SMGs have a median redshift of 2.4, removing any last lingering doubts that the (radio detected) SMGs were indeed high-redshift ULIRGs. The optical spectra themselves show evidence for both starbursts and AGN in most sources, high unobscured star formation rates, and that SMGs are generally metal rich (Swinbank et al., 2004). Very recent optical spectroscopic studies have focused on integral field spectroscopy, though instrumentation is only now becoming capable of integral field studies of distant SMGs. Early results however look very promising; the host galaxies of SMGs are metal rich, and appear to contain surprisingly large numbers of evolved stars and show evidence for ongoing interactions (Tecza et al., 2004; Swinbank et al., 2005).

With accurate redshifts it was possible to use heterodyne instruments (which generally have a very narrow bandpass, making spectroscopic redshifts essential for meaningful observations) to look at far-IR fine structure lines, particularly those of CO. CO surveys of SMGs (Genzel et al., 2003; Neri et al., 2003; Greve et al., 2005) revealed that they have enormous masses (often exceeding $\sim 1 \times 10^{11} M_{\odot}$) of molecular gas, usually unresolved but in some cases extended over 3–5kpc (Genzel et al., 2003), much greater than those masses seen in local ULIRGs and comparable to those seen in high-redshift radio galaxies. HST imaging shows that SMGs often possess disturbed morphologies, consistent with (but not solely supportive of) ongoing major mergers, similar to the local ULIRG population (Smail et al., 1998; Conselice et al., 2003). Very recent results based on deep X-ray observations have provided convincing evidence that the bulk of SMGs harbour a Compton thick AGN in addition to a starburst, implying that a sub-mm galaxy signposts a major period of growth of the central supermassive black hole in these systems, as well as an intense starburst (Alexander et al., 2005). There is some evidence that the periods of black hole growth and stellar mass buildup in these systems may not be coeval (Borys et al., 2005), although there are uncertainties in the derivation of the X-ray obscuring columns, and therefore of the black hole mass (Polletta et al., 2006).

Further results have come from the first observations of SMGs with Spitzer. Results from IRAC and 24μm imaging (Egami et al., 2004; Frayer et al., 2004) have shown that SMGs have a wide range of observed mid-IR colours, and that their rest-frame mid-IR SEDs can in most cases be fitted with a starburst SED template, with the remainder being well fitted by a power-law AGN-dominated template. Photometric redshifts estimated from the near-infrared hump peaking at rest-frame 1.6μm and detected in the Spitzer IRAC bands (3.6–8.0μm) span $1 < z < 3.5$. Ivison et al. (2004) observed 9 MAMBO 1200μm sources in the same region, finding 75% of them to be starburst-dominated using Spitzer mid-infrared spectral shapes (assuming redshifts ~2.5), and concluding that the more AGN-like SEDs are consistent with AGN indications in these objects from UV/optical spectroscopy and X-ray imaging.

Spitzer will be a very powerful tool for finding large populations of ULIRGs at $z > 1.5$, with many hundreds of candidates appearing in the Spitzer surveys such as SWIRE (Lonsdale et al. 2006), the Bootes Shallow field, the First Look Survey, the GTO deep surveys and GOODS (see Fig. 9.1). Moreover Spitzer complements the cold dust selection function of the sub-mm and mm surveys with its mid-IR selection function and enhanced sensitivity to warm systems.

The first source counts and luminosity functions at 24μm, 70μm and 160μm from Spitzer extend the results seen by ISO. Spitzer is revealing large numbers of systems with significantly more extreme infrared/optical luminosity ratio than seen in the local Universe (Rowan-Robinson et al. 2005), a result also known from moderate redshift ISO surveys (Rowan-Robinson et al. 2004; Oliver & Pozzi 2005). Rowan-Robinson et al. (2005) also find a significant population of very cool luminous Spitzer systems, which could be very large disks with quiescent star formation rates rather than starbursts. The 24μm counts in the GOODS fields (Papovich et al., 2004) show evidence for strong evolution, exceeding the 'no evolution' counts predictions by a factor of at least 10, and implying evolution in the comoving IR energy density of the form $(1 + z)^{3.9\pm0.4}$ up to $z \sim 1$ (Le Flóch et al., 2005). The Spitzer counts at 70μm and 160μm (Dole et al., 2004) directly resolve 20% and 7% of the CIB at these wavelengths, respectively, and also show strong evolution, by a factor of ~ 3 over no evolution models, implying that the galaxies responsible for this background mostly lie in the redshift range $0.7 < z < 0.9$. Pérez-González et al. (2005) used find that LIRGs and ULIRGs are increasingly important contributors to the infrared energy density as redshift increases to z~3, being responsible for half of all star formation by z~1.5. The characteristic luminosity of the luminosity function, L^*, increases steadily with z, consistent with the cosmic star formation rate density going as $(1 + z)^4$ to z= 0.8, then flattening somewhat, and with ULIRGs playing a rapidly increasing role above z=1.3. Le Flóch et al. (2005) came to similar conclusions out to z~1, and also compared the UV and IR star formation history since z=1 using deep observations from Spitzer and GALEX, finding that the SFH has evolved much more strongly in the IR than in the UV, thus confirming the ISO-based results of Rowan-Robinson et al. (1997); Flores et al. (1999); Pozzi et al. (2004).

Yan et al. (2005) and Houck et al. (2005) have reported the first Spitzer low resolution IRS spectroscopy for extreme IR/optical z∼2 ULIRG candidates, demonstrating that Spitzer is capable of determining mid-infrared redshifts for the brightest mid-infrared galaxies ($f_{24} > 0.75$mJy). The space density of $L_{ir} > 10^{12.3} L_{\odot}$ z∼2 Spitzer mid-infrared-selected ULIRGs may be similar to that of SMGs (Yan et al.) at this redshift, though, the sample sizes are as yet small and incomplete. Yan et al. infer a higher fraction of starburst-dominated ULIRGs in their sample than Houck et al., which is probably due to the additional selection criterion on high 24/8μm colour (*ie* a rejection of mid-IR warm sources). Lutz et al. (2006) report a low 1.2mm detection rate these systems using MAMBO, and suggest that their mid-infrared-selected sample may have significantly warmer dust than submm-selected samples. Franceschini et al. (2004), Martinez-Sansigre et al. (2005), Donley et al. (2005) and Polletta et al. (2006) have used Spitzer surveys to discover extensive populations of heavily obscured AGN populations that can have very high X-ray obscuring columns, including two z>2, Compton-thick SWIRE HLIRGs with AGN torus-dominated mid-IR SEDs (Polletta et al., 2006). These new Spitzer results indicate that in many ULIRGs highly obscured QSOs are extremely difficult to detect at any wavelength, including hard X-ray and mid-infrared, due to extreme optical depths.

Daddi et al. (2005) have reported the first Spitzer results from the Great Observatories Origins Deep Survey (GOODS) for z∼2 ULIRGs, finding that typical massive ($M_{stellar} \sim 10^{11} M_{\odot}$) star forming galaxies at this redshift are ULIRGs, based on 24μm and radio detections, with a co-moving space density of ULIRGs at z=2 at least 3 orders of magnitude greater than the local one.

In summary it would be fair to say that surveys with SCUBA and other sub-mm/mm observatories, and their followup, have revolutionized our perceptions of ULIRGs, while Spitzer stands poised to contribute its own revolution of knowledge about them. From being little more than an interesting oddity in the local Universe, ULIRGs have become a crucially important population at $z > 1$. Furthermore, whilst $z > 1$ ULIRGs are similar in many ways to their local counterparts - both populations apparently being heavily dust-obscured starbursts and/or AGN triggered by interactions between large, evolved systems - there are important differences. High-redshift ULIRGs are substantially more gas rich than their local counterparts, and some may have extended rather than compact starbursts. Moreover the star formation rates implied (if AGN do not strongly dominate the energetics) are enormous, pushing the limits on theoretical ideas for a "maximal" starburst in a massive galaxy: $M_{gas}/t_{dyn} = 10^{11}$ $M_{\odot}$ / 10^8 yrs ∼1000 $M_{\odot}$/yr (Eggen, Lynden-Bell & Sandage, 1962). To understand these differences, and place these high-redshift ULIRGs into a cosmological context, we must turn to theories of galaxy and large-scale structure formation.

9.6 ULIRGs and Large-Scale Structure

The observation that these distant ULIRGs appear to be very massive systems means that large amounts of raw material are needed to make one, implying that ULIRGs are likely to be found in regions with large concentrations of baryons. Therefore, in order to understand when, where and why ULIRGs exist, we must turn to galaxy and large-scale structure formation models to see where such large concentrations of baryons are predicted to be found. Modern structure formation models describe the evolution of the total mass distribution (comprising both non-baryonic and baryonic 'cold' dark matter) via the evolution of initially Gaussian density fluctuations, which means that the evolution of the total mass distribution should be traced by the formation and evolution of galaxies in some determinable way. Density fluctuations in the dark matter distribution, commonly referred to as 'halos', are predicted to undergo successive mergers over time to build halos of steadily increasing mass. Galaxies are predicted to form from the baryonic matter in these halos, driven either by inter- or intra-halo mergers, or passive collapse of baryons to the halo central regions, or some combination of all three mechanisms (Meza et al., 2003; Kobayashi, 2004). This framework of 'biased' hierarchical buildup (Cole et al., 2000; Granato et al., 2000; Hatton et al. 2003), coupled with a Λ cosmology (Spergel et al., 2003), has proven to be remarkably successful in explaining many aspects of galaxy and large-scale structure formation.

Within this framework, the largest concentrations of baryons are expected to be found within the most massive dark matter halos. A basic verification of this idea can be seen in the local Universe, where the most massive elliptical galaxies are almost invariably found within very rich galaxy clusters; which presumably correspond to large overdensities of baryons in the local Universe. The formation history of these very massive galaxies, even before the advent of sub-mm surveys, was controversial. The 'naive' expectation from hierarchical theory would be that massive galaxies form relatively late and over a long period of time, representing successive mergers between dark matter halos, building up the required large baryon reservoirs (Baugh et al., 1998), and indeed some massive galaxies do appear to form in this way (van Dokkum et al., 1999; Tamura & Ohta, 2003; Bell et al., 2004). There is however observational evidence that many massive galaxies may form at high redshift and on short timescales, for example from the discovery of evolved systems at high redshift (Dunlop et al., 1996; Martini, 2001; Blakeslee et al., 2003), the extremely tight K-band Hubble relation for radio galaxies (De Breuck et al., 2002) implying a very rapid early formation of the most massive galaxies (Rocca-Volmerange et al., 2004), and an inferred early star formation epoch from the colour-magnitude diagram in clusters at $z = 0.5$ (Ellis et al., 1997), implying that the stars in very massive local galaxies formed within a few Gyr of each other at $z > 1$.

The SMGs and Spitzer high z ULIRG samples appear to be perfect candidates for the formation events for those local massive ellipticals that appeared to form very early and very fast; they lie at $z \simeq 2.5$ (Simpson et al., 2004; Chapman et al., 2005), which is a high enough redshift to make stars with the ages seen locally, their star formation rates are high enough to make all the stars in a local massive elliptical fast enough to satisfy the observed colour-scaling relationships, and their comoving number density is comparable to that of local $\gtrsim 3L^*$ ellipticals (Scott et al., 2002; Chapman et al., 2004b). Serious theoretical difficulties remain however. Early CDM-based galaxy formation models lacked the detailed astrophysics to explain the number of observed SMGs and their star formation rates (Granato et al., 2000; Somerville, Primack & Faber, 2001; Guiderdoni et al., 1998). Later models included improved treatments of starbursts, but still had to include further modifications to explain the observed sub-mm counts. These include a top-heavy initial mass function for the stars formed in bursts, so that the same sub-mm flux can be produced with lower star formation rates (Baugh et al., 2005)[8], modified treatments of virialization and the survival of subhalos to solve the 'overmerging' problem in the standard Press-Schechter formalism and make the SFRs high enough (van Kampen et al., 1999, 2005), and modifications to gas cooling and supernova feedback (Granato et al., 2004).

Fundamentally, there are two testable basic predictions of these hierarchical models for high-redshift ULIRGs. Firstly, these ULIRGs should reside within rich environments, that is to say we should see clustering of other galaxies around the ULIRGs on length scales of < 5Mpc, representing galaxy overdensities resulting from the local overdensity of baryons. Testing this observationally is, however, difficult. At $z < 1$ there have been numerous successful measurements of galaxy overdensities around AGN using deep optical/near-IR imaging to find evolved systems (Wold et al., 2001, 2003), but at the mean redshift of high-redshift ULIRGs it is certainly not clear whether their local environments have built up overdensities of evolved galaxies, even if they are overdense in baryons. Secondly, we should observe the high-redshift ULIRGs themselves to cluster together strongly on the sky, or in other words clustering of just the high-redshift ULIRGs on length scales $\gtrsim 25$Mpc. This prediction is a consequence of the assumed form of the evolution of density fluctuations with redshift; rare fluctuations in the underlying mass distribution are predicted to cluster together on the sky, with the strength of clustering depending on the degree of rarity (Bardeen et al., 1986). Very massive halos (those with masses $\gtrsim 10^{14} M_{\odot}$), which are universally predicted to be extremely rare at high redshift, are therefore predicted to cluster together very strongly (e.g. Moscardini et al. 1998; Kauffmann et al. 1999; Mo & White 2002). This clustering should be traced by very massive galaxies, as only the most massive halos contain sufficient baryon reservoirs to form these galaxies. This also means that the clustering strength of massive galaxies should translate straightforwardly to measure of halo mass (Percival et al., 2003). Strong clustering among high-redshift ULIRGs is also predicted if, as some suspect, they are part of a population uniting QSOs and spheroids (Magliocchetti et al., 2001). Measuring this form of clustering observationally is however also difficult, re-

[8]An interesting conundrum from this model is that while the sub-mm counts are predicted to be dominated by bursts, these bursts would be responsible for making only a small fraction of the stellar mass in evolved ellipticals in the local Universe if their IMFs are top heavy.

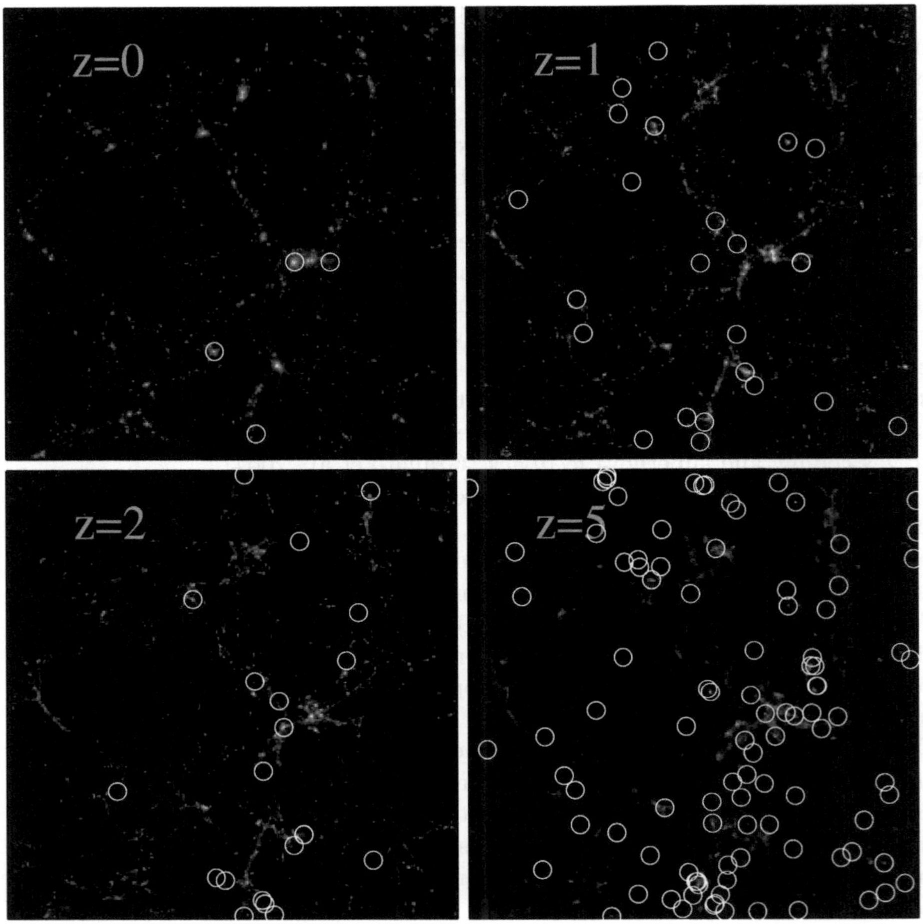

Fig. 9.5. An example of the cosmological hydrodynamic ΛCDM simulation (SPH G6 run) described in Nagamine et al. (2005a,b). Each panel has a comoving size of 143Mpc on a side, and the starforming galaxies with instantaneous SFR > 100 $M_\odot$/yr at each epoch are indicated by the circles (K. Nagamine, priv. comm.)

quiring either $\gtrsim 50$ sources with spectroscopic redshifts so that 'associations' in redshift space between different sources can be searched for, or a minimum of several hundred sources with a reasonably well constrained redshift distribution, and found over a large enough area of sky such that the length scales of interest can be examined.

Despite these difficulties, numerous studies of both the environments of high-redshift ULIRGs, and their clustering, have taken place, and while these studies are not (yet) conclusive, they do generally lend support to the idea that high-redshift ULIRGs form within massive dark matter halos. Considering environment studies first; near-IR imaging studies of the environmental richness of ULIRGs over $0.5 < z < 1.5$ (Farrah et al., 2004b) have shown that these systems reside in richer environments than their local counterparts on average, and that some reside in moderate rich clusters even at $z \gtrsim 1$, though it is not clear whether this represents a genuine increase in environmental richness for ULIRGs, or whether this just reflects the global increase in 'bias' over $0 < z < 1$. At lower redshifts, LIRGs and ULIRGs have also been found in clusters (Duc et al., 2004; Lemonon et al., 1998; Kleinmann et al., 1988). Turning to higher redshifts, then several authors have noted overdensities of sub-mm sources around high-redshift systems that are thought to reside within mass overdensities, including $z \sim 4$ radio galaxies (Stevens et al., 2003; De Breuck et al., 2004) (several of these sub-mm sources are also detected in the X-ray, suggesting the presence of obscured AGN (Smail et al., 2003) and seen as $z \sim 3$ Lyman Break galaxies (Chapman et al., 2001; Geach et al., 2005).

Turning to high-z ULIRG clustering; early efforts to measure angular clustering on the sky came up against the limitations of available instrumentation; at the time of writing there exists no sub-mm array instrument that can map large enough areas of the sky to the required depths (though SCUBA2, APEX and Herschel, are coming soon). No clear picture has therefore emerged; some sub-mm surveys have uncovered tentative hints of clustering on scales of a few arcminutes (Scoville et al., 2000), whereas others show no signs of clustering whatsoever (Borys et al., 2003). An intriguing result from these wide field surveys is that the high-redshift ULIRGs appear to trace moderately bright X-ray survey sources on the sky, even though the direct overlap between the two populations is minimal (Almaini et al., 2003). This could suggest that the high-redshift sub-mm and X-ray populations are tracing the same, overdense dark matter halos. More recent efforts to measure SMG clustering using spectroscopic redshifts have met with greater success, resulting in a reliable detection of clustering for the first time (Blain et al., 2004a). The strength of the clustering, while significant, is not particularly strong, at $r_0 = 6.9 \pm 2.2$ (comoving). This clustering strength is consistent with high-redshift ULIRGs residing in $\sim 10^{13} M_{\odot}$ halos which will eventually become the cores of rich clusters in the local Universe, but the relatively high redshift of the sources in this sample, coupled with the significant error on their clustering and the unknown growth modes of dark matter overdensities (see e.g. Matarrese et al. 1997; Moscardini et al. 1998) leaves other possible interpretations open. More precise measurements of both high-redshift ULIRGs, and other ULIRGs at lower (and higher) redshifts are needed to resolve this issue. The new Spitzer ULIRG surveys, sampling much larger volumes and linear dimensions than current submm/mm surveys, can be expected open up this field of enquiry dramatically.

9.7 Open Questions and New Directions

An incredible amount of energy has been spent trying to understand ULIRGs in the decade since the Sanders & Mirabel (1996) review, and this review has not been able to do adequate justice to all this effort. To summarize the current state of knowledge we address the a few of the most pressing questions from our own perspective.

In the relatively local Universe, ULIRGs are almost all associated with major mergers of gas rich systems. Evolutionary schemes whereby ULIRGs transition into optical QSOs at the end stages of a merger, with an elliptical as the end result, are highly attractive and well supported by simulations. The reality is obviously complex, however, depending strongly on the details of the merger encounter, in particular bulge mass. Observationally it is not necessarily the case that all ULIRGs represent late stages in the merger sequence, however the most luminous systems do seem to represent the late stages, implying that it requires the extraordinary high density conditions of the final merger stages to produce these tremendous levels of power. Evidence that some optical QSOs may have more massive hosts than ULIRGs, or inhabit richer environments, would indicate that the true situation is more complex, but matching AGN/ULIRG samples for such studies is very challenging.

Merger sequences connect starbursts and AGN tightly together via the availability of fuel to build the stellar and SMBH masses, and to fuel the AGN. AGN orientation schemes also tie ULIRGs and QSOs tightly together since off-axis QSOs will be obscured and may appear as ULIRGs. Therefore the empirical determination of the relative power of AGN and starbursts for both infrared-selected ULIRGs and optically-selected ULIRG QSOs is a paramount question. Most ULIRGs are thought to be powered mostly by starbursts however indicators of AGN activity increase with luminosity, recognised from near- and mid-infrared spectroscopy, X-ray detections or VLBI radio imaging. All of these indicators are tricky however and, because nuclear (Starburst) and gravitational accretion (AGN) power often compete for dominance we need more robust estimators of the SFR and AGN activity to compare with merger stage, morphology, mass, etc. to disentangle an evolutionary sequence from other effects. For example, low/high-resolution mid-IR spectroscopic indicators can be obscured; Spitzer will do an excellent job of measuring diagnostic features from 5–40μm, but interpretation will require an understanding of extinction as a function of wavelength throughout the spectrum as well as modeling with full radiative transfer treatment. X-ray columns can be high enough to severely limit AGN detection; separating soft (starburst or reflected AGN) vs hard (AGN or X-ray binary) emission will require sensitivity at higher energies with Con-X and XEUS. High-resolution radio imaging has to be capable of mapping at brightness temperatures above 10^6K to distinguish jet structures from RSN complexes, because starburst and radio-quiet AGN emission cannot be distinguished on the basis of power alone. But we do not yet even understand the complex radio supernovae in the one, closest, well studied system, Arp 220, where questions remain about the nature of the LRSN, the IMF in the luminous starburst, etc. Such imaging observations are time consuming, but need to be expanded to much larger samples. One possible line of inquiry may be to obtain BH mass estimates from the molecular gas kinematics with ALMA ($r \sim 20$pc) as has been done for NGC 4258 in H_2O. Other questions remain, *eg.*: What are the compact masers telling us? Are they pointers to AGN/mass concentrations?

ULIRGs increase dramatically in importance with redshift, from being a very minor component at z∼0 to possibly dominating the energy density at z>2, and ULIRG populations at z>2 are much more extreme beasts than at low redshift, with larger IR luminosities, gas masses and, in some cases, larger ratios of infrared-to-optical luminosity than local ULIRGs, by factors of 10 to 100. They also show an increase in space density of a factor of 100–1000 (Chapman et al., 2005; Le Flóch et al., 2005; Pérez-González et al., 2005), though this number is very uncertain because the low and high redshift LFs barely overlap in luminosity range; at low redshift the number density of HLIRGs is intrinsically extremely small, while at high redshifts, submm surveys are only sensitive *only* to systems above a few $\times 10^{12} L_{\odot}$. Wide area Spitzer luminosity functions can be expected to bridge this gap (Lonsdale et al. 2006b, Babbedge et al. 2006). There is clear evidence that at high redshift many ULIRGs are triggered by mergers of gas rich systems, but there are also some striking differences from local systems: extended radio emission and some evidence for very large cool disk systems, which may not even qualify as starbursts, but rather star formation ocurring in a more-or-less quiescent fashion in these very large disks.

As noted above, the incidence of AGN seems to increase with luminosity, and this is also a strong redshift effect since the most luminous systems are only known at higher redshifts. As also noted above, it remains remarkably difficult to determine the relative power contribution from star formation vs. AGN activity in many ULIRGs. Many lines of argument are used to conclude that star formation powers most ULIRGs at all redshifts. These include the energy reservoir available from nucleosynthesis relative to SMBH accretion; low AGN power deduced from absorption-corrected hard X-ray detections; the number of obscured AGN required to explain the XRB and the correlation between bulge and SMBH masses which suggest bulge growth events must accompany SMBH growth. Arguments for a stronger importance for AGN power include that luminosities above $10^{13} L_{\odot}$ are very extreme for a starburst event while many high-redshift QSOs are known with such luminosities; a large number of QSO-powered ULIRGs is expected in AGN unification-by-orientation schemes; hard X-ray absorption correction is difficult, and Compton thick AGN such as Mrk 231 are very difficult to detect at high redshift.

Perhaps one of the strongest lines of evidence one way or the other comes from the impressive correlation between warm dense molecular gas traced by the HCN molecule with infrared luminosity (Gao & Solomon, 2004), which extends across 4 orders of magnitude in luminosity from 10^9 to 10^{13}. There's little doubt that low luminosity systems are powered by disk star formation, so this relation provides support for the picture that even HLIRGs can be powered by star formation. HCN ULIRG detections are limited at high luminosity/redshift, however, and they tend to be displaced above the lower luminosity redshift relation towards excess L_{IR}/M_{HCN} (Carilli et al., 2005), which is consistent with the presence of an an additional, AGN, power source but is also within the scatter of the lower redshift starburst relation. Spitzer will be able to provide deep insights into this debate because the presence of a significant AGN power source will be evident the mid-infrared spectral shape, as demonstrated by the recent flurry of papers reporting the discovery of large populations of highly obscured QSO/ULIRGs (Martinez-Sansigre et al., 2005; Polletta et al., 2006; Yan et al., 2005; Houck et al., 2005; Donley et al., 2005).

The causes for the dramatic differences in ULIRG number density and nature since z$\sim$3 must lie with the evolving matter density field and galaxy formation processes. Since interactions and mergers are clearly implicated in most ULIRG activity, the most natural assumption to make is that these objects follow 'the action'. They are, after all, the most dramatic active events that occur in the Universe, the prima donnas of the show. At a given epoch they are likely to occur where the most dramatic gas rich mergers are happening, as illustrated in the simulation of K. Nagamine (private communication) in Fig. 9.5. At the earliest epochs of galaxy formation the action is at the sites of future rich clusters, while at current times loose groups are ideal future ULIRG sites since there are a few close galaxies with moderate relative velocity, and gas has not been exhausted yet. At intermediate redshifts, the best ULIRG sites may be the interaction zones in cluster-cluster mergers. These ideas are consistent with the sparse currently available observations of ULIRG environments at various redshifts.

So what are the key unanswered questions and where do we go from here to answer them? So far ULIRG samples are quite sparse - numbering in the hundreds over a huge redshift range. We clearly need much larger, volume-limited samples. Moreover there are serious selection biases since high-redshift systems have been selected at sub-mm wavelengths by their cool dust, and substantially warmer systems will have been missed. Therefore we need to understand the full range of ULIRG SEDs and obtain complete samples with well measured SEDs via selection from the mid-IR to the sub-mm. Spitzer will clearly revolutionize this field, bringing in the warm-samples to complement the sub-mm ones, but also selecting PAH-dominated objects as the mid-IR PAH features redshift into the 24μm band. Larger, deeper sub-mm & mm surveys are also required to reach down the luminosity function at z>1 and combat the effects of cosmic variance.

To complement the ULIRG samples it's obviously important to study complete QSO samples in the same volumes of space and to determine the lifetimes of both types of object, to directly investigate their evolutionary relationships. A tricky aspect to this work will be the fact that ULIRG and QSO luminosities are predicted by the most recent simulations to vary very dramatically on short timescales (as are SMBH and stellar masses), especially as the SMBH building and AGN accretion phase accelerates, so matching samples by luminosity or mass is probably going to match apples and oranges.

Key to understanding the role of ULIRGs in galaxy formation - their connections to other kinds of systems, their progenitors and their descendents, the connections between bulge and SMBH building and the relative importance of starburst *vs* quiescent star formation modes – is to determine the host galaxy properties of ULIRGs and QSOs, especially stellar masses and morphologies, and to determine the richness of their environments and their clustering properties as a function of redshift.

The future holds great promise for answering these questions. Measuring the clustering of ULIRGs requires wide area far-IR/sub-mm surveys, which are or will be forthcoming from Spitzer, SCUBA-2, APEX, ASTRO-F, Herschel and WISE. These surveys will also provide the large samples of ULIRGs needed to measure the range in dust temperatures in the distant ULIRG populations, with detailed followup provided by other groundbased facilities. Measuring sizes, morphologies, and stellar masses, assessing the impact of gravitational lensing and determining environmental richness, on the other hand, require deep pointed observations in the near-IR to radio. Here the key facilities are Spitzer, JWST, ALMA, the EVLA and SKA. Imaging in molecular and other submm/mm lines will also be possible with ALMA.

Acknowledgements

We are deeply indebted to Colin Borys, Jason Surace, Kentaro Nagamine, C. Kevin Xu & Lee Armus who provided original figures from their recent work for this review. The Spitzer IRS ULIRG team who responsible for the spectra shown and described in Sect. 9.3 include B. Thomas Soifer, James Houck, Lee Armus, Vassilis Charmandaris, Henrik Spoon, Jason Marshall, Jeronimo Bernard-Salas & Sarah Higdon. CJL and HES are indebted to Yu Gao and the organizers of the Lijiang Workshop on "Extreme Starbursts" which contributed greatly to the focus of this review.

References

Alexander, D. M., et al. 2005, Nature, **434**, 738
Almaini, O., et al. 2003, MNRAS, **338**, 303
Almaini, O., Dunlop, J. S., Willott, C. J., Alexander, D. M., Bauer, F. E., & Liu, C. T. 2005, MNRAS, **358**, 875
Armus, L., Heckman, T., & Miley, G. 1987, AJ, **94**, 831
Armus, L., Heckman, T., Miley, G. 1988, ApJ, **326**, L45
Armus, L., et al. 2004, ApJS, **154**, 178
Armus, L., et al. 2005, in preparation
Baan, W. A., Wood, P. A. D., & Haschick, A. D. 1982, ApJL, **260**, L49
Baan, W. A. 1989, ApJ, **338**, 804
Baan, W. A., Haschick, A. D., & Henkel, C. 1989, ApJ, **346**, 680
Babbedge, T. S. R., Rowan-Robinson, M., Lonsdale, C. J., et al. 2006, MNRAS, submitted
Baldwin, J. A., Phillips, M. M., & Terlevich, R. 1981, PASP, **93**, 5
Balestra, I., Boller, Th.,Gallo, L., Lutz, D., & Hess, S. 2005, A&A, in press (astro-ph/0507321)
Bardeen, J. M., Bond, J. R., Kaiser, N., & Szalay, A. S. 1986, ApJ, **304**, 15
Barger, A. J., et al. 1998, Nature, **394**, 248
Barger, A. J., Cowie, L. L., Smail, I., Ivison, R. J., Blain, A. W., & Kneib, J.-P. 1999a, AJ, **117**, 2656
Barger, A. J., Cowie, L. L., & Sanders, D. B. 1999b, ApJL, **518**, L5
Barger, A. J., et al. 2005, AJ, **129**, 578
Barnes, J. E. 1989, Nature, **338**, 123
Barnes, J. E., & Hernquist, L. 1992, ARA&A, **30**, 705

Barnes, J. E., & Hernquist, L. 1996, ApJ, **471**, 115
Baugh, C. M., Cole, S., Frenk, C. S., & Lacey, C. G. 1998, ApJ, **498**, 504
Baugh, C. M., et al. 2005, MNRAS, **356**, 1191
Beichman, C. A., et al. 1986, ApJL, **308**, L1
Bell, E. F., et al. 2004, ApJ, **608**, 752
Bertoldi, F., et al. 2000, A&A, **360**, 92
Blain, A. W. & Longair, M. S. 1993, MNRAS, **264**, 509
Blain, A. W., Smail, I., Ivison, R. J., & Kneib, J.-P. 1999a, MNRAS, **302**, 632
Blain, A. W., Kneib, J.-P., Ivison, R. J., & Smail, I. 1999b, ApJL, **512**, L87
Blain, A. W., Chapman, S. C., Smail, I., & Ivison, R. 2004a, ApJ, **611**, 725
Blain, A. W., Chapman, S. C., Smail, I., & Ivison, R. 2004b, ApJ, **611**, 52
Blakeslee, J. P., et al. 2003, ApJL, **596**, L143
Boller, T., Bertoldi, F., Dennefeld, M., & Voges, W. 1998, A&AS, **129**, 87
Boller, T., Gallo, L. C., Lutz, D., & Sturm, E. 2002, MNRAS, **336**, 1143
Bondi, M., Pérez-Torres, M.-A., Dallacasa, D., & Muxlow, T. W. B. 2005, MNRAS, 575
Borne, K. D., Bushouse, H., Lucas, R. A., & Colina, L. 2000, ApJL, **529**, L77
Borys, C., Chapman, S., Halpern, M. & Scott, D. 2003, MNRAS, **344**, 385
Borys, C., Scott, D., Chapman, S., Halpern, M., Nandra, K., & Pope, A. 2004, MNRAS, **355**, 485
Borys, C., et al. 2005, ApJ, in press (astro-ph/0507610)
Braito, V., et al. 2004, A&A, **420**, 79
Bryant P. & Scoville, N. 1999, AJ, **117**, 2632
Burbidge, G. R. & Stein, W. A. 1970, ApJ, **160**, 573
Bushouse, H. A., et al. 2002, ApJS, **138**, 1
Canalizo, G. & Stockton A. 2001, ApJ, **555**, 719
Carilli, C. L., Wrobel, J. M., & Ulvestad, J. S. 1998, AJ, **115**, 928
Carilli, C. L., et al. 2005, ApJ, **618**, 586
Cesarsky, C. J., et al. 1996, A&A, **315**, L32
Chapman, S. C., Lewis, G., Scott, D., Richards, E., Borys, C., Steidel, C., Adelberger, K., & Shapley, A. 2001, ApJL, **548**, L17
Chapman, S. C., Smail, I., Ivison, R. J., Helou, G., Dale, D. A., & Lagache, G. 2002, ApJ, **573**, 66
Chapman, S. C., Smail, I., Windhorst, R., Muxlow, T., & Ivison, R. J. 2004a, ApJ, **611**, 732
Chapman, S. C., Smail, I., Blain, A. W., & Ivison, R. J. 2004b, ApJ, **614**, 671
Chapman, S. C., Blain, A. W., Smail, I. & Ivison, R. J. 2005, ApJ, **622**, 772
Chary, R., et al. 2004, ApJS, **154**, 80
Clavel, J., et al. 2000, A&A, **357**, 839
Clegg, P. E., et al. 1996, A&A, **315**, L38
Clements, D. L., Sutherland, W. J., McMahon, R. G., & Saunders, W. 1996, MNRAS, **279**, 477
Clements, D. L., et al. 2002, ApJ, **581**, 974
Cole, S., Lacey, C. G., Baugh, C. M., & Frenk, C. S. 2000, MNRAS, **319**, 168
Colina, L., et al. 2001, ApJ, **563**, 546
Colina, L., Arribas, S., & Clements, D. 2004, ApJ, **602**, 181
Condon, J. J., Huang, Z.-P., Yin Q. F. & Thuan, T. X. 1991, ApJ, **378**, 65
Conselice, C. J., Chapman, S. C., & Windhorst, R. A. 2003, ApJL, **596**, L5
Cutri, R., Huchra, J., Low, F., Brown, R., & vanden Bout, P. 1994, ApJL, **424**, L65

Cutri, R., Nelson, B., Francis, P. & Smith, P. in "AGN Surveys", ASP Conf. Serire, **284**, 127
Daddi, E., et al. 2005, ApJ, **626**, 680
Darling, J., & Giovanelli, R. 2002, AJ, **124**, 100
De Breuck, C., van Breugel, W., Stanford, S. A., Röttgering, H., Miley, G., & Stern, D. 2002, AJ, **123**, 637
De Breuck, C., et al. 2004, A&A, **424**, 1
de Graauw, T., et al. 1996, A&A, **315**, 49
de Grijp, M. H. K., Miley, G. K., Lub, J. & de Jong, T. 1985, Nature, **314**, 240
de Jong, T., Klein, U., Wielebinski, R. & Wunderlich, E. 1985, A&A, **147**, L6
Dennefeld, M., et al. 2005, A&A, in press (astro-ph/0504344)
Di Matteo, T., Springel, V. & Hernquist, L. 2005, Nature, **433**, 604
Diamond, P. J., Norris, R. P., Baan, W. A., & Booth, R. S. 1989, ApJL, **340**, L49
Diamond, P. J., Lonsdale, C. J., Lonsdale, C. J., & Smith, H. E. 1999, ApJ, **511**, 178
Dole, H., et al. 2001, A&A, **372**, 364
Dole, H., et al. 2004, ApJS, **154**, 87
Donley, J., Rieke, G., Rigby, J. & Perez-Gonzalez 2005, ApJ, in press (astro-ph/0507676)
Dopita, M. A., Kewley, L. J., Heisler, C. A., & Sutherland, R. S. 2000, ApJ, **542**, 224
Dopita, M. A., Groves, B. A., Fischera, J. & Sutherland, R. S. 2005, ApJ, **619**, 755
Downes, D., & Solomon, P. M. 1998, ApJ, **507**, 615
Dubinski, J., Mihos, J. C., & Hernquist, L. 1999, ApJ, **526**, 607
Duc, P.-A., et al. 2002, A&A, **389**, L47
Duc, P., et al. 2004, IAU Colloq. 195: Outskirts of Galaxy Clusters: Intense Life in the Suburbs, 347
Dunlop, J. S., et al. 1996, Nature, **381**, 581
Dunlop, J. S., et al. 2003, MNRAS, **340**, 1095
Dunlop, J. S., et al. 2004, MNRAS, **350**, 769
Dwek, E., et al. 1998, ApJ, **508**, 106
Eales, S., et al. 2000, AJ, **120**, 2244
Efstathiou, A. Rowan-Robinson, M. 2003, MNRAS, **343**, 322
Egami, E., et al. 2004, ApJS, **154**, 130
Eggen, O., Lynden-Bell, D. & Sandage, A. 1962, ApJ, **136**, 748
Elbaz, D., et al. 2002, A&A, **384**, 848
Elbaz, D. 2005, to be published in 'ISO science legacy – a compact review of ISO major achievements', Space Science Reviews, Springer
Ellis, R. S., et al. 1997, ApJ, **483**, 582
Elston, R., Cornell, M. E., & Lebofsky, M. J. 1985, ApJ, **296**, 106
Elvis, M. 2000, ApJ, **545**, 63
Fabian, A. C. 1999, MNRAS, **308**, L39
Farrah, D., et al. 2001, MNRAS, **326**, 1333
Farrah, D., Verma, A., Oliver, S., Rowan-Robinson, M. & McMahon R. 2002a, MNRAS, **329**, 605
Farrah, D., Serjeant, S., Efstathiou, A., Rowan-Robinson M. & Verma A. 2002b, MNRAS, **335**, 1163
Farrah, D., et al. 2003, MNRAS, **343**, 585
Farrah, D., et al. 2004a, MNRAS, **349**, 518
Farrah, D., Fox, M., Rowan-Robinson, M., Clements, D., & Afonso, J. 2004b, ApJ, **603**, 489

Farrah, D., Surace, J. A., Veilleux, S., Sanders, D. B., & Vacca, W. D. 2005, ApJ, **626**, 70
Ferrarese, L. & Merritt, D. 2000, ApJL, **539**, L9
Fischer, J. et al. 1999, Ap&SS **266**, 91
Fixsen, D. J., Dwek, E., Mather, J. C., Bennett, C. L., & Shafer, R. A. 1998, ApJ, **508**, 123
Flores, H., et al. 1999, ApJ, **517**, 148
Fosbury, R. & Wall, J. 1979, MNRAS, **189**, 79
Fox, M. J., et al. 2002, MNRAS, **331**, 839
Franceschini, A., et al. 2003a, MNRAS, **343**, 1181
Franceschini, A., et al. 2003b, A&A, **403**, 501
Franceschini, A., et al. 2004, AJ, **129**, 2074
Frayer, D. T., et al. 2004, ApJS, **154**, 137
Gallagher, S. C., Brandt, W. N., Chartas, G., Garmire, G. P., & Sambruna, R. M. 2002, ApJ, **569**, 655
Gao, Y. & Solomon, P. 2004, ApJ, **606**, 271
Geach, J., et al. 2005, MNRAS, in press (astro-ph/0508357)
Gebhardt, K., et al. 2000, ApJ, **539**, 13
Gehrz, R. D., Sramek, R. A. & Weedman, D. W. 1983, ApJ, **267**, 551
Genzel, R., et al. 1998, ApJ, **498**, 579
Genzel, R., & Cesarsky, C. J. 2000, ARA&A, **38**, 761
Genzel, R., Tacconi, L. J., Rigopoulou, D., Lutz, D., & Tecza, M. 2001, ApJ, **563**, 527
Genzel, R., et al. 2003, ApJ, **584**, 633
Goldader, J. D., et al. 2002, ApJ, **568**, 651
González-Alfonso, E. Smith, H., Fischer, J. & Cernicharo, J. 2004, ApJ, **613**, 247
Granato, G. L., et al. 2000, ApJ, **542**, 710
Granato, G. L., De Zotti, G., Silva, L., Bressan, A., & Danese, L. 2004, ApJ, **600**, 580
Greve, T. R., et al. 2004, MNRAS, **354**, 779
Greve, T. R., et al. 2005, MNRAS, **359**, 1165
Guiderdoni, B., Hivon, E., Bouchet, F. R., & Maffei, B. 1998, MNRAS, **295**, 877
Haas, M., et al. 2003, A&A, **402**, 87
Hacking, P., Houck, J. & Condon, J. 1987, ApJL, **316**, L15
Hatton, S., Devriendt, J. E. G., Ninin, S., Bouchet, F. R., Guiderdoni, B., & Vibert, D. 2003, MNRAS, **343**, 75
Hatziminaoglou, E. & the SWIRE Team 2005, ApJ, **129**, 119
Hauser, M. G., et al. 1998, ApJ, **508**, 25
Hauser, M. G., & Dwek, E. 2001, ARA&A, **39**, 249
Heckman, T. M., Armus L. & Miley G. K. 1987, AJ, **93**, 276
Heckman, T. M., Dahlem, M., Eales, S. A., Fabbiano, G., & Weaver, K. 1996, ApJ, **457**, 616
Heckman, T. M. 2001, in "Gas and Galaxy Evolution", ASP Conf **240**, 345
Helou, G., Soifer, B. T., & Rowan-Robinson, M. 1985, ApJL, **298**, L7
Henkel, C., & Wilson, T. L. 1990, A&A, **229**, 431
Hildebrand, R. H. 1983, QJRAS, **24**, 267
Hines, D., Schmidt, G., Smith, P., Cutri, R. & Low, F. 1995, ApJL, **450**, L1
Holland, W. S., et al. 1999, MNRAS, **303**, 659
Hopkins, P. F., et al. 2005a, ApJL, **625**, L71
Hopkins, P. F., et al. 2005b, ApJ, **630**, 705
Hopkins, P. F., et al. 2005c, ApJ, **630**, 716
Hopkins, P. F., et al. 2005d, ApJ, **631**, in press (astro-ph/0504253)

Hopkins, P. F., et al. 2005e, ApJ, in press (astro-ph/0506398)
Houck, J. R., Schneider, D., Danielson, G., Neugebauer, G., Soifer, B. T., Beichman, C. & Lonsdale, C. J. 1985, ApJL, **290**, L5
Houck, J. R., et al. 2005, ApJL, **622**, L105
Hughes D. H., et al. 1998, Nature, **394**, 241
Hutchings, J. B., & Neff, S. G. 1991, AJ, **101**, 434
Imanishi, M., Terashima, Y., Anabuki, N. & Nakagawa, T. 2003, ApJ, **596L**, 167
Ivison, R. J., et al. 2000, MNRAS, **315**, 209
Ivison, R. J., et al. 2002, MNRAS, **337**, 1
Ivison, R. J., et al. 2004, ApJS, **154**, 124
Iwasawa, K., & Comastri, A. 1998, MNRAS, **297**, 1219
Iwasawa, K., Matt, G., Guainazzi, M., & Fabian, A. C. 2001, MNRAS, **326**, 894
Iwasawa, K., Sanders, D. B., Evans, A. S., Trentham, N., Miniutti, G., & Spoon, H. W. W. 2005, MNRAS, **357**, 565
Joseph, R. D. 1999, Ap&SS, **266**, 321
Joseph, R. D. & Wright G. S. 1985, MNRAS, **214**, 87
Kandalyan, R. A. 1996, Astrophysics, **39**, 237
Kandalyan, R. A. 2003, A&A, **404**, 513
Kashlinsky, A. 2005, Phys. Rep., **409**, 361
Kauffmann, G., Colberg, J. M., Diaferio, A., & White, S. D. M. 1999, MNRAS, **307**, 529
Kaviani, A., Haehnelt, M. G., & Kauffmann, G. 2003, MNRAS, **340**, 739
Kawada, M., et al. 1994, ApJL, **425**, L89
Kessler, M. F., et al. 1996, A&A, **315**, L27
Kewley, L. J., et al. 2000, ApJ, **530**, 704
Kewley, L. J., Dopita, M. A., Sutherland, R., Heisler, C. A., & Trevana, J. 2001, ApJ, **556**, 121
Kim, D.-C. & Sanders, D. B. 1998, ApJS, **119**, 41
Kim, D.-C., Veilleux, S., & Sanders, D. B. 1998, ApJ, **508**, 627
Klaas, U., et al. 2001, A&A, **379**, 823
Kleinmann, D. E. & Low, F. J. 1970, ApJL, **159**, L165
Kleinmann, S. G., et al. 1988, ApJ, **328**, 161
Klöckner, H., Baan, W. A., & Garrett, M. A. 2003, Nature, **421**, 821
Klöckner, H.-R., & Baan, W. A. 2004, A&A, **419**, 887
Kobayashi, C. 2004, MNRAS, **347**, 740
Komossa, S., Burwitz, V., Hasinger, G., Predehl, P., Kaastra, J. & Ikebe, Y. 2005, ApJL, **582**, L15
Kormendy, J. & Sanders, D. B. 1992, ApJL, **390**, L53
Kunth, D. & Sargent, W. L. W. 1981, A&A, **101**, L5
Lagache, G., Puget, J. & Dole, H. 2005, ARA&A, 43
Larson, R. B. & Tinsley, B. M. 1978, ApJ, **219**, 46
Laurent, O., Mirabel, I. F., Charmandaris, V., Gallais, P., Madden, S. C., Sauvage, M., Vigroux, L., & Cesarsky, C. 2000, A&A, **359**, 887
Laurent, G. T., et al. 2005, ApJ, **623**, 742
Lawrence, A. 2001, MNRAS, **323**, 147
Lawrence, A., Rowan-Robinson, M., Leech, K., Jones, D. H. P., & Wall, J. V. 1989, MNRAS, **240**, 329
Le Flóch, E., et al. 2005, ApJ, in press (astro-ph/0506462)
Leech, K. J., Rowan-Robinson, M., Lawrence, A., & Hughes, J. D. 1994, MNRAS, **267**, 253

Leipski, C., et al. 2005, A&A, **440**, L5
Leitherer, C., et al. 1999, ApJS, **123**, 3
Lemke, D., et al. 1996, A&A, **315**, L64
Lemonon, L., et al. 1998, A&A, **334**, L21
Liang, Y. C., Hammer, F., Flores, H., Elbaz, D., Marcillac, D., & Cesarsky, C. J. 2004, A&A, **423**, 867
Lim, J., & Ho, P. T. P. 1999, ApJL, **510**, L7
Lipari, S. 1994, ApJ, **436**, 102
Lipari, S., et al. 2003, MNRAS, **340**, 289
Lo, K.-W. 2005, ARA&A, 43
Lonsdale, C. J., Persson, S. E. & Matthews, K. 1984, ApJ, **287**, 95
Lonsdale, C. J., Hacking, P. B., Conrow, T. P. & Rowan-Robinson, M. 1990, ApJ, **358**, 60
Lonsdale, C. J., Smith, H. E., & Lonsdale, C. J. 1993, ApJL, **405**, L9
Lonsdale, C. J., Diamond, P. J., Lonsdale, C. J. & Smith, H. E. 1994, Nature, **370**, 117
Lonsdale, C. J., Smith, H. E., & Lonsdale, C. J. 1995, ApJ, **438**, 632
Lonsdale, C. J., Lonsdale, C. J., Diamond, P. J., & Smith, H. E. 1998, ApJL, **493**, L13
Lonsdale Carol J., Lonsdale Colin J., Smith, H. E., Diamond P. J. 2003a, ApJ, **592**, 804
Lonsdale, C. J., et al. 2003b, PASP, **115**, 897
Lonsdale, C. J., Diamond, P. J., Lonsdale, C. J., Smith, H. E., Rovilos, E. 2006a, ApJ, in preparation
Lonsdale, C. J., Polletta, M., Farrah, D., et al. 2006, ApJ, submitted
Low, F. J. & Kleinmann, D. E. 1968, AJ, **73**, 868
Luhman, M. L., et al. 2003, ApJ, **594**, 758
Lutz, D., Spoon, H. W. W., Rigopoulou, D., Moorwood, A. F. M., & Genzel, R. 1998, ApJL, **505**, L103
Lutz, D., Veilleux, S. & Genzel, R. 1999, ApJL, **517**, L13
Lutz, D., et al. 2005, ApJL, **625**, L83
Lutz, D., et al. 2006, ApJL, in press (astro-ph/0509058)
MacKenty, J. & Stockton, A. 1984, ApJ, **283**, 64
Madau, P., Ferguson, H. C., Dickinson, M. E., Giavalisco, M., Steidel, C. C., & Fruchter, A. 1996, MNRAS, **283**, 1388
Magliocchetti, M., Moscardini, L., Panuzzo, P., Granato, G. L., De Zotti, G., & Danese, L. 2001, MNRAS, **325**, 1553
Malhotra, S., et al. 1997, ApJL, **491**, L27
Malhotra, S., et al. 2001, ApJ, **561**, 766
Maloney, P. R., & Reynolds, C. S. 2000, ApJL, **545**, L23
Martinez-Sansigre, A. et al. 2005, Nature, in press (astro-ph/0505486)
Martini, P. 2001, AJ, **121**, 2301
Matarrese, S., Coles, P., Lucchin, F., & Moscardini, L. 1997, MNRAS, **286**, 115
McDowell, J. C., et al. 2003, ApJ, **591**, 154
Melnick, J., & Mirabel, I. F. 1990, A&A, **231**, L19
Meza, A., Navarro, J. F., Steinmetz, M., & Eke, V. R. 2003, ApJ, **590**, 619
Mihos, J. C. & Hernquist, L. 1996, ApJ, **464**, 641
Mirabel, I. F. & Sanders, D. 1988, ApJ, **335**, 104
Mirabel, I. F. 2002 in "Lighthouses of the Universe: The Most Luminous Celestial Objects and Their Use for Cosmology", p. 108
Miyoshi, M. et al. 1996, Nature, 373, 127

Mo, H. J., & White, S. D. M. 2002, MNRAS, **336**, 112
Momjian, E., Romney, J. D., Carilli, C. L., Troland, T. H., & Taylor, G. B. 2003, ApJ, **587**, 160
Morel, T., et al. 2001, MNRAS, **327**, 1187
Moscardini, L., Coles, P., Lucchin, F., & Matarrese, S. 1998, MNRAS, **299**, 95
Murphy, T. W., et al. 1996, AJ, **111**, 1025
Naab, T., & Burkert, A. 2003, ApJ, **597**, 893
Nagamine, K., Cen, R.-Y., Hernquist, L., Ostriker, J. & Springel, V. 2005a, ApJ, **618**, 23
Nagamine, K., Cen, R.-Y., Hernquist, L., Ostriker, J. & Springel, V. 2005a, ApJ, **627**, 608
Neff, S. G., Ulvestad, J. S., & Teng, S. H. 2004, ApJ, **611**, 186
Neri, R., et al. 2003, ApJL, **597**, L113
Netzer, H., et al. 2005, ApJ, **629**, 739
Norris, R. P., Kesteven, M. J., Troup, E. R., Allen, D. A., & Sramek, R. A. 1990, ApJ, **359**, 291
Oliver, S. J., Rowan-Robinson, M., & Saunders, W. 1992, MNRAS, **256**, 15P
Oliver, S. J., et al. 1996, MNRAS, **280**, 673
Oliver, S. J., et al. 1997, MNRAS, **289**, 471
Oliver, S. J., et al. 2000, MNRAS, **316**, 749
Oliver, S. J. & Pozzi, F. 2005, to be published in 'ISO science legacy - a compact review of ISO major achievements', Space Science Reviews, Springer
Osterbrock, D. E., & de Robertis, M. M. 1985, PASP, **97**, 1129
Osterbrock, D. E., Tran, H. D., & Veilleux, S. 1992, ApJ, **389**, 196
Papovich, C., et al. 2004, ApJS, **154**, 70
Partridge, R. B., & Peebles, P. J. E. 1967, ApJ, **148**, 377
Peeters, E., Spoon, H., & Tielens, A. 2004, ApJ, **613**, 986
Percival, W. J., Miller, L., McLure, R. J., & Dunlop, J. S. 2001, MNRAS, **322**, 843
Percival, W. J., Scott, D., Peacock, J. A., & Dunlop, J. S. 2003, MNRAS, **338**, L31
Pérez-González, P. G., et al. 2005, ApJ, **630**, 82
Pierre, M., et al. 2001, A&A, **372**, L45
Pihlstrom, Y., Conway, J., Booth, R., Diamond, P. & Polatidis, A. 2001, A&A, **377**, 413
Pihlstrom, Y. M., et al. 2005, ApJ, **618**, 705
Polletta, M., Courvoisier, T., Hooper, E. & Wilkes, B. 2000, A&A, **362**, 75
Polletta, M., et al. 2006, ApJ, submitted
Pope, A., Borys, C., Scott, D., Conselice, C., Dickinson, M., & Mobasher, B. 2005, MNRAS, **358**, 149
Pozzi, F., et al. 2004, ApJ, **609**, 122
Ptak A., Heckman T., Levenson N. A., Weaver K. & Strickland D. 2003, ApJ, **592**, 782
Puget, J.-L., et al. 1996, A&A, **308**, L5
Puget, J. L., et al. 1999, A&A, **345**, 29
Readhead, A., Taylor, G., Pearson, T. & Wilkinson, P. 1996, ApJ, **460**, 634
Rees, M. J., Silk, J., Warner, M. W. & Wickramasinghe, N. 1969, Nature, **223**, 788
Rieke, G. H. & Low F. J. 1972, ApJ, **176**, L95
Rieke, G. H. & Lebofsky, M. J. 1979, ARA&A, **17**, 477
Rieke, G. H. & Lebofsky, M. J. 1986, ApJ, **304**, 326
Rieke, G. H. 1988, ApJL, **331**, L5

Rigopoulou, D., Spoon, H., Genzel, R., Lutz D., Moorwood, A. & Tran Q. 1999, AJ, **118**, 2625
Risaliti, G., Gilli, R., Maiolino, R. & Salvati M. 2000, A&A, **357**, 13
Robertson, B., et al. 2005, ApJL, in press (astro-ph/0503369)
Rocca-Volmerange, B., Le Borgne, D., De Breuck, C., Fioc, M., & Moy, E. 2004, A&A, **415**, 931
Rovilos, E., Diamond, P. J., Lonsdale, C. J., Lonsdale, C. J., & Smith, H. E. 2003, MNRAS, **342**, 373
Rovilos, E., Diamond, P. J., Lonsdale, C. J., Smith, H. E., & Lonsdale, C. J. 2005, MNRAS, **359**, 827
Rowan-Robinson, M. & Crawford, J. 1989, MNRAS, **238**, 523
Rowan-Robinson, M., Hughes, J., Vedi, K., & Walker, D. W. 1990, MNRAS, **246**, 273
Rowan-Robinson, M., et al. 1991, Nature, **351**, 719
Rowan-Robinson, M., et al. 1997, MNRAS, **289**, 490
Rowan-Robinson, M. 2000, MNRAS, **316**, 885
Rowan-Robinson, M. 2001, ApJ, **549**, 745
Rowan-Robinson, M., et al. 2004, MNRAS, **351**, 1290
Rowan-Robinson, M., et al. 2005, AJ, **129**, 1183
Rupke, D. S., Veilleux, S., & Sanders, D. B. 2002, ApJ, **570**, 588
Rupke, D. S., Veilleux, S., & Sanders, D. B. 2005, ApJ, in press (astro-ph/0507037)
Sajina, A., et al. 2003, MNRAS, **343**, 1365
Sakamoto, K., Scoville, N., Yun, M., Crosas, M., Genzel, R. & Tacconi, L. 1999,ApJ, **514**, 68
Sánchez, S. F., et al. 2004, ApJ, **614**, 586
Sanders, D. et al. 1986, ApJL, **305**, L45
Sanders, D. B., et al. 1988a, ApJ, **325**, 74
Sanders, D. B., Soifer, B. T., Elias, J. H., Neugebauer, G. & Matthews K. 1988b, ApJ, **328**, L35
Sanders, D., Scoville, N. & Soifer, B. T. 1991, ApJ, **370**, 158
Sanders, D. B. & Mirabel, I. F. 1996, ARA&A, **34**, 749
Satyapal, S., Sambruna, R. M., & Dudik, R. P. 2004, A&A, **414**, 825
Saunders, W., et al. 1990, MNRAS, **242**, 318
Schlegel, D. J., Finkbeiner, D. P., & Davis, M. 1998, ApJ, **500**, 525
Scoville, N., Yun, M. & Bryant, P. 1997, ApJ, **484**, 702
Scoville, N. Z., et al. 2000, AJ, **119**, 991
Scott, S. E., et al. 2002, MNRAS, **331**, 817
Shier, L. M., Rieke, M. J., & Rieke, G. H. 1996, ApJ, **470**, 222
Simpson, C., et al. 2004, MNRAS, **353**, 179
Smail, I., Ivison, R. J., & Blain, A. W. 1997, ApJL, **490**, L5
Smail, I., Ivison, R. J., Blain, A. W., & Kneib, J.-P. 1998, ApJL, **507**, L21
Smail, I., et al. 1999, MNRAS, **308**, 1061
Smail, I., Ivison, R. J., Blain, A. W., & Kneib, J.-P. 2002, MNRAS, **331**, 495
Smail, I., Scharf, C. A., Ivison, R. J., Stevens, J. A., Bower, R. G., & Dunlop, J. S. 2003, ApJ, **599**, 86
Smail, I., Chapman, S. C., Blain, A. W., & Ivison, R. J. 2004, ApJ, **616**, 71
Smith, H. E., Lonsdale, C. J., & Lonsdale, C. J. 1998, ApJ, **492**, 137
Smith, H. E., Lonsdale, C. J., Lonsdale, C. J. & Diamond P. J. 1998, ApJL, **493**, L17
Soifer, B. T., et al. 1984, ApJL, **278**, L71
Soifer B. T., et al. 1984, ApJL, **283**, L1
Soifer, B. T. & Neugebauer, G. 1991, AJ, **101**, 354

Soifer, B. T., et al. 1987, ApJ, **320**, 238
Soifer, B. T., et al. 1999, ApJ, **513**, 207
Soifer, B. T., et al. 2000, AJ, **119**, 509
Soifer, B. T., Neugebauer, G., Matthews, K., Egami, E., & Weinberger, A. 2002, AJ, **124**, 2980
Solomon, P., Downes, D. & Radford, S. 1992, ApJL, **387**, L55.
Solomon, P., Downes, D., Radford, S. & Barrett, J. 1997, ApJ, **478**, 144
Somerville, R. S., Primack, J. R., & Faber, S. M. 2001, MNRAS, **320**, 504
Spergel, D. N., et al. 2003, ApJS, **148**, 175
Spinoglio, L., Andreani, P., & Malkan, M. A. 2002, ApJ, **572**, 105
Spoon, H. et al. 2002, A&A, **385**, 1022
Spoon, H. et al. 2004, ApJS, **154**, 184
Springel, V., & Hernquist, L. 2005, ApJL, **622**, L9
Springel, V., Di Matteo, T. & Hernquist, L. 2005, MNRAS, **361**, 776
Stanford, S., Stern, D., van Breugel, W. & De Brueck, C. 2000, ApJS, **131**, 185
Stevens, J. A., et al. 2003, Nature, **425**, 264
Strauss, M., Davis, M., Yahil, A., & Huchra, J. 1990, ApJ, **361**, 49
Sturm, E., et al. 2002, A&A, **393**, 821
Sugai, H., & Malkan, M. A. 2000, ApJ, **529**, 219
Surace, J. A., Sanders, D. B., Vacca, W. D., Veilleux, S. & Mazzarella, J. M. 1998, ApJ, **492**, 116
Surace, J. A., Sanders, D. B., & Evans, A. S. 2000, ApJ, **529**, 170
Surace, J. A., and the SWIRE Team 2005, ApJ, in press
Swinbank, A. M., Smail, I., Chapman, S. C., Blain, A. W., Ivison, R. J., & Keel, W. C. 2004, ApJ, **617**, 64
Swinbank, A. M., et al. 2005, MNRAS, **359**, 401
Tacconi, L. J., et al. 2002, ApJ, **524**, 732
Tacconi, L. J., et al. 2002, ApJ, **580**, 73
Tamura, N., & Ohta, K. 2003, AJ, **126**, 596
Taylor, E., et al. 2005, MNRAS, **361**, 1352
Tecza, M., et al. 2004, ApJL, **605**, L109
Teng, S., et al. 2005, ApJ, in press (astro-ph/0508112)
Toomre, A., & Toomre, J. 1972, ApJ, **178**, 623
Tran Q. D., et al. 2001, ApJ, **552**, 527
Turner, T., George, I., Nandra, K., & Mushotzky, R. 1998, ApJ, **493**, 91
Ulvestad, J. S., Wrobel, J. M., & Carilli, C. L. 1999, ApJ, **516**, 127
Vaccari, M. et al. 2005, MNRAS, **358**, 397
van Breugel, W., Fragile, C., Anninos, P. & Murray, S. 2004, IAU Symp, **271**, 472
van der Kruit, P. C. 1971, A&A, **15**, 110
van Dokkum, P. G., Franx, M., Fabricant, D., Kelson, D. D., & Illingworth, G. D. 1999, ApJL, **520**, L95
van Gorkum, J. & Hibbard, 2005, ARA&A, 43
van Kampen, E., Jimenez, R., & Peacock, J. A. 1999, MNRAS, **310**, 43
van Kampen, E., et al. 2005, MNRAS, **359**, 469
Vanden Bout, P. & Solomon, P. 2005, ARA&A, 43
Veilleux, S., & Osterbrock, D. E. 1987, ApJS, **63**, 295
Veilleux, S., Kim, D.-C., Sanders, D. B., Mazzarella, J. M., & Soifer, B. T. 1995, ApJS, **98**, 171
Veilleux, S., Sanders, D. B. & Kim D.-C. 1997, ApJ, **484**, 92
Veilleux, S., Kim, D.-C., & Sanders, D. B. 1999a, ApJ, **522**, 113

Veilleux, S., Sanders, D. B., & Kim, D.-C. 1999b, ApJ, **522**, 139
Veilleux, S., Kim, D.-C. & Sanders, D. B. 2002, ApJS, **143**, 315
Veilleux, S., Cecil, & Bland-Hawthorn, J. 2005, ARA&A, 43
Verma, A., Rowan-Robinson, M., McMahon, R. & Efstathiou, A. 2002, MNRAS, **335**, 574
Verma, A., et al. 2005, to be published in 'ISO science legacy - a compact review of ISO major achievements', Space Science Reviews, Springer
Vignati, P., et al. 1999, A&A, **349**, L57
Voit, G. M., Weymann, R. J. & Koirsta, K. T. 1993. ApJ, **413**, 95
Webb, T. M., et al. 2003, ApJ, **597**, 680
Weiler, K. W., Panagia, N., & Sramek, R. A. 1990, ApJ, **364**, 611
Wilman, R. J., Fabian, A. C., Cutri, R. M., Crawford, C. S., & Brandt, W. N. 1998, MNRAS, **300**, L7
Wilman, R. J., Crawford, C. S., & Abraham, R. G. 1999, MNRAS, **309**, 299
Wilman, R. J., Fabian, A. C., Crawford, C. S., & Cutri, R. M. 2003, MNRAS, **338**, L19
Wisotzki, L., et al. 2000, A&A, **358**, 77
Wold, M., Lacy, M., Lilje, P. B., & Serjeant, S. 2001, MNRAS, **323**, 231
Wold, M., Armus, L., Neugebauer, G., Jarrett, T. H., & Lehnert, M. D. 2003, AJ, **126**, 1776
Worsley, M.A. et al. 2005, MNRAS, **357**, 1281
Xia, X. Y., Xue, S. J., Mao, S., Boller, T., Deng, Z. G., & Wu, H. 2002, ApJ, **564**, 196
Xu, C. K., Lonsdale, C. J., Shupe, D. L., O'Linger, J. & Masci, F. 2001, ApJ, **562**, 179
Xu, C. K., Lonsdale, C. J., Shupe, D. L., Franceschini, A., Martin, C., & Schiminovich, D. 2003, ApJ, **587**, 90
Yan, L., et al. 2005, ApJ, **628**, 604
Yun, M. S. & Scoville, N. Z. 1998, ApJ, **507**, 774
Yun, M. S., Reddy, N. A., & Condon, J. J. 2001, ApJ, **554**, 803
Zou, Z., Xia, X., Deng, Z., & Su, H. 1991, MNRAS, **252**, 593

10 The Role of Galactic Winds in Galaxy Formation

C.L. Martin

Abstract: Astrophysical processes such as gas cooling and feedback, i.e. gas reheating from stars and quasars, render galaxy formation quite complicated compared to large-scale structure formation. Empirical calibrations of the star formation rate and supernova-feedback efficiency have proven critical for simulating the formation and evolution of galaxies in cosmological models. The measured properties of galactic winds, relevant to these recipes, are reviewed here. Observations from X-ray, ultraviolet, optical, and infrared observatories yield important information due to the complex, multiphase nature of galactic outflows. This contribution also examines how these measurements constrain dynamical models of outflows. Possible implications of this, still evolving, dynamical understanding of galactic winds are discussed with regard to shaping the galaxy luminosity function and polluting the intergalactic medium with heavy elements.

10.1 The Astrophysics of Galaxy Formation

Gravity defines the landscape for structure formation in the Universe. By retarding the expansion of overdense regions, gravity amplifies the tiny inhomogeneities present in the early Universe creating sheets and filaments of overdense regions that wrap around the more rapidly expanding voids. The distribution of intergalactic gas clouds (Croft et al. 2002), the galaxy-galaxy correlation function (Percival et al. 2002), and the power spectrum of the microwave background (Spergel et al. 2003) reveal how mass clusters over a range of length and redshift scales. The ΛCDM cosmogony is consistent with the measured spectrum of mass fluctuations and predicts that galaxy halos assemble hierarchically – massive halos assemble at lower redshift than do low mass halos (Lacey & Cole 1993).

These cosmological aspects of structure formation do not, however, describe the colors, sizes, shapes, or even luminosities of galaxies. These artifacts of a galaxy's past star formation history and merger history offer important clues to how galaxies were assembled. Galaxies turn out to present a broad range of star formation histories that depend on both environment and mass. Evolution is accelerated in rich environments where galaxies contain old stellar populations dominated by low mass, and therefore red, stars (Dressler 1980). Many stars in local spiral galaxies formed 6-10 Gyr ago, but the consumption of interstellar gas was slow enough that significant star formation continues today (Kennicutt 1983). In poor environments, the baryons in many dwarf galaxies are still composed primarily of gas, and the star formation rate has probably been fairly constant over cosmic time. Look-back studies show that both star formation and nuclear activity were stronger at $z \sim 1-2$ than today and took place in higher mass objects (Cowie et al. 1996; Heckman et al. 2004). This dependence on galactic mass appears, at first sight, orthogonal to the hierarchical assembly picture and is often called *downsizing.*

This absence of late-forming, massive galaxies in the centers of galaxy groups and clusters is only part of the problem encountered by simple models of star formation in a ΛCDM universe. Figure 10.1 illustrates the number density discrepancy. The data points for the shape of the galaxy luminosity function do not match the shape to the halo mass distribution denoted by the dashed line. The density of galaxies drops off exponentially for luminosities larger than the Milky Way, while the halo abundance falls off more slowly (Somerville et al. 2002; Benson et al. 2003; Croton et al. 2005). And, although more galaxies are found at lower luminosities, the number of dwarf galaxies does not increase as rapidly as

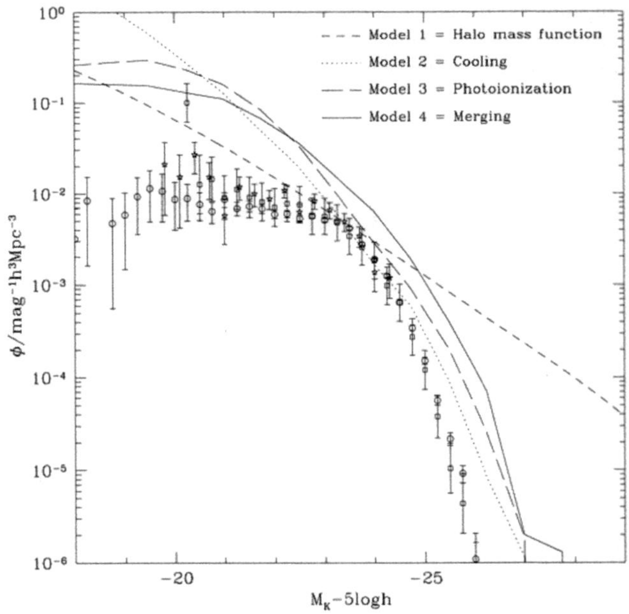

Fig. 10.1. K-band luminosity function of galaxies. Model 1 (dashed line) shows the result of converting the dark matter halo mass function into a galaxy luminosity function by assuming a fixed mass-to-light ratio chosen to match the knee of the luminosity function. Model 2 (dotted line) shows the effect of gas cooling but no feedback or merging. Models 3 and 4 (long-dashed and solid lines, respectively) show the effects of adding photoionization and then galaxy merging. Reproduced from Benson et al. (2003).

the density of low-mass halos in ΛCDM. Choosing an appropriate stellar light to dark matter ratio makes the number of galaxies and halos agree at, and only at, a mass scale typical of the Milky Way. The characteristic cut-off luminosity is called L^*, where $dN = N_0(L/L^*)^{-\alpha}e^{-L/L^*}dL$ and dN is the number density of galaxies per luminosity interval dL.

Taking into account the long cooling time of gas in the most massive halos lessens the discrepancy at high luminosities but does not explain the sharp cut-off in the galaxy luminosity function (Thoul & Weinberg 1995; Benson et al. 2003; Croton et al. 2005). Models that fit the luminosity function include additional gas reheating processes in massive systems (Benson et al. 2003; Croton et al. 2005). In theory, the discrepancy at luminosities below L^* can be rectified by removing a significant amount of gas from the smaller galaxies (Dekel & Silk 1986). *Feedback* refers to physical processes like gas reheating and gas removal that prevent, or at least delay, additional star formation.

In starburst galaxies, the temporal and spatial correlation of supernovae from massive stars is expected to produce more efficient feedback than do lower concentrations of star formation in normal galaxies. As much as 30–50% of the star formation in the Universe occurs in starburst events – thought to be triggered by galaxy interactions. Feedback can have a particularly large effect on galaxy evolution during their assembly phase. Energy sources for this feedback include accretion onto supermassive black holes, supernova explosions, and even ultraviolet radiation from galaxies. This review focuses on how much of the feedback problem is solved by supernova. It examines mass scales from dwarf starbursts to ultraluminous infrared selected galaxies. The recent review by Veilleux et al. (2005) consider AGN feedback in more detail.

10.2 Basic Feedback Physics

The basic unit of feedback in starburst galaxies is the kinetic energy of a supernova explosion from a massive star, or 10^{51} ergs. (The energy carried away by the neutrinos is ignored since it does not couple to the surrounding gas.) The clustering of Type II supernova in time and space leads to a high porosity of hot gas with a long cooling time. Stellar winds make an important contribution to this mechanical power on the timescale of O-star lifetimes – i.e. a few Myr.

Population synthesis models sum the number of supernova per unit time over the stellar mass function. The stellar mass cut-off for supernova is much lower (~ 10 M$_\odot$) than that for hot-star winds (~ 20 M$_\odot$), however, so the supernova completely dominate the total energy injection on the ~ 10 Myr timescale of galactic outflows. It follows that the mechanical feedback on timescales of 10 Myr or more is not predicted to vary strongly with metallicity. For an instantaneous burst of star formation, the increasing number of stars at lower masses is almost exactly canceled by their increasing lifetime leading to a fairly constant $L_w(t)$ for

10.2.1 Feedback from massive stars & population synthesis models

To illustrate the amount of feedback from massive stars, the mechanical power predicted by a population synthesis model is compared to the stellar luminosity and ionizing luminosity – i.e. photons with $h\nu > 13.6$ eV. The results clearly depend on the relative distribution of stellar masses. The low mass stars, say $M < 8\ M_{\odot}$, can be ignored for estimating ionizing luminosity and supernova rate. The low mass cut-off does, however, significantly affect the normalization since most of the mass is in low mass stars. Figure 10.2 shows a model for a Salpeter initial mass function, slope $d\log N/dM = -2.35$, that covers the mass range 1.0 $M_{\odot}$ to 100 $M_{\odot}$. These curves were computed with Starburst 99 (Leitherer et al. 1999) for a constant star formation rate. Equilibrium values are reached after about 40 Myr when the death rate of the least massive stars that undergo core-collapse supernova reach their birth rate. The mechanical power available is only about 1% of the bolometric luminosity. One challenge is to determine what fraction of it is radiated away and unavailable to drive an outflow. Notice that more energy is available in the ionizing continuum, which reaches about 10% of the bolometric luminosity.

Momentum-driven outflows have received attention for just this reason. The momentum supplied from supernova and stellar winds can be estimated from population synthesis models like Starburst 99. The mass flux of stellar debris

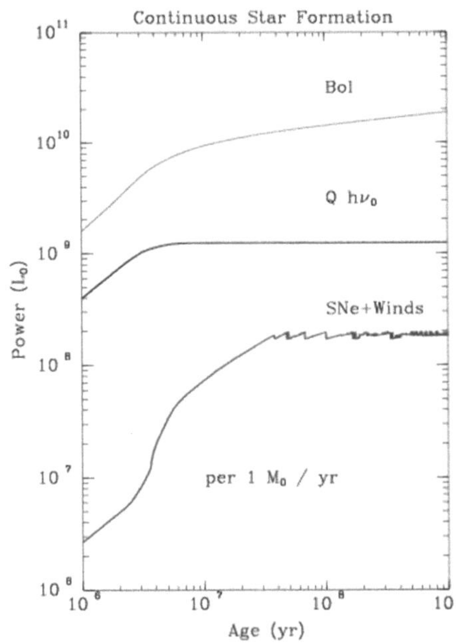

Fig. 10.2. Luminosity, ionizing luminosity, and mechanical power generated by a star formation rate of 1 $M_{\odot}\ yr^{-1}$. The stellar initial mass function has slope $d\log N/dM = -2.35$ and covers the mass range 1.0 $M_{\odot}$to 100 $M_{\odot}$. Model is from Starburst 99 (Leitherer et al. 99). The metallicity is solar.

and the mechanical power determine the ejecta velocity, $v_w^2 = 2L_w/\dot{M}_{ej}$. The total momentum is then $\dot{p}_{ej} = \dot{M}_{ej}v_w$. For the fiducial SB99 model above, this quantity reaches a maximum at 100 Myr which is about 2.5 times the momentum flux at an age of 10 myr. For reference, this value is

$$\dot{p} = 5 \times 10^{33} \text{ dyne}(SFR/1 \text{ M}_\odot \text{ yr}^{-1}). \tag{10.1}$$

The momentum in the radiation field is similar (Murray et al. 2005).

10.2.2 Hydrodynamics

Weaver et al. 1977 described the basic onion-skin structure of a wind-powered bubble. The supersonic outflow from a stellar wind or ensemble of supernova generates a shock wave about the time it sweeps up its own mass in ambient gas. This ambient gas will be compressed into a thin outer shell if it radiates. The hot, shocked supernova/wind ejecta fills most of the volume within this shell. The pressure of this bubble drives the shell outward. Expressing the solutions to this equation of motion in useful units, we have

$$R = 104 \text{ pc } L_{40}^{1/5} n^{-1/5} t_6^{3/5} \tag{10.2}$$

and

$$v = 61 \text{ kms}^{-1} L_{40}^{1/5} n^{-1/5} t_6^{-2/5} \tag{10.3}$$

for the shell radius and velocity, where the mechanical power is in units of 10^{40} ergs s^{-1}; and the age is in units of 1 Myr. An excellent review of more general solutions for blastwaves with power law luminosity $L(t)$ and density $n(r)$ functions can be found in Ostriker & McKee (Rev mod physics).

The mechanics of superbubbles rest on the idea that above a critical star formation rate, supernova remnants will overlap. The combined hot cavity drives an outflow much as an indvidual stellar wind bubble (Tenorio-Tagle & Bodenheimer 1989). The porosity of hot gas can be estimated from the supernova rate per unit volume (S), the lifetime of the SNR before merging with the general ISM (τ), and the size of the SNR at confinement (V), $f_{Hot} = 1 - e^{-Q}$, and $Q = VS\tau$ (McKee & Ostriker 1977). Empirically, outflows are seen when the local density exceeds about 0.1 M$_\odot$ yr^{-1}per square kpc.

The nature of the problem changes when the size of the bubble grows to a scale height (MacLow et al. 1989; de Young & Heckman 1994). This symmetry breaking is quite important. That the shock stalls in the plane of the disk prevents total disruption of the galaxy in many cases where it is energetically feasible (MacLow & Ferrara 1999; Martin 1998). Only after blow-out does a free-flowing wind develop (Chevailer and Clegg 1985). Hydrodynamic instabilities along the wind/disk interface, and at the accelerating shell, generate fragments of interstellar gas that are entrained in the hot wind (Heckman et al. 2000).

10.2.3 Outflows from active galactic nuclei

The properties of the gas accreted by galaxy groups and clusters are well-defined by the standard framework for hierarchical structure formation, but the measured relations reveal excess entropy compared to predictions (Babul et al. 2002). Scannapieco & Oh (2004) point out that the total energy available from all Type II supernovae is not sufficient to explain this cluster reheating – approximately 1 keV per baryon; but the bolometric luminosities of quasars are far above the levels necessary to preheat the ICM. Quasar outflows are potentially more important for structure formation than supernova-driven winds.

The ionizing radiation from quasars clearly has a critical impact on the physical state of the IGM. The prevalence of additional feedback in the form of jets and radiatively-driven winds is complicated by the complex relationships between the numerous classes of AGN; see Antonucci (1993) for a review. Roughly 3-10% of quasars exhibit strong, broad absorption in UV resonance lines, which are blueshifted relative to the center of the broad emission lines (Begelman et al. 1991). These statistics can be interpreted as all radio-quiet quasars hosting outflows with a 10% covering factor. Estimates of their mechanical power vary by a factor of 10 in either direction from L_{bol}. AGN-driven ouflows are common in Sey 1 galaxies, but outflows from the more common Sey 2 galaxies appear to be driven largely by the starburst in many cases (Levenson et al. 2001a,b; Rupke et al. 2005c). It remains uncertain on what spatial scale the AGN energy is deposited in the surrounding gas, something which has a large effect on radiative losses. AGN feedback of this type may be closely associated with starbursts.

Optical AGN and radio-galaxies may be independent phenomenon, triggered by different physical mechanisms (Croton et al. 2005, Best et al. 2005). The outflows from radio-loud quasars form collimated jets, which deposit particles into a cocoon that expands into the surrounding medium. The mechanical power can reach L_{Bol}. The fraction of galaxies that host radio-loud AGN, L(1.4 GHz) greater than 10^{23} WHz^{-1}, rises from nearly zero below a stellar mass of 10^{10} $M_\odot$ to more than 30% at 5×10^{11} $M_\odot$(Best et al. 2005). Considering this strong dependence on black hole mass, it is not surprising that these jets tend to be found in rich clusters. Best et al. (2005) argue that the relation of black hole mass to the radio-loud AGN fraction reflects the gas cooling rate in elliptical galaxies.

10.3 Measured Properties of Galactic Outflows

In a little over a decade, our knowledge of galactic winds has grown from a tunable parameter to a phenomenon highly constrained by observations. The development of high-resolution X-ray spectral imaging has made it possible to constrain the metal content of hot galactic winds, or at least the interfaces between the wind fluid and cooler gas. Much remains to be learned, however, about the relationship between cold, warm, and hot gas in galactic outflows. This knowledge would constrain dynamical models thereby determining how far winds travel. Ideally, the kinematics, composition, density, and temperature would be measured for each phase – something not currently possible. I describe here how a subset of these quantities have been measured using observations from X-ray to radio wavelengths.

One key challenge is to understand how feedback varies with galactic mass. The images in Fig. 10.3 show outflows from the galaxy NGC 1569, $\sim 10^8$ $M_\odot$ of gas, and a lower mass dwarf NGC 3077 ($\sim 10^7$ $M_\odot$ of gas). The ambient gas disks are more effective at restricting the outflow in the larger systems. The trend is even stronger moving up another dex in mass to M82 (see Figure 4 of Veilleux, Cecil, & Bland-Hawthorn 2005). The extended X-ray emission from ultraluminous starbursts is more complicated as shown, for example, in the *Chandra* images of Arp 220 in (McDowell et al. 2003). The ambient medium that the starburst wind encounters is considerably more complicated than that encountered by the M82 or NGC 1569 winds because the tidal interaction propels streams of material outward and generates regions of inflow (Hibbard et al. 2000)

10.3.1 Optical emission lines

Starburst galaxies generally present moderately-bright optical emission-line filaments which extend up to $\sim$ 10 kpc out of the galactic disk (e.g. Heckman, Armus, & Miley 1990; Marlowe et al. 1995; Martin 1998). The optical recombination line luminosity and the collisionally-excited line luminosity are found to be dominated by emission from photoionized clouds (e.g. Lehnert & Heckman 1996; Martin 1997; Miller & Veilleux 2003). Considering the large amount of energy in the ionizing continuum of a normal stellar population, compared to the mechanical power, the dominance of photoionization is not surprising.

Despite the dominance of the emission from photoionized clouds on a global scale, strong evidence for the presense of shocks is seen several kpc from the starburst region in the more extended filaments. The line ratios observed along the outflows evolve toward shock-like line ratios with increasing distance from the starburst region. Figure 2 of Martin (1997) shows several examples in dwarf starburst galaxies. Line ratios become more "shock-like" (high ratios of [N II] $\lambda 6583/\mathrm{H}\alpha$, [S II] $\lambda 6716, 6731/\mathrm{H}\alpha$, and [O I] $\lambda 6300/\mathrm{H}\alpha$) further out along the minor axis. Hybrid models that combine photoionization, shocks, and/or emission from turbulent mixing layers match the data better than photoionization alone. The rough constraints available on the chemical abundances in the dwarf starburst filaments are consistent with them being composed of disk material (Martin 1996, 1997). Non-photoionization processes are responsible for only 3–4% of the Hα emission (Calzetti et al. 2004).

The dynamics of the optical filaments were established in the seminal paper by Heckman, Armus, & Miley (1990). They showed that the network of Hα emitting filaments extending along the minor axis of starburst galaxies presents double-peaked emission profiles characteristic of a moving shell. The pressure gradients indicated the shells were driven by an extended power source, and the starbursts were found to be strong enough to explain the observed shell velocities of 100 – 300 km s^{-1}. Lower velocities were subsequently measured in dwarf galaxies (Marlowe et al. 1995; Martin 1998).

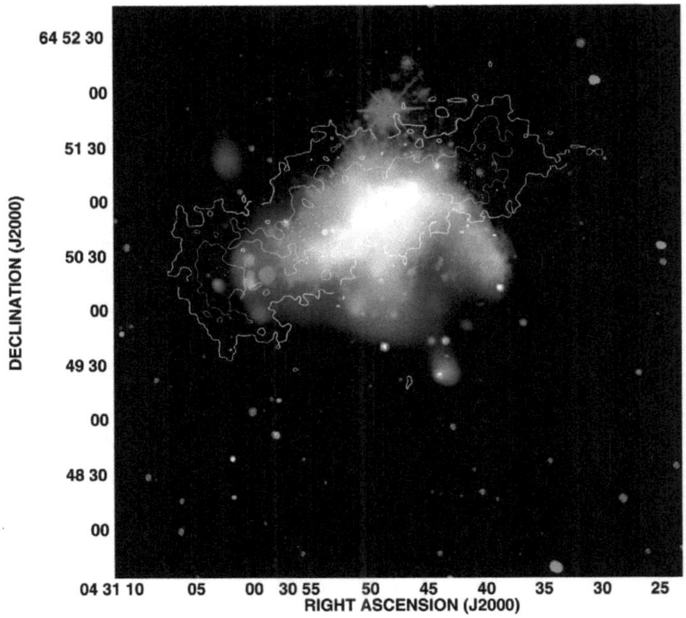

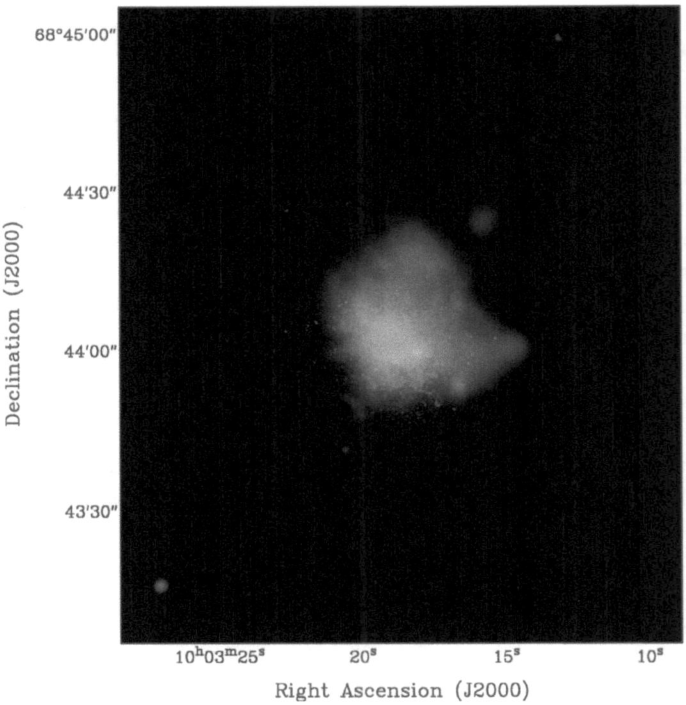

Fig. 10.3. Multi-band pictures of galactic outflows. Outflows become more bipolar, i.e. less spherical, in larger systems. Images are from Ott et al. (2003) and Martin et al. (2002).

For starbursts of all sizes, the estimated mass flux in the warm ionized filaments is similar to the star formation rate (Martin 1999). This assertion rests on a volume filling factor derived from the assumption that the thermal pressure in the filaments is similar to that of the X-ray emitting region. The projected velocities of the warm shells are not much larger than the escape speed however, so some of the warm gas probably remains bound to the parent halo. How much warm gas escapes from the gravitational potential is model dependent.

10.3.2 X-ray emission

The optical observations provide little direct evidence for gas extending more than a few kpc from the plane of the galactic disk, but the expansion renders the wind emission measure, $EM = \int n_e^2 ds$, undetectable due to the steep radial density gradient in the wind. Indeed, the extended wind may only be directly visible when it runs into something as seen, for example, north of M82 (Devine & Bally 1999; Lehnert et al. 1999; Stevens, Read, & Bravo-Guerrero 2003). Similarly, even close to the starburst region, the most sensitive X-ray observations may be blind to the hottest component of the wind which if at $T \sim 10^8$ K would have a volume density about 100 times lower than the detected hot wind (and an emission measure about 10^4 times lower) (Strickland & Stevens 2000). The extraordinary sensitivity of XMM/Newton has very recently detected X-ray emission over a larger region, albeit with a high degree of structure, out to 14 kpc from the disk of M82 (Stevens et al. 2003).

The *Chandra* X-Ray Observatory has revolutionized our understanding of the origins of X-ray emission in starbursts. The arcsecond spatial resolution of the ACIS imaging spectrometer reveals that the brightest X-ray emitting gas lies adjacent to the warm, optical filaments. For example, in the prototypical starburst NGC 253, the X-ray emission was resolved into filaments out to 9 kpc above the disk (Strickland et al. 2000, 2002). Associations between the X-ray emission and the warm-ionized filaments have also been found in NGC 3079 (Cecil et al. 2002) and NGC 1482 and NGC 6240 (Veilleux et al. 2003). Our X-ray view is strongly biased toward the densest regions of the hot wind at its boundary layer with the cooler, optical filaments.

The absence of local enrichment in HII regions suggests that most of the metals from core-collapse supernovae are injected directly into the hot ISM (Kobulnicky & Skillman 1996, 1997; Martin 1996). Measurements of chemical abundances from X-ray spectroscopy are therefore of great interest. In principle, abundance determinations for a collisionally ionized plasma are easier than in the nebular case in which photoionization must be included in the model. Under coronal conditions, the ionization structure is completely determined by the electron temperature, which is fixed by processes like hydrodynamic shocks.

In practice, metal abundances derived from thermal X-ray spectra are often unreliable. The observations have several biases. First, until recently, the spectral apertures were large compared to the thickness of superbubble shells. The X-ray emission within an aperture was the sum of emission from multiple components – i.e. shocked shells, mixing layers, hot bubbles, and point sources –

all weighted disproportionately by the densest regions. Second, the observations were not very sensitive to gas cooler than 10^6 K or hotter than $10^{7.5}$ K. Third, due to a degeneracy between high-metallicity, two-temperature models and low-metallicity, single-temperature models, metallicities fitted to the data tended to be pushed lower than their true value if the temperature range was not fully represented (Buote & Canizares 1994). Fourth, the omission of intrinsic absorption from the spectral models increased the apparent strength of Mg and Si lines relative to Fe L lines. Unless lower energy lines from O were detected, statistically acceptable fits could be obtained with an anomalously high α -element to Fe abundance ratio (e.g. Weaver, Heckman, & Dahlem 2000; Dahlem, Weaver, & Heckman 1998). Generally a volume filling factor of unity and solar abundances were simply assumed leaving significant uncertainties about the mass and composition of the hot wind.

An ultra-deep *Chandra* exposure of the nearby starburst galaxy, NGC 1569, is perhaps the best available for constraining abundances (Martin, Kobulnicky, & Heckman 2002). The metallicity of the ISM in this galaxy is only 0.25 $Z_\odot$, so the contrast with supernova enriched gas is larger than it is in more luminous starbursts. Figure 10.4 shows the ACIS spectrum of NGC 1569. At such low spectral resolution, there is no single spectral model that uniquely fits this complex spectrum. The high-resolution imaging data can break the model degeneracy, however, when combined with data obtained in optical and radio bands. For example, the X-ray color gradients reveal intrinsic absorption which is tightly

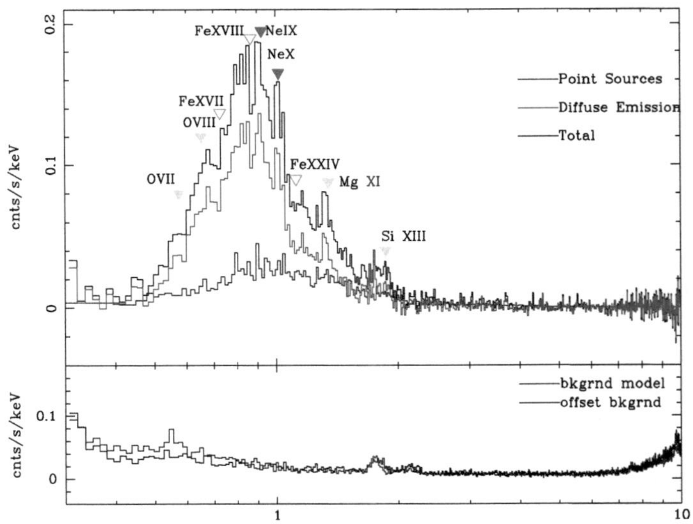

Fig. 10.4. Integrated spectrum of NGC 1569 (top curve). The contribution from point sources (lower curve) is harder than that from diffuse gas (middle curve). Bottom panel shows the folded X-ray background. Although individual lines are not resolved, the Fe L complex and lines from several α-process elements are prominent in the spectrum. See discussion in Martin, Kobulnicky, & Heckman (2002).

correlated with the location of the 21-cm emission from the cold disk gas. The relativity hard point source component is resolved and fitted independently. The X-ray colors show there are at least two thermal emission components, and that they have different mean absorbing columns. The strong spectral lines require an alpha element abundance of at least 0.25 $Z_\odot$ to fit the spectrum. However, the wind metallicity can be increased arbitrarily above this level provided the emission measure, which is proportional to the continuum normalization, is decreased proportionately. Using population synthesis models, normalized to the starburst in NGC 1569, to estimate the mass of supernova ejecta which is a lower limit on the wind mass (i.e. no mass loading), the upper limit on the wind metallicity is about twice solar. The best fit implies α - element abundances in the wind of roughly solar, a mass loading factor ~ 9, and $Z_\alpha/Z_{Fe} \sim 3$. In NGC 1569, most of the metals in the outflow come from the supernovae ejecta not the entrained ISM even though the ISM supplies the majority of the mass in the outflow. It appears likely that the wind carries most of the metals synthesized in the recent burst. The fraction of the total stellar mass formed in such bursts is not yet known.

10.3.3 Interstellar absorption in galaxy spectra

When an outflow passes between the observer and a bright, background light source, the kinematics of the cold gas can be probed in absorption. The surface density of quasars accesible to current telescopes is too low to allow tomographic imaging with an array of distant quasars (Weedman 1986). Single sightlines toward starburst nuclei, and even integrated galaxy spectra, have been used quite effectively. The ultraviolet bandpass contains ground-state transitions for a range of ionization states and elements.

Winds are ubiquitous in the rest-frame-UV spectra of Lyman-break-selected galaxies at $z \sim 3$ (e.g. Pettini et al. 2002; Adelberger et al. 2003; Shapley et al. 2003). Figure 10.5 compares the composite UV spectra of several hundred LBGs to the local composite from Schwartz & Martin (2006). The galaxies that make up the high-z sample are much more luminous than the local UV-selected galaxies, yet the spectra are remarkably similar. The resonance lines are blue-shifted by about 200 km s^{-1} in each case relative to the systemic velocity, marked by stellar lines. One difference is that not all the local galaxies present outflows, so the absorption lines are broader and extend down to zero velocity.

Large samples of nearby galaxies have only been observed in optical lines due to the much higher sensitivity of ground-based telescopes. The strongest lines are from CaII, KI, and NaI, so only fairly cold gas can be probed. Most work has focused on the Na I 5890, 5896 doublet owing to the high depletion of CaII and high density of atmospheric sky lines near KI. This method probes lower column densities than the 21-cm emission observations.

Most starburst galaxies show Na I absorption which is blueshifted relative to the systemic velocity (Heckman et al. 2000; Rupke, Veilleux, & Sanders 2002; Schwartz & Martin 2003; Martin 2005; Rupke et al. 2005a, Rupke et al. 2005b). The absorption is measured against star clusters in the nucleus of the host galaxy,

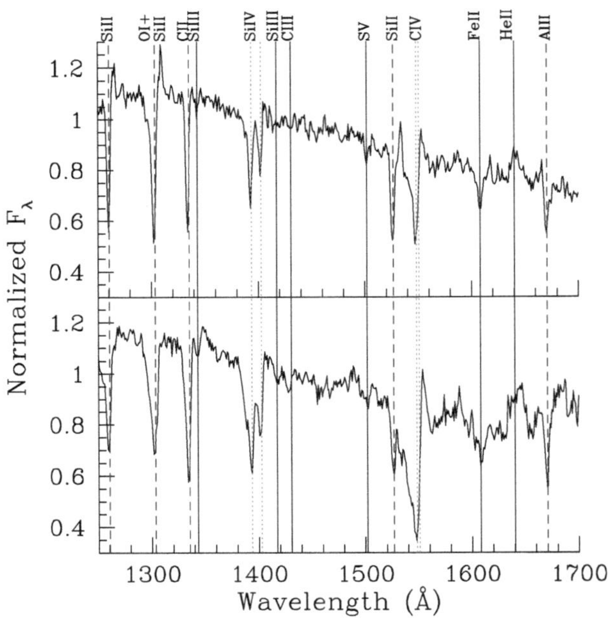

Fig. 10.5. Composite rest-frame ultraviolet spectra of UV-selected galaxies. (*Top*) STIS spectra of 16 star clusters in local galaxies. (*Bottom*) Over 800 Lyman Break Galaxies from Shapley et al. (2003). In spite of a gross difference in luminosity between these samples, the low-ionization resonance lines are blueshifted by a similar amount, 142 ± 80 km s^{-1} and 150 ± 60 km s^{-1} respectively, in the nearby and distant composite spectra. Reproduction of Figure 13 from Schwartz & Martin (2006).

so there is no ambiguity about the direction of the flow. Figure 10.6 summarizes the expansion velocities measured at maximum optical depth for these data sets. Taken together, there is a clear trend for the velocities to increase with increasing burst strength. The outflows that exceed roughly 650 km s^{-1} tend to present active nuclei (Rupke et al. 2005c; Martin 2005).

The hydrogen column is estimated from the NaI column using assumptions about the sodium abundance, depletion, and ionization state. Sub-solar abundances measured along sightlines through the disk and halo of the Milky Way have been interpreted as depletion onto dust (Wakker & Mathis 2000; Savage & Sembach 1996). In the Milky Way, the product of the Na depletion onto dust grains and the NaI ionization fraction presents little variation with N(H). Provided the H column is less than 10^{21} cm^{-2}, their product is ~ 300, within a factor of a few (Wakker & Mathis 2000). The correction may decrease by an order of magnitude at higher columns due to self-shielding of Na from ionizing radiation (Ferlet et al. 1985). Based on their broad range of F_{UV}/F_{FIR} flux ratios, Murray et al. (2006) argue that the ionization correction could vary among infrared selected starbursts by factors of over 100.

The inferred mass flux is also sensitive, however, to the size of the region the sightline represents. Heckman et al. (2000) measured an extent of 1 to 10 kpc

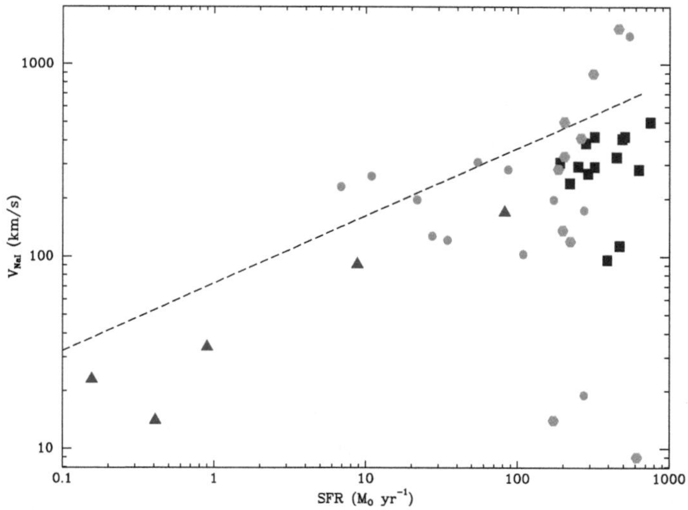

Fig. 10.6. NaI outflow velocity vs SFR. Slower outflow speeds are found in less luminous starbursts. From Martin (2005).

along their slits for the absorbing region, and they argued the cold clouds carried a mass flux similar to the SFR. Rupke et al. (2002) found the mass flux in the ultraluminous starbursts appears to be about an order of magnitude less than the SFR. Of those dwarf galaxies which show interstellar Na I absorption, the area of the outflow would need to be at least 1 kpc to bring the flux of cold material up to that of the SFR (Schwartz & Martin 2003). Martin (2006) measured the run of column density along the slit and found several surprising results. Among these is the large variation in cool outflowing mass among ultraluminous starbursts with similar star formation rates. The neutral clouds carry up to 10% of the interstellar gas mass at rates comparable to the SFR. It is energetically possible that starbursts power these outflows.

10.4 Constraints and Challenges for Dynamical Models

The fate of the gas in these outflows depends on the dynamical model. Figure 10.7 compares the thermal energy of the detected X-ray emitting gas to that needed to escape from the potential of an isothermal sphere. The critical mass below which a lot of hot gas escapes is about 140 km s^{-1} (Martin 1999). The critical mass for this differential feedback is the same as that where the mass – metallicity relation turns over (Tremonti et al. 2004). Since outflows decrease the effective metal yield, one interpretation is that decrease in metallicity with mass is the result of relatively more metal loss in hot winds from smaller galaxies.

The total radiative losses in the halo are also well-constrained by detections of OVI absorption in several dwarfs. Figure 10.8 shows several FUSE spectra of nearby dwarf starbursts. The OVI line is an excellent tracer of cooling gas because stellar radiation is not hard enough to produce OVI. Heckman et al.

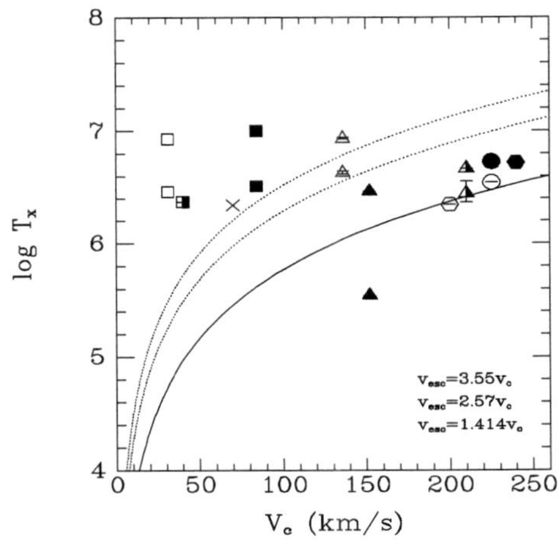

Fig. 10.7. Temperature of the density-weighted diffuse X-ray emission vs. galactic rotation speed. The curves show the escape speed for several halo models. See Figure 1.6 of Martin (2004) for details on individual data points.

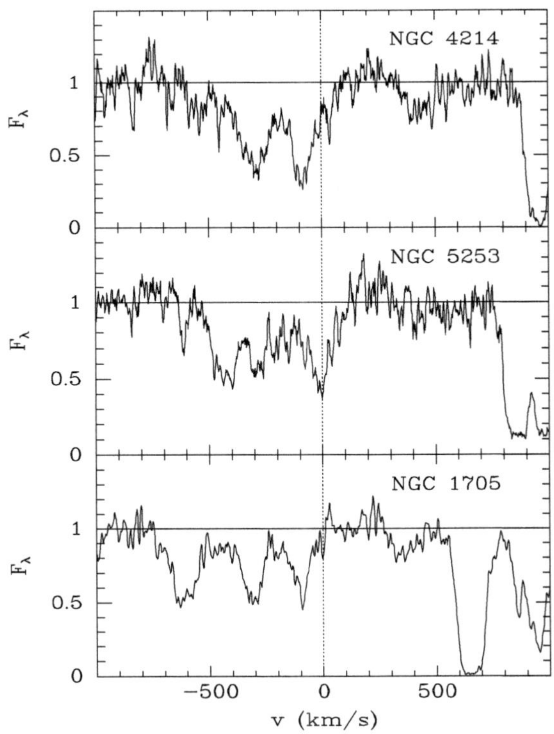

Fig. 10.8. OVI absorption toward nearby dwarf starburst galaxies in FUSE spectra. The OVI traces radiatively cooling gas. From Martin & Heckman, in preparation.

(2001) show that the measured O VI column density is just what is expected for gas that has been heated and then cools radiatively. They set an upper limit on the cooling rate that is 20% of the supernova heating rate. Estimates from optical emission lines give similar values (Martin 1997; Calzetti et al. 2004). The absence of extreme radiative losses indicates that halo gas or a higher ambient pressure in the IGM does not prevent the hot wind from escaping, as suggested by Silich et al. (2001).

The fate of the cool outflow depends on where the absorbing material lies and how it is accelerated. Figure 10.9 shows the measured velocities reach the escape speed. Little material is clearly moving faster than that (Rupke et al. 2002; Martin 2005; Rupke et al. 2005b). Under the assumption that the outflows are driven by the nuclear starburst, the simplest interpretation is that the ram pressure of the hot wind probably accelerates these cool clouds (Heckman et al. 2000, Fujita et al. 2006). The clouds would reach the speed of the hot wind provided the starburst supplies enough momentum to bring the entrained mass up to the speed of the hot wind. One interpretation of Fig. 10.6 is that the velocities rise with increasing SFR up to the point where supernovae supply this critical momentum; but the velocity stops increasing once the SFR is high

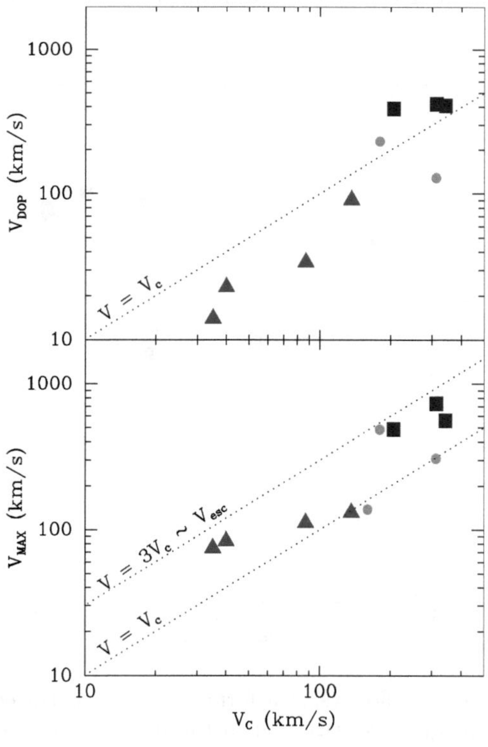

Fig. 10.9. (*Top*), Outflow velocity at NaI line center vs. galactic rotation speed. (*Bottom*), Maximum NaI velocity measured vs. galactic rotation speed. Reproduced from Martin (2005).

enough to accelerate the cool gas to velocity of the hot wind – about $\sqrt{3}c_s \sim$ 650 km s^{-1} (Martin 2005). Although outflows faster than this speed limit are observed, they are nearly always associated with systems containing AGN – and presumably additional acceleration mechanisms (Martin 2005; Rupke et al. 2005c).

The momentum could also be supplied by radiation pressure, especially in ultraluminous systems (Murray, Quartaert, & Thompson 2005; Martin 2005). The abundant Na I lines probe regions where H is neutral – i.e. relatively cool and dense regions. The absence of Na I emission indicates these regions are also dusty. Winds also scatter ultraviolet light indicating significant quantities of halo dust (Hoopes et al. 2005). The velocities continue to increase with luminosity in this model. In principle, the mass loss efficiency (relative to the SFR), $\dot{M}_c/\dot{M}_*$, will fall off slower with halo mass for a momentum-driven wind than for an energy driven wind (Murray et al. 2005). Rough estimates for ULIGs and a few dwarf starbursts are consistent with the momentum-driven wind scenario, and indicate the mass loss rate in the cold phase is of the same order as the SFR for the strongest outflows (Martin 2006).

It is interesting to compare the mass flux of the cold flow to that in the warm and hot gas phases. Figure 10.10 illustrates the enormous range in mass observed among ULIGs with otherwise similar luminosity. It is not currently clear why

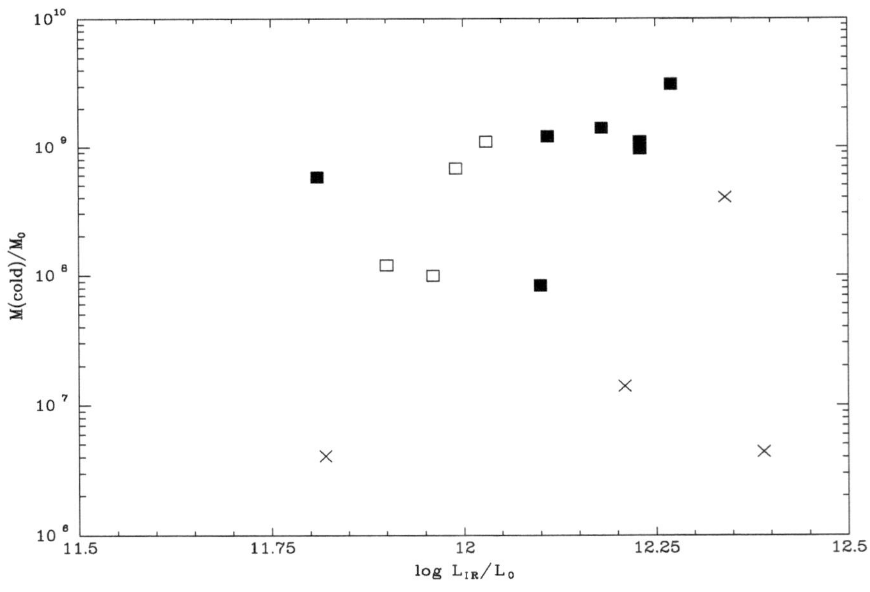

Fig. 10.10. Mass of cool outflow vs. star formation rate, as traced by the far-infrared luminosity. An extremeley large variation in the mass of the cool component of the wind is found among ULIGs with similar star formation rates. Symbol type is related to dynamical age of the merger: double nuclei (solid), single nuclei with prominent tidal tails (open), and single nuclei (crosses). Reproduced from Figure 6 of Martin (2006).

they differ so much. Murray et al. (2006) show that the range could be produced by variations in ionization parameter. Martin (2005) noticed that the systems in the high mass loss state are all relatively young mergers. Their gas disks may have just been shock heated by the encounter, in which case some orbital energy may be tapped to drive the outflow (Cox et al. 2004). Indeed, the large angular extent of these outflows, their large solid angle (as seen from the nucleus), and the presence of significant rotation in a few cases indicate an outflow driven from the entire gas disk, not just the nuclear region (Martin 2006). If merger-driven winds are the answer, the amount of escaping debris will be sensitive to the energy of the encounter.

10.5 Cosmological Impact of Galactic Winds

The maximum extent of galactic winds can be limited by a variety of factors: cosmic time, the pressure of the IGM, and gravitational deceleration. The hot winds from local dwarf starbursts could easily expand to radii of 30 to 50 kpc before being confined by the IGM. In momentum-driven-wind models, gravity wins at large radii in an isothermal halo. Martin (2006) demonstrate that the cool ULIG winds have turn around radii of at least 40 kpc, and possibly much larger. These estimates will undergo revisions as the data converge toward a specific dynamical model for outflows. Hence, the implications for metal dispersal and mass ejection are still preliminary.

10.5.1 The galaxy luminosity function

As described in the Introduction, the faint-end slope of the galaxy luminosity function is shallower than the halo mass function. At luminosities down to roughly $0.01L^*$, the evidence for metal loss from local, dwarf starburst galaxies is strong. The difference from the catastrophic outflows originally described by Larson (1974) and Dekel & Silk (1986) is that bipolar outflows can leave much of the gas disk intact. Numerical simulations, that include axisymmetry, find significant metal loss but not so much mass loss (MacLow & Ferrara 1999). Within just a few Mpc, however, we find two dwarf starbursts – NGC 1569 and NGC 1705 – with much stronger starbursts than MacLow & Ferrara considered. Plausible feedback recipes largely solve the luminosity function – mass distribution on this scale (Benson et al. 2003). Additional mechanisms like photo-ionization supression may be required at the lowest masses.

Theoretical models of galaxy formation in a cold dark matter universe require *tweaking* to reduce the efficiency of cooling in large galaxies, groups, and galaxy clusters. Without energy input, gas cooling proceeds very efficiently in these models, leading to too much star formation in high-mass galaxies. Inspection of the most luminous starbursts in the local Universe suggests that radiation pressure, merger induced shocks, and active nuclei provide additional acceleration. These affects likely play a role in suppressing star formation just above L^*.

On group and cluster scales, the evidence is rapidly growing that feedback from radio-loud quasars is the dominant feedback. In their simulations, Croton et al. (2005) distinguish *radio mode* feedback from optical AGN. Using 10% of the accretion energy, $\dot{M}_{BH}c^2$, for the mechanical power, Croton et al. (2005) demonstrate that this radio model feedback strongly suppresses cluster cooling flows. This prevents the formation of new stars in the most massive halos thereby keeping the color of massive galaxies red – as observed, but not previously produced (Springel et al. 2005). The resulting reduction in the luminosities of bright galaxies induces a sharp break in the luminosity function, which matches observations.

10.5.2 Dispersal of heavy elements & intervening absorption

Intervening absorption line systems (ALS) detected in quasar spectra provide, in principle, a statistically complete census of metals deposited by galactic winds. In practice, the distinction among galactic winds, tidally stripped gas, infalling halo gas, and the general IGM is blurry.

To produce detectable metal-line absorption, the gas must (1) have the proper conditions for the ion to be abundant and (2) be enriched enough to make the line strong. The relevance of three well-studied populations: OVI absorbers, CIV absorbers, and MgII absorbers are reviewed here. Although the ionization of the gas is not necessarily connected to an associated galaxy, galaxies are generally considered the only source of metals after the initial Population III enrichment.

The problem with absorber-galaxy associations is that an unseen, fainter galaxy may exist closer to the sightline than the identified galaxy. Galaxies and absorbers may, independent of each other, be correlated with the same overdensities in large scale filaments. The hydrogen overdensity, $\delta \equiv (\rho - \bar{\rho})/\rho$, related to the HI column density by simulations (Davé et al. 1999), is a useful quantity for identifying similar structures at different epochs.

OVI absorbers

Among these strong, ultraviolet lines, OVI $\lambda\lambda 1031.93, 1037.62$ traces the hottest gas. The ion is produced in shocks, at hot-cold interfaces in the IGM, and via photoionization by the metagalactic background. Cosmological simulations predict that shocks generated by the formation of large-scale structure heat a large fraction of the baryons by the present epoch (Cen et al. 2001). OVI absorption traces only the coolest part of this warm-hot IGM (Nicastro et al. 2005). In collisional-ionization-equilibrium, the maximal fractional ionization for O^{+5} is 20% at $T = 10^{5.45}$ K.

From FUSE detections of OVI from Galactic high-velocity clouds (e.g. Sembach et al. 2003), it is clear that the extended halos of larger galaxies produce OVI absorption. Projecting the Galactic HVCs as if they were observed around another galaxy, TF05 conclude that HVC-like OVI absorbers make a significant contribution to the observed numbers if massive galaxies have large halos or if all galaxies down to $M_r \sim -16$ $(0.01L_r^*)$ have OVI HVCs to 100 kpc.

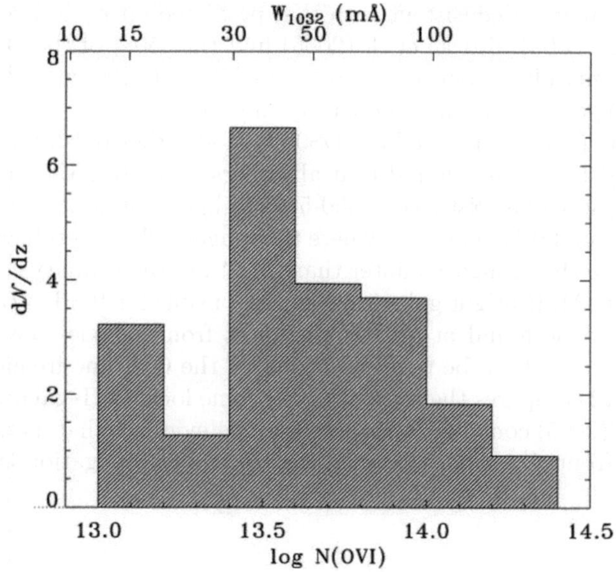

Fig. 10.11. Histogram of observed OVI column density. Forty OVI absorbers found in FUSE spectra are shown. The number of systems rises steeply toward lower columns. Reproduced from DS05 Figure 4.

Figure 10.11 shows the local $d\mathcal{N}/dN_{OVI}dz$ distribution measured from FUSE spectra of AGN (Danforth & Shull 2005). Several properties of this population are relevant to outflows. First, the steep fall-off toward strong systems, $d\mathcal{N}/dN_{OIV} \propto N_{OIV}^{-2.2\pm0.1}$, indicates that numerous weak absorbers contain mass comparable to the strong absorbers. If the deficit of absorbers at $\log N_{OVI} < 13.4$ is real, then metals are not as uniformly distributed throughout the low-δ IGM as simulations assume (Tumlinson & Fang 2005). Second, Danforth & Shull estimate a metallicity $Z_O = 0.09$ $Z_\odot$. This derivation is complicated by the multiphase nature of the absorbing systems but indicates a significant enrichment of the $\delta \geq$ 5 IGM since $z \sim 3$. This result is supported by metallicity estimates for twelve absorbers at $z \sim 0.3$ (Prochaska et al. 2004).

The OVI absorption statistics at $z \sim 2.5$ are summarized by Simcoe et al. (2004). They find a median IGM abundance $[O/H] = -2.82$. The broad spread of values indicates they have not detected a universal metallicity floor. Furthermore, since the median metallicity of systems at $z \sim 3$ is significantly lower than that measured locally, substantial enrichment of the IGM over the past 10 Gyr is indicated. Apparently, the bulk of the intergalactic metals formed much later than the first stars.

Outflows should be more enriched than the general IGM. Other properties that might confirm or refute an outflow origin include kinematic properties and/or direct association with galaxies. The OVI systems from DS05 are kinematically similar to HI systems, but span a much narrower range in column density. For essentially all OVI samples, the OVI column density correlates with line-width in a manner consistent with that predicted for radiatively cooling gas (Heckman et al. 2000). Stocke et al. (2005) find that 50% of the DS05 OVI systems have velocity offsets between Lyα and OVI at 50–200 km s^{-1}, which could represent the speed of the shocks creating the OVI.

Stocke et al. (2005) compared the DS05 OVI absorber redshifts to over a million galaxy redshifts. They found that absorbers with $10 \pm 5\%$ solar metallicity have a median spread in distance of 350-500h_{70}^{-1} kpc around L^* galaxies and 200-270h_{70}^{-1} kpc around $0.1L^*$galaxies, where the ranges reflect metallicity uncertainties. They argue that galaxies fainter than $0.1L^*$ are the primary contributers to metals in the IGM. If bright galaxies were responsible for IGM enrichment, OVI absorbers would be found at greater distances from galaxies. Larger absorber cross sections would then be required to match the OVI line frequency.

Figure 10.12 compares the separations of some local OVI–galaxy pairs. Tumlinson & Fang (2005) conclude that there is direct evidence that metals are spread to $\sim$ 100 kpc from all galaxies. Evidence for further propagation is weaker.

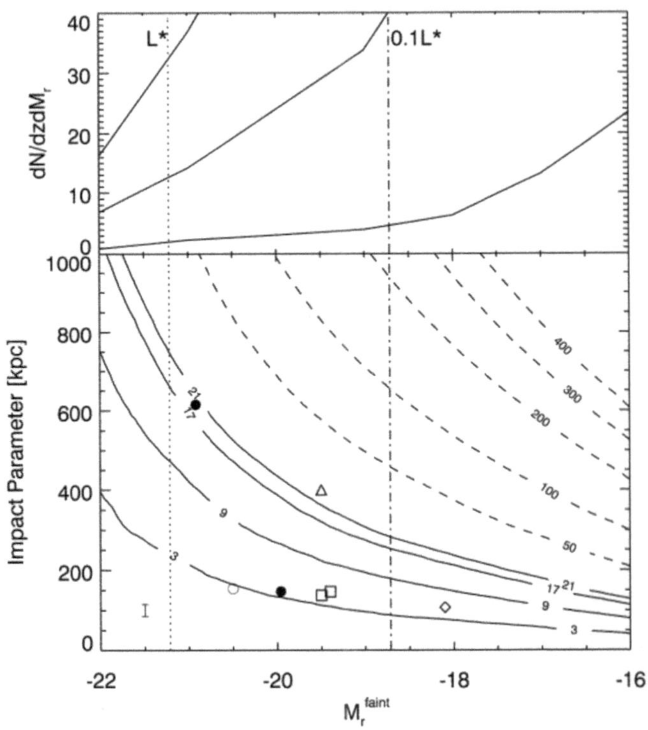

Fig. 10.12. Top: Differential number of OVI absorbers as a function of galaxy luminosity. The model is tied to a local redshift survey. Contours of constant cumulative $d\mathcal{N}_{OIV}/dz$. The symbols mark parameters for the nearest neighbor galaxies for seven well-studied QSO absorbers Reproduced from Tumlinson & Fang (2005) Figure 1.

CIV absorbers

Counting OVI absorbers provides a lower limit on metal production since some gas resides at both hotter and cooler temperatures. For example, the C^{+3} in CIV absorbers is primarily photoionized by the metagalactic background. Higher ionization, i.e. CV, is favored in the most tenuous regions (see Fig. 4 of Simcoe et al. 2004). The addition of these metals compounds the dispersal problem presented by the OVI absorbers.

At $z < 0.9$, CIV $\lambda\lambda 1548, 1550$ absorption systems are strongly clustered around galaxies on velocity scales $v \leq$ 250 km s^{-1} and impact parameter scales of $\rho \leq 100$ h^{-1} kpc (Chen et al. 2001). Adelberger et al. (2003) found a positive correlation between galaxies and CIV systems at redshift 2-3. A typical galaxy is surrounded to radii $r \sim 40$ kpc (proper) by gas that has a large velocity spread, $v > 260$ km s^{-1} and column typical of galaxies, $N(CIV) \gg 10^{14}$ cm^{-2} (Adelberger et al. 2005). At columns $N(CIV) \gg 10^{14}$ cm^{-2}, the extent increases to about 80 kpc. The associated galaxies reside in relatively dense environments and have young stellar populations consistent with prior starburst activity.

The shape of the CIV column density distribution remains essentially invariant, with slope -1.44, from z=1.5 to z=5.5 (Songaila 2005a). At $z \sim 2.5$, high column density CIV systems, $N_{HI} \sim 10^{15} - 10^{16.5}$cm^{-2}, are generally detected in OVI. Simcoe et al. (2005) argue that the absorber geometries resemble thin sheets or bubbles, which are associated with radiative shocks. Their observations and simulations indicate these are dense, $\rho/\bar{\rho} \sim 1000$, metal-rich, $Z \gtrsim 1/10 - 1/3$ $Z_\odot$ regions. Simcoe et al. find a distinct dropoff in the absorber gas density at impact parameters larger than 300 kpc (physical) from galaxies. The strong metal enrichment out to impact parameters reaching 100-200h_{71}^{-1} kpc indicate the shocks are expelled, rather than infalling, gas.

Songaila (2005b) addresses the question of which CIV ALS could be directly associated with galactic outflows. She argues that if CIV is present throughout a bubble, it should be detected from the systemic velocity of the galaxy up to the wind speed. The abundant weak CIV systems, $N(CIV) < 2 \times 10^{13}$ cm^{-2}, have narrow lines that are unlikely to arise from galactic winds. To reproduce the CIV redshift path above 10^{13} cm^{-2}, the cross section per galaxy requires an absorber radius of 200 kpc. Since the galaxies are not more than 1 Gyr old at $z = 3$, the minimum expansion velocity is 200 km s^{-1}. As many as half the stronger CIV systems are wide enough that they could be produced in starburst outflows. Figure 10.13, from Figure 11 of Songaila (2005b), shows that the SiIV/CIV ratios of the $\nabla v(FWTM) > 100$ km s^{-1} CIV systems are larger than for the more quiescent systems. These systems could be ionized by the softer spectrum of the parent galaxy.

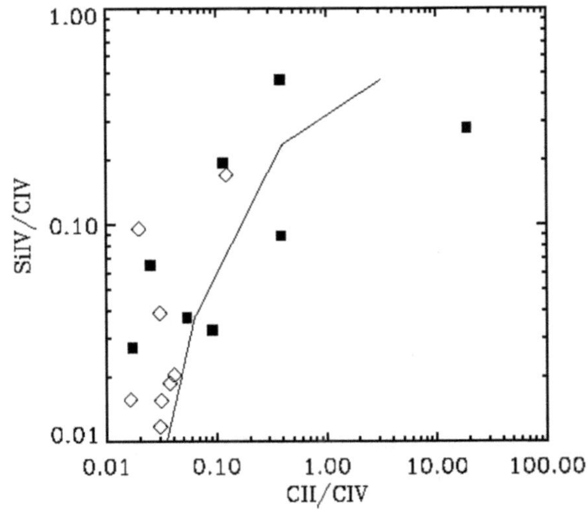

Fig. 10.13. Measured SiIV/CIV and CII/CIV ratios in intervening systems are compared to a simple prediction (AGN power law spectrum and Si/C ratio 2.5 times solar). Filled squares denote systems with $\nabla v(FWTM) > 100$ km s^{-1} . The systems with a large velocity width are also ionized by a softer spectrum. Reproduced from Songaila (2005b).

MgII absorbers

The MgII $\lambda\lambda 2796, 2803$ systems are associated with the photoionized halos of galaxies. At comparable sensitivities, about one-quarter of CIV systems are detected in MgII; and nearly all MgII systems are detected in CIV (Steidel & Sargent 1992). Over the redshift range $0.2 < z < 1.0$, the average MgII system is associated with an $0.7L_B^*$ galaxy (Steidel, Dickinson, & Persson 1994). Bright galaxies, $L > 0.1L_K^*$, within 50 h^{-1} kpc of quasar sightlines produce detectable metal lines of Mg II and/or C IV in the QSO spectrum (Steidel et al. 1997). The origin of the halo gas is debated.

A recent census of strong, $W_r > 1.0Å$ systems, using SDSS spectra, shows a mild evolution with redshift (Prochter et al. 2005). The redshift path density increases as $d\mathcal{N}/dz = 0.094(1+z)^{1.29}$ for $0.35 < z < 2.3$. Bond et al. (2002) proposed previously that a significant fraction of these strong, $W_r > 1.8Å$, absorbers are due to superwinds. They argued that (1) the *symmetric-inverted* structure of many line profiles reflects the two lobes of a bipolar outflow, and (2) the incidence of such systems increases rapidly with redshift over $1 < z < 2$ - reflecting the increase in star formation activity. The velocity splitting of several hundred km s^{-1} in the symmetric-inverted profiles is consistent with measured outflow speeds. Prochter et al. (2005) argue a mixture of galactic winds and accreting gas are required. The cross-section determined from MgII absorption and LBG data, $\sigma \sim 0.01h^{-2}$ Mpc^2 at $z \sim 3$, is consistent with the idea that galactic winds are the primary host of high redshift MgII absorption.

The presence of significant rotation in the halo gas of at least a few systems (Steidel et al. 2002) is not compatible with a wind driven by a nuclear starburst but is similar to the merger-driven outflows discovered recently (Martin 2006). More direct comparisons between absorber velocities host-galaxy kinematics are clearly needed.

10.6 Summary and Future Direction

A substantial amount of gas is entrained in hot galactic winds leading to complicated, multi-temperature flows. These outflows are largely invisible in emission at distances $\gtrsim$ 10 kpc from galaxies. More extended gas contributes, however, to highly-blueshifted absorption lines seen in starburst galaxy spectra. Growing kinematic and physical evidence suggests that a significant subset of intervening absorption line systems are the imprint of galactic winds.

Especially at high redshift, galactic winds appear to be the primary hosts of strong MgII, CIV, and OVI systems. Considering the limited age of high redshift galaxies, the cross section is limited by the velocity of the outflow. Outflow speeds directly measured in cool and warm gas are $\sim$ 200 km s^{-1}, which is just large enough to produce the redshift path density of a significant fraction of CIV and OVI systems. Hence, even though the outflow speeds lie near the halo escape speed, material appears to reach large distances. More outflow modeling is required to determine unique dynamical models that produce a testable cross section for the outflows.

The hot wind clearly escapes from local dwarf galaxies. It carries most of the metals produced in the starburst phase into the IGM. The critical mass for significant metal loss is consistent with the turnover in the mass-metallicity relation for galaxies in that it creates lower effective yields in dwarf galaxies. The relatively recent increase in the metallicity of intervening metal-line systems is consistent with this. And dwarf galaxies appear to make a significant contribution to the absorber – galaxy pairs locally. It remains to be determined whether dwarf galaxies contribute significantly to the absorption cross-section at high redshift. Galaxy downsizing might delay pollution by dwarf galaxies, but it seems just as likely that the absorber – galaxy work at high-redshift is not sensitive enough to select dwarfs. Strong feedback in dwarfs must occur fairly early if it is to solve the luminosity function problem. It appears that it can based on measurements in local dwarf starbursts.

It remains to be determined exactly how radiation pressure from massive stars, AGN, and merger-driven feedback should affect feedback recipes in massive galaxies. A different solution for the missing feedback in very massive halos has been identified (Springel et al. 2005; Croton et al. 2005). While this numerical solution appears to be consistent with many properties of radio-loud quasars (Best et al. 2005), no consensus has been reached regarding the exact physical mechanism driving these outflows. Much remains to be learned about feedback on group and cluster scales too.

Acknowledgements

The author apologies that space and time simply made it impossible to specifically mention many papers in this field. It is hoped that the references provide a gateway for the student to discover additional relevant works. The contributions of my collaborators to the ideas presented here are acknowledged with sincere appreciation. This work was supported by grants from the David and Lucile Packard Foundation and the Alfred P. Sloan Foudation.

References

Adelberger, K. L. et al. 2005, ApJ, 629, 636
Adelberger, K. L., Steidel, C. C., Shapley, A. E., & Pettini, M. 2003, ApJ, 584, 45
Antonucci, R. 1993, ARA&A, 31, 473
Babul, A. et al. 2002, MNRAS, 330, 329
Begelman, M. C. & Nath, B. B. 2005, MNRAS, 361, 1387
Begelman, M. C. et al. 1991, ApJ, 382, 416
Benson, A. J. et al. , MNRAS, **599**, 38 2003
Best, P. N. et al. 2005, MNRAS, 362, 25
Bond, N. A., Churchill, C. W., Charlton, J. C., & Vogt, S. S., 2001, ApJ, 562, 641
Buote, D. A., & Canizares, C. R. 1994, ApJ, 427, 86
Calzetti, D. 2004, AJ, 127, 1405
Cecil, G., Bland-Hawthorn, J., Veilleux, S. 2002, ApJ, 576, 745
Chen, H.-W., Lanzetta, K. M., & Webb, J. K. et al. 2001, ApJ 556, 158
Chevalier, R. A. & Clegg, A. W. 1985, Nature, 317, 44
Cowie, L. L. et al. 1996, ApJ, 112, 839
Croft, R. et al. 2002, ApJ, 581, 20
Croton, D. J. et al. 2005, astro-ph/0508046
Cox, T. J. et al. 2004, ApJ, 607, L87
Dahlem, M., Weaver, K. A., Heckman, T. 1998, ApJS, 118, 401 (DWH98)
Danforth, C.W. & Shull, J.M. 2005, ApJ, 624, 555
Davé, R., Hernquist, L., Katz, N., & Weinberg, D.H. 1999, ApJ, 511, 521
deYoung, D. S. & Heckman, T. M. 1994, ApJ, 431, 598
Dekel, A., & Silk, J. 1986, ApJ, 303, 39
Devine, D. & Bally, J. 1999, ApJ, 510, 197
Dressler, A. 1980, ApJ, 236, 351
Ferlet, R., Vidal-Madjar, A., & Gry, C. 1985, ApJ, 298, 838
Heckman, T. M. et al. 2004, ApJ, 613, 109
Heckman, T., Lehnert, M. D., Strickland D. K., & Lee, A. 2000, ApJS, 129, 493
Heckman, T. M., Armus, L., & Miley, G. K. 1990, ApJS, 74, 833
Hibbard, J. E. et al. 2000, AJ, 119, 1130
Hoopes, C. et al. 2005, ApJ, 619, 99
Kennicutt, R. C. 1983, ApJ, 272, 54
Kobulnicky, H. A. & Skillman, E. D. 1997, ApJ, 489, 636
Kobulnicky, H. A. & Skillman, E. D. 1996, ApJ, 471, 211

Lacey, C. & Cole, S. 1993, MNRAS, 262, 627
Larson, R. B. 1974, MNRAS, 169, 229
Lehnert, M. D., Heckman, T. M., & Weaver, K. A. 1999, ApJ, 523, 575
Lehnert, M. D. & Heckman, T. M. 1996, ApJ, 462, 651
Leitherer, C. et al. 1999, ApJS, 123, 3
Levenson, N. et al. 2001a, ApJ, 550, 230
Levenson, N. et al. 2001b, ApJS, 133, 269
Mac Low, M.-M. & Ferrara, A. 1999, ApJ, 513, 142
Mac Low, M.-M., McCray, R., & Norman, M. L. 1989, ApJ, 337, 14 1
Marlowe, A. T., Heckman, T. M., Wyse, R. F. G., & Schommer, R. 1995, ApJ, 438, 563
Martin, C. L. 2006, ApJ, submitted
Martin, C. L. 2005, ApJ, 621, 227
Martin, C. L. 2004, in Carnegie Observatories Astrophysics Series, Vol. 4: Origin and Evolution of the Elements, ed. A. McWilliam and M. Rauch (Pasadena: Carnegie Observatories)
Martin, C. L., Kobulnicky, H. A., & Heckman, T. M. 2002, ApJ, 574, 663
Martin, C. L. 1999, ApJ, 513, 156
Martin, C. L. 1998, ApJ, 506, 222
Martin, C. L. 1997, ApJ, 491, 561
Martin, C. L. 1996, ApJ, 465, 68
McDowell, J. et al. 2003, ApJ, 591, 154
McKee & Ostriker 1977, ApJ, 218, 148
Miller, S. T. & Veilleux, S. 2003, ApJ, 592, 79
Murray, N., Quataert, E., & Thompson, T. A. 2005, ApJ, 618, 569
Nicastro, F. et al. 2005, Nature, 433, 495
Ostriker, J. P. & McKee, C. F. 1988, Rev Modern Physics, 60, 1
Ott, J. et al. 2003, ApJ, 594, 776
Percival, W. J. et al. 2002, MNRAS, 337, 1068
Pettini, M. et al. 2002, ApJ, 569, 742
Prochter, G. E., Prochaska, J. X., & Burles, S. M. 2005, submitted to ApJ, astro-ph/0411776
Rupke, D., Veilleux, S., & Sanders, D. B. 2005c, ApJ, 632, 751
Rupke, D., Veilleux, S., & Sanders, D. B. 2005b, ApJS, 160, 115
Rupke, D., Veilleux, S., & Sanders, D. B. 2005a, ApJ, 160, 87
Rupke, D., Veilleux, S., & Sanders, D. B. 2002, ApJ, 570, 588
Scannapieco, E. & Oh, S. P. 2003, ApJ, 608, 62
Schwartz, C. & Martin, C. L. 2006, submitted to ApJ
Savage, B. D. & Sembach, K. R. 1996, ARAA, 34, 279
Silich, S. & Tenorio-Tagle, G. 2001, ApJ, 552, 91
Shapley, A. E., Steidel, C. C., Adelberger, K. L., & Pettini, M. 2003, ApJ, 588, 65
Simcoe, R. A., Sargent, W. L. W., Rauch, M., & Becker, G. D., 2005, astro-ph/0508116
Simcoe, R. A. et al. 2004, ApJ, 606, 92
Somerville, R. S. et al., ApJ, 572, L23 (2002)
Songaila 2005b astro-ph/0509821
Songaila 2005a astro-ph/0507649
Spergel, D. N. 2003, ApJS, 148, 175
Springel, V. et al. 2005, ApJ, 620, 79
Steidel, C. C. et al. 2002, ApJ, 570, 526
Steidel, C. C. et al. 1997, ApJ, 480, 568

Steidel, C. C., Dickinson, M., & Persson, S. E. 1994, ApJ, 437, L75
Steidel, C. C. & Sargent, W. L. W. 1992, ApJS, 80, 1
Stevens, I. R., Read, A. M., & Bravo-Guerrero, J. 2003, MNRAS, 343, 47
Stocke, J. T. et al. 2005, astro-ph/0509822
Strickland, D. K. et al. 2002, ApJ, 568, 689
Strickland, D. K. & Stevens, I. R. 2000, MNRAS, 314, 511
Tenorio-Tagle, G., & Bodenheimer, P. 1988, ARA&A, 26, 145
Thoul, A. A. & Weinberg, D. H. 1995, ApJ, 442, 480
Tremonti, C. A., et al. 2004, ApJ, 613, 898
Veilleux, S., Cecil, G., & Bland-Hawthorn, J. 2005, ARA&A, 43, 769
Veilleux, S. et al. 2003, astro-ph/0308330
Wakker, B. P. & Mathis, J. S. 2000, ApJ, 544, L107
Weaver, K. A., Heckman, T. M., & Dahlem, M. 2000, ApJ, 534, 684
Weedman, D. U. 1986, Quasar Astronomy, Cambridge University Press, (Cambridge)

11 Astronomy with Extremely Large Telescopes

F.G. Watson, I.M. Hook & M.M. Colless

Abstract: Optical and infrared astronomy is today characterised by its access to a spectacular array of observing facilities. Most prominent among them are the 8m to 10m class telescopes, which provide an effective complement to space-based instruments. Lessons learned from the construction and operation of large ground-based facilities have enabled astronomers to plan the next generation of instruments with mirrors 20m to 100m in diameter. This chapter introduces the technology of such 'extremely large' telescopes (ELTs), and then highlights some of the scientific problems they might be expected to address. They include the discovery and characterisation of planets beyond the Solar System, the formation of stars throughout the Universe, the growth of galaxies in the early Universe, and the earliest objects and re-ionisation of the Universe. A key aspect of the case for extremely large telescopes is that observations with 20m and 30m instruments occupy a significantly different region of discovery space from those made with larger (50m and 100m) telescopes. Finally, the complementarity between ELTs and other planned facilities is discussed.

11.1 Introduction

What drives the development of new telescopes–technology or science? Throughout history, the dominant force in this symbiotic relationship has ebbed and flowed. Certainly, for the first 175 years of its life, the telescope advanced in a completely empirical fashion, and scientific outcomes were little more than a fortunate spin-off. The discoveries that were made chiefly concerned objects in the Solar System, whose brightness placed relatively minor demands on the aperture of the rudimentary instruments directed towards them.

11.1.1 Back to the future

It was William Herschel who, with the completion of his 'Large Twenty-foot' telescope in 1783, began the process of burgeoning apertures we are confronted with today (see, e.g., Watson 2004, Chapter 10). Herschel perceived that the aperture of a telescope has a crucial bearing on the study of stellar and nebular objects, and wrote at some length on the 'power of penetrating into space' (e.g.,

King 1955, pp. 136–7; Bennett 1976). Moreover, he had definite scientific questions in mind when he embarked on the quest for light-grasp (and funding) that culminated in his 1.2m aperture 'Forty-foot' telescope of 1789. These related principally to the nature of nebulae. Could they all be resolved into stars, or was there real nebulosity out there? And was the Solar System itself embedded in a 'nebula' made of millions of stars?

As telescopes increased in size during the nineteenth and early twentieth centuries, those questions were eventually answered, and the justification for building ever-larger telescopes concentrated on investigations into the newly-discovered 'island universes' (i.e. galaxies). But in the impassioned plea for a 200-inch (5.1m) telescope that led directly to a grant of $6 million from the Rockefeller Foundation, Hale (1928) was remarkably unspecific about the discoveries such an instrument might make, highlighting instead the engineering challenges. He did mention such questions as the spiral nature and rotation of our own Galaxy, but seemed also to suggest the possibility of direct commercial spin-off from large telescopes:

> A far-sighted industrial leader, whose success may depend in the long run on a complete knowledge of the nature of matter and its transformations, would hardly be willing to be limited by the feeble range of terrestrial furnaces. I can easily conceive of such a man adding a great telescope to the equipment of a laboratory for industrial research if the information he needed could not be obtained from existing observatories.

This appeal to the enormous range in states of matter encountered by astrophysicists, while undoubtedly well intentioned, would carry little weight today. Very few contemporary scientists would dare to include an industrial laboratory role in the justification for a large telescope. The nearest might be the technological spin-off from its development and construction. (For the record, the 1928 grant of $6 million for the construction and operation of the Palomar 200-inch is equivalent to a relatively modest $64 million in 2005 dollars; see Friedman 2005).

Today, with two further generations of large optical telescopes having emerged since the Hale Telescope (the 4m class of the 1970s and the current 8-10m instruments), the known horizons of the Universe have expanded by thousands of megaparsecs. It is populated by exotic new classes of objects including neutron stars and black holes, and is completely dominated by components whose nature is unknown–dark matter and dark energy. We stand on the brink of a new generation of telescopes and, with perhaps naive confidence in the ability of technology to deliver them, look to the scientific questions they might answer. Today, the science driver dominates–although engineering comes in a surprisingly close second, and national prestige is not too far behind. No doubt on all those counts William Herschel would have approved.

11.1.2 Scope of this review

One of the more striking aspects of today's generation of 8–10m optical-infrared telescopes is the speed with which they have focused astronomers' attention on the next generation. The proposed new instruments are likely to be both more diverse in size than their predecessors (because of flexibility in the underlying mirror technology) and fewer in number (because of their very high cost). Generically named after a mid-1990s proposal for a 25m telescope, these ELTs (Extremely Large Telescopes) are now understood to be instruments having apertures anywhere between 20m and 100m.

Despite this diversity, there are common threads in the science cases of the new proposals. Without exception, they anticipate exciting and far-reaching results covering all branches of astronomy. In this chapter, we highlight first some of the main scientific drivers for a telescope towards the larger end of the ELT spectrum, with a mirror 50m to 100m in aperture (Figure 11.1). It is representative of the thinking of a large sector of the astronomical community (see, e.g., Hook 2005a), and covers four key observational themes. In order of increasing redshift, these may be summarised as (1) the discovery and characterisation of Earth-like planets beyond the Solar System, (2) the formation of stars through-

Fig. 11.1. OWL (the Overwhelmingly Large Telescope) is a European design for a 100m telescope. It uses large-scale replication of similar optical and mechanical components (Gilmozzi 2004, Dierickx et al. 2004). (*Courtesy European Southern Observatory.*)

out the Universe, (3) the growth of galaxies in the early Universe, and (4) the first objects and the re-ionisation of the Universe.

While a 50m to 100m telescope might represent the highest aspirations of optical and infrared astronomers, perhaps a broader cross-section of the community subscribes to the pragmatic view that technological and financial challenges are likely to be more readily overcome in a telescope of 20m to 30m aperture. Compelling science cases have been developed for such instruments (e.g. GSMT Science Working Group 2003; GMT Science Working Group 2005), and some of their unique capabilities are also discussed in the chapter.

Such ambitious aims need to be placed in the context of other observational facilities, both ground and space based. In particular, how does the scientific impact of an ELT relate to prospects for new radio astronomy facilities such as ALMA (the Atacama Large Millimetre Array), LOFAR (the Low-Frequency Array) or SKA (the Square Kilometre Array)? And how does it compare with new space based facilities like Planck (measuring the cosmic microwave background radiation), GAIA (stellar kinematics) and the near/mid- infrared James Webb Space Telescope? All these facilities are expected to become operational around the turn of the next decade, or soon afterwards.

Finally, ambitious scientific aims do indeed place serious demands on technology, most notably in the design and instrumentation of the ELTs themselves. There are also serious constraints on the location of these gigantic ground-based instruments. We therefore begin by outlining the responses to some of these challenges proposed by the various ELT projects.

11.2 Extremely Large Telescopes

Telescopes in the 4-m class were built with a common mirror technology, namely monolithic ultra-low expansion discs (usually of glass-ceramic material) with aspect ratios (i.e. diameter to thickness) varying from 6 in the earlier examples to around 15 (with active mirror control) by 1989 (see Wilson 1996 chapter 5; 1999 chapter 3). However, their 8–10m successors have utilised a range of technologies.

They can be summarised as follows:

- (a) 'segmented' mirrors; i.e. hexagonal mosaics of many smaller mirrors, as used in the twin 9.8m Keck Telescopes, the 10.4m Gran Telescopio Canarias, the 9.1m Hobby-Eberly Telescope and the 10.0m Southern African Large Telescope;
- (b) thin, monolithic meniscus mirrors (aspect ratio $\geq$ 40) supported on computer-controlled actuators, as in the two 8.1m Gemini telescopes, the Japanese 8.2m Subaru telescope and the four 8.2m unit telescopes of the Very Large Telescope;
- (c) thick, lightweight (hexagonally-webbed) borosilicate glass (Pyrex) mirrors cast in a rotating oven to attain a concave shape before optical surfacing, as in the University of Arizona's 2×8.4m Large Binocular Telescope, the twin 6.5m telescopes of the Magellan project and the 6.5m upgraded MMT telescope.

(The apertures quoted are the maximum effective aperture in the case of the two instruments with variable pupil diameter, the Hobby-Eberly Telescope and the Southern African Large Telescope.) See also Wilson 1999 pp. 346–7 and references therein; Watson 2004, Appendix 4.

11.2.1 ELT mirror technology

The 'original' ELT (a 25m proposal based on an adaptation of the unconventional single-axis Hobby-Eberly telescope; see Bash et al. 1997) envisaged the use of the segmented-mirror technology outlined in (a) above, and this has been almost universally embraced in the ELTs currently planned. (Up-to-date details of all the individual ELT projects can be found on their websites; see References.)

Segmented-mirror ELT concepts have been presented by several groups over a period of some years. Historically, the main projects can be identified as:

- US (AURA) 30m GSMT (Giant Segmented Mirror Telescope), formerly MAXAT, the Maximum Aperture Telescope;
- Caltech/University of California 30m CELT (California Extremely Large Telescope);
- Canadian (ACURA) 30m VLOT (Very Large Optical Telescope);
- European 50m Euro50 (formerly SELT, the Swedish ELT);
- European (ESO) 100m OWL (Overwhelmingly Large Telescope).

Because of the anticipated high cost of ELTs, strategic alliances are now emerging. In 2004, for example, the first three of these projects amalgamated as the TMT (Thirty Meter Telescope) partnership. In the present discussion, however, frequent references are made to the earlier concepts to highlight differing features of their design. Segmented-mirror ELT concepts are also being studied in China, Japan and Russia.

Because there appears to be no possibility of making a monolithic ELT mirror with current technology, option (b) above is not available. However, option (c) offers the possibility of a sparsely-populated segmented mirror using large circular components in an extension of the Large Binocular Telescope concept (two mirrors on a single mounting).

This line of approach is being taken in the GMT (Giant Magellan Telescope; Figure 11.2), a US proposal to combine seven spun-cast 8.4m mirrors in a hexagonal configuration with the equivalent light-grasp of a 21.4m telescope (and the resolving power of a 24.4m). The arrangement contrasts strongly with the hundreds or thousands of segments that need to be made and controlled in a 'conventional' hexagonally-segmented mirror, and this is its principal advantage. It is also expected to present a cleaner point-spread function (i.e. image profile) due to the absence of strong diffraction spikes from the segment interstices. On the other hand, there are difficulties associated with fabricating large mirrors with a deep, off-axis profile, and a substantial mirror-support structure is required. Moreover, 'conventional' segmented mirrors offer significant advantages in that the filled aperture maximises light grasp for a given structure size, and

Fig. 11.2. Concept design for the Giant Magellan Telescope, which uses seven 8.4m mirrors on a common mounting (Gunnells et al. 2004). (© *Todd Mason/Carnegie Observatories.*)

wider-field designs are possible. It seems unlikely that the GMT concept could be expanded to apertures of 50m to 100m.

The optical systems proposed for ELTs cover a wide range of types. Classical Cassegrain (paraboloid-hyperboloid), Gregorian (paraboloid-ellipsoid) and Ritchey-Chrétien (hyperboloid-hyperboloid) configurations have been suggested for the GSMT, GMT and CELT designs respectively. A subtle advantage of the GMT's Gregorian system is that so-called ground layer atmospheric turbulence (at a height of about 150m above the primary mirror) is imaged onto the concave secondary mirror, allowing the possibility of adaptive correction on its surface. For OWL, a complex six-mirror design using a spherical primary and 25.6m diameter flat secondary is proposed. The remaining four elements of this system comprise a spherical aberration corrector and a folding flat mirror to deliver the image to one of six focal stations surrounding the optical axis. OWL's spherical primary mirror offers the enormous advantage that all its 3048 hexagonal segments (each 1.6m across) are identical, saving costs in fabrication.

In general, ELT optics are expected to be made from glass ceramic, although experimentation is taking place with unconventional mirror materials such as silicon carbide.

11.2.2 Structure and sites

All the current proposals for ELTs have altazimuth mountings, many owing their principal design features to large steerable-dish radio telescopes. Most (except GMT and OWL) incorporate Nasmyth foci near the altitude bearings as their main observing stations. GSMT also includes prime and Cassegrain foci; GMT is Cassegrain-only. A generic feature of ELT design is the very fast focal ratio of the primary mirror (for example, OWL is relatively slow at $f/1.25$), which is intended to minimise the separation of the optical components and save weight by simplifying the secondary support structure. In most (GSMT, GMT, Euro50), this is reduced to a basic tripod whose diffraction pattern overlies that of the segment interstices. The TMT design combines a tripod with a 'conventional' hexagonal space-frame tube.

In contrast, OWL's secondary support is a six-vane spider formed as an open lattice of struts, and this mode of construction characterises the telescope itself (Figure 11.1). Using a so-called fractal design, OWL's structure is highly modular, with large numbers of identical components. Most of the telescope parts can be transported in standard 40ft (12.2m) shipping containers. The web-like construction also contributes both to its structural stiffness (resistance to wind-induced vibration) and its relatively light weight. Although the total mass is expected to be around 14,800 tonnes, ESO claim it is only 1–2% of the mass a conventional 8m telescope would have if scaled up to 100m in aperture.

Most ELT designs envisage telescope enclosures similar to those housing today's 8–10m class instruments. A classical dome has been proposed for GSMT, while the VLOT project has explored the possibilities of an over-hemispheric dome with a movable circular aperture. Cylindrical or rectangular enclosures are suggested for GMT and Euro50. All include large apertures to allow free air circulation around the telescope structure. OWL avoids the need for this by using a hemispheric dome that can be completely removed from the telescope, providing total exposure to ambient atmospheric conditions during observing.

Site testing for ELTs has been in progress for some years, and has evolved into a cogent campaign in which data are shared between the various ELT projects. The perspective on possible sites varies from project to project, ranging from the relatively narrow (such as GMT's current focus on Cerro Las Campanas in Chile) to the truly global (as in ESO's stated aim to consider sites anywhere on Earth for OWL). Some projects (such as GSMT) limit their considerations to the Americas plus Hawaii, which includes the best observing sites known today (Mauna Kea and the mountains of northern Chile).

Suitable observing sites are characterised by:

- freedom from cloud cover;
- low atmospheric turbulence;
- high atmospheric transmission in the optical to mid-infrared;
- low wind speed;
- high climatic stability

- low seismic and volcanic activity;
- minimal light and air pollution.

In addition, political, environmental and cost criteria need to be taken into account, together with new considerations such as the thickness of the sodium layer (the 90km high region of the upper atmosphere that will be laser-excited to create artificial guide stars for adaptive optics).

Mauna Kea, as the best site known in the northern hemisphere, faces problems with site availability and the environmental impact of large telescope structures. Other possible northern sites include La Palma and continental mountain sites in the USA, China and, perhaps more speculatively, Uzbekistan (Ehgamberdiev et al. 2000). In the south, well-established Chilean sites include Cerro Tololo, Cerro Pachon, Cerro Las Campanas, Cerro La Silla and Cerro Paranal, and there are possible new sites at Chajnantor and elsewhere. Gamsberg in Namibia has been investigated by ESO, while the high Antarctic plateau (e.g. Dome C) has received considerable attention recently on account of its very low atmospheric turbulence and low infrared background (e.g. Lawrence et al. 2004). Many of these sites would provide logistical challenges on account of their remoteness, but there is a high level of confidence within the engineering community that these can be overcome at a reasonable cost.

11.2.3 Instrumentation

While it remains true that enormous technological challenges face the builders of ELTs, there seems little doubt that they can be overcome given sufficient funding. It is in the field of instrumentation (which is here taken to include adaptive optics systems) that the biggest questions remain (Russell et al. 2004, Taylor 2004).

As we shall see, many of the science drivers for ELTs demand spectroscopy of faint objects, and this brings fundamental problems in instrument design (see, e.g., Watson 2003). For example, if D is the aperture of the telescope, d is the pupil (grating) diameter of the spectrograph and ϕ is the angular spatial resolution element on the sky, then the spectrograph resolution, $R = \lambda/d\lambda$, varies as

$$R \propto \frac{d}{D\phi}.$$

(e.g., McLean and Chaffee 2000). Thus, to achieve a given spectral resolution for a given telescope aperture, either the spectrograph pupil diameter must be made as large as possible (dictating that spectrographs for ELTs will have optics the size of today's medium-class telescope mirrors) or the image diameter must be made as small as possible.

In addition, if $\mathcal{F}$ is the focal ratio of the spectrograph camera and p is the linear pixel size of the detector, then

$$\mathcal{F} \propto \frac{p}{D\phi},$$

which again necessitates a small image diameter in order to secure a workable camera focal ratio (for a given pixel size critically sampling the image). Even today's 4m and 8–10m class instruments utilise spectrograph cameras working at $\sim f/1$, and there would be significant problems of aberration-control in faster designs.

These and other considerations dictate the use of adaptive optics (AO) to keep ϕ small, ideally approaching the diffraction limit of the telescope ($\sim \lambda/D$). It is possible that the best-attainable natural seeing (the image diameter of a point source after its passage through a low-turbulence atmosphere) might be acceptable for certain wide-field applications in an ELT up to 30m in aperture. Such images are rarely less than ~ 0.3 arcseconds in diameter, however.

The diffraction limit of the 100m OWL in the optical waveband is ~ 0.001 arcseconds, and the science case requires that imaging at this level be attained in practice. The technology for achieving it relies on multi-conjugate adaptive optics, which aims to correct for several turbulent layers at differing altitudes in the atmosphere by sensing the wavefront deformation in each layer, and correcting it using multiple deformable mirrors. It is so critical to OWL's performance that 10% of the total cost budget has been allocated to it. It is also widely regarded as potentially the most serious technological show-stopper for the project.

Set against these difficulties are new developments that promise extremely high performance in instruments fed by AO-corrected ELTs. They include wide-field scanning Fabry-Perôt devices (tunable filters) for narrow-band imaging, and advanced focal-plane technologies such as microrobotic positioners for optical fibres and other microcomponents (McGrath et al. 2003). These microcomponents might include deployable integral-field units to spatially-dissect a portion of the telescope focal surface prior to the light being fed to a spectrograph (see, e.g., Watson 2003). Such facilities would have immediate applications in the kind of multi-object spectroscopy described in Section 11.3.3, for example.

Optical fibres themselves, commonly used to transfer and reformat the light from target objects, are about to undergo a revolution. New forms of construction will allow such features as embedded fibre gratings to filter out the forest of narrow atmospheric emission lines in the J and H wavebands at 1.2–1.6 μm (Bland-Hawthorn 2004). These devices have the potential to offer truly spectacular gains in signal-to-noise ratio. On the timescale of ELT development, it is also possible that photonic devices such as highly-efficient 'spectrographs on a chip' (see Watson 1997) could be cheaply mass-produced.

11.2.4 The bottom line

No-one pretends that ELTs will come cheaply. Current estimates put the capital cost of the 30m TMT at $\sim$ \$700 million, with the cost of the 22m GMT estimated at about two thirds of that. Costings for the proposed European facilities are approximately 600 million euros for Euro50 and 1.2 billion euros for OWL. In addition, annual running costs might be up to 10% of these figures, while the budget for a major auxiliary instrument such as a spectrograph will easily run into tens or even hundreds of millions of dollars.

Such ELT capital costs are large compared with existing facilities for ground based astronomy, but are comparable with only medium-sized particle physics facilities. Set against space projects they are extremely modest: the cost to build, launch and fix the 2.4m Hubble Space Telescope, for example, is estimated at \$2 billion (1990 dollars).

The question that must ultimately be asked is how much funding will be available for an ELT? While some of the projects have already received significant amounts for preliminary studies (running in some cases into tens of millions of dollars), it will be the final appropriation that eventually determines what size of ELT achieves first light.

11.3 Some Science Drivers for a 50m to 100m ELT

The order of magnitude increase in aperture from today's 6.5–10m telescopes to the 50–100m promised by this evolving technology will yield an extraordinary advance in knowledge. Not since the time of Herschel has a $\sim$ 100-fold gain in light collecting area been achieved in a single step. The scientific impact of this will be at the very least spectacular, and could utterly revolutionise our view of both the Universe and ourselves.

In this section, the potential of the largest of the currently envisaged ELTs, the 100m OWL, is presented. A lengthy and well-developed science case has been developed for this telescope by the OPTICON ELT Science Working Group (Hook 2005a), but we here limit ourselves to its most exciting aspects. In particular, those applications that demand the full resolution and light gathering power of the 100m aperture are highlighted (Hook 2004, 2005b).

11.3.1 Extra-solar planets

There is no doubt that the most accessible of the roles that will be fulfilled by an ELT is the search for analogues of the Solar System. In the popular imagination, the quest for Earth-like planets, their detailed investigation and perhaps the discovery of life, is the greatest justification for spending public money on such facilities.

This role has been given added impetus in the last decade by the discovery of some 130 extra-solar planets by the so-called Doppler wobble (reflex velocity) method (e.g. Mayor and Queloz 1995; Marcy and Butler 1996; Butler et al. 1999; Tinney et al. 2001; see also IAU Working Group 2005), and it is now known that at least 7% of solar-type stars have giant planets (comparable in mass with Jupiter) in orbits with semi-major axes of 5 AU or less. Because of the high angular resolution offered by a 100m ELT and the possibility of effective coronographic suppression of the central star image, this discovery rate is likely be greatly enhanced by direct imaging. Giant planets of Neptune's mass in orbits down to $\sim$ 1AU will be accessible to distances as great as 100pc.

Terrestrial planets. A terrestrial planet can be defined as one whose mass is less than 10 Earth masses, and whose surface temperature is within the approximate range $270 < T < 320$K. The latter criterion requires that the planet lie in the habitable or 'Goldilocks' zone (where the temperature is just right for liquid water to exist; see Murdin 2000), which means its orbit around a solar-type star will have low eccentricity and a semi-major axis of ~ 1 AU.

These criteria mean the search for terrestrial planets by direct imaging will be one of the greatest challenges facing an ELT. At a distance of 30pc, 1 AU subtends an angle of only 0.033 arcseconds (33 mas). But this is hugely compounded by the fact that an Earth-like planet will be some 25 magnitudes fainter than its parent star in the optical waveband, a luminosity difference of 10^{10}. In order to reveal the planet, a coronograph capable of extreme suppression ratios is required, and stringent demands are placed on the effectiveness of the adaptive optics system. These centre on the Strehl ratio (sharpness) of the delivered image and its freedom from halos and other non-systematic artefacts.

The observing strategy will then be to carry out a survey of ~ 1000 stars in the solar neighbourhood within ~ 30pc. Observations of the Solar System at this distance with such equipment would reveal all the planets except Mercury and Pluto. The survey will provide vital data on the statistics of multiple-planet systems and the frequency with which terrestrial planets occur. Moreover, repeated observations will allow the orbits of individual objects within these planetary systems to be determined.

Physical properties and biospheres. The properties of directly observable planets can be deduced partly from their orbital parameters and the presence of external influences such as neighbouring planets, but more especially from photometry and spectroscopy of the planets themselves. Light curves may reveal rotation and surface features, as well as Saturn- like ring systems. (Such rings give rise to a phase shift of the orbital light curve, as well as low-intensity features induced by ring shadow phenomena; Figure 11.3.)

Spectroscopy is undoubtedly the Holy Grail of exo-Earth studies. In the visible spectrum and near-infrared J band ($\sim 1.25\mu$m), there are features of O_2, O_3, H_2O, CO, CO_2 and CH_4, which are signatures of habitable atmospheres. Water vapour, in particular, is a powerful indicator of habitability for life with a similar carbon-based chemistry to ours. Any H_2O features in the J-band of an extra-solar planet will be distinguishable from the same features in the Earth's atmosphere by the relative Doppler shift induced by the motion of Earth and planet around their respective parent stars.

A similar technique may allow the B band of oxygen (in the far-red at 760nm) in the atmosphere of an extra-solar planet to be disentangled from that in the Earth's atmosphere. The presence of oxygen is highly likely to be indicative of large-scale photosynthesis, and is therefore considered to be a reliable biomarker (i.e. a diagnostic of a biosphere). Another is the spectrophotometric feature at 725nm known as the vegetation red edge (VRE), detectable in the terrestrial atmosphere in the spectrum of lunar Earthshine. The presence of this feature would be a strong indicator of green photosynthetic vegetation; it may even be possible to correlate rotational variability with the presence of an Amazon basin

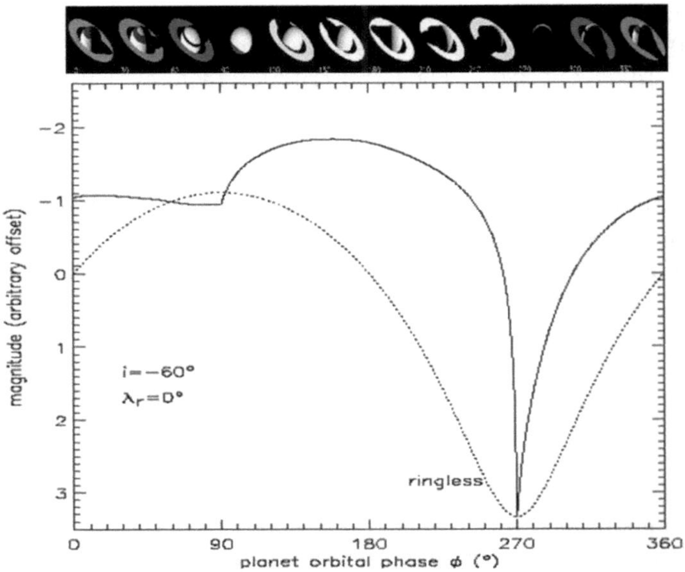

Fig. 11.3. Simulation of the light curve of a ringed planet with orbital phase (solid line) compared with that of a planet with no ring (dotted line). A uniform albedo is assumed. (See, for example, Schneider, Aronold and Borkowski, 2003.) (*Courtesy J. Schneider.*)

analogue, for example. Dry vegetation and soil also give distinctive profiles in the far-red and near-infrared (Clark 1999), and other mineral and geological evidence may come from this spectral region. Most exciting of all is the possibility of quite unexpected markers. While it is probably too far-fetched to expect industrial pollutants to reveal themselves in the atmospheres of extra-solar planets, there may be tracers of exotic chemical and biochemical processes in action. In themselves, these would provide a significant challenge to astrobiologists.

Other aspects of planetary systems. While the search for and study of Solar System analogues is a prime science driver for a 100m ELT, many questions relating to planetary studies will be illuminated by observations with such a telescope. These range from the formation of protoplanetary disks and the detection of emerging planets (via the presence of dark lanes in the disks) to the behaviour of mature planetary systems in diverse environments (around metal-poor stars, high-mass stars, white dwarfs and brown dwarfs, for example).

Protoplanetary disks around young stars do not require high levels of suppression to remove the image of the parent star ($\sim 10^6$ is adequate) but the spatial resolution requirements are not significantly relaxed because of the greater distances at which such stars are typically found. Young gas giant planets are self-luminous, and therefore observable in the near-infrared with similarly lower levels of suppression. Observations of the evolution of such objects with age will allow theories of planetary formation to be tested and refined.

A particularly intriguing prospect is the discovery of Earth-like satellites of giant planets in the habitable zones of solar-type stars. While not able to be resolved directly, such hypothetical objects might be revealed by reflex velocity or astrometric measurements of their parent planets.

Finally, the application of ELT observations to planets is not limited to 'conventional' exo-Solar Systems. The free-floating planetary-mass objects recently discovered in Orion and elsewhere (e.g. Lucas, Roche and Riddick 2003) are potential tracers of the low end of the initial mass function for cloud fragmentation. Their isolation renders them easier to study than conventional planets in orbit around stars. And, of course, ELT observations will supplement and enhance the work of fly-by and planet-orbiting spacecraft in our own Solar System. With 3m resolution available at the Moon's distance and 8km at Jupiter's, there are excellent prospects for the long-term monitoring of planets and satellites. Moreover, studies of asteroids, Kuiper-belt objects (including Pluto and Charon) and comets all stand to gain enormously from the application of ELT studies.

11.3.2 The formation of stars throughout the Universe

Stars have a singularly important function in the evolution of the Universe. They are the furnaces within which elements are forged, and it is only through a broad understanding of the processes involved that we have come to know how the Universe has arrived at its present state. Planets, moons, asteroids and comets–together with the dust from which they formed and any life-forms that might inhabit them–all have their ultimate origin in the hot interiors of stars. And in their later phases, stars make up some of the most exotic objects we find today: white dwarfs, neutron stars and black holes.

Given this pivotal role, it is remarkable that our knowledge of the rates at which stars have formed throughout the Universe's history is very sketchy. Understanding this in greater detail is another key science driver for an ELT, and it is likely that several approaches will be taken to address the problem. Two are suggested here.

Resolved stellar populations. Critical information relating to star formation is encoded in the stars themselves, and can be read by means of their colours ($(B-V)$, $(V-I)$ etc.) and spectra (luminosity class, metallicity [Fe/H], etc.). Stellar atmospheres retain the element abundances that prevailed at the star's formation, and thus provide a marker of age. The difficulty is that for the vast majority of stars in the Universe, we see only their light integrated with that of billions of other stars in their parent galaxies, and the required information is irretrievable.

This problem is further exacerbated by the fact that those stars that can be resolved with present-day facilities are all within the Local Group of galaxies, which is not fully representative of the population of galaxies as a whole. There is, for example, no specimen of a large elliptical galaxy within reach of present-day telescopes for the study of individual stars. That means there is no possibility of contrasting star-formation rates in widely differing environments. In particular, there is no way of constraining theories of galaxy formation by looking at the age distributions of the oldest stars in galaxies throughout the Hubble sequence (tuning-fork diagram).

With a 50–100m ELT, this situation will change dramatically. The stars of Virgo Cluster galaxies (distance modulus 30.9; distance 16 Mpc) will be accessible, even in this confusion-limited application, by virtue of the ELT's high spatial resolution at optical wavelengths, together with its ability to image faint stars (typically $V \sim 35$ at this distance). This will allow the construction of colour-magnitude diagrams for stellar populations and the determination of such evolutionary phases as the main-sequence turn-off. Frayn (2003) has shown that this will be detectable for all stellar ages at the distance of Virgo with a 100m telescope, but not with a 30m.

The identification of distinct stellar populations by means of age and composition can be augmented with kinematic studies to provide information on the merger history of galaxies (Freeman and Bland-Hawthorn 2002). The halo of our own Galaxy shows clear evidence of tidally disrupted remnants of accreted satellite galaxies (shown in simulation in Figure 11.4), and the study of this history is one of the motivations for the RAVE (RAdial Velocity Experiment) survey of stellar radial velocities and metallicities (Steinmetz 2003, Watson et al. 2003). Such studies are also possible in the Galaxy's nearest neighbours, but an ELT will allow them to be extended to Virgo and beyond. Furthermore, the nearest major merging event currently taking place (the Antennae, NGC 4038/9, at a distance of 18 Mpc) will be similarly accessible to kinematic study.

The cosmic star-formation rate from supernovae. Another approach to this problem can be made from an entirely different perspective. Massive stars end their brief lives in supernovae (with 3–8 solar mass stars producing Type Ia supernovae on timescales longer than a billion years, and more massive stars yielding Types Ib/c and II on much shorter timescales). These cataclysmic explosions can be seen over very great distances, and their frequency is directly related to the number of stars that have formed at that epoch. Thus, the measurement of the rate of supernovae over a range of lookback times (corresponding to a redshift, z, of up to $\sim$10, which is possible with a 100m ELT) is a direct way of determining when stars formed, and in what quantities (Della Valle et al., 2005).

Such an experiment requires a well-constructed survey of supernovae, and it is suggested that some 130 nights of ELT time would net a total of $\sim$ 400 supernovae (of which about half would also be observed spectroscopically). This would be sufficient to provide a direct, independent measurement of the star formation rate up to $z \sim 10$ (Madau, Della Valle and Panagia 1998, Miralda-Escude and Rees 1997, Heger et al. 2001).

These observations would also have direct relevance to cosmology, since the supernovae act as standard candles and can be used in an extended Hubble diagram to discriminate between different cosmological models. Type Ia supernovae are reliable distance indicators to $z \sim 4$, and Type II objects are expected to provide cosmological information to $z \sim 10$ (Figure 11.5). Combined with WMAP measurements of the cosmic microwave background and data on the large-scale structure of today's Universe (e.g. from the 2dF Galaxy Redshift Survey; Colless

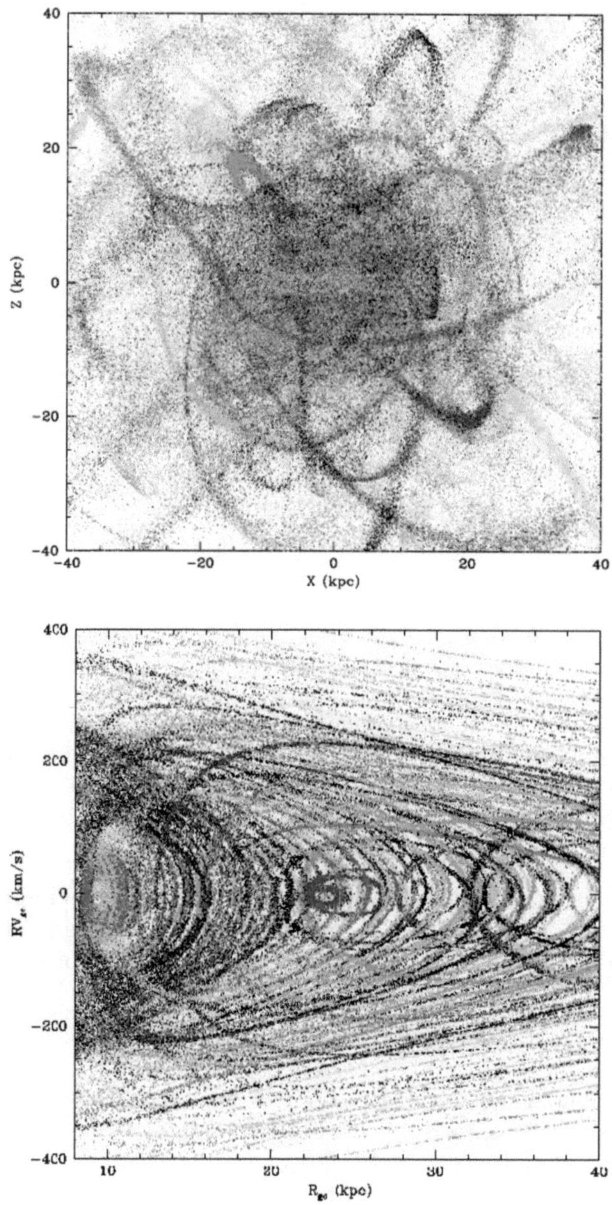

Fig. 11.4. Simulation of a galactic halo made up of the tidally-disrupted remnants of $\sim$ 100 satellite galaxies, plotted in spatial coordinates (upper diagram) and phase space (lower diagram). (*Courtesy P. Harding and H. Morrison.*)

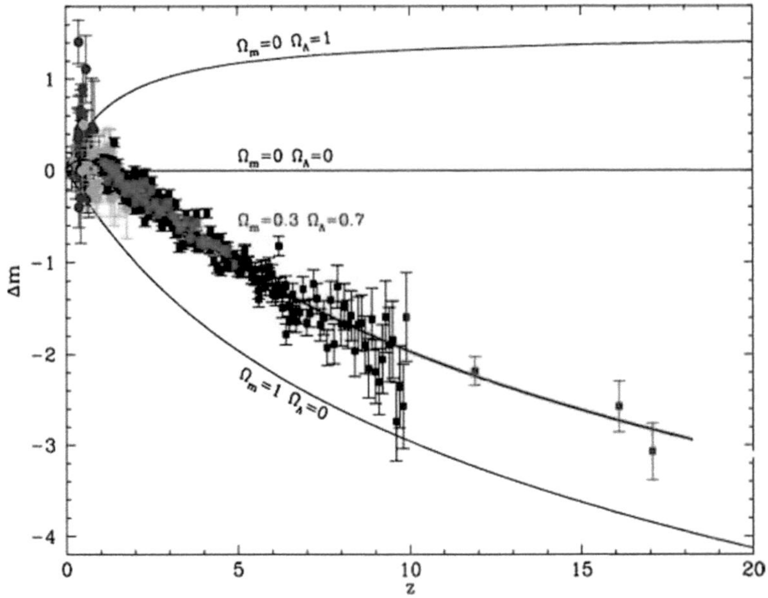

Fig. 11.5. Hubble diagram normalised to a cosmological model for an empty Universe, with simulated observations of supernovae of Type Ia (to $z \sim 5$), Type II and Ib/c (to $z \sim 10$) and from stars of Population III (to $z \sim 17$; see Section 3.4). The simulated observations have been distributed around the track $\Omega_M = 0.3$, $\Omega_\Lambda = 0.7$. The diagram also contains real observations (to $z \sim 2$) of Type Ia supernovae from ground-based telescopes (Perlmutter et al. 1998, 1999; Riess et al. 1998; Knop et al. 2003; Tonry et al. 2003) and from the Hubble Space Telescope (Reiss et al. 2004).

1999, 2001), the supernova observations will constrain various important cosmological parameters. This has particular importance to the study of dark energy, the most abundant (70%) and least-understood component of the Universe.

11.3.3 The growth of galaxies in the early Universe

Just as our present understanding of the star-formation history of the Universe is a broad-brush picture, so it is with our knowledge of the origins of galaxies. While it is clear that the residual temperature fluctuations in the cosmic microwave background radiation relate to the earliest growth of structure in the Universe (e.g. Jaffe 2003, Fairall 2003), the details of that process in terms of the emergence of galaxies remain very uncertain.

The measurement of star formation rates described above clearly relates to the growth of galaxies, but provides no indication of how star formation progressed within any given galaxy or class of galaxies. Moreover, we know that present day structure in the Universe is dominated by dark matter, so simply investigating the baryonic (visible) component of galaxies is unlikely to constrain different models for their growth. We also know that the physical processes that control galaxy growth are complex, involving shocks, feedback from massive stars and active galactic nuclei, merging and the interplay between dark and baryonic matter.

The current best picture of the growth of galaxies involves a hierarchical merging process (Searle and Zinn 1978, Ellis 1998) in which the assembly of mass is separated from the star-formation history. Thus, small galaxies formed first within large dark matter halos. Those that grew into massive galaxies did so by accreting other small protogalaxies, which fed baryonic material into the spheroids and disks of the growing galaxies. The extreme case is the formation of elliptical galaxies, which formed most of their stars (in progenitor galaxies) at high redshift ($z > 3$), but which have undergone major mergers that have significantly increased their mass at lower redshifts ($z < 2$). There is still a need to secure unequivocal evidence that these models are correct, as well as to address questions such as the way in which star formation proceeded in this scenario, and whether some galaxies formed in a non-hierarchical way. (See also Eggen, Lynden-Bell and Sandage 1962; Fukugita, Hogan and Peebles 1998.)

Exploring the growth of galaxies. A 100m ELT provides an appropriate tool for exploring this problem. Again, both the high spatial resolution and high light-gathering power of the instrument are crucial, for both imaging and spectroscopy are essential in addressing the problem. With a diffraction-limited 100m telescope, galaxies appear as extended objects at all cosmological distances (for example, 50pc subtends 6mas at $z = 1.5$; see Figure 11.6). While stars are, of course, unresolved at moderate to high redshifts ($z \sim 1.5$–5.0), the giant HII regions associated with massive hot stars are detectable and can be analysed spectroscopically to reveal element abundances.

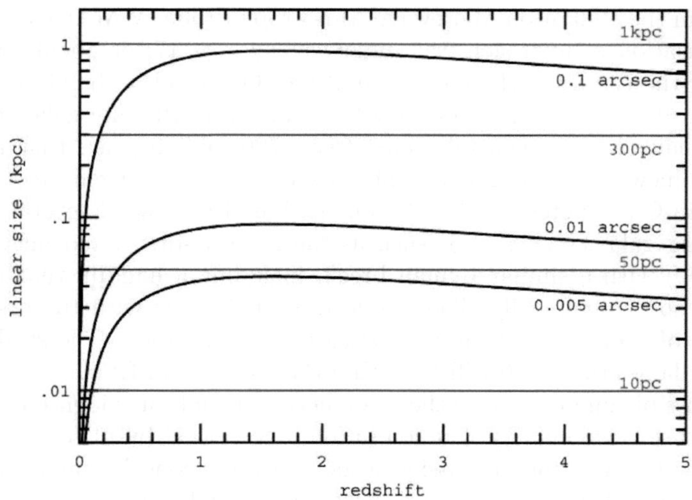

Fig. 11.6. Linear size of objects as a function of redshift, plotted for a cosmology with $H_0 = 65$ kms^{-1} Mpc^{-1}, $\Omega_M = 0.3$, $\Omega_\Lambda = 0.7$. The horizontal lines represent typical linear sizes of a starburst region (1kpc), a giant HII region (300pc), a small HII region (50pc) and a globular cluster (10pc).

More importantly, the velocities of both HII regions and satellite galaxies can be determined from observations of such bright spectral features as H-alpha, [OIII] λ 5007Å and [OII] λ3727Å. Velocity measurements of sufficient numbers of these objects allow the quantity and distribution of dark matter to be traced by kinematic methods. Such studies are among the few means at our disposal for investigating the characteristics of dark matter, but they are difficult to accomplish. The faintness of the targets requires whole nights of integration time, even on a 100m telescope. However, the use of deployable integral field units to allow image-dissected spectroscopy of several objects simultaneously will speed progress.

The evolutionary history of the clumping of dark matter thus obtained will be combined with high-resolution imagery of infant galaxies to present a cogent picture of galaxy formation from its initial phase to maturity. This will provide a critical test of our understanding of the basic evolutionary processes at work in the Universe.

11.3.4 The first objects and the re-ionisation of the Universe

At the epoch of the cosmic microwave background radiation, the cooling of the Universe left atoms of hydrogen and helium in their neutral state. This radiation-absorbing gas then became the raw material from which the first luminous objects evolved. It was ultraviolet emission from these objects that began re-ionising the Universe, so that it again became transparent. One of the problems facing cosmologists is to understand how and when those objects formed, and what was their nature (see, e.g., Keel 2003).

Today, there is conflicting evidence regarding these questions. Analysis of results from the Wilkinson Microwave Anisotropy Probe (WMAP) suggests that the re-ionisation of hydrogen was underway by $z \sim 17$, or about 180 million years after the Big Bang (Bennett et al. 2003). On the other hand, observations of high-redshift ($z \sim 6$) quasars show that enough of the intergalactic medium (IGM) remained in its neutral state after ~700 million years that almost all radiation blueward of the Lyman-alpha line of atomic hydrogen (at λ 1216Å) is absorbed in Gunn-Peterson (1965) troughs. We also know of several candidate star-forming galaxies at $z \sim 6$ (such as the recently-discovered massive spiral HUDF-JD2) with a similar Lyman break, including a handful that have been confirmed spectroscopically. These objects probably vie with the first quasars as sources of radiation for the later stages of re-ionisation. The study of such resolved galaxies to $z \sim 10$ will be within the scope of an ELT.

One possible inference from these conflicting strands of evidence is that there may have been two epochs of re-ionisation, one caused by ultraviolet emission from the first generation of massive stars, and the second by early quasars and galaxies. Or could there have been a slower, more stochastic re-ionisation process that progressed on a time scale spanning the two epochs? Given that our understanding of the process suggests the early IGM was a "Swiss cheese" of ionised bubbles (Strömgren spheres) surrounded by neutral gas, this seem a reasonable possibility. An associated question concerns the enrichment of the early IGM with heavier elements: when did that begin? The answers to these questions lie in observations of highly luminous and extremely remote background objects, whose radiation can act as a tracer of the ionisation state of the IGM through which it travels. Such tomography of the Universe is only possible with an ELT, as smaller ground-based telescopes do not have access to these (relatively rare) background objects in sufficient numbers.

These first bright objects in the Universe are also of great intrinsic interest. Moreover, they allow the identification of early sites of star and black hole formation. The high level of absorption by neutral hydrogen in the IGM dictates that they must be observed in the near infrared. They are all unresolved point sources, and are particularly suitable for observation with a diffraction-limited 100m ELT. In such observations, the contamination from background sky emission is very small, making them vastly superior to space-based observations at wavelengths less than 3μm.

Gamma ray bursts at very high redshifts. The afterglows of gamma-ray bursts (e.g. Vedrenne and Atteia 2003), though short-lived, are sufficiently bright that they are detectable up to $z \sim 15$–20. The more luminous of them at $z \sim 10$ will permit high-resolution ($R \sim 10{,}000$) spectroscopy in the near-infrared (1–2μm) for perhaps 10 days after the burst. It may be noted that similar observations would be possible with a 30m ELT, but for a much shorter time (~ 1 day) after the burst. (See also Lamb and Reichart 2000.).

Population III supernovae. The very first generation of stars to form (the so-called Population III) was by definition extremely metal-weak. Massive stars of this population (140–260 solar masses) explode as very bright supernovae, which are expected to be observable to $z \sim 16$ with a 100m ELT for approximately a month after the event. It is expected that this population of first supernovae will have disappeared by the time the metal enrichment of the interstellar medium reaches 1/10000 of the solar value. The determination of the redshift at which this occurs will itself be a major challenge.

Both Population III supernovae and long period gamma ray bursts (which are thought to be hypernovae at high redshift) may also offer the possibility of extending our knowledge of the cosmic star-formation rate to very early epochs simply by counting them (see Sect. 11.3.2).

Quasars. Understanding when quasars formed is a major science goal of both ground and space based telescope projects. Bright quasars at $z \sim 6$ discovered in the Sloan Digital Sky Survey (York et al. 2000) are presumed to be powered by supermassive black holes, leading to speculation that intermediate luminosity objects with intermediate-mass black holes will be found at $z \sim 10$ and possibly higher redshifts. While these relatively faint objects are expected to be discovered with the James Webb Space Telescope, the high-resolution follow-up spectroscopy needed to investigate the IGM and its ionisation state can only be done with a 100m ELT.

11.4 Science with ELTs up to 30m

The 100m ELT is the largest category of optical-infrared telescope currently envisaged, and it is likely to represent the ultimate aspirations of astronomers for the foreseeable future. But because it presents such extreme technological and financial challenges, many in the community have suggested that there is a need for an intermediate step between 10m and 100m class telescopes. As we have seen, a number of such instruments up to 30m in aperture have been proposed, and have well-developed science cases. Proponents of a 100m telescope, on the other hand, might argue that since the larger aperture merely represents 'more of the same' in terms of adding mirror segments and scaling the structure, one may as well think big.

In many areas of optical-infrared astronomy, there is a continuum improvement of the science that can be achieved as the aperture of the telescope increases. There may also be 'threshold points' at which a telescope of a certain size enables new areas of study to be opened up. Some of these have been identified in the previous section. However, there are also areas in which the 20m and 30m ELTs present a real advantage, and the science cases for these instruments highlight those aspects, as well as the common-ground science up to the limits imposed by their aperture.

11.4.1 Wide field applications

Much of the science case for the 100m class telescope hinges on diffraction-limited ($\sim$ 1 mas) imaging, requiring a very advanced level of adaptive optics correction (Sect. 11.2.3). Because this process requires correction of the turbulence profile at several different levels in the atmosphere, the eventual size of the isoplanatic patch (the angular diameter over which the incoming wavefront is coherent) is small. In the 100m OWL proposal, the area over which diffraction-limited imaging will be achieved is around 2 arcminutes across.

In any area of astronomy requiring large-scale survey observations, field of view is critically important. A simple metric in determining the effectiveness of a survey telescope is its $A\Omega$ product (aperture$^2\times$ field of view^2). It is this quantity that allows the 1.2m UK Schmidt Telescope, for example, to continue producing competitive science in an era dominated by 8m telescopes (e.g. Watson 2003). For the background-limited applications that will dominate much ELT science, the figure of merit can be enhanced to become $A\Omega/\phi^2$ (where ϕ is the angular spatial resolution on the sky). Thus a combination of wide field and high spatial resolution will provide a formidable tool for faint-object survey work.

Both GSMT and GMT highlight wide-field ($\sim$ 20 arcmin) applications in their science cases (GSMT Science Working Group 2003; GMT Science Working Group 2005). They involve simultaneous multi-object spectroscopy of up to 2000 objects (stars or galaxies) within the field to enable the rapid collection of data on survey scales. Microrobotic fibre positioning systems for such instruments are already well-developed (Moore and McGrath 2004). Applications include multi-object tomography of the Universe to determine the ionisation state of the IGM

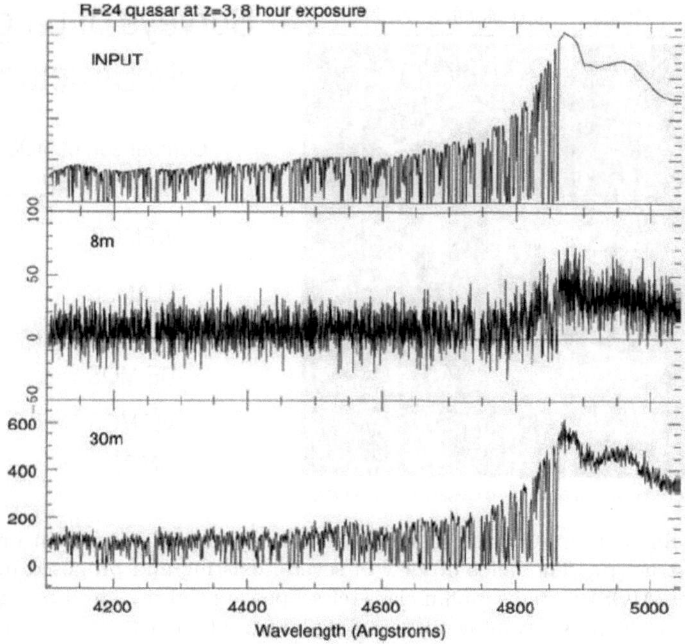

Fig. 11.7. A simulation of the forest of hydrogen and metal absorption lines in the spectrum of a distant quasar caused by discrete clumps of gas in the intergalactic medium (see Sect. 11.3.4). The middle spectrum shows the best that could be expected from an 8m class telescope, while at the bottom is simulated data from a 30m class telescope. The real power of such an instrument lies in its ability to deliver high signal-to-noise spectra for thousands of objects simultaneously in its wide field of view. (*GSMT Science Working Group 2003; courtesy J. Bechtold.*)

and the origin of large-scale structure (compare Sect. 11.3.4), and the evolution of galaxies (compare Sect. 11.3.3).

Some studies are so dependent on the $\sim$ 100-fold advantage in field area over a 100m telescope that they can *only* be carried out with a 20m to 30m ELT. Such investigations include velocity and metallicity surveys of stellar populations in the Andromeda Galaxy over large areas of the bulge, disk and spiral arms, together with related studies on other Local Group members. Similarly, the effects of large-scale structure on galaxy formation will only be revealed by dense tomography of the cosmic web on scales of tens of arcminutes using Lyman Break Galaxies as background sources. While a 100m telescope will resolve details of the gas inflow and outflow in an individual forming galaxy, the exploration of the large-scale picture will be a task much better suited to a 20m to 30m ELT.

Despite its smaller aperture, GMT will be particularly useful for such studies because its optical configuration allows effective ground-layer adaptive optics correction using natural guide stars to give 150 mas imaging over its full 20 arcmin field. The quantity $A\Omega/\phi^2$ is therefore maximised. Furthermore, this is

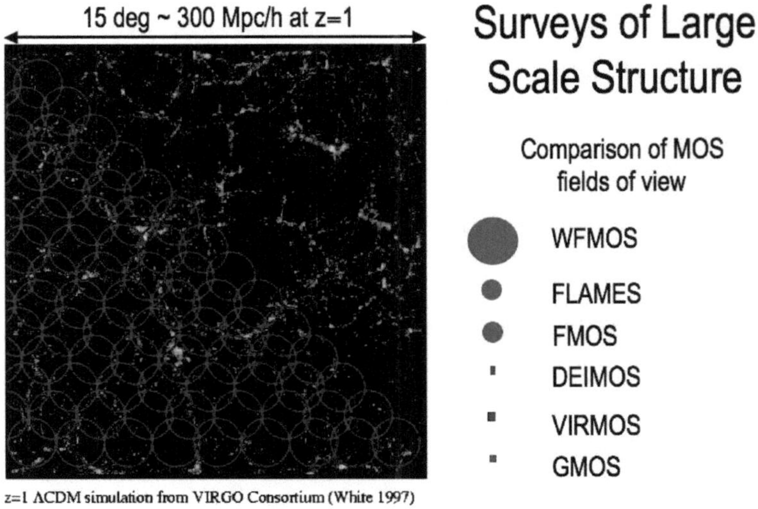

Fig. 11.8. Simulation demonstrating the advantage of field of view in investigating large scale structure. The fields of view of several existing and proposed multi-object spectroscopy (MOS) systems on 8m class telescopes are shown. The 1.5-degree field of WFMOS maximises its advantage in sampling large scale structure at $z = 1$. Similarly, the relatively large fields of 20m and 30m ELTs will make them very suitable for cosmological studies.

achieved using the adaptive Gregorian secondary mirror, reducing the number of optical surfaces in the light-path and, in turn, minimising scattering.

11.4.2 Comparison of scientific capabilities

The Science Case for a 50m to 100m ELT (Hook 2005) includes a summary table comparing the potential capabilities of ELTs of various apertures, and an extended version is presented here in Table 11.1. 'Y' indicates that there is a significant advantage for an ELT of a particular size compared with the next smaller size, while 'N' is shown when that size of ELT cannot make a major contribution in that scientific area. (A temperate mountain site is assumed in all cases.)

The comparison is indicative (rather than quantitative), but does illustrate those threshold points where greater aperture becomes the overriding consideration. It also shows some of those areas in which a 20m to 30m ELT might make a particularly significant contribution. Most noticeable, however, is the fact that almost all areas of astrophysics will derive great benefit from work with an ELT of any aperture.

Table 11.1. Scientific Capabilities of ELTs

Science Case	20m	30m	60m	100m
Solar System	Y	Y	Y	Y
Exo-Planets (direct detection): Gas Giants	Y	Y	Y	Y
Exo-Planets (direct detection): Exo-Earths	N	N	Y?	Y
Proto-Planetary disks	Y	Y	Y	Y
Resolved Stellar Population: Local Group	N?	Y	Y	Y
Resolved Stellar Population: Virgo	N	N	N	Y
Population surveys: Local Group	Y	Y	N	N
Massive Black Holes	Y	Y	Y	Y
Star formation History of the Universe	Y	Y	Y	Y
Physics of Galaxies and Dark Matter, z =1–5	Y	Y	Y	Y
Large-Scale Structure and Galaxy Formation	Y	Y	N	N
Dark Energy	Y	Y	Y	Y
High-z Universe: Sources of re-ionisation	?	?	Y	Y
High-z Universe: $R = 10^4$ on GRB at $z > 10$	N	N	Y	Y
The Unexpected	Y	Y	Y	Y

11.5 Complementing Other Facilities

It is not only ground-based optical-infrared astronomers who have their attention firmly fixed on the prospects for exciting new telescopes. New space-borne facilities have the potential to complement ELTs in the optical-infrared (OIR) waveband, and to exceed their capability in certain areas because of the unique attributes of the space environment for astronomical observations. Likewise, forthcoming radio facilities will also enhance the effectiveness of a proposed ELT.

11.5.1 Space-based astronomy facilities

Space-based astronomy is traditionally associated with wavebands that are inaccessible from the ground (such as the flagship X-ray facilities *Chandra* and XMM-Newton), but the Hubble Space Telescope demonstrated that a modest (2.4m) aperture UV-optical telescope above the atmosphere can revolutionise astrophysics.

The principal advantages offered by a space-based OIR telescope are high spatial resolution, a point spread function that is stable over large field of view, no dependence on atmospheric transmission windows, a low, stable background, continuous access to parts of the sky without day/night limits (from a Sun-Earth L2 orbit), and high sensitivity access to the near- and mid-infrared. It is this last aspect in particular that has prompted the development of the OIR James Webb Space Telescope (JWST), a 6m class instrument with 0.6–27 μm capability, which is expected to be operational around 2012. The sensitivity of this instrument will allow it to outperform an ELT for imaging at wavelengths beyond 3μm. Shortward of this wavelength, however, and particularly in high dispersion spectroscopy, a 100m ELT will do better than the JWST by a factor of perhaps 100 in signal-to-noise. This immediately highlights the main area of complementarity between these two facilities, for without the capabilities of the ELT there will be no means of conducting spectroscopic follow-up on the new objects that the JWST will reveal. The science case for the JWST covers similar areas to those outlined for a 100m ELT, modified by the lower resolution (spatial and spectral) and light grasp. A recent document (GSMT Science Working Group 2005) highlights synergies between JWST and a 30m ELT.

Other space missions that are relevant to ground based ELTs include:

- Planck (mapping the cosmic microwave background radiation), launch date 2007. ELT opportunity: follow-up observations (imaging and spectroscopy) of galaxy clusters revealed by the Sunyaev-Zeldovich effect.
- GAIA (stellar kinematics), 2011. Follow-up ELT observations of planetary systems around other stars.
- TPF and Darwin (terrestrial planet search), 2014-20. Follow-up and complementary ELT observations of planetary systems (see Angel 2003).

In general, the central position of OIR radiation in the electromagnetic spectrum and the fact that stars emit the bulk of their radiation in these bands guarantee ELTs a key ongoing role in space-based observational astronomy.

11.5.2 Radio astronomy facilities

There is a clear symbiosis between ELTs and forthcoming radio astronomy facilities. Interferometric radio telescopes (arrays) provide similar sensitivity and resolution (both spatial and spectral) to diffraction-limited ELTs, and the wavelength baseline defined by the two classes of instrument is of immense astrophysical importance. Some of the most important are:

- ALMA (Atacama Large Millimetre Array). 0.3–10mm milliarcsecond imaging and spectroscopy with R up to 10^7, currently under construction (completion estimated 2011). Complementarity with ELTs in protoplanetary disks, star formation, dust, and imaging black hole accretion disks and jet collimation regions.
- LOFAR (Low-Frequency Array) and related SKA precursors (Allen Array, Mileura Widefield Array, etc.). Complementarity with ELTs in epoch of reionisation studies, IGM at $3.5 < z < 11$.
- SKA (Square Kilometre Array). $>$ 1cm milliarcsecond to arcsecond imaging and spectroscopy with R up to 10^7. (Completion estimated around 2020, although a 10 percent version may be making high-redshift HI observations by 2015). Complementarity with ELT in the early Universe in HI and CO (the 'dark ages'), studies of black holes and other high-energy phenomena, supernovae, star formation history.

Consideration has also been given to the use of ELTs in spectral regions other than their design wavebands, most notably with OWL operating in the submillimetre region (300μm to 2mm). The telescope's collecting area would be effectively twice that of ALMA (although with inferior spatial resolution), and observations with a submillimetre camera (SCOWL; Holland et al. 2003) could complement the array in providing wide-field mapping.

Finally, there are clearly synergies between ELTs and non-electromagnetic facilities. Cosmic ray and neutrino observatories are now opening new windows on the Universe, while gravitational wave instruments such as LIGO (and, eventually, the space-based LISA) are probably on the brink of making their first detections. Collaboration between these widely differing approaches will be mutually beneficial not only to astrophysics, but to physics in general.

11.6 Conclusion

Perhaps the most important lesson in what we might expect to discover from a major new astronomical facility comes, once again, from history. In the early 1840s, Lord Rosse invested some £12000–then a substantial fortune–in the 1.8m reflecting telescope that became known as the Leviathan of Parsonstown (see, e.g. Watson 2004, Chapter 12). His motivation was the same as Herschel's–to see whether all nebulae could be resolved into individual stars. By mid-March 1845, only a matter of days after first-light, he had convinced himself that they could. The fact that he was quite wrong did not emerge until the work of William Huggins nearly twenty years later. But what he discovered next was even more important, a subtle detail that no previous telescope had been large enough to reveal–spiral structure in galaxies.

When the first ELT enters service sometime within the next decade or two, it will almost certainly be the completely unexpected discovery that grabs the headlines. As the GSMT science case declares, 'this telescope will have the light gathering power and angular resolution to open up discovery spaces that virtually assure uncovering of unanticipated phenomena'. Or, put another way, we should welcome serendipity with open minds. Perhaps these discoveries will concern extra-solar planets, perhaps the early Universe. But whatever they are, we can be confident that we will look back on the telescope's science case and smile at just how wide of the mark we were.

Acknowledgements

Both the ELT technology and science case highlights presented here represent the work of a large number of astronomers from a truly international community. The authors are grateful to them all for making available such a wealth of ideas.

References

Websites for individual ELT projects:
22m GMT (Giant Magellan Telescope): http://www.gmto.org/
30m GSMT (Giant Segmented Mirror Telescope): http://www.aura- nio.noao.edu/

30m CELT (California Extremely Large Telescope): http://celt.ucolick.org//
30m VLOT (Very Large Optical Telescope):
`http://www.hia-iha.nrc-cnrc.gc.ca/projects/VLOT_e.html`
30m TMT (Thirty Meter Telescope): http://tmt.ucolick.org/
50m Euro50: http://www.astro.lu.se/ torben/euro50/
100m OWL (Overwhelmingly Large Telescope):
http://www.eso.org/projects/owl/

Angel R., 2003 in: Proceedings of the Conference on Towards Other Earths: DARWIN/TPF and the Search for Extrasolar Terrestrial Planets, 22-25 April 2003, Heidelberg, Germany. Edited by M. Fridlund, T. Henning, compiled by H. Lacoste. ESA SP-539, Noordwijk, Netherlands: ESA Publications Division, 221

Bash, Frank N., Sebring, Thomas A., Ray, Frank B. and Ramsey, Lawrence W., 1997. 'The extremely large telescope: a twenty-five meter aperture for the twenty-first century', in Optical Telescopes of Today and Tomorrow: Following in the Direction of Tycho Brahe (ed. Arne Ardeberg), Proc. SPIE,. **2841**, 576

Bennett, C.L. et al., 2003. 'First Year Wilkinson Microwave Anisotropy Probe (WMAP) Observations: Preliminary Maps and Basic Results', (astro-ph/0302207)

Bennett, J.A., 1976. 'On the Power of Penetrating into Space: the Telescopes of William Herschel', Journal for the History of Astronomy, **7**, 75

Bland-Hawthorn, J., 2004. 'Astrophotonics comes of age: an OH- suppressing infrared fibre', Anglo-Australian Observatory Newsletter, No.106, 4

Butler R. P., Marcy G. W., Fischer D. A., Brown T. M., Contos A. R., Korzennik S. G., Nisenson P. and Noyes R. W., 1999, Astrophys. J., **526**, 916

Clark R. N., 1999, Manual of Remote Sensing, ed. A. Rencz (J. Wiley and Sons, New-York)

Colless, M.M., 1999. Phil. Trans. R. Soc. Lond. A, **357**, 105

Colless, M.M. et al., 2001. Mon. Not. R. Astr. Soc., **328**, 1039

Della Valle, M. et al., 2005. In Exploring the Cosmic Frontier: Astrophysical Instruments for the 21st Century, Berlin, in press

Dierickx, P., et al., 2004. 'OWL Phase A Status Report', Proc. SPIE, **5489**, 391

Eggen, O.J., Lynden-Bell, D. and Sandage, A.R., 1962. Astrophys.J. **136**, 748

Ehgamberdiev, S., et al., 2000. 'The astroclimate of Maidanak Observatory in Uzbekistan', Astronomy and Astrophysics Supplement, **145**, 293

Ellis R., 1998. Nature, **395**, 3

Fairall, A.P., 2003. 'The 2dF and Sloan Surveys', in Astrophysics Update (ed. J. Mason), Springer-Praxis, Chichester, 211

Frayn C., 2003, PhD Thesis, University of Cambridge

Freeman, K.C. and Bland Hawthorn, J., 2002. Ann. Rev. Astron and Astrophys., **40**, 487

Friedman, S.M., The Inflation Calculator, http://www.westegg.com/inflation/, October 2005

Fukugita, M., Hogan, C.J. and Peebles, P.J.E., 1998. Astrophys.J., **503**, 518

Gilmozzi, R., 2004. 'Science and Technology Drivers for Future Giant Telescopes', Proc. SPIE, **5489**, 1

GMT Science Working Group. The Giant Magellan Telescope: Opening a New Century of Cosmic Discovery, http://www.gmto.org/sciencecase, October 2005.

GSMT Science Working Group, 2003. Frontier Science Enabled by a Giant Segmented Mirror Telescope, prepared for the Astronomy Division of the NSF

GSMT Science Working Group, 2005. A Giant Segmented Mirror Telescope: Synergy with the James Webb Space Telescope, University of Hawaii, Honolulu

Gunn, J.E. and Peterson, B.A., 1965. 'On the density of neutral hydrogen in intergalactic space', Astrophys. J., **142**, 1633

Hale, G. E., 1928. 'The Possibilities of Large Telescopes', Harper's Magazine, **156**, 639

Heger, A., Woosley, S.E., Baraffe, I. and Abel, T., 2001. 'Evolution and explosion of very massive primordial stars', (astroph/0112059)

Holland W. et al., 2003. Proc. SPIE, **4840**, 340

Hook, I., for the OPTICON European ELT Science Working Group, 2004. 'Highlights from the Science Case for a 50-100m Extremely Large Telescope', Proc. SPIE, **5489**, 35

Hook, I. (ed.), 2005a. The Science Case for a European Extremely Large Telescope, EC FP6 Infrastructure Coordination Network, OPTICON. (Available at http://www-astro.physics.ox.ac.uk/ imh/ELT/)

Hook, I., for the OPTICON European ELT Science Working Group, 2005b. 'Science with Extremely Large Telescopes', ESO Messenger No. 121, 2

IAU Working Group on Extrasolar Planets, http://www.ciw.edu/boss/IAU/div3/wgesp/planets.shtml, October 2005

Jaffe, A.H., 2003. 'The Cosmic Microwave Background', in Astrophysics Update (ed. J. Mason), Springer-Praxis, Chichester, 89

Keel, W.C., 2003. . 'The Early Universe: From Recombination to Reionisation' in Astrophysics Update (ed. J. Mason), Springer-Praxis, Chichester, 139

King, H. C., 1955. The History of the Telescope, Griffin, London

Knop R., et al., 2003. Astrophys. J., **598**, 102

Lamb, D.Q and Reichart, D.E., 2000. Astrophys.J.Lett. **536**, 1

Lawrence, J.S., Ashley, M.C.B., Tokovinin, A. and Travouillon, T., 2004. 'Exceptional astronomical seeing conditions above Dome C in Antarctica', Nature, **431**, 278

Lucas P. W., Roche P. F., Riddick, F., 2003, in E. Martin, ed., Proc. IAU Symp 211, (ASP: San Francisco), 63

Madau, P., della Valle, M. and Panagia, N., 1998. Mon. Not. R. Astr. Soc., **297**, 17L

Marcy, G. and Butler, R.P. 1996. 'The Planetary Companion to 70 Vir', Astrophys.J. Letters, **464**, L147

Mayor, M. and Queloz, D.,1995. 'A Jupiter-Mass Companion to a Solar- Type Star', Nature, **378**, 355

McGrath, A., for the AAO Instrument Science Group, 2003. 'Directions for future instrumentation development by the AAO', Anglo-Australian Observatory Newsletter, IAU Special Issue, 18

McLean, I.S. and Chaffee, F.H., 2000. In Optical and IR Telescope Instrumentation and Detectors, Proc. Proc. SPIE, **4008**, 2

Miralda-Escude, J. and Rees, M., 1997. Astrophys.J., **478**, 57L

Moore, A. and McGrath, A., 2004. 'The MOMFOS fibre positioner', in Ground-based Instrumentation for Astronomy (ed. A.F.M. Moorwood and M. Iye), Proc. SPIE, **5492**, 1835

Murdin, P., 2000. 'Is there Life in the Universe?', in 2001 Yearbook of Astronomy (ed. Patrick Moore), Macmillan, London, 129

Perlmutter S. et al., 1998. Nature, **391**, 51

Perlmutter S. et al., 1999. Astrophys.J., **517**, 565

Riess A., et al.,1998. Astron.J., **116**, 1009

Riess A., et al, 2004. Astrophys.J., **607**, 665

Russell, A.P.G.et al., 2004. 'Instruments for a European Extremely Large Telescope: the challenges of designing instruments for 30-100m telescopes', in Ground-based Instrumentation for Astronomy (ed. A.F.M. Moorwood and M. Iye), Proc. SPIE, **5492**, 1796

Schneider J., Aronold L. and Borkowski V., 2003. 'New approaches for the search for binary planets and moons of extrasolar planets' in SF2A- 2003: Semaine de l'Astrophysique Franaise, eds. F. Combes, D. Barret, T. Contini, and L. Pagani. EdP-Sciences, Conference Series, 151

Searle L. and Zinn R., 1978. Astrophys.J., **225**, 357

Steinmetz, M., 2003. In GAIA Spectroscopy, Science and Technology (ed. U. Munari), ASP Conference Series, 381

Taylor, K., 2004. 'Instrumentation for the Thirty Meter Telescope', in Ground-based Instrumentation for Astronomy (ed. A.F.M. Moorwood and M. Iye), Proc. SPIE, **5492**, 1818

Tinney, C.G. et al., 2001. Astrophys. J., **551**, 507

Tonry, J. et al., 2003. Astrophys. J., **594**, 1

Vedrenne, G. and Atteia, J.-L., 2003. 'The Story of Gamma Ray Bursts', in Astrophysics Update (ed. J. Mason), Springer-Praxis, Chichester, 255

Watson, F., 1997. 'Waveguide Spectrographs for Astronomy?' In Optical Telescopes of Today and Tomorrow (ed. A. Ardeberg), Proc. SPIE, **2871**, 1373

Watson, F., 2003. 'Optical spectroscopy today and tomorrow', in Astrophysics Update (ed. J. Mason), Springer-Praxis, Chichester, 185

Watson, F., 2004. Stargazer: The Life and Times of the Telescope, Allen and Unwin, Crows Nest, NSW

Watson, F., for the international RAVE team, 2003. 'RAVE hits the Galaxy', Anglo-Australian Observatory Newsletter, No.102, 4

Wilson, R.N., 1996. Reflecting Telescope Optics I, Springer, Berlin

Wilson, R.N., 1999. Reflecting Telescope Optics II, Springer, Berlin

York, D.G. et al., 2000. 'The Sloan Digital Sky Survey: Technical Summary', Astron. J., 120, 1579

Index

MIX
Papier aus verantwortungsvollen Quellen
Paper from responsible sources
FSC® C105338

If you have any concerns about our products,
you can contact us on
ProductSafety@springernature.com

In case Publisher is established outside the EU,
the EU authorized representative is:
Springer Nature Customer Service Center GmbH
Europaplatz 3, 69115 Heidelberg, Germany

Printed by Libri Plureos GmbH
in Hamburg, Germany